Annotated reprints of key works germane to the understanding of the growth of science from the age of antiquity to the twentieth century under the general editorship of HARRY WOOLF, *Willis K. Shepard Professor of the History of Science, The Johns Hopkins University*

A carefully assembled collection of the most significant landmarks in the evolution of scientific development

Presented in scholarly, annotated, and when warranted, translated editions

Commentaries by established scholars whose professional concern embraces the subject, thus assuring analysis and discussion of the highest quality

Ranging from the assembly of scattered articles on a single basic topic, to great synthetic essays, as well as the presentation of unique works of the great writers, comprehensive anthologies of most significant texts published in the past, and encompassing complete runs of certain journals

Distinguished by the association of qualified contemporary judgment with complete and original scientific texts

THE SOURCES OF SCIENCE
Number 11

THE SOURCES OF SCIENCE

Editor-in-Chief: HARRY WOOLF

WILLIS K. SHEPARD PROFESSOR OF THE HISTORY OF SCIENCE
THE JOHNS HOPKINS UNIVERSITY

1. WALLER: Essayes of Natural Experiments
   WITH A NEW INTRODUCTION BY A. RUPERT HALL

2. BOYLE: Experiments and Considerations Touching Colours
   WITH A NEW INTRODUCTION BY MARIE BOAS HALL

3. NEWTON: The Mathematical Works of Isaac Newton, Vol. 1
   WITH A NEW INTRODUCTION BY DEREK T. WHITESIDE

4. LIEBIG: Animal Chemistry
   WITH A NEW INTRODUCTION BY FREDERIC L. HOLMES

5. KEPLER: Kepler's Conversation with Galileo's Sidereal Messenger
   TRANSLATED AND EDITED BY EDWARD ROSEN

6. FARADAY: Achievements of Michael Faraday
   WITH A NEW INTRODUCTION BY L. PEARCE WILLIAMS

7. TAYLOR: Scientific Memoirs Selected from the Transactions of Foreign Academies of Science and Learned Societies, and from Foreign Journals, 7 vols.
   WITH A NEW PREFACE BY HARRY WOOLF

8. CHINCHILLA: Anales Históricas de la Medicina en General y Biográfico-Bibliográficos de la Española en Particular, 4 vols.
   WITH A NEW INTRODUCTION BY FRANCISCO GUERRA

9. MOREJÓN: Historia Bibliográfica de la Medicina Española, 7 vols.
   WITH A NEW INTRODUCTION BY FRANCISCO GUERRA

10. BULLETTINO DI BIBLIOGRAFIA E DI STORIA DELLE SCIENZE MATEMATICHE E FISICHE, 20 vols.
    EDITED BY B. BONCOMPAGNI

11. GREW: The Anatomy of Plants

   WITH A NEW INTRODUCTION BY CONWAY ZIRKLE

12. HALLIWELL: A Collection of Letters Illustrative of the Progress of Science in England from the Reign of Queen Elizabeth to That of Charles II

   WITH A NEW INTRODUCTION BY CARL B. BOYER

13. THE WORKS OF WILLIAM HARVEY, M.D.

   TRANSLATED FROM THE LATIN WITH A LIFE OF THE AUTHOR BY ROBERT WILLIS, M.D.

14. BREWSTER: Memoirs of the Life, Writings, and Discoveries of Sir Isaac Newton

   WITH A NEW INTRODUCTION BY RICHARD S. WESTFALL

THE
ANATOMY OF PLANTS

Nehemiah Grew (1641–1712), engraved after the portrait by R. White, and reproduced in the *Cosmologia Sacra* in 1701.

THE
ANATOMY OF PLANTS

WITH AN IDEA OF
A PHILOSOPHICAL HISTORY OF PLANTS
AND SEVERAL OTHER LECTURES
READ BEFORE THE ROYAL SOCIETY

by Nehemiah Grew

Reprinted from the 1682 edition

With a new introduction by
CONWAY ZIRKLE
PROFESSOR OF BOTANY, DEPARTMENT OF BIOLOGY
UNIVERSITY OF PENNSYLVANIA, PHILADELPHIA

THE SOURCES OF SCIENCE, NO. 11

JOHNSON REPRINT CORPORATION
New York and London

1965

Copyright© 1965 by
JOHNSON REPRINT CORPORATION

Library of Congress Catalog Card Number: 65-18331

PRINTED IN THE UNITED STATES OF AMERICA

INTRODUCTION

NEHEMIAH GREW (1641–1712) may be described as the first microscopist who limited his investigations to the anatomy of plants, or he might just as accurately be classified as the first botanist who used the microscope for studying plant morphology. He was not the first, of course, to examine plants with the microscope, nor was he the first to picture what the microscope revealed. And he was not alone when he began his labors, for at the time—late in the seventeenth century—other microscopists were also looking at plants. In fact, no less than four very able investigators included plants among the objects that excited their curiosity and, naturally, their observations overlapped. Inevitably, several of them noticed the same plant structures at approximately the same time. Thus, at its very beginning, plant anatomy experienced several instances of simultaneous discovery.

This problem of simultaneous discovery has recently attracted the attention of several historians and sociologists of science. The fact that two or more scientists often make the same discovery does call for an explanation. That a single scientist should make a discovery we take for granted, but for two scientists, working independently, to make the same discovery seems odd. It occurs too frequently, however, for it to be a mere chance coincidence. But this duplication of discovery, important as it is, is only a part of a much greater problem. We have to consider also the time when a discovery is made and this, of course, will have to include the time when the discoveries are not made. We shall have to consider not only the duplication of discoveries, but also the discoveries that no one makes. This remark is not an attempted paradox, nor is it even offered as a witticism. To place any discovery in its proper setting, one must at least be aware of the greater problem of the scientific discoveries that could have been made but were not.

When we take full advantage of our hindsight, we can sometimes tell with great accuracy at just what point it was possible for a scientific discovery to have been made, as well as when no one could possibly have made it. We know, for example, exactly which of the recent discoveries in biology could *not* have been made before the electron microscope was invented, and we can measure fairly accurately the time lag between the possibility of such discoveries and the advent of the discoveries themselves. The same hindsight tells us both the time and the type of structures that the light microscope could have revealed. We even know what details the simple glass lenses could have shown if anyone had been curious enough to look at living things through them.

The microscopical investigations of Nehemiah Grew illustrate beautifully both the occurrence of simultaneous discoveries, such as those made by Grew and by Marcello Malpighi, and the discoveries that no one made, even though the technical equipment for making them had been available for over a generation. Grew very wisely took advantage of the equipment he had and very clearly knew how to use it.

It is only recording a truism to state that, to evaluate Grew's contributions we shall have to examine them in the light of what he had to work with, and this means that we shall have to trace briefly the developmental history of the microscope, and try to judge the quality of the instrument he used.

It was not until late in the seventeenth century that anyone explored the finer or microscopic structure of animals and plants. This late beginning of microscopic anatomy is truly remarkable because the microscope had been available for more than half a century, and workable lenses, some of considerable magnifying power, had been in existence some 2500 years. Magnifying lenses were used well before the times of the Greeks and Romans. Layard found a convex lens made of rock crystal in the ruins of the palace at Nimrud, and Seneca described hollow spheres of glass filled with water which were commonly used as magnifiers. But we have no records of these early lenses being used to discover the microscopic anatomy of any living thing, although a single lens, skillfully made, could have revealed many of the details of animal and plant anatomy, and it could have done so at least as well as the early compound microscopes. Indeed, the finely ground single lenses of Leeuwenhoek actually revealed more: it was by means of a single lens mounted in a metal plate that Leeuwenhoek discovered bacteria.

The compound microscopes used by the early plant anatomists were derived from, and were modifications of, the first telescopes. No one knows just when or by whom the first telescope was invented. It was possibly made by Johannes Lippershey some time before 1609, as that was the year that news of the invention reached Galileo Galilei, who was then living in Venice. Galileo at once undertook the task of making a telescope of his own, and with it he examined the heavenly bodies. Immediately, a new world was opened for him to explore. He had only to look through his telescope to see what no man had ever seen before. He was the first to discover the mountains of the moon, the satellites of Jupiter, and the phases of Venus. When the Galilean telescope was constructed so as to have a very short focal length, it became a microscope and, as a microscope, it opened a second new world for exploration.

The Galilean microscope consisted of a convex lens in one end of a tube, and a concave lens—the ocular—in the other. In 1646, Fontana built a microscope with a positive eyepiece, that is, with an eyepiece made of a convex lens, and this instrument is essentially the compound microscope we use today. Naturally, the first microscopes were very crude: their magnification was not great, nor were their images clear. Two hundred years would elapse before their chromatic and spherical aberrations would be corrected. But meanwhile, as defective as they were, they proved to be invaluable for the investigation of the finer structure of living things.

The biologists who used these early instruments—the first microscopists—worked under many handicaps but they worked well. They seem to have used their personal talents to compensate for the imperfections of their equipment. We can appreciate their accomplishments better, perhaps, when we remember that none of their microscopes could give as clear an image at a high magnification as could a single expertly ground lens. But only Leeuwenhoek could grind such lenses.

Astronomical discoveries began almost immediately after the telescope was invented but, as we have stated, a half-century would elapse before anyone would explore the microscopic world. Then, as happens so often in the development of a science, a number of very able men entered the field simultaneously and, in a very short time, they placed the sciences of animal and plant anatomy on an entirely new basis. These men were truly contemporaries; they were all born within a few years, and they died not far apart.

There were in the group two Italians, Marcello Malpighi (1628–1694) and Filipo Buonanni (1638–1725); two Dutchmen, Anthony van Leeuwenhoek (1632–1723) and Jan Swammerdam (1637–1683); and two Englishmen, Robert Hooke (1635–1703) and Nehemiah Grew (1641–1712). These men worked independently but each knew what the others had done and were doing. Four of the six were intimately connected with the Royal Society of London. Hooke was appointed its curator in 1662, and Grew became its secretary in 1677. Leeuwenhoek submitted his numerous papers to the Society, which began their publication in 1674 in the *Philosophical Transactions*. Malpighi was elected a Fellow of the Royal Society in 1668, and in 1669 he submitted the first of his books to the Society for publication. The Royal Society, in fact, published the major works of Hooke, Grew, Leeuwenhoek, and Malpighi.

Hooke, Grew, and Malpighi all contributed to plant anatomy. Hooke published first; his "Micrographia" appeared in 1665. He, of course, has priority in all the plant anatomy that he described, but his contributions to the field were never fundamental. It may seem somewhat heretical but, to the writer, he is reminiscent of an ingenious boy who has been given a microscope and who has become enthralled by what can be done with this new toy. Hooke examined everything he could with his microscope, and the illustrations in his book are remarkably fine. Two of his pictures, one of the flea and another of the louse, have become classics, but his plant illustrations are not so striking, although they were the best that had been made up to the time. He showed the surface markings of many kinds of seed, and he depicted the detailed structure of stinging nettles, fungi, leaves, and sections of cork. In the latter he showed plant cells but, naturally, he did not recognize their importance.

The plant illustrations of Grew and Malpighi are on a somewhat different level. Grew and Malpighi pictured the microscopic structure of plants systematically and in detail. They submitted their work for publication almost simultaneously and this caused some confusion. M. J. Schleiden (1845) has even accused Grew of plagiarism. It is true that Grew, as Secretary of the Royal Society, saw Malpighi's work before it was published, but a careful comparison of the dates of publication shows that the two men made their discoveries independently. It is also true that Grew used Malpighi's results in his later work but whenever he did, as Julius von Sachs (1875) has pointed out, he gave Malpighi full credit. It is well to discard this canard of Schleiden's before we examine the work of Grew in detail.

The publications of Grew and Malpighi could hardly have overlapped more. Malpighi's "Anatome Plantarum Idea" is dated Bologna, November 1, 1671, but it was presented to the Royal Society on December 7. It was at this meeting that Grew presented his "The Anatomy of Plants Begun," a small book that had just appeared in print. Its manuscript, however, had been tendered to the Society in May. Grew published his "Anatomy of Vegetables" the next year (1672) and his "Idea of a Phytological History, etc.," in 1673. His "Comparative Anatomy of Trunks" appeared in two parts in 1674 and in 1676 and finally, in 1682, he published the work that is reproduced here, "The Anatomy of Plants." Malpighi's "Anatome Plantarum" was published in London, the first part in 1675, the second in 1679.

There has been some speculation about how Grew became interested in the microscopic structure of plants. Miall has suggested that Grew was a pious man who thought that he should demonstrate the wisdom of God as manifested in His works of creation. At the time, of course, nearly all biological thinking had teleological overtones and it was assumed almost universally that the best reason for studying nature in all of her details was to show how carefully and skillfully everything had been planned. Grew *was* pious, and he came from a pious—even a pig-headed—family.

Nehemiah Grew was the son of Obadiah Grew, a Puritan clergyman of St. Michaels, Coventry. His grandfather, Francis Grew, was a Nonconformist who lost most of his substance as a result of the Star Chamber proceedings. When the Civil War broke out Obadiah, who had entered the Church, sided with the Parliamentary party. For him the Restoration was something of a calamity as he had to resign his living in 1662. Of course it was not the Restoration alone, but the Restoration plus his own troublesome conscience, that caused his adversity. His conscience would not let him comply with the Act of Conformity. And his troubles continued. In 1682, he was imprisoned for six months in the Coventry jail. This was the year in which his son's (Nehemiah's) "Anatomy of Plants" was published, and dedicated by its author to "His most sacred Majesty Charles II." The incarceration of the father and the dedication by the son, however, need not have been connected. Nehemiah was only following one of the established customs of the times. He was probably both pious and nonpolitical, a true son of his father but not very much concerned with affairs of state.

Nehemiah Grew had been an undergraduate at Cambridge. Later, he studied medicine at Leiden where, at the age of thirty, he received his degree. He practiced medicine very successfully in London. John Wilkins, one of the founders of the Royal Society and later the bishop of Chester, brought Grew's work to the attention of his colleagues in the Society, and Grew was elected a Fellow. He became secretary of the Society in 1677.

Professionally, Grew was a doctor of medicine and it could have been that his first interest in plants was in their medicinal properties. After he had become acquainted with them, however, he could well have found in their structure evidence for the existence of some divine plan. Ultimately, he seems to have become interested in them for their own sake. His interest was very broad, extending to all their structures and all their activities. While he called his great work an "anatomy," we should remember that, in his day, anatomy did not have its modern meaning. Then, it included more than the microscopic structures; it covered much of what we would label morphology—even gross morphology. Some of his illustrations, especially those of seeds, are shown at a magnification that a good modern hand lens could give. He showed few seeds as large as those illustrated in Hooke's "Micrographia" but his was still an anatomy of seeds.

Grew also described many of the activities of plants—their physiology—although, at the time, very little was known of the subject. Nearly half a century would elapse before Stephen Hales would publish his "Vegetable Statics," and plant physiology could be organized into a separate discipline. Grew was especially interested in the germination of seeds, the way they absorbed water, and the way they extruded the radicle. He described clearly how the stem growing point with its surrounding leaves—the "plume"—emerged. He called the cotyledons the "lobes," but noted how some of the seed lobes, such as those in the bean, emerged from the ground, turned green, and became "dissimilar" leaves. He described seeds that had two lobes (the dicotyledons), seeds that had one lobe (the monocotyledons), and seeds that had more than two lobes. But here he was not describing the gymnosperms (the polycotyledons) but the common cress (*Lepidum sativum*), whose seed leaves are trilobed.

In his description of the mature plant, Grew invented a number of terms that were never widely adopted and have now been superseded by their modern names. Any botanist reading his works, however, will have no difficulty in equating his terms with their modern equivalents. For example, Grew's "attire" is no longer used and the word stamens has taken its place. Grew's "parenchyma" was more fortunate, however, and is still in use. Its etymology gives us perhaps our best clue to Grew's concept of the plant cell. Parenchyma is, of course, the unaltered Greek *parenchyma* and means anything poured in beside (*para*, beside, and *enchyma*, an infusion). Grew considered the parenchyma to be a form of solid foam. Hooke had described the pith as a "heap of bubbles;" Grew called them "bladders" and concluded that his bladders formed both the pith and the cortex. He likened his parenchymous tissue to "the Froth of *Beer* or *Eggs*." The cell walls, however, he thought of as woven fibers that extended far beyond the individual cells, nor did he believe that these fibers always enclosed cells. But even when he described cells specifically he seems not to have considered that their contents were of any real importance. A century and a half would elapse before Schleiden and Swann would enunciate the cell theory.

Grew did notice, however, that his little bladders sometimes fused to form vessels or tubes. A single row or file of bladders could "by the shrinking up of their Horizontal *Fibres*, all regularly break one into another and so make one *continued Cavity;* or a *Tube* whose *Diametre* is the same with that of the *Bladders*, . . ." (p. 118, §35). Here he also tells us how "Aer-Vessels" originate from "*Parenchymous Parts*" where, through an "alteration of the *Fibres*," they are transformed much as "Caterpillars are into *Flies*." Earlier (p. 110, §18–20), he had described in some detail how the turpentine ducts were formed in the bark of the pine, and how the milk ducts were produced in the fig tree.

A glance at the plates reprinted in this volume will reveal Grew's overwhelming interest in the various kinds of stems. He contrasted the scattered vascular bundles and the lack of bark in the monocotyledons with the regularly arranged bundles and well-developed bark in the dicotyledons. He was impressed with the effective use made of the supporting tissue where, arranged in a hollow cylinder, it gives the stem a strength and rigidity it would not otherwise have. He had a clear concept of the medullary rays in the dicotyledonous stem (Pls. 3, 37) and perhaps some notion of their function, and he noted that they connected the cortex with the pith.

Grew was convinced that plants, like animals, possessed vessels and he proved this very simply. He cut a section of root and pressed its sides gently. Where the vessels were cut, water was extruded, but when he withdrew the pressure the water was sucked back in. When the pith was cut, however, this did not occur. The only water from the pith came from the cells that had been cut, and this water was not sucked in again.

Grew was especially interested in buds and he described how they were sheltered in an axillary position, and how they were protected by the bud scales. He showed how the bud leaves were folded up or rolled together and how they unfolded during their vernation. He mentioned the fact that these bud leaves had been formed months before they emerged. He concluded correctly that a bulb was, in fact, a great underground bud.

Grew described, although sometimes rather hastily, all of the plant structures that he was able to see. He described the tendrils used for climbing, the plant hairs that grow on and protect the surface of the mature leaves, and the stomata that he found only on the "upper" surface of the leaves (p. 153). This latter observation has been cited as an oversight, or as an outright error, on Grew's part. But this would be too simple-minded an error for a man such as Grew.

Certainly he was not mistaken if by "upper" he meant "dorsal." It is the ventral side of the leaf that has few or no stomata. The modern critics of Grew apparently forgot that the leaf turns over in vernation and that it is the ventral side of the mature leaf that is "up."

Grew's description of the flower and its many parts is excellent, although the terms he used have become archaic. His "empalement" is the calyx, his "foliature" is the corolla, and his "attire" the stamens. He considered the pistil not a part of the flower but the beginning of the fruit; in this, however, he was only following the convention of the sixteenth- and seventeenth-century herbalists. It is worth recording that our modern definition of the fruit is that it is the mature ovary and consists of the pericarp and seed. Grew described the anthers (the semets) and noted correctly the function of pollen. He described the pollen grains as "congeries, usually, of so many perfect globules or globulets, sometimes of other figures but always regular." He also recognized that the pollen grains were the granules that were found in beebread, and this was something that Swammerdam himself had missed.

In recognizing the role of pollen, of course, Grew became aware that the flowering plants reproduced sexually, and for recording this fact the historians of science have given him full credit. Grew's publication preceded Camerarius' (1691) by nine years. His contribution to our knowledge of sex in plants was actually very minor, but it did involve a number of oddities; so many, in fact, that we would do well to consider this part of his work in some detail. We can set the stage best by quoting directly from his "Anatomy of Vegetables" (1672). In recording the use of pollen, Grew wrote (pp. 147–148):

"Yet the finding out of Food is but in order to enjoy it: which, that is provided for a vast number of little Animals in the *attire* of all Flowers, observation persuades us to believe. For why else are they evermore here found? Go from one flower to another, great and small, you shall meet with none untaken up with these guests. In some, and particularly the *Sunflower*, where the parts of the Attire, and the *animals* for which they provide, are larger, the matter is more visable. We must not think that God Almighty hath left any of the whole Family of his Creatures unprovided for; but as the Great Master, somewhere or other carveth out to all; and that for a great number of these little Folk, he hath stored up their peculiar provisions in the *Attires* of Flowers; each *Flower* thus becoming their Lodging and their Dining-Room, both in one.

"Wherein the particular parts of the *Attire* may be the more distinctly serviceable, this to one Animal, and that to another, I cannot say: Or to the same Animal, as a *Bee*, whether this for the *Honey*, another for their *Bread*, a third for the *Wax:* Or whether all only suck from hence some *Juices;* or some may not also carry some of the Parts, as of the Globulets, wholly away: Or lastly, what may be the primary use of the *attire* (for even this abovesaid, though great, is yet but secondary). I now determine not."

When Grew wrote the above, it is practically certain that he knew the primary function of the pollen, but it appears that, at the time, he just did not care to commit himself and tell what that function was. The secondary function may even have seemed to be the more respectable as it called attention to the kindness of nature and to the providence of God. Grew's emphasis, in his treatment of the subject, is certainly an exercise in piety. Obviously God provided for all His creatures; He created large fruit for the large animals to eat, and small fruit—the pollen grains—for the small.

But the primary function of pollen was still there and sooner or later it would have to be announced. Grew himself announced it in a paper he read to the Royal Society on November 9, 1676, but the paper was not printed until 1682, when he included it in "The Anatomy of Plants" (from p. 171):

"3. In discourse hereof with our Learned *Savilian* Professor Sir *Thomas Millington,* he told me, he conceived, That the *Attire* doth serve, as the *Male* for the *Generation* of the *Seed.*

"4. I immediately reply'd, That I was of the same Opinion; and gave him some reasons for it, and answered some *Objections,* which might oppose them. But withall, in regard to every Plant's αῤῥενόϑηλυς or *Male* and *Female,* that I was also of Opinion, That it servith for the *Seperation* of some *Parts,* as well as the *Affusion* of others. The sum therefore of my Thoughts, concerning this *Matter,* is as follows."

The above is perhaps the most quoted passage in all of Grew's works. It is followed (p. 172) by a mass of speculations, which to us seems almost meaningless. In spite of the fact that Grew cited no experimental evidence, and apparently he himself never experimented, his endorsement of the idea that the flowering plants reproduced sexually was exceptionally influential. Camerarius (1694) cited it prominently in the classic paper in which he recorded his careful and controlled experiments that supplied the proof that pollen was in fact the male element. All the historians of botany have emphasized this contribution of Grew's.

The prominence given Grew in the history of the discovery of sex in plants, however, seems to be little more than an accident. The sexuality of at least some of the flowering plants was well known at the time he wrote. It had, in fact, been described by many botanists. Knowledge of the sexuality of the date palm was really very old. The custom of pollinating it by hand antedates history. It is illustrated graphically on numerous Babylonian cylinder seals and on several Assyrian monuments. In classical times, Herodotos described the sexuality of the date palm but his description is very inaccurate. Theophrastus, however, described it correctly and recorded the real role of pollen. Almost every other classical author who wrote on botanical topics also recorded the sexuality of this one plant. As we would expect, the sex of the date palm was common knowledge in the Arabic world, for there the production of dates was of great economic importance. Even in Europe, the fact that date trees were male and female was well known and was recorded routinely in the sixteenth- and seventeenth-century herbals. In general, however, the recognition of sex in the date palm did not lead to the discovery that the other flowering plants also reproduced sexually, although Adam Zaluziansky (1592) did make the obvious generalization and recognized the universal role of pollen.

It could have been the lack of prominent dioecious plants in the flora and cultivated crops of Europe that limited popular knowledge of plant sexuality. The date palm was strictly a tropical plant, and few Europeans had ever seen it pollinated by hand. With the discovery of the New World, however, a second dioecious plant was introduced to the Europeans, and its sexuality was also described in the herbals. This plant was a small tree, *Carica papaya.* The Spaniards named it "mamoera" because its pendulous fruit reminded them of the mammary gland. The English changed the pronunciation to "mammy" and called it the "mammy tree" but, through some strange anomaly of popular nomenclature, the mammy tree has now become the "papaw." Sexual reproduction in the papaw was described by Linschoten (1596), l'Ecluse (1607), Bauhin (1620), Parkinson (1640), and later by Grew's friend, John Ray (1688). Neither

Grew nor his contemporaries could have remained in ignorance of the sexual reproduction of at least two plants. Thus Grew's hesitancy in announcing the "primary" function of pollen is a little puzzling.

Grew was interested also in the reproduction of ferns, especially in the "seed cases" he found on the lower surface of the fronds. His verbal description of these bodies is far superior to the pictures (Pl. 72) he used to illustrate them. He described these sporangia and the action of the annulus accurately and vividly. The "seed case" (p. 200) is "girded about with a sturdy *Tendon* or *Spring* of the *Colour* of *Gold:* the whole *Machine* looking not much unlike a little *Padlock*. The *Surface* of the *Spring* resembles a fine *Screw*, or some of the *Aer-Vessels* in the *Wood* of a *Plant*. So soon as by the *Innate Aer* of the Plant or otherwise, this *Spring* is become stark enough, it suddenly breaks the *Case* into two halfs, like two little *Cups,* and so flings the *Seed*."

After such a description, the picture is an anticlimax, which brings up a point that needs our consideration. By what standards are we to judge Grew's illustrations? How effective was his microscope? By modern standards, of course, it was crude. A sophisticated microtechnique was also lacking. Grew certainly could have depicted the annulus of the fern sporangium more accurately if he had mounted it in water and viewed it in transmitted light. Without depreciating Grew and his discoveries in the slightest—he was, of course, a genius—we have to recognize the many inaccuracies in the finer details of his drawings. Most of these can be traced to the limitations imposed by his microscope but there is also a basic human limitation that Sachs has discussed in detail. We can all see more in a microscopic field if we know what to look for, but Grew had no one to tell him what to look for. Sachs pointed out that we can see more detail in an object with the naked eye once we have examined it under a magnifying glass. We could doubtless see more than Grew did, using his instrument, simply because we have used superior microscopes and know what details to expect.

There is no need, of course, for us to belabor Grew's imperfections. His contributions were outstanding and more than a century would pass before his work would be superseded.

Within the limits imposed by the magnification at his disposal, Grew is superb, and the value of his contributions was recognized immediately by his contemporaries, as is shown by the popularity of his publications. His first contribution to plant morphology was "The Anatomy of Plants Begun" (London, 1671). The next year he published a small book, "The Anatomy of Vegetables," which was translated into French and published in Paris as "Anatomie des Plantes" (editions in 1675 and 1679). A French edition was also published in Leiden in 1685 and another in 1691. A Latin translation, "Anatomiae Vegetabilium Primordia," was published in 1678 in the Appendix to the *Ephemerides Acad. Nat. Cur.* (Dec. I: Ann. 8, pp. 287–379). Pritzel also lists an Italian edition, translated by F. M. Nigrisoli, but he lists neither the place nor the date of publication.

Grew's second contribution to plant anatomy, "An Idea of a Phytological History Propounded," was published in London in 1673. This was translated into Latin as "Idea Historiae Phytologicae" and published in 1679, also in the Appendix of the *Ephemerides* (Dec. I: Ann. 9 et 10, pp. 99–218). His third botanical work, "The Comparative Anatomy of Trunks," was published in London in 1675. This appeared in Latin as "Comparativa Anatomia Truncorum" in the same volume of the *Ephemerides* (Appendix pp. 219–293).

Grew revised all these works when he republished them in the volume that is here reprinted, that is, in his definitive "Anatomy of Plants" (London, 1682). They are included here on pages 1 to 140. The remaining portions of the "Anatomy of Plants" were presented to the Royal Society at several meetings held during the years 1675, 1676, and 1677, but they were printed for the first time in this volume.

Grew's nonbotanical publications are not of major importance and are mentioned here only because they give some indication of the range of his interests and activities. The best known of these is his "Musaeum Regalis Societatis" London, 1681. This is a somewhat over-detailed catalog of the belongings of the Royal Society though it does include items of general biological interest. Appended to this work, and almost concealed by it, is Grew's "The Comparative Anatomy of the Stomach and Guts," and this short contribution is a real comparative anatomy. In it, Grew described the gullets, stomachs, and intestines of fifteen mammals, including those of both carnivores and herbivores. He described also the alimentary track of thirteen birds and five fish. This work was first presented as a series of lectures before the Royal Society during the year 1676. It certainly deserves to be better known.

Grew's last book was "Cosmologia Sacra," London, 1701. This was a defence of Christianity, and it illustrates Grew's view of the universe, his teleology, and his over-all piety. His papers published in the *Philosophical Transactions* are minor and contain no plant anatomy. Here we merely list them: "On the nature of snow," **8**, 5193; "Description and use of the pores in the skin of the hands and feet," **14**, 566; "Observation on a diseased spleen," **17**, 543; "Description of the American Tomineius, or humming bird," **17**, 760; "On the food of the humming bird," **17**, 815; "A demonstration of the number of acres contained in England, or South Britain, and the use that may be made of it," **27**, 266.

The chief value of Grew's contributions lies in his description of the anatomy of plants. He also published some physiological data, but his contributions to physiology were only incidental and in no way systematic although his physiological observations were often very shrewd. His account of the ejection of spores from the fern sporangium, for example, deserves our recognition and our respect, as do a number of his other incidental records of plant activities and of the physical principles involved. He was concerned with the rise of sap in trees and with the forces that pushed it up. He noted that water would rise a little way in a glass tube, how far it rose depending upon the bore of the tube. He inferred that this capillarity would also force water up the water-ducts of a tree, but when he considered the height to which the sap was raised he concluded very reasonably that other forces were also involved.

Grew noted that olive oil would dissolve the green color from a plant leaf but that the solution of the pigment would appear red or a deep yellow when it was held up against a candle (p. 273). Here, of course, he may have observed the fluorescence of chlorophyll, but he does not tell us just how he held his solution "up against a candle." He may have discovered that chlorophyll in solution is red in reflected light, or that the last rays to be absorbed by a chlorophyll solution are in the red end of the spectrum. He did not pursue this matter any further. He maintained his interests, however, in the changes he could induce in the solutions he made of the flower pigments, and he noted especially the changes that occurred when the solutions were made acid or basic. He was thus one of the first to recognize the possibility of a color test for acidity, though he was preceded in this by both J. Wray (*Phil. Trans.*, **4**, 963, 1670) and M. Lister (*Phil. Trans.*, **6**, 2132, 1671).

Color tests in the seventeenth century, however, presented problems of their own. Published charts, in which the various shades were permanent, apparently were nonexistent. But in his talks before the Royal Society, Grew had to describe his colors so that the fellows would know what he meant, and here he showed himself to be very ingenious. Knowing the habits and customs of the Fellows, he knew the colors that they were familiar with. In the following passage he spoke their language (p. 274):

"So the *Green Leavs* of *Cinquefoyl*, give a *Tincture* no higher than to resemble *Rheinish Wine;* those of *Hyssop, Canary;* of *Strawberries, Malaga;* of *Mint, Muscadine;* of *Wood-Sorrel, Water* and some drops of *Claret;* of *Blood-Wort, Water* and a dash of *Claret;* and those of *Bawm* make a Tincture near as *red* as ordinary *Claret* alone."

Grew's interests were miscellaneous and varied, at least by modern standards, but by the standards of the late seventeenth century they would have appeared concentrated and somewhat limited. When we consider that he had a microscope, and that the whole world lay before him, we can appreciate his self-control. In one respect he reminds us of Warren Hastings; when we consider his opportunities, we can marvel at his restraint. In 1684 (*Phil. Trans.* 14, 566) he gave an excellent account of the sweat pores and over-all pattern of the human hand. He observed the sweat ooze out of the pores and his illustration depicts the patterns formed by the ridges. He could even have established the technique of fingerprinting.

The less we say about Grew's chemical interpretations and philosophical speculations, however, the better. Our kindest course of action here is to note that they were no worse than those of his contemporaries. Chemistry and philosophy do not seem to us too closely related but, in the late seventeenth century, the connection was still quite pronounced. Grew's greatness lay not in his speculations but in his observations, not in his philosophical orientation but in his hard-headed, empirical search for facts. Here he had no superiors. More than a century would elapse before his microscopic observations of plants would be surpassed.

Conway Zirkle

REFERENCES

Arber, Agnes. "Nehemiah Grew (1641–1712)" *in* "Makers of British Botany" (F. W. Oliver, ed.), Cambridge, 1913.
Disney, Alfred N., ed. "Origin and Development of the Microscope," London, 1928.
Greene, E. L. "Landmarks of Botanical History," *Smithsonian Miscellaneous Collections* **54**, No. 1870, Washington, 1909.
Hawks, Ellison. "Pioneers of Plant Study," London, 1928.
Locy, William A. "Biology and Its Makers," New York, 1910.
Miall, L. C. "The Early Naturalists: Their Lives and Work (1530–1789)," London, 1912.
Nordenskiöld, Erik. "The History of Biology," New York, 1936.
Sachs, Julius von. "History of Botany (1530–1860)," Oxford, 1890.
Singer, Charles. "A History of Biology," New York, 1950.
Sirks, M. J., and Conway Zirkle. "The Evolution of Biology," New York, 1964.

THE
ANATOMY OF PLANTS

At a Meeting of the Council of the ROYAL SOCIETY, *Feb.* 22. 168½.

DR. *Grew* having read several *Lectures* of the *Anatomy* of *Plants*, some whereof have been already printed at divers times, and some are not printed; with several other *Lectures* of their *Colours, Odours Tasts*, and *Salts*; as also of the *Solution* of *Salts* in *Water*; and of *Mixture*; all of them to the satisfaction of the said *Society*: It is therefore Ordered, That He be desired, to cause them to printed together in one Volume.

*C*HR. *W*REN **P.R.S.**

THE ANATOMY OF PLANTS.

WITH AN

IDEA

OF A

Philoſophical Hiſtory of Plants.

And ſeveral other

LECTURES,

Read before the

ROYAL SOCIETY.

By *NEHEMJAH GREW* M.D. Fellow of the *ROYAL SOCIETY*, and of the *COLLEGE* of *PHYSICIANS*.

Printed by *W. Rawlins*, for the Author, 1682.

TO HIS MOST
Sacred Majesty
CHARLES II.
King of Great Britain, &c.

May it pleaſe Your Majeſty,

HE Dedication of one Part of the following Anatomy having been very graciouſly received by Your Majeſty: I am now emboldened moſt humbly to preſent the Whole into Your Royal Hands.

By which Your Majeſty will find, That there are Terræ Incognitæ in Philoſophy, as well as Geography. And for ſo much, as lies here, it comes to paſs, I know not how, even in this Inquiſitive Age, That I am the firſt, who have given a Map of the Country.

Your

The Epistle Dedicatory.

Your Majesty will here see, That there are those things within a Plant, *little less admirable, than within an* Animal. *That a* Plant, *as well as an* Animal, *is composed of several* Organical Parts; *some whereof may be called its* Bowels. *That every* Plant *hath* Bowels *of divers kinds, conteining divers kinds of* Liquors. *That even a* Plant *lives partly upon* Aer; *for the reception whereof, it hath those* Parts *which are answerable to* Lungs. *So that a* Plant *is, as it were, an* Animal *in* Quires; *as an* Animal *is a* Plant, *or rather several* Plants *bound up into one* Volume.

Again, that all the said Organs, Bowels, *or other* Parts, *are as artificially made; and for their* Place *and* Number, *as punctually set together; as all the* Mathematick Lines *of a* Flower *or* Face. *That the Staple of the Stuff is so exquisitely fine, that no* Silk-worm *is able to draw any thing near so small a* Thred. *So that one who walks about with the meanest* Stick, *holds a* Piece *of* Natures Handicraft, *which far surpasses the most elaborate* Woof *or* Needle-Work *in the World.*

That by all these Means, the Ascent *of the* Sap, *the* Distribution *of the* Aer, *the* Confection *of several sorts of* Liquors, *as* Lympha's, Milks, Oyls, Balsames; *with other parts of* Vegetation, *are all contrived and brought about in a* Mechanical *way.*

In

The Epistle Dedicatory.

In sum, Your Majesty will find, that we are come ashore into a new World, whereof we see no end.

It may be, that some will say, into another Utopia. *Yet not I, but Nature speaketh these things: the only true* Pallas, *wherewith it is treasonable for the most couriously handed* Arachne *to compare. In whose Name, I, the meanest of her Pupils, do in all humility crave Your Majesties Gracious Patronage Whereof I cannot doubt, since Your Majesty hath been pleased to be the Founder, and to style Your Self the Patron of that Society, of which I have the honour to be a Member. Your Majesty deeming it to be a more Noble Design, To enlarge the Territories of Knowledge, than those of Dominion: and the Highest Pitch of Human Glory, not to rule, in any sort, over many; but to be a Good Prince over Wise Men. I am*

Your Majesties

most humble

and

most obedient

Subject

NEHEMJAH GREW.

THE PREFACE.

T is a *Politick* or *Civil Virtue* in every prudent mans Eye, To set himself an example, in what he doth, unto others. And in so doing, he looks upon himself as accountable, in some sort, to all Men. To those therefore, who may either expresly, or tacitly, expect the Reasons, upon which I first undertook the *Anatomy* of *Plants*, and also made the after-progress therein; I shall summe them up as follows.

The first occasion of directing my Thoughts this way, was in the *Year* 1664, upon reading some, of the many and curious Inventions of Learned Men, in the *Bodies* of *Animals*. For considering, that both of them came at first out of the same *Hand*; and were therefore the *Contrivances* of the same *Wisdom*: I thence fully assured my self, that it could not be a vain Design; to seek it in both. And being then newly furnished with a good stock of *Seeds*, in order to raise a *Nursery* of *Plants*; I resolved, besides what I first aimed at, to make the utmost use of them for that purpose: that so I might put somewhat upon that side the *Leaf* which the best *Botanicks* had left bare and empty. And in which, notwithstanding some other Learned Men had inserted somewhat of this nature; as Dr. *Highmore* in his *Book* of *Generation*, Dr. *Sharrock* of the *Propagation* of *Plants*, and Mr. *Hook* in his *Micrography*: yet but collaterally, and whithout shewing any purpose of managing this *Part* of *Natural History*. And although it seemed at first an Ob-

The Preface.

jection in my way, That the first projectors seldome bring their business to any good end: yet I also knew, That if Men should stay for an Example in every thing; nothing extraordinary would ever be done.

But notwithstanding the reasonableness of the Design; yet I did not forget, that, in respect of the Undertaker, there might be *Impar congressus.* And therefore, before I had ventured very far, in the Year 1668, I imparted it to my Brother-in-Law, the Learned Dr. *Henry Sampson*, now *Fellow* of the *Colledge of Physicians* in *London*. Who not only very well liked the same; but also excited me to a vigorous and accurate prosecution of it. Which he did, partly, by mentioning a very pertinent passage of Dr. *Glisson*, in the *Preface* to his *Book de Hepate*, (*a*) which I had not then read.

(*a*) Ch.1.

Plantæ quoque in hunc censum (sc. Anatomicum) *veniunt; variâ enim Partium texturâ, & differentiis constant: & proculdubio, ex acurata earundem dissectione, utiles valde observationes nobis exurgerent: præstaretque in illis (inferioris licet ordinis) rebus examinandis operam impendere, quam in transcribendis ut sæpe sit, aliorum laboribus, inutiliter ætatem transigere. Quippe hoc pacto, ignavarum apum more, aliena duntaxat alvearia expilamus, nihilque bono publico adjicimus.*

After I had finished the *First Book*, that I might know the sense also of other Learned Men, whether the steps I had already taken, would warrant me to proceed any further: I put some part of it into the same Hand; who, in the Year 1670, communicated the same to Mr. *Oldenburge*, then *Secretary* to the *Royal Society*: and after he had read it over, it was, upon his motion, delivered to that excellent Person Dr. *John Wilkins* then *Bishop* of *Chester*; who produced it at a Meeting of the *Royal Society*, and desired, they might see the rest. Which, or the greatest part, being also presented to them, the Right Honourable the Lord Vicount *Brouncker*, then *President* of the *Royal Society*, was pleased to peruse the same. Presently, after which, at a Meeting of the Council of

The Preface.

of the said *Society*, the following *Order* was made, and entred in their Council-Book with this Date, and in these words:

May 11th 1671.

Then was Licensed Dr. Nehemjah Grew's *Book, Entituled,* The Anatomy of Vegetables *begun; together with an account of* Vegetation *grounded thereupon. And Ordered to be Printed by the Printer to the* Royal Society.

Hereupon, I was obliged to send the Book to the Press. And upon the 9th of *November* following in the same Year 1671, when it was near being printed, my Lord *Brouncher* signed the forementioned *Order*: the Printer, whose Name was to be inserted therein, not having received his *Diploma* till that time.

The Book being quickly after printed off; I ordered it to be Presented to the *Royal Society*; which was accordingly done at one of their Meetings December 7, 1671. And also to be sent to the Bishop of *Chester*; who was pleased to signifie his acceptance thereof by a Letter dated at *Chester, December* 26th 1671. now filed amongst others in the Custody of the *Royal Society*: part whereof, in regard it relates to matter of Fact, I shall here recite.

Sir,

I did yesterday receive your Book; and am very sensible of the Honour you have done me in the Dedication of it. You was very happy in the choice

The Preface.

of this Subject to write upon; one of the most Noble and the most Copious parts of Philosophy; *and such an one, as hath hitherto lain uncultivated. And you have been very successful in your first Attempt about it, in so many remarkable Observations and Discoveries, as you have made already. I could heartily wish, that you would still apply your self to this kind of Enquiries. You will find that Additionals will come in more copiously and easily. And it is not fit, that any one should, by his Superstructions, carry away the praise from him, who was the first Inventor, and who laid the Foundations, wherein the greatest difficulty doth consist,* &c.

Having thus submitted my self to the Judgment of many Learned Men; I saw that my Journey must not here end. So that, like one who is got into a Wood, I thought I might as fairly find my way out, by going on, as by making a retreat. Whereupon, I began to draw up a *Scheme* of the whole *Design*.

While I was doing this, I received news from *London*, that the same day, *December* 7. 1671, in which my Book, then printed, was presented to the *Royal Society*: there was also presented a *Manuscript* (without *Figures*) from Seignior *Malpighi*, upon the same Subject; dated at *Bononia, November,* 1ˢᵗ 1671. the same, which Mr. *Oldenburge*, when it came to be printed, calleth his *Idea*. And of this, entry was made in their Journal Book. So that the *Royal Society* having now a Prospect of the good service of an Ancient *Member*, and one, who had highly merited by his Works then extant; from thence forward, I looked upon my self to be excused.

But soon after, receiving another Letter from the Bishop of *Chester*, dated at *London, Febr.* 18. 1672. I

found

The Preface.

found the matter otherwife; and that the *Society* were pleafed to engage me to proceed. Whereof entry was made by the *Secretary* in their Journal Book, at one of their Meetings, *April,* 18. 1672, in thefe words:

The Society *was made acquainted with one particular lately paffed in the Council;* fc. *That the Bifhop of* Chefter *had there propofed Dr. Grew to be a Curator to the* Royal Society *for the Anatomy of Plants: and that the Council had approved of that* Propofal. *Upon which, it was Ordered, That the Thanks of the* Society *be returned to the Lord Bifhop of* Chefter, *for this* Propofal, *and to the Council for their Approbation of the fame.*

This they might be induced to do; upon confidering, that it would be no difadvantage to the credit of thofe matters, which were fo new and ftrange, to be offered to the World from a double Authority. For one, although he may have no mind to deceive; yet is it more likely for one, than for two, to be deceived. Likewife, that the fame Subject, being profecuted by two Hands, would be the more illuftrated by the different Examples produced by both. And that, as in other matters, fo here, the defects of both, would mutually be fupplyed.

Whether for thefe, or other Reafons alfo, they were pleafed to pafs the forementioned *Order;* that being done, it had been very ill manners in me, not to have anfwered their expectation therein. And therefore reaffuming the Defign I had laid by, and having reduced it to fome intelligible *Idea,* it was fubmitted to the Cenfure of the *Royal Society:* and it was thereupon ordered it fhould be printed.

Not

The Preface.

Not long after, I received a Curious and Learned Book from Monſ. *Dodart*, *Archiater* to the Prince of *Conde*, and *Fellow* of the *Royal Academy* at *Paris*; in perſuance of whoſe Order, it was by him compoſed and publiſhed. Which being a Deſign of a like Import, I was glad to ſee it ſo far juſtify'd by that Illuſtrious *Society*, as well as by our own.

In this *Idea*, one principal Thing I inſiſt upon, for a *Philoſophical Hiſtory* of *Plants*, is *Anatomy*. And, agreeing to the *Method* therein propoſed, all the Obſervations conteined in the *Firſt Book*, except one or two, were made with the *Naked Eye*. To the end, I might firſt give a proof, How far it was poſſible for us to go, without the help of *Glaſſes*: which many Ingenious Men want; and more, the patience to manage them. For the Truth of theſe Obſervations, Seignior *Malpighi*, having procured my Book to be tranſlated into Latin for his private uſe, ſpeaks his own ſenſe, in ſome of his Letters to Mr. *Oldenburge*, printed at the end of his *Anatomy* of *Plants*. And ſome of them, have ſince been confirmed, both by our Learned Country-men Dr. *Wallis*, and Mr. *Liſter*; and by the Ingenious Mr. *Lewenhoeck*, abroad.

Having thus begun with the bare Eye; I next proceeded to the uſe of the *Microſcope*. And the Obſervations thereby made, firſt on *Roots*, and afterwards on *Trunks* and *Branches*, together with the *Figures*, were all exhibited to the *Royal Society* at ſeveral times from *May* 15. 1672. to *April* 2. 1674; being the *Materials* for the *Second* and *Third Parts*: and hereof *Memorials* were inſerted in their Journal Books.

After this, the *Royal Society* received from Seignior *Malpighi* his *Second Part* of the *Anatomy* of *Plants*, together with the *Figures* therein deſcribed, and his *Letters* to their *Secretary*, dated at *Bononia Aug.* 20th of the ſame year 1674. when, and not before, he gave leave that the two ſaid *Parts* ſhould be printed.

So

The Preface.

So soon as I had finished the *Second* and *Third Parts*, I proceeded to the *Last*, *sc.* of *Leaves, Flowers, Fruits* and *Seeds*: and those Things I met with, more remarquable, were presented to the said *Society* in the Years 1676 & 1677. And the publishing of the former *Parts* successively, as well as of all together, hath been done in pursuance of their several *Orders* for the same.

Having concluded the *History* of *Perfect Plants*; I intended to have subjoyned the *Description* of those which are *Imperfect*. As also of *Parasitical, Marine*, and *Sensitive Plants*. And lastly, a view of the chief Particulars, wherein the *Mechanisme* of a *Plant*, is different from that of an *Animal*. But these things I leave to some other Hand.

The *First Book*, a little after it came forth; was translated into the French *Tongue*, by Monf. Le *Vasseur* an Ingenious Gentleman in *Paris*; elegantly, and in the Judgment of those who are well skilled in that *Language*, with much exactness, as to the sense. He having taken special care, to have all the difficulties of our own, by Me, cleared to him. And in a late Book Entituled, *Philosophia vetus & nova* printed at *Noriberg* 1682. the Learned Author seems to have made use of this Translation, for all that he hath taken notice of in that my *First Book*.

By the Ingenious Collectors of the *German Ephemerides*, both my *First, Second*, and *Third Books*, are all published in *Latine*. But their unskilful Interpreter doth often fail of the *Grammatical Sense*. Whose Errors, many of them very gross, I desire may be imputed neither to them, nor to my self.

Besides these, the *Second Lecture* of *Mixture* is also translated into *French*, by Monf. *Mesmin* a Learned Physician in *Paris*: whose *Version* is very well approved by those who are competent Judges hereof.

This,

The Preface.

This, and the rest which follow, are placed, not in the order of *Time*; but more according to their Nature or Relation one to another. All of them intended as a Commentary upon some particulars mentioned, either in the *First Lecture*, or in the *Idea*.

In the *Plates*, for the clearer conception of the *Part* described, I have represented it, generally, as entire, as its being magnified to some good degree, would bear. So, for instance, not the *Barque*, *Wood*, or *Pith* of a *Root* or *Tree*, by it self; but at least, some portion of all three together: Whereby, both their *Texture*, and also their Relation one to another, and the *Fabrick* of the whole, may be observed at one *View*. Yet have I not every where magnify'd the *Part* to the same degree; but more or less, as was necessary to represent what is spoken of it. And very highly, only in some few Examples, as in *Tab.* 40. which may suffice to illustrate the rest. Some of the *Plates*, especially those which I did not draw to the *Engravers* hand, are a little hard and stiff: but they are all well enough done, to represent what they intend.

AN IDEA
OF A
Philosophical History
OF
PLANTS.

Read before the
ROYAL SOCIETY,
January 8. and *January* 15. 1672.

By *NEHEMJAH GREW* M.D. Fellow of the
Royal Society, and of the *College* of *Physicians.*

The Second Edition.

LONDON,

Printed by *W. Rawlins,* 1682.

TO THE

Most Illustrious

THE

ROYAL SOCIETY,

The following

IDEA

Is most HUMBLY

PRESENTED.

AND,

In their NAMES also

PROPOSED

TO THE

CONSIDERATION

Of other

Learned Men.

By the AUTHOR

NEHEMJAH GREW.

THE CONTENTS.

UNTO what Degree the knowledge of Plants is arrived, §. 1. Wherein defective, §. 2. Why concluded to be so, §. 3. Yet capable of Improvement, §. 4. And worthy of it, §. 5.

Divers Instances given, wherein; first of the Organical Parts, as to their external Accidents and Oeconomical Uses, 6. Then of their Contents, Qualities, and Powers, 7. And an Improvement of this Part, will further that of divers other parts of knowledge; whereof Instances are given, 8.

In order whereto, Five General Means are propounded, 9. The First, a particular and comparative Survey of whatever is of more External consideration about Plants, 10. Instanced as to their Figures, 11, 12. Proportions, 13. Seasons, 14. Places, 15. Motions, 16.

The Second, A like Survey of the Organical Parts by Anatomy, as that which is very necessary, 17. In what manner to be prosecuted, both without, and with the Microscope, 18. What thereupon to be observed, 19. And what, from observantion made, probably attainable, 20.

The Third, A like Survey of the Contents of Plants; their several Kinds, 21. Of all which, their Receptacles, 22. Motions, 23. Qualities, 24. Consistence, 25. Colours, Smells, and Tastes, 26. Where also the same Qualities are to be inquired into, as generally belonging to Plants, 26. As their Colours, 27. Odours, 28. Tastes, 29. Also their Faculties, 30. All these to be further examined, 31. By Contusion, 32. Agitation, 33. Frigifaction, 34. Infusion, 35. Subsession, 36. Digestion, 37, 38. Decoction, 39. Destillation, 40. Arefaction, 41.

Assation,

The Contents.

Assation, 42. *Ustion*, 43. *Calcination*, 44. *By Composition with other Bodies*, 45. *And by Compounding the Experiment it self*, 46. *What hence attainable*, 47.

The Fourth, *A like* Survey *of the* Principles, *as well as of the* Contents, *of the* Organical Parts, 48. *The Difficulty hereof, in some respects cleared*, 49. *Further, by two Instances*, 50, 51. *Some Remarques hereupon, of the* Principles of Plants, 52. *From hence will be attainable a further knowledge of the* Modes of Vegetation, 53. *Of the* Qualities *of* Vegetables, 54. *And of their* Powers, 55, 56.

The Fifth, *A like* Survey *of those* Bodies, *either from which these* Principles *are derived, or wherewith they have any communion*, 57. *Which are* Four *in general*, scil. Earth, *and all solid Receptacles*, 58. Water, *and all liquid Receptacles*, 59. Aer, 60. *And* Sun, 61.

A Sixth General Inquiry, *only hinted*, 62.

The Conclusion, 63.

A N

AN IDEA OF A Philosophical History OF PLANTS.

F WE take an account of the *Degrees* whereunto the Knowledge of *Vegetables* is Advanced, it appeareth, That besides the great Varieties, which the Succesful *Arts* of *Florists*, or Transplantations from one *Climate* to another, have produced; we have very many *Species* brought to light, especially Natives of the *Indies*, which the *Ancients*, for any thing that appears in their *Writings* now extant, were ignorant of. In which particular *Clusius, Columna, Bauhinus, Boccone,* and others, have performed much. Withall, That their *Descriptions* (of all Parts above ground) their *Places* and *Seasons*, are with good diligence and preciseness set before us. Likewise their Order and Kindred: for the adjusting whereof our Learned Countryman Mr. *Ray*, and Dr. *Morrison*, have both taken very laudable pains. As also the ordering of them with respect to their *Alimental* and *Mechanick* Uses; for which, amongst others, Mr. *Evelyn* and Dr. *Beal* have deserved many thanks, and great praise. We are also informed, of the *Natures* and infallible *Faculties* of *many* of them. Whereunto so many as have assisted, have much obliged their *Posterity*.

2. §. By

2. §. By due Reflection upon what hath been *Performed*; it also appears, what is left *Imperfect*, and what *Undone*. For the *Virtues* of most *Plants*, are with much *uncertainty*, and too *promiscuously* ascribed to them. So that if you turn over an *Herbal*, you shall find almost every *Herb*, to be good for every *Disease*. And of the *Virtues* of many, they are altogether silent. And although, for the finding out, and just appropriation of them, they have left us some *Rules*, yet not all. The *Descriptions* likewise of many, are yet to be perfected; especially as to their Roots. Those who are very curious about the other *Parts*, being yet here too remiss. And as for their *Figures*, it were much to be wished, That they were all drawn by one *Scale*; or, at most, by Two; one, for *Trees* and *Shrubs*; and another for *Herbs*. Many likewise of their *Ranks* and *Affinities*, are yet undetermined. And a great number of *Names*, both *English* and *Latine*, not well given. So what we call *Goat's-Rue*, is not at all of kin to that Plant, whose *Generical Name* it bears. The like may be said of *Wild-Tansy*, *Stock-July-Flowers*, *Horse-Radish*, and many more. So also when we say *Bellis Major*, *& Minor*, as we commonly do, these *Names* would intimate, That the *Plants* to which they are given, differ (as the great double *Marigold*, doth from the less) only in Bulk: whereas, in truth, they are two *Species* of *Plants*. So we commonly say, *Centaurium Majus & Minus*, *Chelidonium Majus & Minus*, and of others in like manner, which yet are distinct *Species*, and of very different *Tribes*. But for the *Reason* of *Vegetation*, and the *Causes* of all those infinite *Varieties* therein observable (I mean so far as *Matter*, and the various *Affections* hereof, are instrumental thereunto) almost all Men have seemed to be unconcerned.

3. §. That Nothing hereof remaineth further to be known, is a Thought not well Calculated. For if we consider how long and gradual a *Journey* the *Knowledge* of *Nature* is; and how short a Time we have to proceed therein; as on the one hand, we shall conclude it our ease and profit, To see how far Others have gone before us: so shall we beware on the other, That we conceive not unduly of *Nature*, whilst we have a just value for Those, who were but her *Disciples*, and instructed by Her. Their Time and Abilities both, being short to her; which, as She was first Designed by *Divine Wisdom*; so may Her vast Dimensions best be adjudged of, in being compared Therewith. It will therefore be our Prudence, not to insist upon the Invidious Question, Which of Her *Scholars* have taken the fairest measure of Her; but to be well satisfied, that as yet She hath not been Circumscribed by Any.

4. §. Nor doth it more behove us to consider, how much of the Nature of *Vegetation* may lie before us yet *unknown*; Than, to believe, a great part thereof to be *knowable*. Not concluding from the acknowledged, much less supposed Insuccessfulness, of any Mens Undertakings: but from what may be accounted Possible, as to the Nature of things themselves; and from *Divine Providence*, by Infinite Ways conducting to the knowledge of them. Neither can we determine how great a part This may be: Because, *It is impossible to Measure, what we See not*. And since we are most likely to under-measure, we shall hereby but intrench our Endeavours, which we are not wont to carry beyond the *Idea*, which we have of our *Work*.

5. §. And

5. §. And how far foever this kind of Knowledge may be attainable, its being fo far alfo worthy our attainment will be granted. For beholding the Many and Elegant Varieties, wherewith a Field or Garden is adorned; Who would not fay, That it were exceeding pleafant to know what we See: and not more delightful, to one who has *Eyes*, to difcern that all is very fine; than to another who hath *Reafon*, to underftand *how*. This furely were for a Man to take a True Inventory of his Goods, and his beft way to put a price upon them. Yea it feems, that this were not only to be *Partaker of Divine Bounty*; but alfo, in fome degree, *To be Copartner in the Secrets of Divine Art*. That which were very defireable, unlefs we fhould think it impertinent for us to defign the *Knowing of That*, which *God* hath once thought fit to *Do*.

6. §. If for thefe, and other *Reafons*, an inquiry into the Nature of *Vegetation* may be of good Import; It will be requifite to fee, firft of all, What may offer it felf to be enquired of; or to underftand, what our *Scope* is: That fo doing, we may take our aim the better in making, and having made, in applying our Obfervations thereunto. Amongft other Inquiries therefore, fuch as thefe deferve to be propofed. Firft, by what means it is that a *Plant*, or any *Part* of it, comes to *Grow*, a *Seed* to put forth a *Root* and *Trunk*; and this, all the other *Parts*, to the *Seed* again; and all thefe being *formed*, by continual Nutrition ftill to be *increafed*. How the Aliment by which a *Plant* is fed, is duly prepared in its feveral *Parts*; which way it is *conveyed* unto them; and in what manner it is *affimilated* to their refpective Natures in them all. Whence this Growth and Augmentation, is not made of one, but many differing Degrees, unto both extremes of *fmall* and *great*; whether the comparifon be made betwixt feveral *Plants*, or the feveral *Parts* of one. How not only their *Sizes*, but alfo their *Shapes* are fo exceeding various; as of *Roots*, in being Thick or Slender, Short or Long, Entire or Parted, Stringed or Ramified, and the like: of *Trunks*, fome being more Entire, others Branched, others Shrub'd: of *Leaves*, which are Long or Round, Even-edg'd or Efcallop'd, and many other ways different, yet always Flat: and fo for the other *Parts*. Then to inquire, What fhould be the reafon of their various *Motions*; that the *Root* fhould *defcend*; that its defcent fhould fometimes be *perpendicular*, fometimes more *level*: That the *Trunk* doth *afcend*; and that the afcent thereof, as to the fpace of *Time* wherein it is made, is of different *meafures*: and of divers other *Motions*, as they are obfervable in the *Roots*, *Trunks*, and other *Parts* of *Plants*. Whence again, thefe *Motions* have their Different, and Stated *Terms*; that *Plants* have their fet and peculiar *Seafons* for their *Spring* or *Birth*, for their *Full Growth*, and for their *Teeming*; and the like. Further, what may be the Caufes as of the *Seafons* of their *Growth*; fo of the *Periods* of their *Lives*; fome being *Annual*, others *Biennial*, others *Perennial*; fome *Perennial* both as to their *Roots* and *Trunks*; and fome as to their *Roots* only. Then, as they pafs through thefe feveral *Seafons* of their *Lives*, in what manner their convenient *feeding*, *houfing, cloathing* or *protection* otherwife, is contrived; wherein, in this kind and harmonious *Oeconomy*, one *Part*, may be officious to another, for the prefervation of the health and life of the *whole*. And laftly, what care is taken, not only for themfelves, but for their *Pofterity*; in

what manner the *Seed* is prepared, formed and fitted for *Propagation*: and this being of so great concernment, how sometimes the other *Parts* also, as *Roots*, in putting forth *Trunks*; *Trunks* in putting forth *Roots*; yea in turning oftentimes into *Roots* themselves; whereof, in the *second Book of the Anatomy of Plants*, I shall give some instances. With other *Heads* of Inquiry of this kind.

7. §. Nor are the *Natures*, *Faculties*, and *Contents* of *Vegetables* less various, or a particular Inspection hereinto, of less concernment. For since All, or Most, seem to grow in the same manner, with one *Sun*, one *Rain*, indifferently well upon one *Soil*, and, to outward appearance, to have the same *Common Parts*; it may be asked, How it comes to pass, that their *Liquors*, or other *Contained Parts*, are of such different *Kinds*; one being Watry, another Winy, a third Oily, a fourth *Milky*, and the like. How also there is such a variety in their *Sensible Qualities*, as their *Colours*, *Tastes*, and *Smells*; what those *Materials* are, which are necessary to the *Being* of these *Qualities*; and those *Formalities*, wherein their *Essence* doth consist; as what it is that makes a *Plant*, or *Flower*, to be *white* or *red*; *fragrant* or *fetid*; *bitter* or *sweet*; or to be of any other *Colour, smell*, or *Taste*. In like manner, their *Faculties* and *Powers*, what that is, or those things are, by which they are constituted; as whence one becomes *Purgative*, another *Vomitory*, a third *Diaphoretick*, &c. These, I say, with many other particular Inquiries depending hereupon; as they cannot but much oblige the Reason of Man to be obsequious to them, so by bringing in, at least, some satisfaction, will no less reward it. Especially, if it be withal considered, that besides our satisfaction as to the *Nature* of *Vegetation*; some further Light, to divers other parts of Knowledge, may likewise hence arise.

8. §. For since the present Design will ingage us, to an accurate and multifarious Observation of *Plants*; we may hereby be enabled to *range* and *sort* them with more certainty, according to the Degrees of their *Affinity*. And all *Exoticks*, *Plants* or *Parts* of *Plants*, may probably be reduced to some such *Domesticks*, unto which they may bear the best Resemblance. Again, it may frequently conduct our minds to the consideration of the *State* of *Animals*; as whether there are not divers material Agreements betwixt them both; and what they are. *Wherein* also they may considerably differ, and what those things are which are more essential to their distinguishment. And *besides*, not only to compare what is already known of both; but also, by what may be observed in the *one*, to suggest and facilitate the finding out of what may yet be unobserved in the *other*. So *also* the consideration of the *Colours*, *smells* and *Tastes* of *Vegetables*, may conduce to the Knowledge of the same *Qualities* in General; or of what it is, that constitutes them such, in any other Body: not as they are *actually* received by *Sense*; but so far, as such *Materials* or external *Circumstances*, are requisite to their becoming the *Adequate Objects* thereof. It *may* lead us also to inquire into further Ways of *Cultivation*, with respect to the whole *Plant*, or to the *Flower*, *Fruit*, or other *Part*: To amend them as to their *Sizes*, *Colours*, *Tastes*, *Fruitfulness*, or otherwise: To think of other Ways of *Propagation*; or to apply those already known to other *Plants* than hath been used. Likewise the Knowledge of their *Mechanical* Uses may hereby be enlarged; both as to the

Rea-

Reason of their use, in such particular *Trades* and *Manufactures*, already known; and the discovery of other uses yet unknown. As also their *Alimental*, with respect both to *Meats* and *Drinks*; the preparation of some, and the finding out of others. But especially their *Medicinal*; some *Plants* which have hitherto been neglected, may be applied to use; the *Perverted* uses of some, and the *Confused* uses of others, may be rectified. What may best correct their *Malignancies*, or inforce their *Virtues*; When needful to add the preparations of *Art* to That of *Nature*; How to Enlarge those of *Art*, and Rectifie those which are indeed Inartificial, may hereby be better conjectured. The knowledge of all which, that we may know how far it is accessible, and what probable Approaches may be made towards it; those several Means I have thought of, and suppose necessary thereunto, are next to be proposed.

9. §. Reflecting then upon the present Design, and seeing this to lie wide; we shall, in the first place, conclude the *Means* attending thereon, should do so likewise. Wherefore, although some may present themselves unto us as more promising; yet let us suppose what several Persons, were they hereunto engaged, each according to his Sense and Genius, would possibly make choice of. Believing, that although Considering Men may vary, in the approval of their own Sense and Notion; yet not always mearly, because it is their own; but because each, may probably see somewhat more in his own, than others do. Wherefore it will be our surest Logick to conclude, Not because no *Mean* may be approved by all Men, that all *Means* should be rejected; but rather, because each may be approved by some, that therefore, all be made choice of. And these, I think, may be comprehended under Five General *Heads* of Enquiry. *First*, Of those Things, which are of more *External* Consideration about *Plants*, as their *Figures*, &c. *Secondly*, Of their Compounding *Parts*, as *Vessels*, &c. *Thirdly*, Of their *Liquors*, and other *Contents*. *Fourthly*, Of their *Principles*, as *Salts*, &c. *Fifthly*, Of their *Aliment*, as *Water*, and other *Means* of Growth.

10. §. AND FIRST of all, whatever is of more *External* Consideration, as the *Figures, Proportions, Motions, Seasons, Situations* of *Vegetables*, and of their several *Parts*, should be observed. In doing which, a particular survey of all their Varieties should be taken. And then a Comparison made betwixt these, and the several *Plants*, or *Parts* of *Plants*, whereof they are the *Properties*. To the end, We may, if possible, be thereby conducted to find out, what other, either sensible, or more recluse *Property*, any of them may agree together in. For it is not more certain, that the three Angles of every *Rectilinear Triangle*, because always equal to two *Right Angles*, are therefore, if put together, always the same: than that *one Property*, agreeing to divers *Vegetables*, should have one *Cause*: For although the *Scope* and *End* may vary; yet the *Cause*, as it is the *Cause* of that *Property*, must be *one*: and consequently, must also import some *Identity* in the *Nature* of all those *Vegetables* wherein it Acts. Wherefore by thus comparing of them, we shall be able more exactly to state the *Orders* and *Degrees* of their *Affinities*; Better to understand both the *Causes* and *Ends* of their *Varieties*: And more probably to conjecture of their *Natures* and *Vertues*.

The First General Mean.

11. §. First

11. §. First then the various *Figures* of their several *Parts* should be observed; and that with respect both to the *Forms*, and the *Positions*, by which their *Roots, Trunks, Branches, Leaves, Flowers, Fruits,* and *Seeds* may vary, or agree; and those several *Lines,* by which both the said *Varieties* are determin'd. In which of these *Parts,* the agreement chiefly lies; this being both more observable, and more material in some of them; less in the Root, more in the Flower, or Seed. And in how many of these *Parts* together; whether one, more, or all. By both which, the *Orders* and *Degrees* of *Affinity,* which are many, may be accounted; either as to what we strictly call *Kindred,* or else *Analogy.* For there are found, not only *Herbs* accounted of several *Tribes,* which are ally'd; and some of the Smallest, which are of kin to the Greatest: But there are also, probably, some *Herbs,* which have a particular Relation, to many Kinds of *Shrubs*; and some *Shrubs,* to many Kinds of *Trees.* Thus the several sorts of *Letuce,* are of Kin, together in the *First Degree*; with *Endive,* in the *Second.* The several *Clarys,* amongst themselves in the *First*; with *Horehound,* in the *Second*; with *Lamium,* in the *Third.* All *Strawberries* agree together, in the *First Degree*; with *Cinquefoyl,* in the *Second*; with *Tormentil* in the *Third*; and with *Avens, &c.* in other Degrees more remote. So *Agrimony,* hath alike Analogy unto *Strawberry*; as *Goats-Rue,* hath to *Claver*: And *Strawberry,* the like unto the *Rasp*; as *Gooseberry* to the *Vine*; or *Burnet,* to the *Rose.* Amongst the several *Sorts* of *Grass,* there are some which match all those of *Corn*; which is but a greater kind of *Grass.* So again all *Pulse,* are not only of kin, in their several Degrees, to one another; but likewise, to almost all kinds of *Trefoyls,* as *Melilot, Fœnugreek,* and the common *Clavers* themselves; as by comparing not only their *Leaves,* but *Flowers, Seeds,* and *Cods* together, may be evident. For the several *parts* of the *Flower* of a *Trefoyl,* are so many more *Flowers,* containing so many *Cods* of small *Seeds,* all, in shape, agreeable to the *Flowers, Cods,* and *Seeds* of *Pulse.* The same Relation, which *Trefoyls* have to the *Peas* or other *Pulse*; *Colts-foot,* hath to *Buttyr-Bur*; *Chickweed* to *Leucanthemum*; *Groundsell,* to *Jacobæa*; or *Scorodonia,* to *Foxglove*: Or, to go higher, as the *Leguminous* Kinds of *Herbs,* have to *Sena,* or some other of the *Lobed Shrubs* and *Trees.* And, as among *Animals,* there are some which connect several Kinds; as the *Batt* doth *Beasts* and *Birds*: So, among *Plants,* there are some also, which seem to stand between two *Tribes*; as *Lappa,* between *Knapweeds* and *Thistles*; *Lampsana,* between the *Intybaceous* Kind, and the *Mouse-ears.*

12. §. From hence likewise, the *Natures* of *Plants* may be conjectured. For in looking upon divers *Plants,* though of different *Names* and *Kinds*; yet if some affinity may be found betwixt them, then the *Nature* of any one of them being well known, we have thence ground of conjecture, as to the *Nature* of all the rest. So that as every *Plant* may have somewhat of *Nature individual* to it self; so, as far as it obtaineth any *Visible Communities* with other *Plants,* so far, may it partake of *Common Nature* with those also. Thus the *Wild,* and *Garden Cucumers,* have this difference; that the one *purgeth* strongly, the other, *not at all*: yet in being *Diuretick,* they both agree. The *Natures* of *Umbelliferous Plants,* we know, are various; yet 'tis most probable, that they all agree in this one, *scil.* in being *Carminative.*

The

The several sorts, both of *Corn* and *Grass*, are all akin; there is no doubt therefore, but that the *Seeds* of *Grass* themselves (of *Rye* and *Oats* it is tryed) if it were worth the while to order them, as *Barley*, would yield an inflammable Spirit. So likewise the several Kinds of *Pulse*, have some one community in their Form, as is said: for which reason, I question not, but that in some Cases, wherein *Cicers* are esteemed a good *Medicine*; a *Decoction* of the better sort of *Pease*, especially that we call the *Sugar-Pease*, may go beyond them. As doth also the *Flower* or *Meal* of *Beans*, that of the *seeds* of *Fœnugreek*; even there, where they are accounted excellent. So *Tulips, Lillies, Crocuses, Jacynths,* and *Onions* themselves, with meny others, in their several Degrees, are all *allied*. If therefore *Crocuses, Onions, Lillies,* agree in one or more *Faculties*, then why may not all the rest? as in being *Anodyne*; or in some other *Common Nature*; whereby, in their *Vegetation*, their *Parts* are Governed and Over-ruled, to one Common or *Analogous Form.*

13. §. The *Proportions* likewise, amongst the several *Parts* of *Vegetables*, for the same Reasons, deserve to be observed; the comparison being made, both betwixt the *Parts* of several *Plants*, and the several *Parts* of one. And here again, either betwixt any Two of the *Parts*, or any One of them, and the Whole besides, or all the rest put together. So some larger *seeds*, produce a small *Root*; as those of *Cucumer*: and others smaller, produce one very great; as those of *Bryony*. Some *Plants*, as the *Melon*, though themselves but very slender, yet have a vast and bulky *Fruit*; others again, as *Thistles*, and many yet more substantial, have no other *Fruit*, besides their *Seed*. So the *Seeds* of all *Pulse*, and especially, the *Garden Bean*, though large, yet produce but a small *Plant:* but those of *Foxglove, Mullen, Burdock, Sun-flower, &c.* being themselves much less, do yet produce a far greater. And especially, those *Seeds*, which are inclosed in the Thicker sort of *Cover,* (analogous to that I have elsewhere called the *Secondine*) as that of *Peony*; whose *seed,* so called, is only the *Nest* wherein the true and real *Seed* is lodged, no bigger than a little Pins head: which is also observable of the *Seeds* of divers other *Plants*. These, and the like *Proportions*, as they lie betwixt the several *Parts*, should be noted: and to what *Plants* or *Parts* especially, any of them may agree: comparing also in what other kind of *Properties* an agreement betwixt the said *Parts* may be found: that so doing, we may, if possible, amongst all their *Individual Natures,* be instructed to single out those *Common Ones,* which are concomitant to such Agreeing *Properties*.

Anat. Plant. Book 1. Chap. ult.

14. §. The several *seasons* also of *Plants*, and of their *Parts*, should be considered. Observing at what particular Times of the Year, any of them chiefly *Spring,* Early or Late. The Times wherein they *Germinate*; whether for some Space only, or all the Year long. Wherein they *Spring,* after Sowing; or *Flower,* after Springing, sooner, or flower. Which Flower, the *first* Year, or not till the *second.* Which *after* the *Leaves* are put forth, or *before* them; for so, some do, as the *Crocus Vernus, Bears-foot, Hepatica aurea,* and others; all the *Leaves,* at the time of their flowering, being old, or of the foregoing Year's growth. So likewise the *Maturation* of the *Fruit* or *Seed*; how long after the *Flower,* and the like. All or some of which *Varieties,* being

laid

laid together, we may probably conjecture the *Causes* thereof; and the *Natures* of the *Plants* in which they are seen: *scil.* as such a Degree of Heat may be necessary for the Fermentation, or the better Distribution of the *Sap* of such a *Plant*; or for the Impregnation of the *Aer*, to be mixed therewith; or the due Disposing of the *Soil*, to render the most convenient Aliment thereunto. So the *Principles* of such *Plants*, which flower all the Year, may be more equally proportion'd. Those which flower before the *Leaves* put forth, as the *Crocus Vernus*, and those which flower in *Spring*, may be accounted *Rank*, and full of *Volatile Salt*. But *Autumn Plants* especially, to abound with a Fixed: and the like.

15. §. The proper *Places* also of *Plants*, or such wherein they have, from their *Seeds*, or other way of *Propagation*, a Spontaneous growth, should be considered. And that as to the *Climate*; whether in one Colder, Temperate, or more Hot. The *Region*; Continent, or Island. The *Seat*; as Sea, or Land, Watry, Boggy, or Dry; Hills, Plains, or Vallies; Open, in Woods, or under Hedges; Against *Walls*, rooted in them, or on their Tops: and the like. And perhaps the *Seeds* of some *Plants*, as of *Mosses*, (which, through their smallness, will ascend like Moths in the Sun) may fly or swim for some time, in the Aer, *viz.* till they begin to shoot, and so become heavy enough, to fall down upon the Ground. From whence, in like manner, as from their *Seasons*, their particular *Natures* may be directed unto. In that, so far as we may conjecture the nature of such an *Aer*, *Soil*, or *Seat*, we may also of such a *Plant*, to which they are *congenial*.

16. §. So likewise, those many Varieties observable in the *Motions* of *Plants*, and of their Parts, both Kinds and Degrees; *Ascending*, *Descending*, and *Horizontal*; *Rectilinear*, and *Spiral Motions*, should be noted; to what *Plants* they agree, and wherein any of these Motions may be analogous to those of *Animals*. And in a word, any other *Forensick Properties* of *Plants*. And then, to Compare them all together; both being necessary. For *Thoughts* cannot work upon nothing, no more than *Hands*. He that will build an House, must provide Materials. And on the contrary, the Materials will never become an House, unless, by certain Rules, we joyn them all together. So, it is not, *simply*, the Knowledge of *many things*, but a multifarious Copulation of them in the Mind, that becomes prolifick of further Knowledge. And thus much for the first General *Mean*.

The Second General Mean.

17. §. THE NEXT which I propose, and that a most necessary one, is *Anatomy*. For when upon the Dissection of *Vegetables*, we see so great a difference in them, that not only their Outward *Figures*, but also their Inward *Structure*, is so Elegant; and in all, so Various; it must needs lead us thus to Think, That these Inward *Varieties*, were either to no *End*; or if they were, we must assign to what. To imagine the first, were exceeding vain; as if *Nature*, the Handmaid of *Divine Wisdom*, should with Her fine *Needle* and *Thred*, stitch up so many several *Pieces*, of so difficult, and yet so groundless a Work. But if for some *End*, then either only to be looked upon, or some other besides. If for this only, then this must be such as in respect whereof, Her Work is at no time, nor in any degree frustrate; the contrary whereunto, is most manifest. For although Men do every where, with frequent pleasure, behold the Outward Elegancies of

Plants

Plants; yet the Inward Ones, which, generally, are as Precise and Various as the Outward; we see, how usual it is, for the beholding of These, to be omitted by them. And besides, when we have observed *Nature's* Work, as well as we can; it may be no impediment to our best Endeavours, to believe, That some Parts of it, will still remain behind, *Unseen.* So that if to be *Seen,* were the only End of it, it must needs be wholly frustrate, as to the greater number of Men; and, in some part, as to all. Wherefore, we must suppose some other *Ends* of the said Varieties, which should have their *Effect*, and so These, not be in vain, whether Men beheld them or not; which, are, therefore, such as have respect to *Vegetation:* That the *Corn* might grow, *so*; and the *Flower, so*, whether or no Men had a mind, leisure, or ability, to understand *how.*

18. §. If then the *Anatomy* of *Vegetables* be so useful a *Mean*, we ought not to streighten it; but to force this, as well as the rest, to its utmost Extent. And therefore, first of all, To go through all the *Parts*, with equal care; examining the *Root, Trunk, Branch, Leaf, Flower, Fruit,* and *Seed.* Then to Repeat or Retrograde the Dissection, from *Part* to *Part*: in that, although the best Method of Delivery, for clear Discourse, can be but one, according to that of *Nature*, from the *Seed* forward, to the *Seed*: yet can it not but be useful, for That of Dissection, to proceed *to* and *fro*; somewhat or other being more Visible in each several *Part*, from whence still an Hint may be taken, for the ushering in the observation of it in the others. To examine, again, not only all the *Parts*, but *Kinds* of *Vegetables*, and comparatively, to observe divers of the same *size, shape, motion, age, sap, quality, power,* or any other way the *same*, which may also agree, in some one or more particulars, as to their *Interiour Structure*: and to make this comparison, throughout all their *Parts* and *Properties.* To observe them likewise, in several *Seasons* of the Year, and in several *Ages* of the *Plants*, and of their *Parts*; in both which, divers of them may be noted to change, not only their *Dimensions*, but their *Natures* also; as *Vessels*, do into *Ligaments*; and *Cartilages*, into *Bones*, sometimes, in *Animals*. And to do all this by several Ways of *Section*, Oblique, Perpendicular, and Transverse; all three being requisite, if not to Observe, yet the better to Comprehend, some Things. And it will be convenient sometimes to Break, Tear, or otherwise Divide, without a *Section.* Together with the *Knife* it will be necessary to joyn the *Microscope*; and to examine all the *Parts*, and every Way, in the use of That. As also, that both Immediate, and Microscopical Inspections, be Compared: since it is certain, That some things, may be demonstrated by Reason and the Eye conjunct, without a Glass, which cannot be discovered by it; or else the discovery is so dark, as which, alone, may not be safely depended on.

19. §. By these several Ways of Inspection, it will be requisite, To observe their Compounding *Parts*; as *Simply* considered, and as variously *Proportioned*, and *Disposed*. As *simply* considered, to note their *Number*; what, and whether the same, in all: their *Kinds*, wherein different in the same, or divers *Vegetables*: their *Original*, in part, or in whole: *Structure*, as to their *Contexture* and their *Cavities*; Their *Contexture*, within themselves severally, and as joyned together: their *Cavities*, as to their *Size, Shape,* and *Number*; in which a great variety

riety will be found. Next their *Positions* one amongst another, which are also various; as Anterior, Posterior, Collateral, Surrounding, Mediate, Immediate, Near, Remote; both as they respect the several *Parts*, and the several portions of one: And all these, as few, or more; these or others of them, may be diversly Compounded together. And then the *Proportions* they bear one to another; whether as to Minority, Equality, or Excess; each *Part* compared with each, and that as to the several Degrees appearing in the said *Proportions*; the Varieties whereof may be exceeding numerous. For if we should suppose but *Four* considerable *Parts* generally constitutive of a *Vegetable*: These *Four*, produce a Variety *Four* ways. First, when One is Unequal; and then it produceth only *Four* Varieties: and those two ways, *scil.* when one is Greater, and the other three, Equal and Less; or when one is Less; and the other three, Equal and Greater. Secondly, when Two be Unequal; and then they produce *Six* Varieties. Thirdly, when Three be Unequal, which produceth *Twelve* Varieties. Or lastly, when all Four be Unequal; which produceth *Twenty four*: which general Varieties, may be further multiplied by their several Degrees.

20. §. From all which, we may come to know, what the *Communities* of *Vegetables* are, as belonging to all; what their *Distinctions*, to such a Kind; their *Properties*, to such a Species; and their *Peculiarities*, to such Particular ones. And as in *Metaphysical*, or other Contemplative Matters, when we have a distinct knowledge of the *Communities* and *Differences* of Things, we may then be able to give their true *Definitions*: so may we possibly, here attain, to do likewise: not only to know, That every *Plant* Inwardly differs from another, but also wherein; so as not more surely to Define by the Outward *Figure*, than by the Inward *Structure*, What that is, or those things are, whereby any *Plant*, or Sort of *Plants*, may be distinguished from all others. And having obtained a knowledge of the *Communities* and *Differences* amongst the *Parts* of *Vegetables*; it may conduct us through a *Series* of more facile and probable *Conclusions*, of the ways of their *Causality*, as to the *Communities* and *Differences* of *Vegetation*. And thus much for the Second General *Mean*.

<small>The Third General Mean.</small>

21. §. HAVING THUS far examined the *Organical* and *Containing Parts* of *Vegetables*; it will be requisite, more designedly, to observe those also which are *Fluid*, or any others Contained in them: and that, for our better understanding both of the *Nature of Vegetation*, and of the said *Contained Parts*. And to make inquiry, *First* of their *Kinds*; as *Spirits*; both such as agree, in general, in being *Vinous*; and those that are Special, to particular *Plants*. *Aers* and *Vapours*; for the existence whereof, in all *Vegetables*, there are Arguments certainly concluding. And for the difference of their *Natures*, in being more dry, or moist, more simple or compounded, as they are existent in several *Parts*, there are probable ones. *Lympha's* or clear and watry *Saps*; which most *Plants*, in one *Part* or other, at some time of the Year, do Bleed *Mucilages*; as in *Mallow* and *Violet Leaves*; in many *Seeds*, as of *Quinces*, *Clary*; *Fruits*, as in *Cucumers*; distinct from the watry *Sap*, as by permitting it to stand and gelly upon the *Vessels* from whence it issues, is plain: And in the young *Berrys* of White *Bryony*, when about the bigness of a *Pepper-Corn*;

Corn; the juyce whereof is fo Vifcous, that the twentieth part of a *Grain*, will draw out above a *Yard* in length. *Oyles*; not only in *Seeds*, and fome *Fruits*, but other *Parts*; as in certain little cavities in the *Leaves* of *Savine*, vifibly collected while they are growing. *Gumms* or *Refines*; as in *Pine*, *Fir*, and others of this Kind. *Milks*; as in a vaft number of *Plants*, and amongft them, many not fufpected to yield any. For, of *Herbs*, not only moft of the *Umbelliferous Kind*, are *Milky*; but all or moft of the *Intybous*; *Poppys*; *Trachelium*s; *Perwinkles*; divers *Thiftles*; and even *Onions*, if cut at the bottome; with a great many more. Of *Trees*, not only the Little *Maple*, but the young *Shoots* of *Lawrel*, efpecially being crufhed; as alfo thofe of *Elder*, and fome others. To which may be added, fuch *Mucilages*, which though not fo properly contained *within* the *Parts*, yet are found lying *over* them; as over the firft *Spring-leaves* of all kinds of *Docks*; betwixt the *Leaves* and the *Veil* wherein they are involved. That fine white Flower or Powder, which lies over the *Leaves* of fome *Plants*, as of *Bears-Ear*: And in *Princes-Feather*, about certain *Apertures* only on the edges of the *Leaves*.

22. §. Of all thefe fhould be obferved, *firft* their *Receptacles*; fome of them, being proper to one; others, common to two or more of them: fince it is certain, that fome of them do Tranfmigrate from one, into another *Receptacle*, or that the fame *Receptacle* is filled with Fluid Bodies, of a quite different Nature, at the different *Seafons* of the *Year*, and *Ages* of the *Vegetable*. And it is alfo very probable, That two of fome of them, may, fometimes, be contained in one *Receptacle*, at the fame time; as in *Animals*, the *Lympha* in the D. *Thoracicus*, and that, and the *Chyle*, in the *Sanguineous Veffels*.

23. §. Then their *Motions*; both *Natural*, and fuch as may be effected by *Art*: and thofe either by Defcent or Afcent; And in afcending, through what different *Chanels* or *Parts* of the *Trunk*; fince it is certain, That there is a variety, both in refpect of the *Seafon*, and of *Vegetables*. Where it will fall in, To obferve the *Tapping of Trees*. As alfo their *Bleeding*: to what *Trees* it is proper to *bleed*: in thofe to which it is, with what difference of *Celerity*: and when their peculiar *Seafon*: for none will bleed at all times; neither will all *bleed* at the fame. And then their Collateral *Motion*, together with the Mode of their Tranfition from one *Organical Part* to another.

24. §. Next their *Quantities*, either of *one*; as the Comparifon is made betwixt feveral *Plants*, or betwixt the *Parts* of the fame. So the true *Seed* of all *Plants*, containeth more *Oyl*, in proportion, than any of the other *Parts*. Or elfe of *divers*, as coexiftent and bearing fuch a proportion one to another in the fame *Part*: of moft of which, it may be known by their refpective *Receptacles*. Yet the Computation muft not be made from the number of the faid *Receptacles*, *fimply*; but as that is in conjunction with their *Capacity*; and as their *Capacity* is proportioned to their furrounding *Sides*; the *Sides* of thofe of the leaft *Capacity*, being ufually as thick, as thofe of the greateft: fo that fuppofe Ten leffer, to lye within the compafs of One greater; the *Content* of thefe altogether, would fcarce be equal to half the *Content* of that One.

25. §. Also their *Consistence*; *scil*: of so many of them as are discriminable by Touch; in being Soft or Hard; Thin or Thick; Mucilaginous, Gummous, Glutinous, Friable, &c. And these in their several Degrees; in which there is a Variety, as in the *Milks* of some *Plants*, which are more *Dilute*, than that of others: *Mucilages*; which in some, are very thick and *Viscous*, in others, more diluted and coming nearer to a *watry Sap*. And by This, to be compared in the same manner, as by their *Quantity*.

26. §. Likewise their *Colours, Smells*, and *Tastes:* The general and particular Kinds of all which should be noted. And to what *Contained Parts*, and in what Variety, they appertain. So most *Resinous Gumms* are Tinctur'd, some, not; as that which drops from the *Domestick Pine*, is as clear as Rock-water. The *Milks* of some *Plants* are *Paler*, as in *Burdock*; of others *Whiter*, as in *Dandelyon, Scorzonera*; *Citrine*, as in the Root of *Trachelium, Angelica*; *Yellow*, as in *Lovage*. In some *Plants, Odorous*, as in *Umbelliferous*; in others not, as in *Cichoraceous*. That of Little *Maple, Tastless*; of Garden *Chervil, Sweet*; of *Fenil, Hot*; of *Scorzonera, Astringent*; of *Dandelion, Bitter*; and generally, in other *Plants*; but with many Degrees of Strength, and in conjunction with other *Tasts*. But most *Mucilages*, have little either *Colour, Taste*, or *Smell*; and the like. Here also the same *Qualities* are to be inquired into, as, in general speaking, they are said to belong to a *Vegetable*. Since it is more than probable, that all *Colours* (excepting *White*, which is sometimes common both to *Containing* and *Contained Parts*) all *Odours*, and *Tastes*, which are more immediately, and without a resolution of their *Essential Principles*, perceptible in a *Plant*; are not ascribable either to the *Organical*, or *Containing Parts*; but only to Those, Contained in them; as from divers reasons hereafter may appear.

27. §. And *first*, their *Colours*; where, with respect to several *Plants* and *Parts*, they are more *Changeable*; as Red, in *Flowers*; or *Constant*, as Green, in *Leaves*. Which, with respect to several Ages of one *Part*, are more *fading*, as Green in *Fruits*; or *durable*, as Yellow in *Flowers*. In what *Parts* more *Single*, as always in the *Seed*; or more *Compounded*, as in the *Flower*; and in what *Plants* more especially, as in *Pancy*. Which proper to *Plants* that have such a *Taste* or *Smell*, as both, in *White Flowers*, are usually less strong. To *Plants* that flower in such a *Season*, as a *Yellow Flower*, I think, chiefly, to *Spring Plants*. And to *Plants* that are natural to such a *Soil* or *Seat*, as to *Water-plants*, more usually, a *white Flower*. What, amongst all *Colours*, more Common to *Plants*, as *Green*; or more Rare, as *Black*. And what all these Varieties of *Colours* are upon *Cultivation*, but chiefly, in their natural *Soil*. To observe also with their superficial *Colours*, those within: so the *Roots* of *Docks*, are *Yellow*; of *Bistort, Red*; of *Avens*, Purple; but of most, White. Where the Inward, and Superficial *Colours* agree; as in the *Leaves*; or vary, as in the other Parts frequently. And in what manner they are *Situated*; some universally spreading, others running only along with the *Vessels*, as in the *Leaves* of *Red Dock*, and the *Flowers* of *Wood-sorrel*.

28. §. Next their *Odours*; what may be their principal *Seat*; whether one or divers *Seats* in the same *Plant*. What the chief *Matter* out of which they are continually bred. What similitude betwixt

the *smells* of divers *Vegetables*; as betwixt *Baume*, and a *Limon*; the Green *Leaves* of *Meadow-sweet*, and the green Rinds of *Walnuts*. Or betwixt those of *Plants* and *Animals*; as the *Smell* of green and well-grown *Carduus*, is like to that rank *scent, ab aliquorum axillis spiranti*. Which have a more sensible *Smell*; as most have; and which have less, as *Corn*. Where the green *Leaf* is the most Fragrant *Part*, as in *Musk-Cranesbill*; where the *Flower*, as in *Roses*; the *Root*, as in sweet *Calamus*. Where all the *Parts* have some *Odour*, where some, or one, only; as in *Scurvy-grass*, only the *Flowers*, unless the *Leaves* are bruis'd; and in *Arum*, the *Pestil* only; for neither the *Leaf*, nor *Root* hath any *smell*, unless cut; but this is strong enough, not much unlike to *Humane Excrements*.

29. §. But especially their *Tastes*, which it much importeth us more precisely to distinguish; *First*, by their general *Kinds*; for the number, even of these, may be computed greater than usually it is. I remember not, that *Heat* and *Acritude*, with respect to *Taste*, are distinguished; yet *Arum-Root* is very *Pungent*, without any proper *Heat*; and *Cloves*, are very *Hot*, without any proper *Pungency*. So the White *Roots* of *Yarrow*, have a *Taste*, hardly any other way perceptible, than by causing a gentle *glowing* and *continued Warmth* upon the Tongue. Also their *Respondencies* one to another; as that of *Zedoary*, and of the lesser *Cardamoms*, is somewhat like to *Camphire*. Likewise their *Degrees*; in which there is a great latitude, and may be extended from *One* to *Ten*, or with easie distinction, from *One* to *Five*: So the Root of *Sorrel*, is Bitter in the *first*; of *Dock*, in the *second*; of *Dog-Rose*, in the *third*; of *Dandelyon*, in the *fourth*; of *Gentian*, in the *fifth*: observing them, not only as they vary in several *Kinds* of *Plants*, but the several *Species* of one, as in *Cichory*, *Hawkweed*, *Dandelyon*. And then their *Compositions*; for *Tastes* are as truly *conjunct* in one *Part*, as *Colours*: by which, the latitude is still greater; In that all Kinds of *Tastes*, in all their Degrees, and in differing Numbers, may be variously Compounded together: For the most part, *Two*, as in the *Leaves* of *sharp-pointed Dock*, *Astringent*, and *Sowre*; in *Sorrel-Roots*, *Astringent* and *Bitter*; and in *Aloes*, *Bitter* and *sweet*; the one in the *fifth*, the other, in the *first* Degree; as upon an unprejudiced tryal may be perceived: and yet more evidently in the *Gall* of any *Land-Animal*. Sometimes *three*, as in *Agrimony*, *Bitter*, *Rough*, and *Sowrish*; and in *Agarick*, *Bitter*, *Rough*, and *Sweet*. And sometimes, perhaps more. The Sensible distinctions of all which, may lye almost as wide, as of *Plants* themselves. Wherefore, although it may be thought rashness, to take away the distinctions of *Hot*, *Cold*, *Moist*, *Dry*, *Thin*, *Gross*, and other *Qualities*, in their several Degree, which the *Ancients* have affixed to particular *Plants*: yet since they have done it, to many of them, with much uncertainty; and that, withal, they are, more properly, the *Effects* and *Operations* of *Plants*, than their *Qualites*; Practical Observation, may therefore approve it useful, to add these Sensible Ones of various *Tastes*, precisely distinguishing their *Conjugations* and *Degrees*. Lastly, their several Varieties and Mutations, with respect to the Subject wherein they reside, should also be noted, As, of all *Tastes* found in *Plants*, *Bitter* and *Sowr*, are most common; *Sweet* and *Salt*, most rare. Which latter, is not only perceptible in some *Sea-Plants*; but upon some others, as upon the fresh

Leaves of *Tamarisk*; which being licked while they grow, or when immediately gathered, are plainly *saltish*. How they vary with the *Age* of the *Plant*, or *Part*; as the *Roots* of *Radishes*, growing up to *Seed*, lose the strength of their *Tast*; so most *Fruits* are first *Sowre*, then *Sweet*. What proper to the several *Parts* of any one *Plant*; so the *Leaves* of *Wormwood* are extraordinary *Bitter*; the *Root* scarcely so at all; of an *Hot*, but quite different *Taste*. What more Common, or Rare, to any *Part*; so no *Root* that I ever tasted, is *Sowre*. And how they Alternate in several *Plants*; as the *Root* of *Stock-July-flower* is biting, not the *Leaves*; on the contrary, the *Leaves* of the *Water-Arsmart*, are *Biting*; but not the *Root*; and the like. To which we may add the difference of Time wherein the *Tastes* of *Plants* are perceived; as those of *Arum*, and *Rape-Crowfoot*, are both *Biting*; but that of the first, as it is slowly perceived, so it continues long; that of the other, quickly comes, and quickly goes.

30. §. Amongst the other Adjuncts of the *Contained Parts*, though not of these only, the *Faculties of Vegetables* are to be reputed. For so the *Rosin* of *Jalap*, which is Purgative, is as truly contained in the *Organical Parts* of that *Root*, as Blood is in *Veins*: It will be requisite therefore to make particular observation of these also. And first, what *Faculties* chiefly may reside in *Plants*, above others: so there is none of known use in *Salivation*, except by holding in the mouth: Although we may ask, Why some amongst them, may not (being Taken inwardly) have a power to evacuate by This, as well as other Violent ways? Where the *Faculty* is more universally spread over all the *Parts* of a *Vegetable*, as in *Asarum*. Where belonging chiefly or wholly to any particular *Parts* or *Part*; as chiefly to the *Root* of *Rhubarb*; and only to the true and proper *Seed* of *Barbado Nuts*. Whether some *Faculties*, may be proper to some *Parts* especially. What conjunction they may have with any sensible *Qualities*. So, many *Purgers*, are not only *Resinous* and *Gummous*; But also *Mucilaginous*; as *Bryony*, wild *Cucumer*, *Lapathum Sativum*; and therefore probably *Rhubarb*, when growing; *Mallows*, *Violets*, &c. Such as are Purging and Vomitory, though some of them have a strong *Taste*, yet the greater part, and of those, many of the stronger sort, have no *Taste*, or not Great; as *Senna*, *Jalap*, *Scammony*, *Hellebore*, *Asarum*, and others. Amongst which, although *Hellebore* hath a very *Durable Taste*, yet is it not very *High* or *Great*. So also, those that are most sensibly tasted, are, I think, for the most part, more or less *Bitter*; either simply, as *Colocynthis*; or *Bitter* and *Astringent*, as *Rhubarb*; or *Bitter* and *Sweet*, as *Aloe*; or *Bitter*, *Astringent*, and *Sweet*, as *Agarick*. Few are *Hot*, as *Iris*. Or *simply* Sweet. And though some may be Subacid, that are Mollifying or Lenitive, yet no proper Purge or Vomit is *Sowre*. Such *Plants* as are of a soft and sweetish *Taste*, without Viscosity, may be accounted good *Antiscorbuticks*, especially against the *Sea*, or other *Salt-Scurvey*; as are good sweet Pease: And sometimes the *Water* or *Spirit* of the *Shells*; which may easily be drawn from them, being first duly fermented, and hath a true *Vinous Taste*; but very mild, and not unpleasant. Those *Plants*, whose Parts are not only *Hot* but *Volatile*, as *Onions*, are generally good for *Burns*. Such as have a *Balsamick Taste* or *smell*, with a little *Astringency*, as *Hypericum*, *Golden-Rod*, *Lamium Luteum*, &c. the best *Wound-Herbs*. And such as are gently

Bitter,

Bitter, and *Penetrant* upon the *Tongue*, or in the *Throat*, as *Daify, Anagallis*, good *Cleanfers*. That fuch *Bodys*, principally, are *Anodyne*, which are *Yellow*, I think, is more than a conceit; Yelks of Eggs, Fœnugreek Seeds, Lint-feed Oyl, May-Butyr, Marrow, *Pinguedo Humana, Hyofcyamus luteus*, Safron, Sulphur, Opium, all *Anodyne* and *Yellow*. How likewife their *Faculties* and *Qualities* may vary their Degrees, either differently or together: fo *Aloe* and *Colocynthis*, are both *Bitter* in the higheft Degree; yet *Aloe*, which is alfo *Sweet*, Purgeth more moderately; *Colocynthis*, which is *Bitter*, but not *Sweet*, moft Violently. How far the *Faculties* of *Vegetables*, as well as their *Qualities*, may be Compounded; where, and which chiefly; as Aftrictive and Purgative in *Rhabarb*. Where this Queftion may be put, Whether divers other, and yet more extreme *Faculties*, as well as thefe of Aftrictive and Purgative, may not fomewhere or other be alfo found, or made, to meet: whereby the fame *Plant*, or fome *Preparation* of it, may be moft Potent, and yet moft Innocent; the *Malignity* thereof exerting its Power, and the *Virtue* its Soveraignty at the fame time. And laftly, what *Affinity* there may be betwixt them; as moft *Plants*, that are ftrong Purgatives, and efpecially Vomitories, I think, are alfo *Sternutatory*; as white *Hellebore, Jalap, Tobacco* : and on the contrary, fuch as are *Sternutatory*, are fome of the moft proper and moft potent *Medicines* for the Head, Brain, and *Genus Nervofum*, Taken inwardly, as *Lilium convalle, &c*. and the like.

31. §. Thus far a particular obfervation of the *Qualities* and *Faculties* of the *Contents* of *Vegetables* may proceed, as they are exiftent in their *Natural Eftate*. From which, although fome probable Conjectures may be made, of their Material and Formal *Effences*, and of the *Caufes* of their determinate Varieties, or the *Modes* of *Vegetation* neceffary thereunto : yet will our Conceptions hereof be more facile, clear, and comprehenfive, if by all other Ways of Obfervation, they be likewife examined, according as *Experiment* may be applicable to any of them.

32. §. As by *Contufion*; fo fome *Plants* give their *Smell*, not without Rubbing, or not fo well; as the green *Leaves* of *Stramonium, Scurvygrafs*, and many more: others lofe it by Rubbing, as the *flowers* of *Violets, Carnations, Borage, &c.* others yield it both ways, as *Rofemary, &c.* So fome *Apples* mend their *Tafte*, by Scoaping, and *Pears* by Rowling, efpecially that called the *Rowling Pear*.

33. §. By *Agitation*; which doth that, fometimes, by Force, which *Digeftion*, doth by Heat : fo any cold *Oyl* and a *Syrup* being, in a due manner, agitated together, of two Fluid bodies will become one Confiftent, as is known.

34. §. By *Frigifaction*; how far the *Juyces* of *Plants*, either without or within them, may be any of them, or fome more than others, fubject to *Cold* : and thereby to be deprived of their *Motion* or natural *Confiftence*, or may fuffer alteration in their *Colour, Tafte*, or *Smell*.

35. §. By *Infufion*; where I mean *Infufion* only in Common Water; So both *Caffia Lignea*, and *Cinnamon* are a little Mucilaginous; but the former moft. Some of the *Contents* of *Plants*, may be wholly diffolved in Common Water; fome but in part, others not at all; or very little; which is proper to fome *Milks*, as well as *Gums*. The *Colours, Smells* or *Taftes* they hereupon yield, are found various; and in
fome

some very unexpected: So the green *Leaves* of *Bawm*, being duly infused in common Water, without any other Body added, tincture it with a clear and deep Red, near that of *Claret Wine*, as I have often tryed.

36. §. By *Subsiding*; So the Juyce of *Sorrel*, being ordered as that of *Grapes*, will, in time, let fall a kind of *Tartar* or *Essential Salt*. And so perhaps will that of many other *Plants*, without any previous *Decoction*; although that be commonly thought to be necessary.

37. §. By *Digestion* with *Fermentation*; either of the entire *Vegetables*, or of the *Juyces*, or other *Contents*; and these by themselves, or with common Water. And hereby to note, what difference may be in the Strength, Celerity, or Continuance of the *Fermentation*. Likewise, how their *Qualities* may thereby be altered; as the Smell of *Violet-flowers*, from a most excellent *Fragrancy*, may, by *Digestion*, be reduced to an odious and abominable *stink*, like that of the black Mud of *Gutters*.

38. §. By *Digestion* with *Calefaction*; so the *Colour* of the *Juyce* of *Limons*, from Transparency (if that be a Colour) may be turned to a perfect Red. Whence it is that many are deceived in the *Preparation* called the *Tincture* of *Corals*; supposing the *Corals* to give the *Menstruum* its Colour. Whereas the *Menstruum* will obtain it, only by *Digestion*, without any *Corals*, mixed with it.

39. §. By *Decoction*; either of *Vegetables* themselves, or of their *Liquors*; and to observe what alterations follow. So *Turpentine* boiled becometh friable; *Sugar*, *Bitter*, and of a Brown Red. *Turneps* lose their Biting *Taste*; *Onions*, their Picquancy; yet neither of them convey those self same *Qualities* to the Water. The same may be observed in the *Decoction* of *sweet-Fennel-seeds*, *Aniseeds*, and others, losing much of their *Tastes* themselves, and yet conveying very little of them to the *Liquors* wherein they are boiled; the greater portion of their Volatile parts, and so their *Virtue* and *Taste* therewith, flying away. Whereof therefore it is much better to make an *Emulsion*, than to *decoct* them; or to make an *Emulsion* from them, with their own *Decoction*, especially if the *Medicine* be intended to be *Carminative*, as I have frequently observed. The *Decoction* should also be carried on throughout all degrees to that of an *Extract*; by which the *Qualities* thereof, sometimes, are much altered; as the *Colour* of all or most green *Leaves*, from a kind of Yellow, deepens at last into a dark one, as Black as Pitch.

40. §: By *Distillations*; both with the cold *Still*, *Alembick*, *Chappel-* and open *Furnace*: and to note what *Vegetables* thus give their *Smell* or *Taste*, and in what Degrees of strength, either under, or over their natural ones; as *Mint*, *Pennyroyal*, and the like, which are *Aromatick* and *Hot*, give their *Tastes* perfect: but *Wormwood*, which is *Aromatick* and *Bitter*; gives it but by halfs, pretty fully as *Aromatick*, little as *Bitter*. And *Carduus*, though also so exceeding *Bitter*, yet not being *Aromatick*, yieldeth a much weaker *Taste*. Also what *Vegetables* yield *Oyl* most plentifully; and what difference may be in those *Oyls*, as to their *Colour*, *Weight*, or otherwise; as that of *Cloves* is sometimes Red; of *Cinnamon*, limpid; both Ponderous. So to distil *Juices*, *Gums*, or other *Contents*, with an hot *fire*; and to see, what Bodies they yield, and of what *Qualities*; as *Turpentine* is known to yield, besides

its

its *Oyl*, a fubacid *Water*; *Vinegar*, an Eager *Spirit*; as that part may be called, which *Chymists* are wont to call the *Phlegm*.

41. §. By *Arefaction*; fo *Milks* which are Liquid, and White in their Natural Eftate, in Standing, grow Gummous, Yellow, and otherwife different; fo doth that of *Scorzonera*; and that of *Fenil* becomes a Balfamical, but Limpid *Oyl*. The *Roots* of *Angelica*, being dry'd, and cut by the length, exhibit their fmall *Veins* fill'd with an Aromatick *Rofin*. In the whiter parts of *Rhubarb*, is gathered a kind of *Saline Concret*; by which, this *Root*, in chewing, feems as if it were a little gritty. *Cabbage-stalks*, fliced, and laid in the Shade to dry, gather on them a kind of *Nitrous* Hoar. *Raifins* and *Corins* contain, not only a fweet Juyce, but alfo a true *Sugar*, which lies *curdled* in the *Pulp*, as the more *Saline* parts do in Green Soap. And the like is gather'd on the out-fide of a *Fig*; faving, that it is more *Nitrous*, as lying next the *Aer*. The *Roots* of *Arum*, upon drying, lofe much of the ftrength of their *Tafte*; but the contrary may be noted of many other *Roots*, which, upon drying, increafe it. Some, being cut and laid by, change their Natural *Colours*, into Red, Pnrple, Yellow, Green, or White; as *Liquorifh*, into White, in fome places; and *Peony*, into Red: and fometimes into two; as *Patience*, into Yellow and Red.

42. §. By *Affation*; thus *Apples*, by roafting, eat more Sowre. The *Root* of *Horfe-Radifh*, toafted, tafteth like a *Turnep*. *Potatoes*, *Onions*, and many other *Roots*, and *Parts*, have their *Taftes*, either Altered or Refracted; which chiefly, and in what manner, fhould be obferved. There is one alteration, as remarkable, as commonly known; and is that which followeth upon roafting or baking in one kind of the *Waldenfian* Pears, which, for a *Walden*, we corruptly call a *Warden*.

43. §. By *Uftion*; wherein fome *Plants*, or *Parts* of them, burn very quietly; others, not without violent motions; fo *Fenil-Seeds*, held in the flame of a Candle, will fpit and fpurtle, like the *Serum* of *Blood*. Some *Vegetables* lofe their *Smell*, as *Rofes*; others, keep it, as *Rofemary*; and others, mend it, as *Lignum Aloes*, To note, not only the alteration of their *Qualities*, but what they yield; as *Turpentine*, which, in *Diftillation*, yieldeth *Oyl* and *Water*, both limpid; upon *Uftion*, fheweth nothing but a black *Soot*. So *Benzoine*, by *Diftillation*, *Oyl*; by *Uftion*, white *Flowers*, as is known.

44. §. By *Calcination*; and here to obferve, wherein the *Caput Mortuum* of one, may differ from, or agree in Nature with that of another; and alfo to compare thefe with thofe of *Animal Bodies*. As alfo in their *Quantities*. And to compare them with what they yield by *Diftillation* and *Uftion* as to both. Thus far they have been tryed *fingly*, or by themfelves. They fhould alfo be examined,

45. §. By *Compofition*; not only with *Water*, as in fimple *Infufions*, *&c.* but with any other Bodies, which may have a power of acting upon them, or upon which, thefe may have a power to act. And fo to make *Infufions*, *Deftillations*, *Decoctions*, *Digeftions*, in divers kinds of *Liquors*, as *Vinegar*, *Urine*, *Spirit* of *H. H. Wine*, *Blood*, *Milk*, or others. So in *Infufions*, fome Red Colours are heightned by *Acids*; Blews, turned Purple. So fetid *Spirits* (as of *H.H.*) may be rendred much more grateful, by being *Rectified*, once or twice, with frefh *Aromaticks*, To obferve alfo what follows, upon mixing the *Liquors*, or other Parts

of *Plants* together; as *Oyl* of *Turpentine*, by *Digestion* with a *Lixivial Salt*, extracteth thence a Red *Tincture*. Or with *Salts, Earths, Metals*, or any other Bodies; as the *Juyce* of the green *Leaves* of *Rasberry, Primrose*, and divers other *Plants* (I think principally such as are Astringent) expressed upon *Steel*, as it drieth, becometh of a *Purple Colour*.

46. §. Lastly, by *Compounding* the *Experiment* it self, or joyning two or more of them, upon the same matter: as *Fermentation* and *Destillation*, as is used for some *Waters*. *Infusion* and *Fermentation*, as in making of *Beer*. *Fermentation* and *Coction*, or rather *Assation*, as in making of *Bread*. *Arefaction* and *Destillation*, as may be tryed upon some *Herbs*; and with what difference from what may be noted, upon their being distilled, moist.

47. §. Having proceeded thus far, by all the above particular Ways of Observation; a Comparative Prospect must be taken of them: by which, at last, the *Communities* and *Differences* of the *Contents* of *Vegetables*, may be discerned; the manner of their *Causation* and *Original*, partly, be judged of; and wherein it is, that the *Essence* of their several *Natures* and *Qualities* doth consist, in some measure comprehended. And consequently, both from the knowledge of their particular *Natures*, and the Analogy found betwixt them; we may be able, better to conjecture, and try, what any of them are, or may be good for. For certainly, we shall then know, more readily, to apply things unto, and more fitly to prepare them for, their Proper Uses, when we first know, *what they are*. Notwithstanding, since the *Faculties* of *Plants*, do often lie more recluse; it is best, therefore, not wholly to acquiesce in such Conjectures, as their *Tastes*, or other *Sensible Properties* may suggest; but to subjoyn *Experiment*. In making of which, and in passing a Judgment thereupon, many Cautions, both in respect of the *Plant* whereof, and the *Subject* whereupon it is made, are requisite to be attended. Which yet, in regard they result not so directly from the Matter at present in hand; I shall not, therefore, here insist upon them, And thus much for the Third General *Mean*.

The Fourth General Mean.

48. §. THE *Contents* of the *Organical Parts* of *Vegetables*, having been thus duly Examined: it will be requisite to make the like Inquiry, into their *Principles*; or the *Bodys*, immediately concurrent and essential to their Being. And of these, we are to observe, First, their *Number*; whether well reducible to *five, six, seven*, or *more*, or *fewer*: and the Special Differences observable under any one General; since there are many Bodies, of very different Natures, confounded under one Name. Next their *Conjugation*; which they are, that either under or over those observable in *animal*, or other *Bodies*, are here joyned together in a *Plant*; How far common to the *Organical Parts* of divers *Plants*; or to the several *Organical Parts* of one; or how far different in them. So the predominant *Principle* of the *Parenchymous Parts* of a *Plant*, that it is an *Acid*, seems evident, From the general Nature of *Fruits*; and of *Corn*; and most *Parenchymous Roots*, which are either Spirituous, or Sower, or by Digestion, do easily become such. Likewise their *Proportions*; which stand in the greatest, which in the least, or in the meaner *Quantities*, and in what Degrees; both in divers *Vegetables*, and in the several *Organical Parts* of one. And then the *Concentration* and *Union* of them altogether; as to the degrees

grees of their Closeness or Laxity; or the manner of their Implication and Coherency; or as to their Location, one being more Central, another more Exposed and Rampant over the rest; or otherwise different. To examine these *Principles*, by their *Colour, Taste, Smell, Consistence, Fixedness, Volatility, Weight, Figures*, or other *Accidents*. And to these purposes, to go through the formentioned Ways of *Experiment*; as *Ustion, Calcination, Destillation, &c.* as any of them may appear applicable hereunto. So the *Essential Salt* of *Wormwood*, which may be obtained from the *Lixivial*; is *Bitter*, transparent, and commonly, of a *Cylindrick* figure: whereas that which is obtained by *Coction*, or from the *Extract*, is *tastless*, greyish, and almost *Cubick*: and that in the Extract of the Green *Leaves* of *Violets*, appears in fine transparent *Shoots*, like so many little Needles. And it is probable, That the *Salts* of most Kinds of *Plants*, whether Lixivial or Essential; and of these, whether obtained by *Decoction*, or otherwise, have either their *Figure*, or other *Qualities*, proper to themselves, whereby they are all distinguished one from another. And lastly, to make *Experiment* upon these *Principles*, mixing them with one another, or with other Bodies, or otherwise.

49. §. I know it will be difficult to make observations of this kind upon the *Organical Parts* of *Plants*, severally. Yet I have thought of some Ways, whereby true and undeceivable ones may be made. And the better to illustrate what I mean, I shall give one or two Instances of Tryal to this purpose. For the making of which, and some others of the like nature, I considered, That upon the *Anatomical Analysis* of all the *Parts* of a *Plant*, I had certainly found, (and shall hereafter shew) That in all *Plants*, there are *Two, and only Two Organical Parts Essentially distinct*, viz. The *Pithy Part*, and the *Lignous Part*, or such others as are analogous to either of These. So that, if we can think of any *Plants*, which will afford us either of these two, though not perfectly, yet in some good measure, simple and unmixed: We may then see, by putting them to a *Chymical, Test*, what *Principles* and *Proportion* of *Principles*, concur to *specifie* their *Substantial Forms*.

50. §. To the *Pithy Part, Starch*, or pure *Manchet* is analogous, as having very little of the *Lignous* mixed with them. I therefore ordered ℔ ij of *Starch* to be put into a *Retort*, and with a *Receiver* affixed, to be set in a *Sand Furnace*; and that all it would yield, should, by degrees, be forced over; which, besides what was evaporated at the Neck of the *Receiver*, was about ℔ j. of an acid and eager *Liquor*, of a heavy and blackish *Oyl* ℥ ss, and of a light *Oyl* ʒ j. The *Caput Mortuum* could not be reduced to Ashes, by the strongest heat which a *naked fire* in that Furnace would produce.

51. To the *Lignous Part, Hemp* or *Flax* is analogous, having very little of the *Pithy* mixed with them. I caused therefore ℔ ij of *Flax* to be put into a *Retort*, and manag'd as the *Starch*: whereupon, it yielded a Liquor, as I remember, somewhat like the former, and about the same quantity; no *Oyl* which remained liquid, when cold; but instead of that a *Butyr*, almost of the Consistence and Colour of the *Oyl* of *Mace*; and of this above ℥ iij, or near six times the quantity of the *Oyl* which was yielded by the *Starch*. The *Caput Mortuum* being burned to a white *Ash*, yielded some portion of a *Lixivial Salt*.

52. §. From whence, I shall, at present, only make these two Remarques; *First*, That although the chief portion, as to quantity, in both these *Bodys*, (as in most *Plants*) is an Acid Liquor; yet the latter, yields also some of an Alkaly, which the other doth not. So that they are the *Lignous Parts* of a *Plant*, generally, which yield the *Alkalick Salt*, or at least in the greatest Proportion. *Secondly*, That the *Sulphurious or Oleous Principle*, is also much more predominant in the *Lignous Part*, than in the *Pithy*. To these, the like Tryals upon other *Plants*, should be added; and other ways. So, in regard the *Soot* of most *Woods*, yields a *Volatile Alkaly*; it were fit to examine, Whether the *Soot* which is made of the *Pithy Parts* and that, of the *Lignous*, afford the said *Alkaly*, in equal qantity; or whether, as is most likely, that of the *Lignous* doth afford it in a far greater: and the like.

53. §. The prosecution of what is here proposed, will be requisite, To a fuller and clearer view, of the *Modes* of *Vegetation*, of the *Sensible Natures* of *Vegetables*, and of their more Recluse *Faculties* and *Powers*. First, of the *Modes* of *Vegetation*. For suppose we were speaking of a *Root*; from a due consideration of the *Properties* of any *Organical Part* or *Parts* thereof; 'tis true, that the real and genuine *Causes* may be rendred, of divers other dependent *Properties*, as spoken generally of the whole *Root*. But it will be asked again, What may be the *Causes* of those *first* and Independent ones? Which, if we will seek, we must do it by inquiring also, What are the *Principles* of those *Organical Parts*? For it is necessary, that the *Principles* whereof a Body doth consist, should be, if not all of them the *active*, yet the *capacitating Causes*, or such as are called *Causæ sine quibus non*, of its becoming and being, in all respects, both as to *Substance* and *Accidents*, what it is: otherwise, their Existence, in that Body, were altogether superfluous; since it might have been without them: which if so, it might then have been made of any other; there being no necessity of putting any difference, if neither those, whereof it is made, are thought necessary to its Being. Wherefore if we will allow a Body, and so the *Organical Parts* of a *Vegetable* to have *Principles*, we must allow these *Principles* their necessary Use; and that the Shapes or other Properties of the said *Parts*, are as much dependant upon the *Nature* of These; as is the Roundness of a Drop of Ink, upon the Fluidity of Water, ingredient to it.

54. §. Again, the *Principles* of the *Organical Parts* being known, we may from thence obtain a further knowledge of the *Natures*, and *Causation* or *Original* of their *Contents*; since these *Contents* are not only included in the said *Organical Parts*, but also Created by them: and must needs be so, whether we will suppose the *Principles* of these *Contents* to be præ-existent to their reception thereinto, or not. For, if not præ-existent, what can be clearer, than that the said *Parts* give them their Existence? And if præ-existent, yet in regard they are distinguished, and such only of them admitted in such sort into an *Organical Part*, from amongst others, as are apt to combine and mix together in such a *Form*, and so to constitute such a *Liquor*; it is as clear, that the Existence, if not of those *Principles*, yet of that *Liquor*, is dependent on the said *Part*.

55. §. And if by means of the said *Organical Parts*, it is, that their *Contents* become *such* and *such* peculiar *Mixtures*; it is hence also manifest, That, by the same means, they are of *such* distinct *Faculties* and *Powers*: Because the *Faculty* or *Power* of a Body, lieth not in any of its *Principles* apart; but is a Resultance from them all; or from their being, in such peculiar sort and manner, United and Combined together. So the *Principles* of the Purgative Parts of a *Root*, as of *Rhubarb*, although we should suppose them to be existent in the surrounding *Earth*, yet we cannot say, That *that Earth*, or the *Principles* therein contained, are Purgative; but only that they are such, as by being combined together, in such a peculiar way, may become *so*. So the several parts of a *Clock*, although they are and must be all præ-existent to it, and it is their *Form*, by which they are, what they are; yet is it the *setting together* of such *Parts*, and in such a way only, that makes them a *Clock*. And since we see that the *Mixture* of two Bodies of two different *Qualities*, as of Two *Colours*, will produce a Third *Colour*, differing from them both; as Blue and Red, do a Murrey: Why should not Two or More Bodies of different *Natures*, be so combined together, as to produce a Third *Nature*? Or wherefore may not that be allowed to be performed by *Nature*, which by Artificial Compounding of *Medicines*, or other Bodies, is designed, and oftentimes effected? I'll give but one Instance; *Water*, *Grease*, and an *Alcalizate Salt*, may be easily so ordered as to be invested with new *Qualities*, Nature, and *Powers*; the *Salt*, to lose its extreme fiery Pungent *Taste*; the *Tallow*, its *Smell*; and being before unsociable with the *Water*, to mingle therewith: neither *Tallow, Salt*, nor *Water* alone, will fetch out a spot of *Grease*; but all united easily do it: the same Three Bodys united, are, in some Cases, as in the *Jaundies*, no ill *Medicine*; any of which, given alone, may rather prove prejudicial, than a cure: and all this done, only by duly boiling them together into one Body, which we call *Sope*.

56. §. Whence again, if it be such an *Union*, and *Proportion*, of such a *Sort* of *Principles*, which produceth such a *Faculty*; and that we may, by any means, come to know what these are; we may, possibly, also attain to the knowledge of such *Rules*, whereby any kind of *Faculty* may be made; as to Compound such Bodies, which are neither Purgative nor Vomitory, so together, as to be Invested with those *Faculties*. And if to Make them, then consequently, to Mend, Exalt, Strengthen, and Enoble them, with greater ease and certainty. And thus much for the Fourth General *Mean*.

57. §. HITHERTO, We have considered the *Materials* of a *Vegetable*, only as Ingredient to it: there yet remains a *Fifth Story* to be ascended; which is, to consider these *Materials* as they are derived from *abroad*: or as, after they are received and naturalized, they may, with others yet abroad, have any kind of correspondence. And these are *Four* in general, *scil. Earth, Water, Aer*, and *Sun*; all which, in that they contribute so universally to *Vegetation*, and to whatsoever is contained in a *Vegetable*, it is therefore requisite, that of These likewise, Particular Observation should be made.

The Fifth General Mean.

58. §. And First, of the *Earth*, and of all Solid Receptacles of *Plants*. Where we are to consider their several *Kinds*; as Mellow, Sandy, Clayie, Chalky, and others. Their *Ingredients*; as Rank and

Mellow *Earth*, with Sand, or with Clay; or Sand with Clay; or altogether; and in what Proportions. The *Principles* whereinto any one of these *Ingredients*, separated from the rest, and put to the Test of *Distillation, Ustion, Calcination*, or other, either alone, or by mixture with other Bodies, may be Resolved. And by their *Qualities*, as *Colour, Smell, Taste, &c.* both *Ingredients* and *Principles* to be examined. To make tryal of the growth of *Plants*, in all kinds of *simple Soils*; either *Earthy* or *Mineral*, as Clay, Marl, Oker, Fullers Earth, Bole Armeniac, Vitriol, Allum, *&c.* or *Vegetable*, as Rotten Wood, Brans, Starch, or Flower, *&c.* or *Animal*, as Dungs, pounded Flesh, dried and powdered Blood, and the like; that it may appear, how far any of these may contribute to the growth of a *Plant*; or to one, above another.

59. §. Next of the *Water*, and of all Liquid Receptacles. Where the several kinds of *Water*, from Wells, Springs, Rain, and Rivers are, by their *Qualities* and *Faculties*, to be examined; as these, and by these, their *Principles*, either in their Natural State, or upon Digestion, or otherwise, may be observable: since Common *Water* it self, is undoubtedly compounded of several *Principles*; the simplicity thereof, not being argued, from its Clearness and Transparency; for a Solution of *Alum*, though it containeth a considerable quantity of *Earth*, is yet very Clear: nor from its seeming to have neither *Smell* nor *Taste*; for *Water-drinkers* will tell you of the varieties of both in different *Waters*. Besides, if these *Qualities* should be accounted rather Phansie, than Sense; the difference of *Waters* is yet more manifest, from their different Effects, observed by *Cooks, Laundresses, Brewers,* and others, that have occasion to use them: for not to mix with *Sope*, without curdling; not to boil Meat tender, or without colouring it red; and the like, are the vices of some *Waters*, not of others, which yet would seem, in Colour, Taste, and Smell, to be the same. Tryal should also be made of the growth of *Plants* in all kinds of Liquid Receptacles, as *Common Water, Snow Water, Sea Water, Urine, Milk, Whey, Wine, Oyl, Ink, &c.* Or any of these, with a solution of *Salt, Nitre, Sal prunellæ, Sope,* or other body. And hereby to observe what follows, either in the *Liquor*, or in the *Plant* it self: as if any *fixed* Body, being weighed before its dissolution in *Water*; and if the *Plant*, set herein, groweth; the *Water*, being then evaporated; whether the quantity of that dissolved body, continue the same, or is lessened. So, whether any *Vegetable* will become *Opiate*, by growing a considerable time in a plain Solution or Water-tincture of *Opium*; and the like. Which *Experiments*, what event soever they have, yet at least, for our further instruction in the *Nature of Vegetation*, may be of use.

60. §. Next of *Aer*, where it will be requisite to inquire, what sort of Bodies may be herein contained: It being probable, from the variety of *Meteors* formed herein; and of *Vapours* and *Exhalations* continually advanced hereinto; that some or other of them, may bear an Analogy, to all Volatile Bodys, whether *Animal, Vegetable,* or *Mineral*. The flourishings also of *Frozen Dew*; and the Green Colour, which the *Aer* gives the Ground or Water, when, for some time exposed to it; and other effects; seem to argue, that it is Impregnated with *Vegetable Principles*. To consider also the peculiar Nature of that Body, which is strictly called, *Aer,* And of that true *Aerial Salt*, which to me, seemeth probable,

bable, that it is diffolved in the Æther, as other *Salts* are in *Water*, or in the Vaporous parts of the *Aer*. As alfo to try, what different Effects, a diverfity of *Aer* may have upon a *Vegetable*; as by fetting a *Plant*, or *Seed*, either exceeding Low, as at the bottom of a deep Well; or exceeding High, as on the top of a Steeple. Or elfe by expofing fome *Soil* to the *Aer*, which is affuredly free from any *Seed*, and fo, as no *Seed* can light upon it; and to obferve, whether the *Aer* hath a power of producing a *Vegetable* therein, or not : and the like.

61. §. Laftly of the *Sun*; as to which, it may be confidered, What Influence it may have upon the *Plant* it felf ; upon the *Soil* ; Or upon the *Aer*. Whether that *Influence* is any thing elfe befides Heat : or may differ from that of a *Fire*, otherwife, than by being Temperate, and more Equal. That it doth, feems evident from an *Experiment* fometime fince given us, in one of the *Parifian Journals des Scavans*, and which I therefore think very applicable to our prefent purpofe. If you hold a *Concave* at a due diftance, againft a Fire, it will collect and caft the Heat into a burning *Focus*: but if you put a peice of plain Glafs between them, the Glafs will fcatter the Heat, and deftroy the *Focus*. Whereas the *Sun-Beams*, being gathered in like manner, will pafs through the interpofed Glafs, and maintain their *Focus*. As for That, of the Collection of the *Sun-beams*, by the help of *Glaffes*, in the form of a *Magiftery*, or of *Flowers*, and fuch like, I defire to fufpend my thoughts of them, till I fee them. I will only fay thus much further at prefent, That I do not underftand why the *Sun* fhould not have fome Influence upon Bodies, befides by Heat, if it may be granted, That the *Moon* hath ; for which, it fhould feem, there are fome good Arguments.

62. §. WE HAVE thus far examined the *Principles* neceffary to *Vegetation*. The *Queftion* may be put once more, In what manner are thefe *Principles* fo adapted, as to become capable of being affembled together, in fuch a *Number*, *Conjugation*, *Proportion*, and *Union*, as to make a *Vegetable Body* ? For the comprehenfion whereof, we muft alfo know, What are the *Principles* of thefe *Principles*. Which, although they lie in fo great an abyfs of obfcurity ; yet, I think, I have fome reafon to believe, that they are not altogether undifcoverable. How far they may be fo, I am fo far from Determining, that I fhall not now Conjecture. *A Sixth General Mean. Only hinted.*

63. §. THIS is the *Defign*, and thefe the *Means* I propofe in order thereunto. To which, I fuppofe, they may all appear to be neceffary. For what we obtain of *Nature*, we muft not do it by commanding, but by courting of Her. Thofe that woo Her, may poffibly have her for their Wife ; but She is not fo common, as to proftitute her felf to the beft behaved *Wit*, which only practifeth upon it felf, and is not applied to her. I mean, that where ever Men will go beyond Phanfie and Imagination, depending upon the Conduct of *Divine Wifdom*, they muft Labour, Hope and Perfevere. And as the *Means* propounded, are all neceffary, fo they may, in fome meafure, prove effectual. How far, I promife not ; the Way is long and dark : and as Travellers fometimes amongft Mountains, by gaining the top of one, are fo far from their Journeys end ; that they only come to fee another lies before them : fo the Way of *Nature*, is fo impervious, and, as I may fay, down Hill and up Hill, that how far foever we go, yet the furmounting of one difficulty, is wont ftill to give us the profpect of another. We may *The Conclufion.*

there-

therefore believe, our attainments will be imperfect, after we have done all: but because we cannot attain to all, that therefore we should endeavour after nothing; is an Inference, which looks so much awry from the Practical Sense of Men, that it ought not to be answered. Nor with better Reason, may we go about determining, what may be done. The greatest Designs that any Men undertake, are of the greatest uncertainty, as to their Success: which if they appear to be of good Import, though we know not how far they are attainable, we are to propound the *Means*, in the utmost use whereof only, we can be able to judge: A *War* is not to be quitted, for the hazards which attend it; nor the *Councils* of *Princes* broken up, because those that sit at them, have not the Spirit of Prophecy, as well as of Wisdom. To conclude, If but little should be effected, yet to design more, can do us no harm: For although a Man shall never be able to hit *Stars* by shooting at them; yet he shall come much nearer to them, than another that throws at *Apples*.

F I N I S.

THE ANATOMY OF PLANTS,

BEGUN.

WITH A General Account OF VEGETATION,

Grounded thereupon.

The FIRST BOOK.

Prefented in Manufcript to the ROYAL SOCIETY, Sometime before the 11th. of *May*, 1671.

And afterwards in Print, *December* 7. of the fame Year 1671.

By *NEHEMJAH GREW* M.D. Fellow of the *Royal Society*, and of the *College* of *Phyficians*.

𝕿𝖍𝖊 𝕾𝖊𝖈𝖔𝖓𝖉 𝕰𝖉𝖎𝖙𝖎𝖔𝖓.

LONDON,

Printed by *W. Rawlins*, 1682.

TO THE
RIGHT HONOURABLE

WILLIAM

Lord Vi-Count Brouncker,

THE

PRESIDENT,

And to the

Council and Fellows

OF THE

ROYAL SOCIETY,

The following

ANATOMY

Is moſt HUMBLY

PRESENTED

By the AUTHOR

NEHEMJAH GREW.

TO THE
Right Reverend
JOHN
Lord Bishop of
CHESTER.

MY LORD,

 Hope your pardon, if while you are holding *That best of Books* in one Hand, I here present some Pages of that of *Nature* into your other: Especially since *Your Lordship* knoweth very well, how excellent a *Commentary* This is on the *Former*; by which, in part, *GOD* reads the World his own Definition, and their Duty to him.

But if this Address, *my Lord*, may be thought congruous, 'tis yet more just; and that I should let *Your Lordship*, and others know, how much, and how deservedly, I resent *Your* extraordinary Favours. Particularly, that you were pleased, so far to animate my Endeavours, towards the Publishing the following *Observations*. Ma-

F 2 ny

Epistle Dedicatory.

ny whereof, and moſt belonging to the *Firſt Chapter,* having now lain dormant, near ſeven years; and might ſtill, perhaps, have ſo continued, had not *Your Lordſhips* Eye, at length, created Light upon them. In doing which, *You* have given one, amongſt thoſe many Tokens, of as well *Your* readineſs to promote Learning and Knowledge by the hands of others; as *Your* high Abilities to do it by *Your* Own: Both which, are ſo manifeſt in *Your Lordſhip,* that, like the firſt Principles of *Mathematical Science,* they are not ſo much to be aſſerted, becauſe known and granted by all.

The Conſideration whereof, *my Lord,* may make me not only *Juſt,* in owning of your Favours; but alſo moſt *Ambitious* of your *Patronage:* Which yet, to beſpeak, I muſt confeſs, I cannot well. Not that I think, what is Good and Valuable, is always its own beſt Advocate: for I know, that the Cenſures of Men, are humorous, and variable; and that one *Age,* muſt have leave to frown on thoſe *Books,* which another, will do nothing leſs than kiſs and embrace. But, chiefly, for this Reaſon, Leſt I ſhould ſo much as ſeem deſirous, of *Your Lordſhips* Solliciting my Cauſe, as to all I have ſaid. For as it is your Glory, that you like not ſo to ſhine, as to put out the leaſt Star; ſo were it to *Your* Diſhonour, to borrow *Your* Name, to illuſtrate the Spots, though of the moſt conſpicuous. I am,

My Lord,

    Your Lordſhips

        Moſt Obliged,

           And

        Moſt Humble Servant

           NEHEMJAH GREW.

Coventry,
June 10. 1671.

THE CONTENTS.

CHAP. I.

Of the Seed in its State of Vegetation.

THE Method propounded, §. 1. The Garden-Bean, *dissected*, 2. The two Coats *Described*, 3, 4. The Foramen *in the outer* Coat, 5, 6. *What generally observable of the Covers of the* Seed, 7. *The Organical* Parts *of the* Seed, 8. *The Main Body*, 9, 10. The Radicle *in the* Bean, 11. *In other* Seeds, 12. *The Plume*, 13, 14. *The Similary* Parts, 15. *The* Cuticle, 16, 17. *The* Parenchyma, 18, 19, 20. *The* Inner Body, 21, *to* 29. *No solid Account yet given, of* Vegetation, 30. *The* Coats *how in common subservient to the* Vegetation *of the* Seed, 31. *The* Foramen, *of what use herein*, 32. *The use of the* Inner Coat, 33. *Of the* Cuticle, 34. *Of the* Parenchyma, 35. *Of the* Seminal Root, 36. *How the* Radicle *first becomes a* Root, 37. *By what means, the* Plume *all this while preserved*, 38. *How after the* Root *the* Plume *vegetates*, 39. *How the* Lobes, 40. *But not in all* Seeds, 41. *That they do in most, demonstrated*, 42, 43, 44. *What hence resolvable*, 45. *The use of the* Dissimilar Leaves, 46, *to the end.*

CHAP. II.

Of the Root.

THis *also to be Dissected*, §. 1. *The* Skin *hereof, its Original*, 2. *The* Cortical Body, *its Original*, 3. Texture, 4. Pores, 5. Proportions, 6. *The* Lignous Body, *its Original*, 7. Texture, 8. Proportions, 9. *The* Insertment, *its Original*, 10. Pores, 11. Number and Size, 12. *A fuller description hereof, with that of the Osculations of the* Lignous Body, 13, 14, 15. *The* Pith, *its original sometimes from the* Seed, 16. *Sometimes from the* Barque, 17. *Its* Pores, 18, 19. Proportions, 20. Fibres *of the* Lignous Body *therein*, 21. *The* Pith *of those* Fibres, 22. *How the* Root *grows, and the use of the* Skin, Cortical

The Contents.

tical *and* Lignous Body *thereto*, 23. *How it groweth in length*, 24. *By what means it descends*, 25. *How it grows in breadth*, 26. *And the* Pith, *how thus framed*, 27. *The use of the* Pith, 27. *Of the* Insertment, 28. *The joynt service of all the Parts.* 29, 30, 31.

CHAP. III.

Of the Trunk.

THE Coarcture, §. 1. *The* Skin, *its original*, 2. *The original of the* Cortical Body, 3. *Of the* Lignous, 4. *Of the* Insertment *and* Pith, 5. *The Latitudinal Shooting of the* Lignous Body, *wherein observable*, 6, 7. *The* Pores *of the* Lignous Body, *where and how most remarkable*, 8. *A lesser sort of* Pores, 9. *A third sort only visible through a* Microscope. *Observed in Wood or* Charcoal, 10. *Observed in the Fibres of the Trunks of Herbs*, 11. *The* Insertions, *where more visible*, 12, 13. *Their Westage with the* Lignous Body, 14. *The smaller* Insertions, *only visible through a* Microscope, 15. *No* Valves *in a* Plant, 16. *The Ranks of the Pores of the* Insertions, 17. *The Pores of the* Pith, 18, 19, 20. *How the* Trunk *ascends*, 21. *The disposition of its Parts consequent to that Ascent*, 22. *Consequent to the different Nature of the Sap*, 23. *The effects of the said Differences*, 24, *to* 28. *Which way, and how the Sap ascends*, 29, *to the end.*

The Appendix.

Of Trunk-Roots and Claspers.

Trunk-Roots *of two kinds*, §. 1, 2.   Claspers *of one kind*, 3. *The Uses of both*, 4, *to the end.*

CHAP.

The Contents.

CHAP. IV.

Of the Bud, Branch, and Leaf.

THE *Parts of the* Germen *and* Branch *the same with those of the* Trunk, §. 1, 2. *The manner of their growth,* 3. *How nourished,* 4. *And the use of* Knots, 5. *How secur'd* 6. *The Parts of a Leaf,* 7. *The Positions of the* Fibres *of the Stalks of Leaves,* 8. *For what Uses,* 9, 10. *The visible cause of the different circumference of Leaves,* 11. *And of their being flat,* 12. *And filamentous,* 13. *The Foulds of Leaves, their Kinds and Use,* 14, 15, 16. *The Protections of Leaves,* 17. *The use of the Leaf,* 18, *to the end.*

The Appendix.

Of Thorns, Hairs and Globulets.

Thorns *of two Kinds; the* Lignous, §. 1. *The* Cortical, 2. *An argument of the* Magnetick Descent *of the* Cortical Body, 3. Hairs *of divers Kinds,* 4, 5. *Their Use,* 6. Globulets *of two Kinds,* 7, 8.

CHAP. V.

Of the Flower.

ITs *three Parts,* §. 1. *The* Impalement, *of divers kinds,* 2. *Their use,* 3, 4. *The* Foliation, *its nature,* 5. *Foulds,* 6. *Protections,* 7. Downs, 8, 9. Globulets, 10. *Its Use,* 11, 12. *The Attire of two kinds. The Description of the first,* 13, 14, 15, 16. *Of the other,* 17, 18, 19, 20, 21. *Their use,* 22, *to the end.*

CHAP. VI.

Of the Fruit.

THE *Vital Parts of all, the same,* §. 1. *The Number, Description, and Original of the Parts of an* Apple, 2. *Of a* Pear, 3, 4. *Of a* Plum, 5, 6, 7. *Of a* Nut, 8. *Of a* Berry, 9. *The use of the Fruit,* 10, *to the end.*

CHAP. VII.

Of the Seed in its State of Generation.

WHat here further observed, not in the First Chapter, §. 1. *The* Case, *its Figures*, 2. *The outer Coat, its Figures*, 3. *Various Surface*, 4. *And Mucilages*, 5. *The nature of the outer Coat*, 6. *Its Apertures*, 7. *Next to which the Radicle usually placed*, 8. *The Original of the Outer Coat*, 9. *The Original of the Inner*, 10. *Its Nature*, 11, 12. *The Essential Parts of a* Plant, 13, 14. *The Secondine*, 15. *The* Colliquamentum *herein*, 16. *The Navel Fibres*, 17. *In the Generation of the Seed, the Sap first prepared in the Seed-Branch*, 18, 19. *Next in the inner Coat*, 20. *With the help of the Outer*, 21, 22. *The use of the Secondine*, 23. *Of the Ramulets of the Seed-Branch*, 24. *Of their Inosculation*, 25. *How the* Colliquamentum *becometh a* Parenchyma, 26, *to the end.*

THE ANATOMY OF PLANTS,

BEGUN.

With a General Account of *Vegetation*,

Founded thereupon.

CHAP. I.

Of the Seed in its State of Vegetation.

EING to speak of *Plants* ; and, as far as Inspection, and consequent Reason, may conduct, to enquire into the visible *Constitutions*, and *Uses* of their several *Parts* : I choose that Method, which, to the best advantge, may suit with what we have to say hereon. And that is the Method of Nature her self, in her continued Series of *Vegetations* ; proceeding from the *Seed sown*, to the formation of the *Root, Trunk, Branch, Leaf, Flower, Fruit,* and last of all, of the *Seed* also to be *sown again*; all which, we shall, in the same order, particularly speak of.

2. §. The *Essential Constitutions* of the said *Parts* are in all *Plants* the same : But for Observation, some are more convenient ; in which I shall chiefly instance. And first of all, for the *Seed*, we choose the great Garden-Bean.

3. §.

3. §. If then we take a *Bean* and dissect it, we shall find it cloath‐ed with a doubled *Vest* or *Coat*. These *Coats*, while the *Bean* is yet green, are separable, and easily distinguished. Or in an old one, after it hath lay'n two or three days in a mellow Soil; or been soaked as long a time in Water: as in *Tab.*1. When 'tis dry, they cleave so closely together, that the Eye not before instructed, will judge them but one; the inner *Coat* (which is of the most rare contexture) so far shrinking up, as to seem only the roughness of the outer, somewhat resembling *Wafers* under *Maquaroons*.

Tab. 1. *f.* 2.

4. §. The Inner *Coat*, in its Natural State, is every where twice, and in some places, thrice as thick, as the Outer. Next to the *Radicle*, which I shall presently describe, it is six or seven times thicker; and encompasses the *Radicle* round about, as in the same *Figure* appears.

*Tab.*1. *f.* 2.

5. §. At the thicker end of the *Bean*, in the outer *Coat*, a very small *Foramen* presents it self, even to the bare Eye. In Dissection 'tis found to terminate against the point of that *Part* which I call the *Radicle*. It is of that capacity, as to admit a small *Virginal* Wyer; and is most of all conspicuous in a green *Bean*. Especially, if a little magnified with a good *Spectacle-Glass*. This *Foramen* is not a hole casually made, or by the breaking off of the Stalk; but designedly formed, for the uses hereafter mentioned. It may be observed not only in the great *Garden-Bean*, but likewise in the other *kinds*; in the *French-Bean* very plainly; in *Pease*, *Lupines*, *Vetches*, *Lentiles*, and other *Pulse* 'tis also found; and in many *Seeds* not reckoned of this kind‐red, as in that of *Fœnugreek*, *Medica Tornata*, *Goats-Rue*, and others: In many of which, 'tis so very small, as scarcely, without the help of *Glasses* to be discovered; and in some, not without cutting off part of the *seed*, which otherwise would intercept the sight hereof.

*Tab.*1. *f.*1--a

6. §. That this *Foramen* is truly permeable, even in old *Setting-Beans*, and the other *Seeds* above named, appears upon their being soaked for some time in Water. For then, taking them out, and crushing them a little, many small bubles will alternately arise and break upon it.

7. §. Of all *Seeds* which have thick or hard *Covers*, it is also observable, That they have the same likewise *Perforated*, as above said, or in some other manner. And accordingly, although the *Coats* of such *Seeds* as are lodg'd in *Shells* or *Stones*, being thin, are not visibly perforated; yet the *Stones* and *shells* themselves always are; as in *Chap.* 7. shall be seen how. To which *Chapter*, what is farther observable, either as to the nature and number of the *Covers* of the *Seed*, I also refer.

8. §. The *Coats* of the *Bean* being stripp'd off, the proper *Seed* shews it self. The parts whereof it is composed, are three; *sc.* the *Main Body*, and two more, appendant to it; which we may call, the Three *Organical Parts* of the *Bean*.

9. §. The *Main Body* is not one entire piece, but always divided, lengthwise, into two halves or *Lobes*, which are both joyn'd together at the *Basis* of the *Bean*. These *Lobes* in dry *Beans*, are but difficultly separated or observ'd; but in young ones, especially boil'd, they easily slip asunder.

*Tab.*1. *f.*2,3.

10. §.

Book I. of Plants. 3

10. §. Some very few *Seeds* are divided, not into two *Lobes*, but into more; as that of *Cresses* into Six. And some are not at all divided, but entire; as the Grains of *Corn*. Excepting which few, all other *Seeds*, even the smallest, are divided, like as the *Bean*, into just two *Lobes*. Whereof, though in most *Seeds*, because of their minuteness, we cannot by dissection be inform'd; yet otherwise, we easily may; as in this *Chapter* shall be seen. *Tab.* 1. *f.* 4. *f.* 5.

11. §. At the *Basis* of the *Bean*, the two other *Organical Parts* stand appendent; by mediation whereof, the two *Lobes* meet and join together. The greater of these two *Parts* stands without the two *Lobes*, and upon divesting the *Bean* of its *Coats*, is immediately visible. 'Tis of a white colour, and more glossie than the *Main Body*, especially when the *Bean* is young. In the *Bean*, and many other *Seeds*, 'tis situated somewhat above the thicker end, as you hold the *Bean* in its most proper posture for growth. In *Oak-Kernels*, which we call *Acorns*, *Apple-Kernels*, *Almonds*, and many other *Seeds*, it stands prominent just from the end; the *Basis* and the *End* being in these the same, but in the *Bean* divers. *Tab.* 1. *f.* 2, & 3---a *Tab.*1.*f.*6---c

12. §. This *Part* is found not only in the *Bean*, and the *Seeds* above mentioned; but in all others: being that, which upon the Vegetation of the *Seed*, becomes the *Root* of the *Plant*; which therefore may be called the *Radicle*: by which, I mean the Materials, abating the Formality, of a *Root*. In *Corn*, it is that *Part*, which *Malsters*, upon its shooting forth, call the *Come*. 'Tis not easie to be observed, saving in some few *Seeds*, amongst which, that of the *Bean* is the most fair and ample of all I have seen. But that of some other *Seeds*, is, in proportion, greater; as of *Fœnugreek*, which is full as big as one of its *Lobes*. *Tab.*1.*f.*7---e

13. §. The lesser of the two said Appendents lies occult between the two *Lobes* of the *Bean*, by separation whereof only it is to be seen. 'Tis enclos'd in two small *Cavities*, form'd in the *Lobes* for its reception. Its colour comes near to that of the *Radicle*; and it is founded upon the *Basis* thereof, having a quite contrary production, *sc.* towards the *Cone* of the *Bean*; as being that very *Part*, which, in process, becomes the *Body* or *Trunk* of the *Plant*. In *Corn*, it is that *Part*, which after the *Radicle* is sprouted forth, or come, shoots towards the smaller end of the *Grain*; and by many *Malsters*, is called the *Acrospire*. *Tab.*1.*f* 3--b

14. §. This *Part* is not, like the *Radicle*, an entire Body, but divided, at its loose end, into divers pieces, all very closely couched together, as Feathers in a Bunch; for which reason it may be called the *Plume*. They are so close, that only two or three of the outmost are at first seen: but upon a nice and curious separation of these, the more interiour still may be discovered. In the *Bean*, this may be done: but in very few other *Seeds*; because of the extreme smallness of the *Plume*. Now as the *Plume* is that *Part* which becomes the *Trunk* of the *Plant*, so these pieces are so many true, and already formed, though not display'd *Leaves*, intended for the said *Trunk*, and foulded up in the same plicature, wherein upon the sprouting of the *Bean*, they afterwards appear. In a *French Bean*, and especially in the larger white Kind, or in the great *Indian Phaseolus*, the two outmost are very fair and elegant. In the great *Garden-Bean* two extraordinary small *Plumes* often, if not always, stand one on either side the great one now describ'd: From *Tab.*1.*f.*8--b

G 2 which,

which, in that they differ in nothing save in their size, I therefore only here just take notice of them. And these three Parts, *sc.* the *Main Body*, the *Radicle*, and the *Plume*, are concurrent to the making up of a *Seed*; and no more than these

15. §. Having thus taken a view of the *Organical Parts* of the *Bean*, and other *Seeds*; let us next examine the *Similary, sc.* those whereof the *Organical* are compos'd: a distinct observation of which, for a clear understanding of the *Vegetation* of the *Seed*, and of the whole *Plant* arising thence is requisite: To obtain which, we must proceed in our *Anatomy*.

16. §. Dissecting a *Bean* then, the first *Part* occurring is its *Cuticle*. The Eye and first Thoughts, suggest it to be only a more dense and glossy Superficies; but better enquiry discovers it a real *Cuticle*. 'Tis so exquisitely thin, and for the most part, so firmly continuous with the Body of the *Bean*, that it cannot, except in some small Rag, be distinctly seen, which, by carrying your Knife aslant into the *Bean*, and then very gently bearing upward what you have cut, will separate, and shew it self transparent. This *Cuticle* is not only spread upon the *Convex* of the *Lobes*, but also on their *Flats*, where they are contiguous, extending it self likewise upon both the *Radicle* and *Plume*, and so over the whole *Bean*.

17. §. This *Part*, though it be so far common with the *Coats* of the *Bean*, as to be like those, an *Integument*; yet are we in a quite different Notion to conceive of it: For whereas the *Coats*, upon setting the *Bean*, do only administer the *Sap*, and, as being superseded from their Office, then die; as shall be seen: this, on the contrary, with the *Organical Parts* of the *Bean*, is nourished, augmented, and by a real *Vegetation* co-extended.

18. §. Next to the *Cuticle*, we come to the *Parenchyma* it self; the *Part* throughout which *the Inner Body*, whereof we shall speak anon, is disseminated; for which reason I call it the *Parenchyma*. Not that we are so meanly to conceive of it, as if (according to the stricter sense of that word,) it were a meer concreted Juyce. For it is a Body very curiously *organiz'd*, consisting of an infinite number of extreme small *Bladders*; as in *Tab.* 1. is apparent. The *Surface* hereof is somewhat dense, but inwardly, 'tis of a laxer Contexture. If you view it in a *Microscope*, or with a very good *Spectacle-Glass*, it hath some similitude to the *Pith*, while *sappy* in the *Roots* and *Trunks* of *Plants*; and that for good reason, as in *Ch.* 2. shall be seen. This is best seen in green *Beans*.

Tab. 1. *f.* 9.

19. §. This *Part* would seem by its colour to be peculiar to the *Lobes* of the *Bean*; but as is the *Cuticle*, so is this also, common both to the *Radicle* and *Plume*; that is, the *Parenchyma* or Pulp of the *Bean*, as to its essential substance, is the same in all three. The reason why the colour of the *Plume*, and especially of the *Radicle*, which are white, is so different from that of the *Lobes*, which are green, may chiefly depend upon their being more compact and dense, and thence their different Tinctures. And therefore the *Lobes* themselves, which are green while the *Bean* is *young*; yet when it is *old* and *dry*, become whitish too. And in *many* other *Seeds*, as *Acorns*, *Almonds*, the *Kernels* of *Apples*, *Plums*, *Nuts*, *&c.* the *Lobes*, even *fresh* and *young*, are pure white as the *Radicle* it self.

20. §.

20. §. But although the *Parenchyma* be common, as is said, to all the *Organical Parts*; yet in very differing proportions. In the *Plume*, where it is proportionably least, it maketh about three *Fifths* of the whole *Plume*; in the *Radicle*, it maketh above five Sevenths of the whole *Radicle*; and in each *Lobe*, is so far over-proportionate, as to make at least nine Tenths of the whole *Lobe*.

21. §. By what hath been said, that the *Parenchyma* or *Pulp* is not the only constituting *Part*, besides the *Cuticle*, is imply'd: there being another *Body*, of an essentially different substance, embosom'd herein: which may be found not only in the *Radicle* and *Plume*, but also in the *Lobes* themselves, and so in the whole *Bean*.

22. §. This *Inner Body* appears very plain and conspicuous in cutting the *Radicle* athwart, and so proceeding by degrees towards the *Plume*, through both which it runneth in a large and strait *Trunk*. In the *Lobes*, being it is there in so very small proportion, 'tis difficultly seen, especially towards their *Verges*. Yet if with a sharp Knife you smoothly cut the *Lobes* of the *Bean* athwart, divers small *Specks*, of a different colour from that of the *Parenchyma*, standing therein all along in a Line, may be observ'd; which *Specks* are the Terminations of the *Branches* of this *Inner Body*. *Tab.*1. *f.* 10, 11, & 12.

*Tab.*1. *f.* 13.

23. §. For this *Inner body*, as it is existent in every *Organical Part* of the *Bean*; so is it, with respect to each *Part*, most regularly distributed. In a good part of the *Radicle* 'tis one entire *Trunk*; towards the *Basis* thereof, 'tis divided into three main *Branches*; the middlemost runneth directly into the *Plume*; the other two on either side it, after a little space, pass into the *Lobes*; where the said *Branches* dividing themselves into other smaller; and those into more, and smaller again, are terminated towards the Verges of each *Lobe*; in which manner the said *Inner Body* being distributed it becomes in each *Lobe* a true and perfect *Root*. *Tab.*1. *f.* 14.

24. §. Of this *Seminal Root*, as now we'll call it, from the Description here given, it is further observable; That the two main *Branches* hereof; in which the several *Ramifications* in each *Lobe* are all united, are not committed into the *Seminal Trunk* of the *Plume*, nor yet stand at right angles with *That* and the *Radicle*, and so with equal respect towards them both: but being produced through part of the *Parenchyma* of the *Radicle*, are at last united therein to the main *Trunk*, and make acute Angles therewith: as may be seen in the same. *f.* 14. *Tab.*1. *f.* 14.

25. §. This *Seminal Root* being so tender, cannot be perfectly excarnated, (as may the *Vessels* in the *Parts* of an *Animal*) by the most accurate Hand. Yet by dissection begun and continu'd, as is above declared, its whole frame and distribution may be easily observ'd. Again, if you take the *Lobe* of a *Bean*, and lengthwise pare off its *Parenchyma* by degrees, and in extreme thin slices, many *Branches* of the *Seminal Root*, (which by the other way of Dissection were only noted by so many *Specks*) both as they are fewer about the Basis of the *Bean*, and more numerous towards its Verges, in some good distinction and entireness will appear. For this you must have new *Beans*: or else soaked in Water, or buryed for some time.

26. §. As the *Inner Body* is branched out in the *Lobes*, so is it in the *Plume*: For if you cut the *Plume* athwart, and from the *Basis* proceed along the Body thereof, you'll therein find, first, one large *Trunk*

or *Branch,* and after four or five very small *Specks* round about it, which are the terminations of so many lesser *Branches* therewith distributed to the several parts of the *Plume.* The distribution of the *Inner Body,* as it is continuous throughout all the *Organical Parts* of the *Bean,* is represented, *Tab.* 1. *f.* 14.

*Tab.*1.*f.* 11-c

27. §. This *Inner Body* is, by dissection, best observable in the *Bean* and great *Lupine.* In other larger *Pulse* it shews likewise some obscure Marks of it self. But in no other *seeds,* which I have observed, though of the greatest size: as of *Apples, Plums, Nuts,* &c. is there any clear appearance hereof, upon dissection, saving in the *Radicle* and *Plume;* the reason of which is partly from its being, in most *seeds,* so extraordinary little; partly from its Colour, which in most *Seeds,* is the same with that of the *Parenchyma* it self, and so not distinguishable from it.

*Tab.*1.*f.*15.c

28. §. Yet in a *Gourd-seed,* the whole *Seminal Root,* not only its *Main Branches,* but also the Sub-divisions and *Inosculations* of the lesser ones, are without any dissection, upon the separation of the *Lobes,* on their contiguous Flats immediately apparent.

And as to the existence of this *Seminal Root,* what Dissection cannot attain, yet an ocular inspection in hundreds of other *Seeds,* even the smallest, will demonstrate; as in this *Chapter* shall be seen how.

29. §. In the mean time, let us only take notice; That when we say, every *Plant* hath its *Root,* we reckon short. For every *Plant* hath really two, though not contemporary, yet successive *Roots;* its *Original* or *Seminal-Root* within the *Lobes* or *Main Body* of its *Seed;* and its *Plant-Root,* which the *Radicle* becometh in its growth: the *Parenchyma* of the *Seed,* being in some resemblance, that to the *Seminal Root* at first, which the Mould is to the *Plant-Root* afterwards; and the *Seminal Root* being that to the *Plant-Root,* which the *Plant-Root* is to the *Trunk.* For our better understanding whereof, having taken a view of the several *Parts* of a *Bean,* as far as Dissection conducts; we will next briefly enquire into the Use of the said *Parts,* and in what manner they are the Fountain of *Vegetation,* and concurrent to the being of the future *Plant.*

An Account of the *Vegetation* of the *Seed.*

30. §. THE GENERAL Cause of the growth of a *Bean,* or other *Seed,* is *Fermentation.* That is, the *Bean* lying in the Mould, and a moderate access of some moisture, partly dissimilar, and partly congenerous, being made, a gentle *Fermentation* thence ariseth. By which, the *Bean* swelling, and the *Sap* still encreasing, and the *Bean* continuing still to swell, the work thus proceeds: as is the usual way of explicating. But that there is simply a *Fermentation,* and so a sufficient supply of *Sap* is not enough: but that this *Fermentation,* and the *Sap* wherein 'tis made, should be under a various Government, by divers *Parts* thereto subservient, is also requisite; and as the various preparation of the *Aliment* in an *Animal,* equally necessary: the particular process of the Work according whereto, we find none undertaking to declare.

31. §. Let us look upon a *Bean* then, as a piece of Work so fram'd and set together, as to declare a Design for the production of a *Plant;* which, upon its lying in some convenient Soil, is thus effected. First of all, the *Bean* being enfoulded round in its *Coats,* the *Sap* wherewith it is fed, must of necessity pass through these: By which means, it is

not

not only in a proportionate quantity, and by degrees; but also in a purer body; and possibly not without some *Vegetable Tincture*, transmitted to the *Bean*. Whereas, were the *Bean* naked, the *Sap* must needs be, as over-copious, so but crude and immature, as not being *filtred* through so fine a *Cotton* as the *Coats* be. And as they have the use of a *Filtre* to the transient *Sap*; so of a *Vessel* to that which is still deposited within them; being alike accommodated to the securer *Fermentation* hereof, as Bottles or Barrels are to Beer, or any other *Fermentative Liquor*.

32. §. And as the *Fermentation* is promoted by some *Aperture* in the Vessel; so have we the *Foramen* in the upper *Coat* also contrived. That if there should be need of some more *Aiery* Particles to excite the *Fermentation*; through *this*, they may obtain their Entry. Or, on the contrary, should there be any such *Particles* or *Steams*, as might damp the genuine proceeding thereof, through this again, they may have easie issue. Or if, by being over copious, they should become too high a *Ferment*; and so precipitate those soft and slow degrees, as are necessary to a due *Vegetation*. The said *Aperture* being that, as a common Pasport, here to the *Sap*, which what we call the *Bung-hole* of the Barrel, is to the new tunn'd Liquor.

33. §. And the *Radicle* being designed to shoot forth first, as presently shall be shew'd how; therefore is it distinctly surrounded with the Inner and more succulent *Coat*. That being thereby suppled on every side, its eruption may be the better promoted.

34. §. The *Sap* being passed through the *Coats*, it next enters the Body of the *Bean*; yet not indiscriminately neither; but, being filtred through the *Outer Coat*, and fermented in the Body of the *Inner*, is by mediation of the *Cuticle*, again more finely filtr'd, and so entereth the *Parenchyma* it self under a fourth Government.

35. §. Through which Part the *Sap* passing towards the *Seminal Root*, as through that which is of a more spatious content; besides the benefit it hath of a farther *percolation*, it will also find room enough for a more free and active fermenting and *maturation* herein. And being moreover, part of the true Body of the *Bean*, and so with its proper *Seminalities* or *Tinctures* copiously repleat; the *Sap* will not only find *room*, but also *matter* enough, by whose Energy its *Fermentation* will still be more advanced,

36. § And the *Sap* being duly prepared here, it next passeth into all the *Branches* of the *Seminal Root*, and so under a fifth Government. Wherein how delicate 'tis now become, we may conceive by the proportion betwixt the *Parenchyma* and this *Seminal Root*; so much only of the best digested *Sap* being discharged from the whole Stock in that, as this will receive. And this, moreover, as the *Parenchyma*, with its proper *Seminalities* being endowed; the *Sap* for the supply of the *Radicle*, and of the young *Root* from thence, is duly prepared therein, and with its highest *Tincture* and *Impregnation* at last enriched.

37. §. The *Sap* being thus prepared in the *Lobes* of the *Bean*, 'tis thence discharg'd; and either into the *Plume*, or the *Radicle*, must forthwith issue. And since the *Plume* is a dependent on the *Radicle*; the *Sap* therefore ought first to be dispenced to this: which accordingly, is ever found to shoot forth before the *Plume*: and sometimes an inch or two in length. Now because the primitive course of the *Sap* into the

the *Radicle*, is thus requisite; therefore, by the frame of the *Parts* of the *Bean* is it also made necessary. The two main *Branches* of the *Seminal Root*, being produced, as is before observed, not into the *Plume*, but the *Radicle*. Now the *Sap* being brought as far as the *Seminal Root*, in either *Lobe*; and according to the conduct thereof continuing still to move: it must needs immediately issue into the same *Part*, *Tab.*1. *f.* 14. whereinto the main *Branches* themselves do; that is, into the *Radicle*. By which *Sap*, thus bringing the several *Tinctures* of the *Parts* aforesaid with it, being now fed; it is no longer a meer *Radicle*, but is made also *Seminal*, and so becomes a perfect *Root*.

38. §. The *Plume*, all this while, lyes close and still. For the sake of which, chiefly it is, that the *Bean* and other *Seeds* are divided into *Lobes*, *viz.* That it might be warmly and safely lodged up between them, and so secur'd from the Injuries so tender a *Part* would sustain from the Mould; whereto, had the *Main Body* been entire, it must, upon the cleaving of the *Coats*, have lay'n contiguous.

39. §. But the *Radicle* being thus impregnated and shot into a *Root*; 'tis now time for the *Plume* to rouze out of its *Cloysters*, and germinate too: In order whereto, 'tis now fed from the *Root*, with laudable and sufficient *Aliment*. For as the Supplies and Motion of *Sap* were first made from the *Lobes*, towards the *Root*: so the *Root* being well shot into the Mould, and now receiving a new and more copious *Sap* from thence; the motion hereof must needs be stronger, and by degrees proceed in a contrary course, *sc.* from the *Root* toward the *Plume*: and, by the continuation of the *Seminal Root*, is directly conducted thereinto; by which being fed, it gradually enlarges and displays it self.

40. §. The course of the *Sap* thus turned, it issues, I say, in a direct Line from the *Root* into the *Plume*: but collaterally, into the *Lobes* also; *sc.* by those two aforesaid *Branches* which are obliquely transmitted from the *Radicle* into either *Lobe*. By which *Branches* the said *Sap* being disbursed back into all the *Seminal Root*, and from thence, likewise into the *Parenchyma* of the *Lobes*, they are both thus fed, and for some time augmenting themselves, really grow: as in *Lupines* is evident.

41. §. Yet is not this common to all *Seeds*. Some rot underground; as *Corn*; being of a laxer and less Oleous Substance, differing herein from most other *Seeds*; and being not divided into *Lobes*, but one entire thick Body. And some, although they continue firm, and are divided into *Lobes*, yet rise not; as the great *Garden Bean*. In which, therefore, it is observable, That the two Main *Branches* of the *Lobes*, in comparison with that which runs into the *Plume*, are but *Tab.*1. *f.* 14. mean; and so insufficient to the feeding and vegetation of the *Lobes*; the *Plume*, on the contrary, growing so lusty, as to mount up without them.

42. §. Excepting a few of these Two Kinds, all other *Seeds* whatsoever, (which I have observed) besides that they continue firm; upon the *Vegetation* of the *Plume*, do mount also upwards, and advance above the Ground together with it; as all *Seeds* which spring up with one or more *Dissimilar Leaves*: These *Dissimilar Leaves*, for the most part *Two*, which first spring up, and are of a different shape from those that follow, being the very *Lobes* of the *Seed*, divided, expanded, and thus advanced.

43. §.

43. §. The Impediments of our apprehension hereof are the Colour, Size and Shape of the *Dissimilar Leaves*. Notwithstanding, that they are nothing else but the *Main Body* of the *Seed*, how I came first to conceive, and afterwards to know it, was thus. First, I observed in general, that the *Dissimilar Leaves*, were never jagg'd, but even edg'd: And seeing the even verges of the *Lobes* of the *Seed* hereto respondent, I was apt to think, that those which were so like, might prove the same. Next descending to particular *Seeds*, I observed, first, of the *Lupine*; that, as to its Colour, advancing above the Ground, (as it useth to do) it was always changed into a perfect Green. And why might not the same by parity of Reason be inferred of other *Seeds*? That, as to its size, it grew but little bigger than when first set. Whence, as I discern'd (the Augmentation being but little) we here had only *Tab. 2. f. 1.* the two *Lobes*: So, (as some augmentation there was) I inferr'd the like might be, and that, in farther degrees, in other *Seeds*.

44. §. Next of the *Cucumber-seed*, That, as to its Colour, often appearing above ground, in its primitive white, from white it turns to yellow, and from yellow to green; the proper colour of a *Leaf*. That, as to its size, though at its first arise, the *Lobes* were little bigger than upon setting; yet afterwards, as they chang'd their Colour, so their di- *Tab. 2. f. 2.* mensions also, growing to a three-four-five-fold amplitude above their primitive size. But whereas the *Lobes* of the *Seed*, are in proportion, narrow, short and thick: how then come the *Dissimilar Leaves*, to be so exceeding broad, or long, and thin? The Question answers it self: For the *Dissimilar Leaves*, for that very reason are so thin, because so very broad or long; as we see many things, how much they are extended in length or breadth, so much they lose in depth, or grow more thin; which is that which here befalls the now effoliated *Lobes*. For being once dis-imprisoned from their *Coats*, and the course of the *Sap* into them, now more and more encreased; they must needs very considerably amplifie themselves: and from the manner wherein the *Seminal Root* is branched in them, that amplification cannot be in thickness, but in length or breadth. In both which, in some *Dissimilar* *Tab. 2. f. 3.* *Leaves*, 'tis very remarkable; especially in length, as in those of *Lettice*, *Thorn-Apple*, and others; whose *Seeds*, although very small, yet the *Lobes* of those *Seeds* growing up into *Dissimilar Leaves*, are extended an Inch, and sometimes more, in length. Though he that shall attempt to get a clear sight of the *Lobes* of *Thorn-Apple*, and some others, by *Dissection*, will find it no easie Task; yet is that which may be obtained; and in the Last *Book* shall be shew'd. From all which, and the observation of other *Seeds*, I at last found, that the *Dissimilar Leaves* of a young *Plant*, are nothing else but the *Lobes* or *Main Body* of its *Seed*. So that, as the *Lobes* did at first feed and impregnate the *Radicle* into a *perfect Root*; so the *Root*, being perfected, doth again feed, and by degrees amplifie each *Lobe* into a perfect *Leaf*.

45. §. The Original of the *Dissimilar Leaves* thus known, we understand, why some *Plants* have none; because the *Seed* either riseth not, as *Garden-Beans*, *Corn*, &c. Or upon rising, the *Lobes* are little alter'd, as *Lupines*, *Pease*, &c. Why, though the proper *Leaves* are often indented round; the *Dissimilar* like the *Lobes* are even-edg'd. Why, though the proper *Leaves* are often hairy, yet these are ever smooth. Why some have more *Dissimilar Leaves* than two, as *Cresses*,

H which

Hist. of the Prop. of Vege.

which have six, as the Ingenious Mr. *Sharrock* also observes. The reason whereof is, because the *Main Body* is not divided into Two, but Six, distinct *Lobes*, as I have often counted. Why *Radishes* seem at first to have four, which yet after appear plainly two: because the *Lobes* of the *Seed*, have both a little Indenture, and are both plaited, one over the other. To which, other Instances might be added.

46. The use of the *Dissimilar Leaves* is, first, for the protection of the *Plume*; which being but young, and so but soft and tender, is provided with these, as a double Guard, one on either side of it. For this reason it is, that the *Plume*, in *Corn*, is trussed up within a membranous *Sheath*: and that of a *Bean*, cooped up betwixt a pair of *Surfoyls*: But where the *Lobes* rise, there the *Plume* hath neither of them, being both needless.

Tab. 3.

47. §. Again, since the *Plume*, being yet tender, may be injur'd not only by the *Aer*, but also for want of *Sap*, the supplies from the *Root* being yet but slow and sparing; that the said *Plume* therefore, by the *Dissimilar Leaves*, may have the advantage likewise of some refreshment from Dew or Rain. For these having their *Basis* a little beneath that of the *Plume*, and expanding themselves on all sides of it, they often stand after Rain, like a Vessel of Water, continually soaking and suppling it, lest its new access into the *Ayr*, should shrivel it.

48. §. Moreover, that since the *Dissimilar Leaves* by their *Basis* intercept the *Root* and *Plume*, the greater and grosser part of the *Sap*, may be, by the way, deposited into those; and so the purest proceed into the yet but young and delicate *Plume*, as its fittest Aliment.

49. §. Lastly, we have here a demonstration of the being of the *Seminal Root*: which, since through the colour or smalness of the *Seed*, it could not by Dissection be observ'd, except in some few; Nature hath here provided us a way of viewing it in the now effoliated *Lobes*, not of one or two *Seeds*, but of hundreds; the *Seminal Root* visibly branching it self towards the Cone or Verges of the said *Lobes*, or now *Dissimilar Leaves*.

CHAP.

CHAP. II.

Of the ROOT.

AVING Examin'd and pursu'd the Degrees of *Vegetation* in the *Seed*, we find its two *Lobes* have here their utmost period: and, that having conveyed their *Seminalities* into the *Radicle* and *Plume*; these therefore, as the *Root* and *Trunk* of the *Plant*, still survive. Of these, in their order, we next proceed to speak; and first, of the *Root*: whereof, as well as of the *Seed*, we must by Dissection inform our selves.

2. §. In Dissection of a *Root* then, we shall find it with the *Radicle*, as the *Parts* of an Old Man with those of a *Fœtus*, substantially, one. The first *Part* occurring is its *Skin*, the Original whereof is from the *Seed*: For that extreme thin *Cuticle* which is spread over the *Lobes* of the *Seed*, and from thence over the *Radicle*, upon the shooting of the *Radicle* into a *Root*, is co-extended, and becomes its *Skin*.

3. §. The next *Part* is the *Cortical Body*. Which, when it is thin, is commonly called the *Barque*. The Original hereof, likewise is from the *Seed*; or the *Parenchyma*, which is there common both to the *Lobes* and *Radicle*, being by *Vegetation* augmented and prolonged into the *Root*, the same becomes the *Parenchyma* of the *Barque*. *Tab.*2. *f.* 4.

4. §. The Contexture of this *Parenchyma* may be well illustrated by that of a *Sponge*, being a Body Porous, Dilative and Pliable. Its *Pores*, as they are innumerable, so, extream small. These *Pores* are not only susceptive of so much Moisture as to fill, but also to enlarge themselves, and so to dilate the *Cortical Body* wherein they are: which by the shriv'ling in thereof, upon its being expos'd to the Air, is also seen. In which dilatation, many of its Parts becoming more lax and distant, and none of them suffering a solution of their continuity; 'tis a Body also sufficiently pliable; that is to say, a *most exquisitely finewrought Sponge*.

5. §. The Extention of these *Pores* is much alike by the length and breadth of the *Root*; which from the shrinking up of the *Cortical Body*, in a piece of a cut *Root*, by the same dimensions, is argu'd.

6. §. The proportions of this *Cortical Body* are various: If thin, 'tis, as is said, called a *Barque*; and thought to serve to no other end, than what is vulgarly ascrib'd to a *Barque*; which is a narrow conceit. If a Bulky Body, in comparison with That within it, as in the young *Roots* of *Cichory*, *Asparagus*, &c. 'tis here, because the fairest, therefore taken for the prime *Part*; which, though, as to Medicinal use, it is; yet, as to the private use of the *Plant*, not so. The Colour hereof, though it be originally white, yet in the continued growth of the *Root*, divers *Tinctures*, as yellow in *Dock*, red in *Bistort*, are thereinto introduced.

7. §. Next within this Part ſtands the *Lignous Body:* This *Lignous Body*, lyeth with all its parts, ſo far as they are viſible, in a Circle or *Ring.* Yet are there divers extreme ſmall *Fibres* thereto parallel, uſually mixed with the *Cortical Body*; and by the ſomewhat different colour of the ſaid *Cortical Body* where they ſtand, may be noted. Theſe *Fibres* the *Cortical Body,* and *Skin,* altogether, properly make the *Barque.* The Original of this *Lignous Body,* as of the two former, is from the *Seed*; or, the *Seminal Roots* of both the *Lobes,* being united in the *Radicle,* and with its *Parenchyma* co-extended, is here in the *Root* of the *Plant,* the *Lignous Body.*

Tab.2.f 4.b.

8. §. The Contexture hereof, in many of its parts, is much more cloſe than that of the *Cortical*; and their *Pores* very different. For whereas thoſe of the *Cortical* are infinitely numerous, theſe of the *Lignous* are in compariſon nothing ſo. But theſe, although fewer, yet are they, many of them, more open, fair and viſible: as in a very thin Slice cut athwart the young *Root* of a *Tree,* and held up againſt the light, is apparent. Yet not in all equally; in *Coran*-Tree, *Gooſberry*-Tree, &c. leſs, in *Oak, Plums,* and eſpecially *Damaſcens,* more; in *Elder, Vines, &c.* moſt conſpicuous. And as they are different in number and ſize, ſo alſo (whereon the numerouſneſs of the *Pores* of the *Cortical Body* principally depends) in their ſhape. For whereas thoſe of the *Cortical Body* are extended much alike both by the length and breadth of the *Root*; theſe of the *Lignous,* are only by the length; which eſpecially in *Vines,* and ſome other *Roots* is evident. Of theſe *Pores,* 'tis alſo obſervable, that although in all places of the *Root* they are viſible, yet moſt fair and open about the *filamentous Extremities* of ſome *Roots,* where about, the *Roots* have no *Pith*; as in *Fenil.* And in many *Roots,* higher.

Tab. 2. f. 5.

Tab.2.f.7 a.b.

9. §. The proportion betwixt this *Lignous Body* and the *Cortical,* is various, as was ſaid; yet in this, conſtant, *ſc.* that in the *filamentous* and ſmaller Parts of the *Root,* the *Lignous Body* is very much the leſs; running like a ſlender *Wyer* or *Nerve* through the other ſurrounding it. Whereas in the upper part, it is often times of far greater quantity than the *Cortical,* although it be encompaſs'd by it. They ſtand both together pyramidally, which is moſt common to *Infant Roots,* but alſo to a great many others.

10. The next *Part* obſervable in the *Root,* is the *Inſertment.* The exiſtence hereof, ſo far as we can yet obſerve, is ſometimes in the *Radicle* of the *Seed* it ſelf; I cannot ſay always. As to its ſubſtantial nature, we are more certain; that it is the ſame with that of the *Parenchyma* of the *Radicle*; being always at leaſt augmented, and ſo, in part, originated from the *Cortical Body,* and ſo, at ſecond hand, from the ſaid *Parenchyma.* For in diſſecting a *Root,* I find, that the *Cortical Body* doth not only environ the *Lignous,* but is alſo wedg'd, and in many Pieces *inſerted* into it; and that the ſaid inſerted Pieces make not a meer Indenture, but tranſmit and ſhoot themſelves quite through as far as the *Pith*: which in a thin Slice cut athwart the *Root,* as ſo many lines drawn from the Circumference towards the Center, ſhew themſelves.

Tab. 2. f. 5.

11. §. The *Pores* of the *Inſertment* are ſometimes, at leaſt, extended ſomewhat more by the breadth of the *Root,* as about the top of the *Root* of *Borage* may be ſeen; and are thus different from thoſe of

the

the *Cortical Body*, which are extended by the length and breadth much alike; and from thofe of the *Lignous*, being only by its length.

12. §. The number and fize of thefe *Infertions* are various. In *Hawthorn*, and fome others, and efpecially *Willows*, they are moft extream fmall; in *Cherries* and *Plums* they are Biger; and in the *Vine* and fome other *Trees*, very fairly apparent. In the *Roots* of moft *Herbs* they are generally more eafily difcoverable; which may lead to the obfervation of them in all. *Tab.* 2. *f.* 5. & 6.

13. §. Thefe *Infertions*, although they are continuous through both the length and breadth of the *Root*; yet not fo in all Parts, but by the feveral fhootings of the *Lignous Body* they are frequently intercepted. For of the *Lignous Body* it is (here beft) obfervable; That its feveral *Shootings*, betwixt which the *Cortical* is inferted, are not, throughout the *Root*, wholly diftinct, ftrait and parallel: but that all along being enarch'd, the *Lignous Body*, both in length and breadth, is thus difpofed into *Braces* or *Ofculations*. Betwixt thefe feveral *Shootings* of the *Lignous Body* thus ofculated, the *Cortical* fhooting, and being alfo ofculated anfwerably *Brace* for *Brace*, that which I call the *Infertment* is framed thereof. *Tab.* 2. *f.* 8.

14. §. Thefe *Ofculations* are fo made, that the *Pores* or *Fibres* of the *Lignous Body*, I think, notwithftanding, feldom or never run one into another; being, though contiguous, yet ftill diftinct. In the fame manner as fome of the *Nerves*, though they meet, and for fome fpace are affociated together, yet 'tis moft probable, that none of their *Fibres* are truly inofculated, faving perhaps, in the *Plexures*.

15. §. Thefe *Ofculations* of the *Lignous Body*, and fo the interception of the *Infertions* of the *Cortical*, are not to be obferv'd by the traverfe cut of the *Root*, but by taking off the *Barque*. In the *Roots* of *Trees*, they are generally obfcure; but in *Herbs* often more diftinctly apparent; and efpecially in a *Turnep*: the appearance whereof, the *Barque* being ftripp'd off, is as a piece of clofe-wrought Network, fill'd up with the *Infertions* from thence.

16. §. The next and laft diftinct *Part* of the *Root* is the *Pith*. The fubftantial nature thereof, is, as was faid of the *Infertment*, the fame likewife with that of the *Parenchyma* of the *Seed*. And according to the beft obfervation I have yet made, 'tis fometimes exiftent in its *Radicle*; in which, the two main *Branches* of the *Lobes* both meeting, and being ofculated together, are thus difpos'd into one round and tubular *Trunk*, and fo environing part of the *Parenchyma*, make thereof a *Pith*; as in either the *Radicle*, or the young *Root* of the great *Bean* or *Lupine*, may, I think, be well feen.

17. §. But many times the Original hereof is immediately from the *Barque*. For in diffection of divers *Roots*, both of *Trees* and *Herbs*, as of *Barberry* or *Mallows*; it is obfervable, That the *Cortical Body* and *Pith*, are both of them participant of the fame Colour; in the *Barberry*, both of them tinged yellow, and in *Mallows*, green. In cutting the fmaller Parts of the *Roots* of many *Plants*, as of *Borage*, *Mallows*, *Parfley*, *Columbine*, &c. 'tis alfo evident, That the *Lignous Body* is not there, in the leaft Concave, but ftandeth Solid, or without any *Pith* in the Center; and that the *Infertions* being gradually multiplied afterwards, the *Pith*, at length, towards the thicker parts of the *Root*, fhews and enlarges it felf. Whence it appears, that in all fuch *Roots*, *Tab.*2.*f.*9.--a

the

the *Pith* is not only of the same substantial nature, and by the *Insertions* doth communicate with the *Barque*; and that it is also augmented by it; which is true of the *Pith* of all *Roots*; but is moreover, by mediation of the said *Insertions*, wholly originated from it; that is to say, from the *Parenchymous* Part thereof. The various appearances of the *Insertions* and *Pith* from the filamentous *Parts* to the top of the *Root*, see in *Tab.* 2. The *Pores* of the *Lignous Body*, as it stands entire in the said filamentous *Parts*, are best seen when they have lain by a night to dry, after cutting.

Tab. 2. *f.* 9.

18. §. A farther evidence hereof are the Proportions betwixt the *Cortical Body* and *Pith*. For as about the inferiour Parts of the *Root*, where the *Pith* is small, the *Cortical Body* is proportionably great; so about the top, where the *Pith* is enlarged, the *Cortical Body* (now more properly becoming a *Barque*) groweth proportionably less, *sc.* because the *Insertions* do still more and more enlarge the *Pith*. Likewise the peculiar frame of some *Roots*, wherein besides the *Pith*, the *Lignous Body* being divided into two or more *Rings*, there are also one or more thick *Rings*, of a white and soft substance, which stand betwixt them; and are nothing else but the *Insertions* of the *Cortical Body* collected into the said *Rings*; but, towards the top of the *Root*, being inserted again, thus make a large and ample *Pith*; as in older *Fennel*-Roots, those of *Beet*, *Turnep*, and some other *Herbs*, is seen.

Tab. 2. *f.* 8.

19. §. The *Pores* of the *Pith*, as those of the *Cortical Body*, are extended both by the breadth and length of the *Root*, much alike; yet are they more or less of a greater size than those of the *Cortical Body*.

20. §. The Proportions of the *Pith*, are various; in *Trees*, but small; in *Herbs*, generally, very fair; in some making by far the greatest part of the *Root*; as in a *Turnep*: By reason of the wide circumference whereof, and so the finer Concoction and Assimilation of its Sap; that *Part* which in most old *Trunks* is a dry and harsh *Pith*, here proves a tender, pleasant meat.

21. §. In the *Roots* of very many *Plants*, as *Turneps*, *Carrots*, &c. the *Lignous Body*, besides its main utmost *Ring*, hath divers of its *osculated Fibres* dispersed throughout the Body of the *Pith*; sometimes all alike, and sometimes more especially in, or near, its Center; which *Fibres*, as they run towards the top of the *Root*, still declining the Center, at last collaterally strike into its Circumference; either all of them, or some few, keeping the Center still. Of these principally, the Succulent part of the *Lignous Body* of the *Trunk* is often originated.

22. §. Some of these *Pith-Fibres*, although they are so exceeding slender, yet in some *Roots*, as in that of *Flower de liz*, they are visibly concave, each of them, in their several Cavities also embosoming a very small *Pith*; the sight whereof, the *Root* being cut traverse, and laid in a Window for a day or two to dry, may without *Glasses* be obtain'd. And this is the general account of the *Root*; the declaration of the manner of its growth, with the use and service of its several *Parts*, we shall next endeavour.

An Account of the Growth of the Root.

23. §. I SAY THEN, That the *Radicle* being impregnate, and shot into the Moulds, the contiguous moisture, by the *Cortical Body*, being a Body laxe and Spongy, is easily admitted: Yet not all indiscriminately, but that which is more adapted to pass through the surrounding *Cuticle.*

Cuticle. Which tranfient *Sap*, though it thus becomes fine, yet is not fimple; but a mixture of *Particles*, both in refpect of thofe originally in the *Root*, and amongft themfelves, fomewhat heterogeneous. And being lodg'd in the *Cortical Body* moderately laxe, and of a Circular form; the effect will be an eafie Fermentation. The *Sap* fermenting, a feparation of *Parts* will follow; fome whereof will be impacted to the Circumference of the *Cortical Body*, whence the *Cuticle* becomes a *Skin*; as we fee in the growing of the Coats of Cheefes, of the Skin over divers Liquors, and the like. Whereupon the *Sap* paffing into the *Cortical Body*, through this, as through a *Manica Hippocratis*, is ftill more finely filtred. With which *Sap*, the *Cortical Body* being dilated as far as its *Tone*, without a folution of Continuity, will bear; and the fupply of the *Sap* ftill renew'd: the pureft part, as moft apt and ready, recedes, with its due *Tinctures*, from the faid *Cortical Body*, to all the parts of the *Lignous*; both thofe mixed with the *Barque*, and thofe lying within it. Which *Lignous Body* likewife fuper-inducing its own proper *Tinctures* into the faid *Sap*; 'tis now to its higheft preparaton wrought up, and becomes (as they fpeak of that of an Animal) the Vegetative *Ros* or *Cambium*: the nobleft part whereof is at laft coagulated in, and affimilated to the like fubftance with the faid *Lignous Body*. The remainder, though not united to it, yet tinctur'd therein, thus retreats, that is, by the continual appulfe of the *Sap*, is in part carried off into the *Cortical Body* back again, the *Sap* whereof it now tinctures into good *Aliment*. So that whereas before, the *Cortical Body* was only relaxed in its Parts, and fo dilated; 'tis now increas'd in real quantity or number of parts, and fo is truly nourifh'd. And the *Cortical Body* being faturate with fo much of this Vital *Sap* as ferves it felf; and the fecond Remainders difcharged thence to the *Skin*; this alfo is nourifh'd and augmented therewith. So that as in an *Animal Body* there is no inftauration or growth of Parts made by the *Bloud* only, but the *Nervous Spirit* is alfo thereunto affiftant; fo is it here: the *Sap* prepared in the *Cortical Body*, is as the *Bloud*, and that part thereof prepared by the *Lignous*, is as the *Nervous Spirit*; which partly becoming Nutriment to it felf, and partly being difcharged back into the *Cortical Body*, and diffufing its Tincture through the *Sap* there, that to the faid *Cortical Body* and *Skin*, becomes alfo true Nutriment, and fo they all now grow.

24. §. In which growth, a proportion in length and breadth is requifite: which being rated by the benefit of the *Plant*, both for firm ftanding and fufficient *Sap*, muft therefore principally be in length. And becaufe it is thus requifite, therefore by the conftitution of one of its *Parts, fc.* the *Lignous Body*, it is alfo made neceffary. For the *Pores* hereof, in that they are all extended by its length, the *Sap* alfo according to the frame and fite of the faid *Pores* will principally move; and that way as its *Sap* moves, the fame way will the generation of its Parts alfo proceed; *fc.* by its length. And the *Lignous Body* firft (that is by a *priority caufal*) moving in length it felf; the *Cortical* alfo moves therewith. For that which is nourifh'd, is extended: but whatever is extended, is mov'd: that therefore which is nourifh'd, is mov'd: The *Lignous Body* then being firft nourifh'd, 'tis likewife firft mov'd, and fo becomes and carries in it the Principle of all Vegetative motion in the *Cortical*; and fo they both move in length.

25. §.

25. §. Yet as the *Lignous Body* is the *Principle* of Motion in the *Cortical*; so the *Cortical* is the *Moderator* of that in the *Lignous*: As in Animal Motions, the *Principle* is from the *Nerves*; yet being once given to the *Muscle* or *Limb*, and that moving proportionably to its structure, the *Nerves* also are carried in the same motion with it. We suppose therefore, that as the principal motion of the *Lignous Body* is in length, so is its *proper tendency* also to *Ascend*. But being much exceeded both in Compass and Quantity by the *Cortical*, as in the smaller parts of the *Root* it is; it must needs therefore be over-born and governed by it; and so, though not lose its motion, yet make it that way wherein the *Cortical Body* may be more obedient to it; which will be by descent. Yet both of them being sufficiently pliable, they are thus capable, where the Soyl may oppose a direct descent, there to divert any way, where it is more penetrable, and so to descend obliquely. For the same reason it may also be, that though you set a *Bean* with the *Radicle* upward; yet the *Radicle*, as it shoots, declining also gradually, is thus arch'd in form of an Hook, and so at last descends. For every declination from a perpendicular Line, is a mixed motion betwixt Ascent and Descent, as that of the *Radicle* also is, and so seeming to be dependent upon the two *Contrary Tendencies* of the *Lignous* and *Cortical Bodies*. What may be the cause of those *Tendencies* (being most probably external, and a kind of *Magnetisme*) I shall not make my Task here to enquire.

26. §. Now although the *Lignous Body*, by the position and shape of its *Pores*, principally groweth in length; yet will it in some degree likewise in breadth: For it cannot be supposed that the purest *Sap* is all received into the said *Pores*; but that part thereof likewise, staying about its *Superficial Parts*, is there tinctur'd and agglutinated to them. And because these *Pores* are prolonged by its length; therefore it is much more laxe and easily divisible that way; as in flitting a Stick, or cleaving of Timber, and in cutting and hewing them athwart is also seen. Whence it comes to pass, that in shooting from the Center towards the Circumference, and there finding more room, its said original *Laxity* doth easily in divers places now become greater, and at length in open *Partments* plainly visible. Betwixt which *Partments*, the *Cortical Body*, being bound in on the one hand, by the surrounding *Skin* and *Moulds*, and pressed upon by the *Lignous* on the other, must needs insert it self, and so move contrary to it, from the Circumference *towards* the *Center*. Where the said contrary motions continued as begun, they at last meet, unite, and either make or augment the *Pith*. And thus the *Root* is fram'd, and the *Skin*, the *Cortical* and *Lignous Bodies*, so as is said, thereunto concurrent. We shall next shew the use of the two other *Parts, sc.* the *Insertment* and *Pith*; and first of the *Pith*.

27. §. ONE true use of the *Pith* is for the better Advancement of the *Sap*, whereof I shall speak in the next Chapter. The use I here observe, is for the quicker and higher Fermentation of the *Sap*: For although the Fermentation made in the *Cortical Body* was well subservient to the first *Vegetations*, yet those more perfect ones in the *Trunk* which after follow, require a Body more adapted to it, and that is the *Pith*; which is so necessary, as not to be only common to, but considerably large in the *Roots* of most *Plants*; if not in their inferiour

parts,

parts, yet at their tops. Where though either deriv'd or amplify'd from the *Cortical Body,* yet being by its *Insertions* only, we may therefore suppose, as those, so this, to be more finely constituted. And being also from its coarctation, while inserted, now free; all its *Pores,* upon the supply of the *Sap,* will more or less be amplified: Upon which accounts, the *Sap* thereinto received, will be more pure, and its fermentation therein more active. And as the *Pith* is superiour to the *Cortical Body* by its *Constitution,* so by its *Place.* For as it thus stands central, it hath the *Lignous Body* surrounding it. Now as the *Skin* is the Fence of the *Cortical Body,* and that of the *Lignous*; so is the *Lignous* again a far more preheminent one unto the *Pith*; the *Sap* being here a brisk Liquor, *tunn'd up* as in a wooden *Cask.*

28. §. And as the *Pith* subserves the higher Fermentation of the *Sap*; so do the *Insertions* its purer Distribution; that separation which the parts of the *Sap,* by being fermented in the *Pith,* were dispos'd for; being, upon its entrance into the *Insertions,* now made: So that as the *Skin* is a *Filtre* to the *Cortical Body,* so are the *Insertions* a more preheminent one to the *Lignous.* And as they subserve the purer, so the freer and sufficient distribution of the *Sap*: For the *Root* enlarging, and so the *Lignous Body* growing thicker, although the *Cortical* and the *Pith* might supply *Sap* sufficient to the nutrition of its *Parts* next adjacent to them; yet those more inward, must needs be scanted of their *Aliment*; and so, if not quite starv'd, yet be uncapable of equal growth: Whereas the *Lignous Body* being through its whole breadth frequently disparted, and the *Cortical Body* inserted through it; the *Sap* by those *Insertions,* as the *Bloud* by the disseminations of the *Arteries,* is freely and sufficiently convey'd to its intimate Parts, even those, which from either the *Barque* or from the *Pith,* are most remote. Lastly, as the consequent hereof, they are thus assistant to the *Latitudinal growth* of the *Root*; as the *Lignous Body* to its growth in Length; so these *Insertions* of the *Cortical,* to its better growth in Breadth.

29. §. Having thus seen the solitary uses of the Several *Parts* of the *Root,* I shall lastly propound my Conjectures of that Design whereto they are altogether concurrent, and that is the *Circulation* of the *Sap.*

30. §. That the *Sap* hath a Double, and so a *Circular* Motion, in the *Root*; is probable, from the proper Motion of the *Root,* and from its Office. From its Motion, which is Descent: for which, the *Sap* must likewise, some where, have such a Motion proper to it. From its Office, which is, To feed the *Trunk*: for which, the *Sap* must also, in some *Part* or other, have a more especial Motion of Ascent.

31. §. We may therefore suppose, That the *Sap* moving in the *Barque,* towards the *Pith,* through the *Insertions,* thereinto obtains a pass, Which passage, the upper *Insertions* will not favour; because the *Pith* standing in the same heigth with them, is there large, the fermenting and course of the *Sap* quick, and so its opposition strong. But through the lower it will much more easily enter; because there, from the smalness of the *Pith,* the opposition is little, and from the shortness of the *Insertions,* the way more open. So that the *Sap* here meeting with the least opposition, here it will bestow it self (feeding the *Lignous Body* in its passage) into the *Pith.* Into which, fresh *Sap* still entring, this being yet but crude, will subside: that

first receiv'd, and so become a Liquor higher wrought, will more easily mount upwards. And moving in the *Pith*, especially in the *Sap-Fibers* there dispers'd, as in the *Arterys*, in equal altitude with the upper-*Insertions*; the most volatile parts of all will still continue their direct ascent towards the *Trunk*. But those of a middle nature, and, as not apt to ascend, so being lighter than those beneath them, not to descend neither; they will tend from the *Pith* towards the *Insertions* in a Motion betwixt both. Through which *Insertions* (feeding the *Lignous Body* in its passage) it is, by the next subsequent *Sap*, discharged off into the *Cortical Body*, and so into the *Sap-Fibres* themselves, as into the *Veins*, back again. Wherein, being still pursu'd by fresh *Sap* from the Center, and more occurring from the Circumference, towards the lower *Insertions*, it thus descends. Through which, together with part of the *Sap* afresh imbib'd from the Earth, it re-enters the *Pith*. From whence, into the *Cortical Body*, and from thence into the *Pith*, the cruder part thereof, is reciprocally disburs'd; while the most *Volatile*, not needing the help of a *Circulation*, more directly ascendeth towards the *Trunk*.

CHAP.

CHAP. III.

Of the TRUNK.

HAVING thus declar'd the degrees of *Vegetation* in the *Root*; the continuance hereof in the *Trunk* shall next be shew'd: in order to which, the Parts whereof this likewise is compounded, we shall first observe.

1. §. That which without dissection shews it self, is the *Coarcture*: I cannot say of the *Root*, nor of the *Trunk*; but what I choose here to mention, as standing betwixt them, and so being common to them both; all their *Parts* being here bound in closer together, as in the tops of the grown *Roots* of very many *Plants*, is apparent.

2. §. Of the *Parts* of the *Trunk*, the first occurring is its *Skin*: The Formation whereof, is not from the Air, but in the *Seed*, from whence it is originated; being the production of the *Cuticle*, there investing the two *Lobes* and *Plume*.

3. §. The next *Part* is the *Cortical Body*; which here in the *Trunk* is no new substantial Formation; but, as is that of the *Root*, originated from the *Parenchyma* of the *Plume* in the *Seed*; and is only the increase and augmentation thereof. The *Skin*, this *Cortical Body* *Tab*. 3. *f*. 1, or *Parenchyma*, and (for the most part) some Fibers of the *Lignous* & 4. mixed herewith, alltogether make the *Barque*.

4. §. Next, the *Lignous Body*, which, whether it be visibly divided into many softer *Fibers* or small *Threads*, as in the *Bean*, *Fen-* *Tab*. 3. *f*. 1. *nel*, and most *Herbs*; or that its Parts stand more compact and close, shewing one hard, firm and solid piece, as in *Trees*; it is, in all, one and the same *Body*; and that not formed originally in the *Trunk*, but in the *Seed*; being nothing else but the prolongation of the *Seminal Root* distributed in the *Lobes* and *Plume* thereof.

5. §. Lastly, The *Insertions* and *Pith* are here originated likewise from the *Plume*, as the same in the *Root*, from the *Radicle*: So that as to their *Substantial Parts*, the *Lobes* of the *Seed*, the *Radicle* and *Plume*, the *Root* and *Trunk* are all one.

6. §. Yet some things are more fairly observable in the *Trunk*. First, the *Latitudinal* shootings of the *Lignous Body*, which in *Trunks* of several years growth, are apparent in so many *Rings*, as is commonly known. For several young Fibers of the *Lignous Body*, as in the *Tab*. 3. *f*. 5, *Root*, so here, shooting in the *Cortical* one year, and the spaces be- & 8. twixt them being after fill'd up with more (I think not till) the next, at length they become altogether a firm compact *Ring*; the *Perfection* of one *Ring*, and the *Ground-work* of another, being thus made concomitantly.

7. §. From these Annual younger *Fibers* it is, that although the *Cortical Body* and *Pith* are both of the same substantial nature, and their *Pores* little different; yet whereas the *Pith*, which the first year is green, and of all the *Parts* the fullest of *Sap*, becomes afterwards white and dry: The *Cortical Body*, on the contrary, so long as the *Tree* grows, ever keepeth green and moist, *sc.* because the said *Sap-Fibers*, annually grow therein, and so communicate with it.

8. §. The *Pores* likewise of the *Lignous Body*, many of them, in well-grown Timber, as in Oaken boards, are very conspicuous, in cutting both lengthwise and traverse. They very seldom, if ever, run one into another, but keep, like so many several *Vessels*, all along *Tab. 3. f. 2.* distinct; as by cutting, and so following any one of them as far as *& 3.* you please, for a Foot or half a Yard, or more together, may be observ'd. And so, the like, in any *Cane*.

9. §. Besides these, there are a lesser sort; which, by the help only of a good *Spectacle Glass* may be observ'd.

10. §. And these are all the *Pores* visible without a *Microscope*. The use of which, excepting in some few particulars, I have pur-
Micrography. posely omitted in this first *Book*. Mr. *Hook* sheweth us, besides these, a third, and yet smaller Sort; and (as a confirmation of what, in *C. 2. §. 8.* the Second *Chapter*, I have said of the *Pores* of the *Lignous Body* in general) that they are all continuous and prolonged by the length of the *Trunk*, as are the greater ones: whereof he maketh Experiment, by filling up, in a piece of *Char-coal*, all the said *Pores* with *Mercury*: which appears to pass quite through them, in that by a very good *Glass* it is visible in their Orifices at both ends; and without a *Glass*, by the weight of the Coal alone, is also manifest. All these I have seen, with the help of a good *Microscope*, in several *Tab. 3. f. 7.* sorts of *Woods*. As they all appear in a piece of *Oak*, cut transversely, See *Tab. 3*.

11. §. Upon further Enquiry, I likewise find, That the *Pores* of the *Lignous Body* in the *Trunks* of *Herbs*, which at first I only supposed, by the help of good *Glasses*, are very fairly visible: each *Fibre* being sometimes perforated by 30, 50, 100, or hundreds of *Pores*. Or what I think is the truest notion of them, That each *Fibre*, though it seem to the bare eye to be but *one*, yet is, indeed, a great number of *Fibres* together; and every *Pore*, being not meerly a space betwixt the several parts of the Wood, but the *Concave* of a *Fiber*. So that if it be asked, what all that Part of a *Plant*, either *Herb* or *Tree*, which is properly called the *Woody-Part*; what all that is, I suppose, That it is nothing else but a *Cluster* of innumerable and most extraordinay *Tab. 3. f. 6.* small *Vessels* or *Concave Fibers*: as in a Slice of the Trunk of *Burdock* is apparent.

12. §. Next the *Insertions* of the *Cortical Body*, which in the *Trunk* of a *Tree* saw'd athawrt, are plainly discerned as they run from the Circumference toward the Center; the whole Body of the *Tree* being visibly compounded of two distinct Substances, that of the several *Rings*, and that of the *Insertions*, running cross; shewing *Tab. 3. f. 5.* that in some resemblance in a *Plain*, which the *Lines* of *Latitude* and *& 8.* of the *Meridian* do in a *Globe*. The entrance of the *Insertions* into the *Wood*, is also, upon striping off the Barque, very apparent; as in the same *Fig.* 8.

13. §.

13. §. These *Insertions* are likewise very conspicuous in Sawing of *Trees* length-ways into Boards, and those plain'd, and wrought into *Leaves* for *Tables, Wainscot, Trenchers,* and the like. In all which, as in course *Trenchers* made of *Beech,* and *Tables* of *Oak*, there are many parts which have a greater smoothness than the rest; and are so many *inserted Pieces* of the *Cortical Body*; which being by those of the *Lignous,* frequently intercepted, seem to be discontinuous, although in the *Trunk* they are really extended, in continued Plates, throughout its Breadth. *Tab.* 3. *f.* 2. *& Tab.* 4. *f.* 1.

14. §. These *Insertions,* although as is said, of a quite distinct substance from the *Lignous Body,* and so no where truly incorporated with it, yet being they are in all parts, the one as the *Warp,* the other as the *Woof,* mutually *braced* and *interwoven* together, they thus constitute one strong and firmly coherent Body; as the Timber of any *Tree*. *Tab.* 4. *f.* 1.

15. §. As the Pores or Vessels are greater or less, so are the *Insertions* also: To the bare eye usually the greater only are discernable: But through an indifferent *Microscope* there are others also, much more both numerous and small, distinctly apparent, as in a transverse piece of *Oak*. *Tab.* 3. *f.* 7.

16. §. In none of all the *Pores* can we observe any thing which may have the true nature and use of *Valves,* which is, Easily to admit that, to which they will by no means allow a regress. And their non-existence is enough evident, from what in the first *Chapter* we have said of the *Lobes* of the *seed:* in whose *Seminal Root,* were there any *Valves,* it could not be, that by a contrary *Course* of the *Sap,* they should ever grow; which yet, where-ever they turn into *Dissimilar Leaves,* they do. Or if we consider the growth of the *Root,* which oftentimes is *upward* and *downward* both at once. And being cut transversely, will bleed, both the same ways, with equal freedom. *C.* 1. §. 42.

17. §. The *Insertions* here in the *Trunk* give us likewise a sight of the position of their *Pores.* For in a plained piece of *Oak,* as in *Wainscot, Tables, &c.* besides the larger *Pores* of the *Lignous Body,* which run by the *length* of the *Trunk*; the *Tract* likewise of those of the *Insertions* may be observed to be made by the breadth, and so directly cross. Nor are they continuous as those of the *Lignous Body,* but very short, as those both of the *Cortical Body* and *Pith,* with which the *Insertions,* as to their substance, are congenerous. Yet they all stand so together, as to be plainly ranked in even *Lines* or *Rows* throughout the breadth of the *Trunk :* As the *Tract* of those *Pores* appears to the naked Eye, see in *Tab.* 3. *Fig.* 9. The *Pores* themselves may be seen in the *Root* of a *Vine* described and figured in the Second *Book*, as it appears through a good *Microscope*. *Tab.* 3. *f.* 2. *Tab.* 17.

18. §. The *Pores* of the *Pith* likewise being larger here in the *Trunk*, are better observable than in the *Root:* the width whereof, in comparison with their *sides* so exquisitely thin, may by an *Hony-Comb* be grossly exemplified; and is that also which the vast disproportion betwixt the Bulk and Weight of a dry *Pith* doth enough declare. In the *Trunks* of some *Plants,* they are so ample and transparent, that in cutting both by the length and breadth of the *Pith,* some of them through the transparency of the *Skins* by which they are bounded, or of which they consist, would seem to be considerably

bly extended by the length of the *Pith*; but are really difcontinuous and fhort, and as 'tis faid, fomewhat anfwerable to the *Cells* of an *Hony-Comb*. This is the neareft we can come to them, by the bare Eye without the affiftance of a good *Microfcope*. Mr. *Hook* fheweth in his *Micrography*, That the *Pores* of the *Pith*, particularly of *Elder-Pith*, fo far as they are vifible, are all alike difcontinuous; and that the *Pith* is nothing elfe but (as he calls them) an heap of *Bubbles*. Although, in regard they are not fluid, but fixed Parts, I fhall choofe rather to call them, *Bladders*. As they appear through a good *Tab. 3. f. 6.* *Glafs*, in a piece of *Burdock*, See in *Tab.* 3. But a more particular Defcription of the *Sizes*, *Figures*, and admirable *Textures* hereof, I have given in feveral places in the following *Books*.

C.2. §.3,16, 17. 19. §. Befides what this Obfervation informs us of here, it farther confirms what in the Second *Chapter* we have faid of the Original of the *Pith* and *Cortical Body*, and of the famenefs of both their natures with the *Parenchyma* of the *Seed*: which is no-
C. 1. §. 18. & Tab.1. f.9. thing elfe but a Mafs of *Bladders*; as in the Firft *Chapter* hath been faid.

20. §. In the *Piths* of many *Plants*, the greater *Pores* or *Bladders* have fome of them leffer ones within them, and fome of them are divided with crofs Membranes: And betwixt their feveral fides, have, I think, other fmaller *Bladders* vifibly interjected. However, that they are all permeable, is moft certain. They ftand together not confufedly, but in even *Ranks* or *Trains*; as thofe of the *Infertions* by the breadth, fo thefe by the length of the *Trunk*. And thus far there is a general correfponding betwixt the parts of the *Root* and *Trunk*. Yet are there fome confiderable Difparities betwixt them; wherein, and how they come to pafs, and to what efpecial Ufe and End, fhall next be faid.

An Account of the Growth of the *Trunk*. 21. §. WE SAY then, that the *Sap* being in the *Root* by Filtrations, Fermentations (and in what *Roots* needful, perhaps by Circulation alfo) duly prepar'd; the prime part thereof paffing through the intermediate *Coarcture*, in due moderation and purity is entertain'd at laft into the *Trunk*. And the *Sap* of the *Trunk* being purer and more volatile, and fo it felf apt to afcend; the motion of the *Trunk* likewife will be more noble, receiving a difpofition and tendency to afcend therewith. And what by the *Sap* the *Trunk* is in part dipos'd to, by the refpective pofition and quantity of its Parts it is effectually enabled. For whereas in the *Root* the *Lignous Body* being in proportion with the *Cortical*, but little, and all lying clofe within its Center; it muft therefore needs be under its controul: on the contrary, being here comparatively of greater quantity, and alfo more dilated, and having divers of its Branches ftanding more abroad towards the Circumference, as both in the *Leaves* and Body of the young *Trunk* and *Plume*, is feen; it will in its own *magnetical* tendency to afcend, reduce the *Cortical Body* to a compliance with it.

22. §. And the *Trunk* thus ftanding from under the reftraint of the Ground in the open Air, the difpofition of its *Parts*, originally different from that of the *Parts* in the *Root*, will not only be continued, but improved. For by the force and preffure of the *Sap* in its collateral Motion, the *Lignous Body* will now more freely and farther be dilated.

lated. And this being dilated, the *Cortical Body* also, must needs be *inserted*; and is therefore in proportion always, more or less, smaller here in the *Trunk*, than in the *Root*. And as the *Cortical Body* lessens, so the *Pith* will be enlarged, and by the same proportion is here greater. And the *Pith* being enlarged it self, its *Pores* (the *Lignous Body*, upon its dilatation, as it were tentering and stretching out all their sides) must needs likewise be enlarged with it; and accordingly, are ever greater in the *Pith* of the *Trunk*, than of the *Root*. And the dilatation of the *Lignous Body* still continued, it follows, that whereas the *Pith* descendent in the *Root*, is not only in proportion less and less, but also in the smaller extremities thereof, and sometimes higher, altogether absent: Contrariwise, in the *Trunk*, it is not only continued to its top and smallest *Twigs*, but also there, in proportion, equally ample with what it is in any other inferiour part.

23. §. But although the openness of the Aer permitting, be allways alike; yet the Energy of the *Sap* effecting, being different; as therefore that doth, the dilatation of the *Trunk*, will also vary. If that be less, so is this; as in the *Trunks* of most *Trees*: If that be greater, so this; as in *Herbs* is common; the *Lignous Body* being usually so far dilated, that the *utmost Shootings* thereof may easily be seen to jut out, and adjoyn to the *Skin*. And if the *Sap* be still of greater energy, it so far dilates the *Lignous Body*, as not only to amplifie the *Pith* and all its *Pores*; but also so far to stretch them out, as to make them tear. Whereupon either running again into the *Cortical Body*, or shrinking up towards it, the *Trunk* thus sometimes becomes an *hallow Stalk*, the *Pith* being wholly, or in part voided. But generally it keeps entire; and where it doth, the same proportion and respect to the *Lignous* and *Cortical Bodies*, as is said. The Consequences of all which will be, the *Strength* of the *Trunk*, the *Security* and *Plenty* of the *Sap*, its *Fermentation* will be quicker, its *Distribution* more effectual, and its *Advancement* more sufficient.

24. §. First, the Erect Growth and Strength of the *Trunk*; this being, by the position of its several Parts, effected: for besides the slendering of the *Trunk* still towards the top, the *Circumferential* position of the *Lignous Body*, likewise is, and that eminently, hereunto subservient. So that as the *Lignous, Body*, in the smaller parts, of the *Root* standing Central, we may thence conceive and see their pliableness to any oblique motion; so here, on the contrary, the *Lignous Body* standing wide, it thus becomes the Strength of the *Trunk*, and most advantageous to its Perpendicular Growth. We see the same Design in *Bones* and *Feathers*: The strongest *Bones*, as those in the Legs, are hollow. Now should we suppose the same *Bone*, to be contracted into a Solid; although now it would be no heavier, and in that respect, as apt for motion; yet would it have far less strength, than as its Parts are dilated to a *Circumferential* posture. And so for *Quills*, which, for the same Reasons, in subserviency to flying, as they are exceeding light; So, in comparison with the thinness of their *Sides*, they are very strong, and much less apt to bend, than if contracted into a Solid *Cylinder*. We see it not only in *Nature*, but *Art*. For hence it is, that *Joyners* and *Carpenters* unite and set together their Timberpieces and several Works oftentimes with double Joynts; which, although

though they are no thicker, than a single one might be made; yet standing at a distance, have a greater strength than That could have. And the same Architecture, will have the same use, in the *Trunks* of *Plants*; in most whereof 'tis very apparent; as for instance, in *Corn*. For *Nature* designing its *Sap* a great Ascent; for its higher maturity, hath given it a tall *Trunk*: But to prevent its ravenous despoiling either of the *Ear*, or *Soyl*; although it be tall, yet are its sides but thin: And because again, it should grow not only tall and thriftily, but for avoiding propping up, strongly too; therefore, the same proportion as its heigth bears, to the thinness of its sides, doth the greatness of its Circumference also; being so far dilated as to parallel a *Quil* it self.

25. §. Besides the position of the *Lignous Body* within the compass of a *Ring*, there are some *Shootings* thereof, often standing beyond the Circumference of the said *Ring*, making sometimes a triangular, oftner a quadrangular Body of the *Trunk*. To the end, that the *Ring*, being but thin, and not self-sufficient, these, like *Splinters* to *Boxes*, might add strength and stability to it.

26. §. Next, the security and plenty of the *Sap*. For should the *Lignous Body*, as it doth in the smaller Parts of the *Root*, stand Central here also, and so the *Cortical* wholly surround it: the greater part of the *Sap* would thus be more immediately expos'd to the *Sun* and *Aer*; and being lodged in a laxe Body, by them continually be prey'd upon, and as fast as supplied to the *Trunk*, be exhausted. Whereas, the *Pith* standing in the Center, the *Sap* therein being not only most remote from the *Aer* and *Sun*, but by the *Barque*, and especially the *Wood*, being also surrounded and doubly immur'd, will very securely and copiously be convey'd to all the Collateral Parts, and (as shall be said how) the top of the *Trunk*.

27. §. And the *Sap* by the amplitude, and great porosity of the *Pith*, being herein more copious, its Fermentation also will be quicker; which we see in all Liquors, by standing in a greater quantity together, proceeds more kindly: And being *tun'd up* within the *Wood*, is at the same time not only secur'd from loss, but all extream mutations; the Day being thus, not too hot; nor the Night, too cold for it.

28. §. And the Fermentation hereof being quicker, its motion also will be stronger, and its Distribution more effectual, not only to the dilatation of the *Trunk*, but likewise the shooting out of the *Branches*. Whence it is, that in the Bodies of *Trees*, the *Barque* of it self, though it be Sappy, and many *Fibres* of the *Lignous Body* mixed with it, yet seldom sendeth forth any; and that in *Herbs*, those with the least *Pith* (other advantages not supplying this defect) have the fewest or smallest Branches, or other collateral Growths: and that *Corn*, which hath no *Pith*, hath neither any Branches.

29. §. Lastly, the Advancement of the *Sap* will hence also be more ready and sufficient. For the understanding where, and how, we suppose, That in all *Trunks* whatsoever there are two Parts joyntly hereunto subservient. In some, the *Lignous Body* and the *Cortical*, as in older *Trunks*; the *Pith* being either excluded, or dried: But in most, principally, the *Lignous Body* and *Pith*; as in most Annual Growths of *Trees*; but especially *Herbs*, where the *Cortical Body* is usually much and often wholly Inserted.

30. §.

30. §. Of the *Lignous Body* it is so apparent by its *Pores*, or rather by its *Vessels*, that we need no farther Evidence. For to what end are *Vessels*, but for the conveyance of Liquor? And is that also, which upon cutting the young Branch of a Sappy *Tree* or *Herb*, by an accurate and steady view may be observed. But when I say the *Vessels* of the *Lignous Body*, I mean principally them of the *younger shootings*, both those which make the *new Ring*, and those which are mixed with the *Cortical Body* in the *Barque*: that which ascendeth by the *Pores* or *Vessels* of the Wood, being probably, because in less quantity, more in form of a *Vapour*, than a *Liquor*. Yet that which drenching into the sides of its *Pores*, is with all thereunto sufficient Aliment; as we see *Orpine, Onions, &c.* only standing in a moyster Aer will often grow. And being likewise in part supplied by the *Insertions* from the younger *Shoots*: But especially because as it is but little, so (considered as Aliment) it serveth only for the growth of the *Wood*, and no more; whereas, the more copious Aliment ascendent by the younger *shoots*, subserves not only their own growth, but the generation of others; and is besides with that in the *Cortical Body* the Fountain of *Perspirations*, which we know even in *Animals* are much more abundant than the *Nutritive Parts*; and doubtless in a *Vegetable* are still much more.

31. §. But these *Pores*, although they are a free and open way to the ascending *Sap*; yet that meer *Pores* or *Vessels* should be able of themselves to advance the *Sap* with that speed, strength and plenty, and to that height, as is necessary, cannot probably be supposed. It follows then, that herein we must grant the *Pith* a joynt service. And why else is the *Pith* in all Primitive *Growths* the most *Sappy* part, why hath it so great a stock of *Sap*, if not, after due maturation within it self, still to be disbursed into the *Fibres* of the *Lignous Body*? Why are the Annual *Growths* of all both *Herbs* and *Trees*, with great *Piths*, the quickest and the longest? But how are the *Pores* or *Bladders* of the *Pith* permeable? That they are so, both from their being capable of a repletion with *Sap*, and of being again wholly emptied of it, and again, instead thereof fill'd with *Aer*, is as certain as that they are *Pores*. That they are permeable, by the breadth, appears from the dilatation of the *Lignous Body*, and from the production of *Branches*, as hath been, and shall hereafter be said. And how else is there a Communion betwixt *This* and the *Cortical Body*? That they are so also, by the length, is probable, because by the best *Microscope* we cannot yet observe, that they are visibly more open by the breadth, than by the length. And withall are ranked by the length, as those of the *Insertions* by the breadth of the *Trunk*. But if you set a piece of dry *Elder-Pith* in some tinged Liquor, why then doth it not penetrate the *Pores*, so as to ascend through the Body of the *Pith*? The plain reason is, because they are all fill'd with *Aer*. Whereas the *Pith* in a Vegetating *Plant*, as its Parts or *Bladders* are still generated, they are at the same time also fill'd with *Sap*; which, as 'tis gradually spent, is still repaired by more succeeding, and so the *Aer* still kept out; as in all Primitive *Growths*, and the *Pith* of *Elder* it self: Yet the same *Pith*, by reason of the following Winter, wanting a more copious and quick supply of *Sap*, thus once become, ever after keeps dry. And since in the aforesaid Trial the Liquor only ascends by the sides of the *Pith*, that

K is

is of its broken *Bladder*, we should thence by the same reason conclude that they are not penetrable by the breadth neither, and so no way; and then it need not be ask'd what would follow. But certainly the *Sap* in the *Bladders* of the *Pith* is discharged and repaired every moment, as by its shriv'ling up, upon cutting the *Plant*, is evident.

32. §. We suppose then, that as the *Sap* ascendeth into the *Trunk* by the *Lignous Body*, so partly also by the *Pith*. For a piece of *Cotton* with one end immers'd in some tinged Liquor, and with the other erect above, though it will not imbibe the Liquor so far as to over-run at the top, yet so as to advance towards it, it will. So here, the *Pith*, being a porous and spongy Body, and in its Vegetating state, its *Pores* or *Bladders* being also permeable, as a curious *Filtre* of *Natures* own contrivance, it thus advanceth, or as people use to say, sucks up the *Sap*. Yet as it is seen of the Liquor in the *Cotton*; so likewise are we to suppose it of the *Sap* in the *Pith*; that though it riseth up for some way, yet is their some term, beyond which it riseth not, and towards which the motion of the ascending *Sap* is more and more broken, weak and slow, and so the quantity thereof less and less. But because the *Sap* moveth not only by the length, but breadth of the *Pith*; at the same time therefore as it partly ascendeth by the *Pith*, it is likewise in part pressed into the *Lignous Body* or into its *Pores*. And since the motion of the *Sap* by the breadth of the *Pith* not being far continued, and but collateral, is more prone and easie, than the perpendicular, or by its length; it therefore follows, that the collateral motion of the *Sap*, at such a height or part of the *Pith*, will be equally strong with the perpendicular at another part, though somewhat beneath it; and that where the perpendicular is more broken and weak, the collateral will be less; and consequently where the perpendicular tendency of the *Sap* hath its term, the collateral tendency thereof, and so its pressure into the *Pores* or *Vessels* of the *Lignous Body*, will still continue. Through which, in that they are small, and so their sides almost contiguous, the *Sap* as fast as pressed into them will easily run up; as in very small Glasse Pipes, or betwixt the two halves of a Stick first slit, and then tyed somewhat loosely together, may also any Liquor be observed to do. By which Advantage the facility and strength of that ascent will be continued higher in the said *Vessels*, than in the *Pith*. Yet since this also, as well as that in the *Pith* will have its term; the *Sap*, although got thus far, would at last be stagnant, or at least its ascent be very sparing, slow and feeble, if not some way or other re-inforced. Wherefore, as the *Sap* moving by the breadth of the *Pith*, presseth thence into the *Vessels* of the *Lignous Body*; so having well fill'd these, is in part by the same Collateral motion disbursed back, into a yet higher Region of the *Pith*. By which partly, and partly, by that portion of the *Sap*, which in its perpendicular ascent was before lodged therein; 'tis thus here, as in any inferiour place equally replenished. Whereupon the force and vigour of the perpendicular motion of the *Sap* herein, will likewise be renew'd; and so its Collateral motion also, and so its pressure into the *Vessels* of the *Lignous Body*, and consequently its ascent therein: and so by a pressure, from these into the *Pith*, and from the *Pith* into these, reciprocally carried on; a most ready and copious ascent of the *Sap* will be continued, from the bottom to the top, though of the highest *Tree*.

An Appendix.

Of Trunk-Roots and Claspers.

THE diſtinct *Parts* whereof theſe are compoſed, are the ſame with thoſe of the *Trunk*, and but the continuation of them.

1. §. *Trunk Roots* are of two kinds: Of the one, are thoſe that vegetate by a direct deſcent: The place of their Eruption is ſometimes all along the *Trunk*; as in *Mint*, &c. Sometimes only at its utmoſt point, as in the *Bramble*.

2. 2. The other ſort are ſuch as neither aſcend nor deſcend, but ſhoot forth at right Angles with the *Trunk*; which therefore, though as to their *Office*, they are true *Roots*, yet as to their *Nature*, they are a *Middle Thing* betwixt a *Root* and a *Trunk*.

3. §. *Claspers*, though they are but of one kind, yet their Nature is double; not a mean betwixit that of the *Root* and that of the *Trunk*, but a compound of both; as in their Circumvolutions, wherein they often mutually aſcend and deſcend, is ſeen.

4. §. The uſe of theſe *Parts* may be obſerved as the *Trunk* Mounts, or as it Trails. In the mounting of the *Trunk*, they are for Support and Supply. For Support, we ſee the *Claspers* of *Vines*: the *Branches* whereof being very long, fragile and ſlender; unleſs by their *Claspers*, they were mutually contain'd together, they muſt needs by their own weight, and that of their Fruit, undecently fall; and be alſo liable to frequent breaking. So that the whole care is divided betwixt the Gardener and Nature; the Gardener, with his Ligaments of Leather, ſecures the main Branches; and Nature, with theſe of her own finding, ſecures the Leſs. Their Conveniency to which end, is ſeen in their *Circumvolutions*, a motion, not proper to any other *Part*: As alſo in their toughneſs, though much more ſlender than the *Branches* whereon they are appendent.

5. §. The *Claspers* of *Bryony* have a retrograde motion about every Third *Circle*, to the form a Doublet-Claſp. Probably for the more certain hold; which, if it miſs one way, it may be ſure to take another.

6. §. For Supply, we ſee the *Trunk-Roots* of *Ivy*. For mounting very high, and being of a cloſer or more compact Subſtance than that of a *Vine*; the *Sap* could not be ſufficiently ſupplied to the upper *Sprouts*, unleſs theſe, to the *Mother-Root*, were joyntly aſſiſtant. Yet ſerve they for ſupport likewiſe; whence they ſhoot out, not as in *Creſſes*, *Brook-lime*, &c. reciprocally on each ſide, but commonly, all on one; that ſo they may be faſtned at the neareſt hand.

7. §. In the Trailing of the *Trunk*, they ſerve for ſtabiliment, propagation and ſhade. For ſtabiliment, the *Claspers* of *Cucumbers* are of good uſe. For the *Trunk* and *Branches* being long and fragile, the Bruſhes of the Winds would injuriouſly hoiſe them to and fro, to the dammage both of themſelves and their tender Fruits, were

they not by thefe Ligaments brought to good Affociation and Settlement.

8. §. As for this end, fo for Propagation, the *Trunk-Roots* of *Chamæmile* do well ferve. Whence we have the reafon of the common obfervation, that it grows better by being trod upon: the Mould, where too laxe, being thus made to lie more conveniently about the faid *Trunk-Roots* newly bedded therein; and is that which is fometimes alfo effected in Rowling of *Corn*.

9. §. For both thefe ends, Serve the *Trunk-Roots* of *Strawberries*; as alfo for fhade; for in that all *Strawberries* delight; and by the trailing of the *Plant* is well obtain'd. So that as we are wont to tangle the Twigs of *Trees* together to make an *Arbour Artificial*; the fame is here done to make a *Natural one:* as likewife by the *Clafpers* of *Cucumbers*. For the *Branches* of the one by the Linking of their *Clafpers*, and of the other by the Tethering of their *Trunk-Roots*, being couched together; their tender Fruits thus lie under the Umbrage of a *Bower* made of their own Leaves.

CHAP. IV.

Of the GERMEN, BRANCH, *and* LEAF.

THE Parts of the *Germen* and *Branch*, are the fame with thofe of the *Trunk*; the fame *Skin*, *Cortical* and *Lignous Bodies*, *Infertment* and *Pith*, hereinto propagated, and diftinctly obfervable herein.

2. §. For upon Enquiry into the Original of a *Branch* or *Germen*, it appears, That it is not from the *Superficies* of the *Trunk*; but fo deep, as to take, with the *Cortical*, the *Lignous Body* into it felf: and that, not only from its Circumference, but from in *Inner* or *Central Parts*; So as to take the *Pith* in alfo. Divers of which *Parts* may commonly be feen to fhoot out into the *Pith*; from which *Shoots*, the furrounding and more fuperiour *Germens* are originated; in like manner as the Succulent Part of the *Lignous Body* of the *Trunk* is fometimes principally from thofe Fibrous *Shoots* which run along the *Pith* in the *Root*.

3. §. The manner wherein ufually the *Germen* and *Branch* are fram'd, is briefly thus: The *Sap* (as is faid, *Chap.* 3.) mounting in the *Trunk*, will not only by its length, but by its breadth alfo, through the *Infertions* partly move. Yet, its Particles being not all alike qualified, in different degrees. Some are more grofs and fluggifh; of which we have the formation of a Circle of Wood only, or of an *Annual Ring*. Others are more brifk; and by thefe, we have the *Germen* propagated. For by the vigour of their own motion from the Center, they imprefs an equal tendency on fome of the inner *Portions* of the *Lignous Body* next adjacent

cent to the *Pith*, to move with them. And fince the *Lignous Body* is not entire, but frequently difparted; through thefe *Difpartments*, the faid interiour *Portions*, upon their Nutrition, actually fhoot; not only towards the Circumference, fo as to make part of a *Ring*; but even beyond it, in order to the production of a *Germen*. And the *Lignous Body* thus moving, and carrying the *Cortical* along with it; they both make a force upon the *Skin*. Yet their motion being moft even and gradual, that force is fuch likewife; not to caufe the leaft breach of its parts, but gently to carry it on with themfelves; and fo partly, by the extenfion of its already exiftent parts, as of thofe of *Gold* in drawing of Guilded *Wyer*; and partly, by the accretion of new ones, as in the enlarging of a *Bubble* above the Surface of the Water; it is extended with them to their utmoft growth. In which growth, the *Germen* being prolonged, and fo difplaying its feveral parts, as when a *Profpective* or *Telefcope* is drawn out, thus becomes a *Branch*.

4. §. The fame way as the propagation of the *Parts* of a *Germen* is contriv'd, is its due nutrition alfo. For being originated from the inner part of the *Lignous Body*, 'tis nourifhed with the beft fermented *Sap* in the *Trunk*, *fc.* that next adjacent to it in the *Pith*. Befides, fince all its *Parts*, upon their fhooting forth, divaricate from their perpendicular, to a crofs Line, as thefe and the other grow and thrive together, they bind and throng each other into a *Knot*: through which *Knot* the *Sap* being ftrain'd, 'tis thus, in due moderation and purity delivered up into the *Branch*.

5. § And for *Knots*, they are fo neceffary, as to be feen not only where collaterl *Branches* put forth; but in fuch *Plants* alfo, as fhoot up in one fingle *Trunk*; as in *Corn*. Wherein, as they make for the ftrength of the *Trunk*; fo by fo many percolations, as they are *Knots*, for the trafmiffion of the *Sap* more and more refined towards the *Ear*. So that the two general ufes of *Knots* are, For *fimer ftanding*, and *finer growth*.

6. §. Laftly, as the due Formation and Nutrition of the *Germen* are provided for, fo is its fecurity alfo; which both in its pofition upon the *Trunk*, and that of its *Parts* among themfelves, may be obferved. The pofition of its *Parts* fhall be confidered in fpeakimg of the *Leaf*. As to its ftanding in the *Trunk*, tis alwayes betwixt the *trunk* or *older Branch*, and the *Bafis* of the Stalk of *a Leaf*; whereby it is not only guarded from the Injuries of any contingent Violence; but alfo from the more piercing affaults of the Cold; fo long, till in time 'tis grown larger, and more hardy. The maner and ufes of the pofition of every *Germen*, confidered as after it becomes a *Branch*; hath already been, by the Ingenious Mr. *Sharrock* Hift. of the very well obferved; to whom I refer. Prop.of Veget.

7. §. UPON THE prolongation of the *Germen* into a *Branch*, its *Leaves* are thus difplay'd. The *Parts* whereof are fubftantially the fame with thofe of a *Branch*. For the *Skin* of the *Leaf*, is only the ampliation of that of the *Branch*; being partly by the accretion of new, and partly the extention of its already exiftent parts, dilated (as in making of *Leaf-Gold*) into its prefent breadth. The *Fibres* or *Nerves* difperfed through the *Leaf*, are only the Ramifications of the *Branch's* Wood, or *Lignous Body*. The *Parenchyma* of the *Leaf*,

which

which lies betwixt the *Nerves*, and as in Gentlewomens Needle-works, fills all up, is nothing else, but the continuation of the *Cortical Body*, or *Parenchymous* part of the *Barque* from the *Branch* into it self, as in most *Plants* with a thick *Leaf*, may easily be seen.

8. §. The *Fibers* of the *Leaf* neither shoot out of the *Branch*, or the *Trunk*, nor stand in the *Stalk*, in an *even* Line; but alwayes in either an *Angular* or *Circular* posture; and usually making either a *Triangle*, or a *Semi-Circle*, or *Chord* of a Circle; as in *Cichory*, *Endive*, *Cabbage*, &c. may be observed. And if the *Leaf* have but one main *Fiber*, that also is postur'd in a bowed or *Lunar* Figure; as in *Mint* and others. The usual number of these *Vascular Threds* or *Fibres* is 3, 5, or 7.

Tab. 4. f. 2. to f. 11.

9. §. The reason of the said Positions of the *Fibers* in the *Stalk* of the *Leaf*, is for its more *Erect* growth, and greater *Strength*: which, were the position of the said *Fibers* in an *even Line*, and so the *Stalk* it self, as well as the *Leaf*, flat; must needs have been defective; as from what we have said of the Circumferential posture of the *Lignous Body* in the *Trunk*, we may better conceive.

C. 3. §. 24.

10. §. As likewise for the security of its *Sap*: For by this means it is, that the several *Fibers*, and especially the main or middle *Fiber* of the *Leaf*, together with a considerable part of the *Parenchyma*, are so disposed of, as to jut out, not from its upper, but its back, or neither Side. Whence the whole *Leaf*, reclining backward, becomes a Canopy to them, defending them from those Injuries which from colder Blasts, or an hotter Sun, they might otherwise sustain. So that by a mutual benefit, as These give *suck* to all the *Leaf*, so that again *protection* to These.

11. §. These *Fibers* are likewise the immediate Visible Cause of the Shape of the *Leaf*. For if the nethermost *Fiber* or *Fibers* in the *Stalk* (which thence runs chiefly through the length of the *Leaf*) be in proportion greater, the *Leaf* is long; as in *Endive*, *Cichory*, and others: If all of a more equal size, it spreads rounder, as in *Ivy*, *Doves-foot*, *Colts foot*, &c. And although a *Dock-Leaf* be very long, whose *Fibers* notwithstanding, as they stand higher in the *Stalk*, are disposed into a *Circle* all of an equal size; yet herein one or more peculiar *Fibres*, standing, in or near the *Center*, betwixt the rest, and running through the length of the *Leaf*, may be observed.

Tab. 4.

12. §. In correspondence also to the size and shape of these *Fibres*, is the *Leaf* flat. In that either they are very small, or if larger, yet they never make an entire Circle or *Ring*; but either half of one, as in *Borage*, or at most three parts of one, as in *Mullen*, may be seen. For if either they were so big, as to contain, or so entire, as perfectly to include a *Pith*, the Energy of the *Sap* in that *Pith*, would cause the said *Lignous Ring* to shoot forth on every side, as it doth in the *Root* or *Trunk*: But the said *Fibers* being not figur'd into an entire *Ring*, but so as to be open; on that hand therefore where open they cannot shoot any thing directly from themselves, because there they have nothing to shoot; and the *Sap* having also a free vent through the said opening, against that part therefore which is thereunto opposite, it can have no force; and so neither will they shoot forth on that hand; and so will they consequently, that way only, which the force of the *Sap* directs, which is only on the right and left.

Tab. 4.

13. §.

13. §. The several *Fibers* in the *stalk*, are all Inosculated in the *Leaf*, with very many Sub-divisions. According as these *Fibers* are Inosculated near, or at, or shoot directly to the edge of the *Leaf*, is it Even, or Scallop'd. Where these *Inosculations* are not made, there we have no *Leaves*, but only a company of *Filaments*; as in *Fennel*.

14. §. To the *Formations* of *Leaves*, the *Fouldings* immediately follow. And sometimes they have one Date, or are the contemporary works of *Nature*; each *Leaf* obtaining its distinct shape, and proper posture together; both being perfect, not only in the outer, but Central and minutest *Leaves*, which are five hundred times smaller than the outer: both which in the Cautious opening of a *Germen* may be seen.

15. §. Nor is there greater Art in the *Forms*, than in the *Foulds* or *Postures* of *Leaves*; both answerably varying, as this or that way they may be most agreeable. Of the *Quincuncial* posture, so amply instanc'd in by the Learned Sir *Thomas Brown*, I shall omit to speak. Others there are, which though not all so universal, yet equally necessary where they are, giving two general advantages to the *Leaves*, *Elegancy* and *Security*, *sc.* in taking up, so as their *Forms* will bear, the least room; and in being so conveniently couch'd, as to be capable of receiving protection from other Parts, or of giving it one to another; as for instance, [Treat. of the *Quincunx*.]

16. §. First, **T**here is the *Bow-Lap*, where the *Leaves* are all laid somewhat convexly one over another, but not plaited; being to the length, breadth and number of *Leaves* most agreeable; as in the Buds of *Pear-tree*, *Plum-tree*, &c. But where the *Leaves* are not so thick set, as to stand in the *Bow-Lap*, there we have the *Plicature*, or the *Flat-Lap*; as in *Rose-Tree*, *Strawberry*, *Cinquefoyl*, *Burnet*, &c. For the *Leaves* being here plaited, and so lying in half their breadth, and divers of them thus also collaterally set together; the thickness of them all, and half their breadth, are much alike dimensions; by which they stand more secure within themselves, and in better consort with other *Germen-Growths* in the same *Truss*. If the *Leaves* be much indented or jagg'd, now we have the *Duplicature*; wherein there are divers *Plaits* in one *Leaf*, or *Labels* of a *Leaf*, but in distinct *Sets*, a lesser under a greater; as in *Souchus*, *Tansey*, &c. When the *Leaves* stand not collaterally, but single; and are moreover very broad; then we have the *Multiplicature*; as in *Gooseberries*, *Mallows*, &c. the *Plaits* being not only divers in the same *Leaf*, but of the same *set* continuant, and so each *Leaf* gather'd up in five, seven or more *Fouids*, in the same manner as our Gentlewomens Fans. Where either the thickness of the *Leaf* will not permit a *Flat-Lap*, or the fewness of their number, or the smalness of their *Fibers*, will allow the *Rowl*, there This may be observed. Which is sometimes single, as in *Bears-Ears*, *Arum*, *Flammula*, *Jerusalem Cowslip*, &c. Sometimes double, the two *Rowls* beginning at each edge of the *Leaf*, and meeting in the middle. Which again, is either the *Fore-Rowl*, or the *Back-Rowl*. If the *Leaf* be design'd to grow long, now we have the *Back-Rowl*, as in *Docks*, *Sorrels*, and the rest of this Kindred: as also in *Primrose*, and other like *Plants*. For the main *Fibers*, and therewith a considerable part of the *Cortical Body* standing prominent from the *Back-side* of the *Leaf*, they thus stand securely couch'd up betwixt

twixt the two *Rowls*; on whose security the growth of the *Leaf* in length depends. But those of *Bears-Ears, Violets, Doves Foot, Warden,* and many more, upon contrary respects, are rowled up inwards. Lastly, there is the *Tre-Rowl,* as in *Fern*; the *Labels* whereof, though all rowled up to the *main Stem,* yet could not stand so firm and secure from the Injuries either of the *Ground* or *Weather,* unless to the *Rowls* in breadth, that by the length were super-induc'd; the *stalk* or *main Stem* giving the same Protection here, which in other *Plants* by the *Leaves,* or some particular *Mantling,* is contriv'd. These, and other *Foulds,* See in the *Figures* belonging to the *First* Part of the *Fourth* Book.

17. §. According to the *Form* and *Foulding* of every *Leaf* or *Germen,* is its Protection order'd; about six ways whereof may be observ'd; *sc.* by *Leaves, Surfoyls, Interfoyls, Stalks, Hoods* and *Mantlings.* To add to what we have above given, one or two Instances. Every *Bud,* besides its proper *Leaves,* is covered with divers Leafy *Pannicles* or *Surfoyls*; which, what the *Leaves* are to one another, are that to them all: For not opening except gradually, they admit not the *Weather, Wet, Sun* or *Aer,* to approach the *Leaves,* except by degrees respondent, and as they are gradually inur'd to bear them. Sometimes, besides *Surfoyls,* there are also many *Interfoyls* set betwixt the *Leaves,* from the Circumference to the Center of the *Bud*; as in the *Hasel.* For the *Fibres* of these *Leaves* standing out so far from a plain surface; they would, if not thus shelter'd, lie too much expos'd and naked to the *Severities* of the *Weather.* Where none of all the Protections above-named, are convenient, there the *Membranes* of the *Leaves* by continuation in their first forming (together with some *Fibres* of the *Lignous Body*) are drawn out into so many *Mantles* or *Veils*; as in *Docks, Snakeweed,* &c. For the *Leaves* here being but few, yet each *Leaf* and its *Stalk* being both exceeding long; at the bottom whereof the next following *Leaf* still springs up; the form and posture of all is such, as supersedes all the other kinds of Protection, and so each *Leaf* apart is provided with a *Veil* to it self. These, and other *Protections,* See in the *Figures* belonging to the *First* Part of the *Fourth* Book.

18. §. The *Uses* of the *Leaves,* I mean in respect of their service to the *Plant* it self, are these: First, for Protection; which, besides what they give one to another, they afford also to the *Flower* and *Fruit.* To the *Flower* in their *Foulds*; that being, for the most part, born and usher'd into the open *Aer* by the *Leaves.* To the *Fruit,* when afterwards they are display'd, as in *Strawberries, Grapes, Rasps, Mulberries,* &c. On which, and the like, should the Sun-Beams immediately strike, especially while they are young, they would quite shrivel them up; but being by the *Leaves* screened off, they impress the circumjacent *Aer* so far only as gently to warm the said *Fruits,* and so to promote their *Fermentation* and *Growth.* And accordingly we see, that the *Leaves* above-named are exceeding large in propotion to the *Fruits*: whereas in *Pear-trees, Apple-trees,* &c. the *Fruit* being of a solider *Parenchyma,* and so not needing the like protection, are usually equal with, and often wider in Diameter than the *Leaves.*

19. §. Another use is for Augmentation; or, the capacity for the due spreading and ampliation of a *Tree* or other *Plant*, are its *Leaves*. For herein the *Lignous Body* being divided into small *Fibres*, and these running all along their lax and spongie *Parenchyma*; they are thus a *Body* fit for the imbibition of *Sap*, and easie *Growth*. Now the *Sap* having a free reception into the *Leaves*, it still gives way to the next succeeding in the *Branches* and *Trunk*, and the voyding of the *Sap* in these, for the mounting of that in the *Root*, and ingress of that in the *Ground*. But were there no *Leaves* to make a free reception of *Sap*, it must be needs be stagnant in all the *Parts* to the *Root*, and so the *Root* being clogg'd, its fermenting and other Offices will be voyded, and so the due *Growth* of the whole. As in the motion of a *Watch*, although the original term thereof be the *Spring*, yet the capacity for its continuance in a due measure throughout all the *Wheels*, is the free and easie motion of the *Ballance*.

20. §. Lastly, As the *Leaves* subserve the more copious advancement, so the higher purity of the *Sap*. For this being well fermented both in the *Root*, and in its Ascent through the *Trunk*, and so its *Parts* prepar'd to a farther separation; the grosser ones are still deposited into the *Leaves*; the more elaborate and essential only thus supplied to the *Flower*, *Fruit* and *Seed*, as their convenient *Aliment*. Whence it is, that where the *Flowers* are many and large, into which the more odorous *Particles* are copiously receiv'd, the green *Leaves* have little or no smell; as those of *Rose-tree*, *Carnations*, *French-Marigold*, *Wood-bind*, *Tulips*, &c. But on the contrary, where the *Flowers* are none, or small, the green *Leaves* themselve are likewise of a strong savour; as those of *Wormwood*, *Tansie*, *Baum*, *Mint*, *Rue*, *Geranium Moschatum*, *Angelica*, and others.

An Appendix.

Of Thorns, Hairs and Globulets.

THorns are of two kinds, *Lignous* and *Cortical*. Of the first are such as those of the *Hawthorn*, and are constituted of all the same substantial *Parts* whereof the *Germen* or *Bud* it self, and in a like proportion: which also in their Infancy are set with the resemblances of divers minute *Leaves*. Of affinity with these are the *Spinets* or *Thorny Prickles* upon the Edges and Tops of divers *Leaves*, as of *Barbery*, *Holly*, *Thistle*, *Furze*, and others; all which I think are the filamentous extremities of the *Lignous Body* sheathed in the *Skin*. But this principal differnce betwixt a *Bud* and these *Lignous Thorns*, is observable; That the *Bud* hath its Original from the Inner part of the *Lignous Body*, next the *Pith*: But these *Thorns*, from the outer, and less fecund Part; and so produceth no *Leaves*, but is, as it were, the *Mola* of a *Bud*.

2. §. *Cortical Thorns* are such as those of the *Rasberry* Bush, being not, unless in a most extraordinary small and invisible proportion propagated from the *Lignous Body*, but as, it seems, wholly from the *Cortical* and *Skin*, or from the exteriour part of the *Barque*.

C. 2. §. 25.

3. §. The *Growth* of this *Thorn* may farther argue what in the second *Chapter* we supposed; *sc.* That as the proper *Tendency* of the *Lignous Body*, is to *Ascend*; so of the *Cortical* to *Descend*. For as the *Lignous Thorn*, like other *Parts* of the *Trunk*, in its *Growth* ascends; This, being almost wholly *Cortical*, pointeth downward. The use of *Thorns* the Ingenious Mr. *Sharrock* hath observed.

Hist. of the Prop. of Veget.

4. §. Upon the *Leaves* of divers *Plants* two *Productions* shew themselves, *sc. Hairs* and *Gloubulets*. Of *Hairs*, only one kind is taken notice of; although they are various. Ordinarily they are of a *Simple Figure*; which when fine and thick set, as on most *Hairy Buds*; or fine and long, as on those of the *Vine*, we call them *Down*.

5. §. But sometimes they are *Branched out*, from the bottom to the top, reciprocally on every side, in some resemblance to a *Stags Horn*; as in *Mullen*. And sometimes they are *Astral*, as upon *Lavender*, and some other *Leaves*, and especially those of *Wild Olive*; wherein every *Hair* rising in one round entire *Basis* a little way above the surface of the *Leaf*, is then disparted, Star-like, into several, four, five or six *Points*, all standing at right Angles with the said perpendicular *Basis*.

6. §. The *Uses of Hairs* are for *Distinction* and *Protection*. That of *Distinction* is but secondary, the *Leaves* being grown to a considerable size. That of *Protection* is the prime, for which they were originally form'd together with the *Leaves* themselves, and whose service they enjoy in their Infant-estate: For the *Hairs* being then in form of a *Down*, always very thick set, thus, give that *Protection* to the *Leaves*, which their exceeding tenderness then requires; so that they seem to be vested with a Coat of *Frize*, or to be kept warm, like young and dainty *Chickens*, in *Wool*.

7. §. *Globulets* are seen upon *Orach*, both Garden and Wild; and yet more plainly on *Mercury* or *Bonus Henricus*. In these, growing almost upon the whole *Plant*, and being very large, they are by all taken notice of.

8. §. But strict Observation discovers, that these *Globulets* are the natural and constant Off-spring of very many other *Plants*. Both these *Globulets*, and likewise the diversity of *Hairs*, I find that Mr. *Hook* hath also observed. I take notice, that they are of two kinds; *Transparent*, as upon the *Leaves* of *Hysop*, *Mint*, *Baume*, and many more *White*, as upon those of *Germander*, *Sage*, and others. All which, though the naked Eye will discover, yet by the help of *Glasses* we may observe them most distinctly. The use of these we suppose the same, in part, with those of the *Flower*, whereof we shall speak.

Micrography.

CHAP.

CHAP. V.

Of the FLOWER.

E next proceed to the *Flower*. The general *Parts* whereof are most commonly three; *sc.* the *Empalement*, the *Foliation*, and the *Attire*.

2. §. The *Empalement*, whether of one or more pieces, I call that which is the utmost *Part* of the *Flower*, encompassing the other two. 'Tis compounded of the three general *Parts*, the *Skin*, the *Cortical* and *Lignous Bodies*, each *Empaler* (where there are divers) being as another little *Leaf*; as in those of a *Quince-Flower*, as oft as they happen to be overgrown, is well seen. As likewise in the *Primrose*, with the *green Flower*; commonly so call'd, though by a mistake: For that which seems to be the *Flower*, is only the more flourishing *Empalement*, the *Flower* it self being *White*. But the continuation of all the three aforesaid *Parts* into each *Empaler*, is discoverable, I think, no where better than in an *Artichoke*, which is a true *Flower*, and whose *Empalers* are of that amplitude, as fairly to shew them all: As also, that the Original of the *Skin* of each *Empaler* or *Leaf* is not distinct from that of the rest; but to be all one piece, laid in so many *Plaits* or *Duplicatures*, as there are *Leaves*, from the outermost to the inner and most Central ones.

3. §. The Design of the *Empalement*, is to be *Security* and *Bands* to the other two *Parts* of the *Flower*: To be their *Security* before its opening, by intercepting all extremities of *Weather*: Afterwards to be their *Bands*, and firmly to contain all their *Parts* in their due and most decorous posture: so that a *Flower* without its *Empalement*, would hang as uncouth and taudry, as a *Lady* without her *Bodies*.

4. §. Hence we have the reason why it is various, and sometimes wanting. Some *Flowers* have none, as *Tulips*; for having a fat and frim *Leaf*, and each *Leaf* likewise standing on a broad and strong *Basis*, they are thus sufficient to themselves. *Carnations*, on the contrary, have not only an *Empalement*, but that (for more firmitude) of one piece: For otherwise, the Foot of each *Leaf* being very long and slender, most of them would be apt to break out of compass: yet is the top of the *Empalement* indented also; that the *Indentments*, by being lapp'd over the *Leaves* before their expansion, may then protect them; and by being spread under them afterwards, may better shoulder and prop them up. And if the Feet of the *Leaves* be both long and very tender too, here the *Empalement* is numerous, though consisting of several pieces; yet those in divers *Rounds*, and all with a counterchangeable respect to each other (which also the Learned Sir *Thomas Brown* observes) as in all *Knapweeds*, and other *Flowers*; whereby, how commodious they are for both the aforesaid ends, may easily be conceiv'd; and well enough exemplified by the Scales of *Fishes*, whereunto, as to their position, they have not an unapt resemblance. *Treat. of the Quincunx.*

5. §. THE FOLIATION alfo, is of the fame fubftantial *Nature* with the green *Leaf*; the *Membrane*, *Pulp*, and *Fibres* whereof, being, as there, fo here, but the continuation of the *Skin*, the *Cortical* and *Lignous Bodies*.

6. §. The *Foulds* of the *Flower* or *Foliation* are various, as thofe of the green *Leaf*; but fome of them different. The moft general are, Firft, The *Clofe-Couch*, as in *Rofes*, and many other double *Flowers*. Then the *Concave-Couch*, as in *Blattaria flore albo*. Next the *Plait*, as in fome of the *Leaves* of *Peafe-Blooms*, in the *Flowers* of *Coriander*, &c. which is either fingle, as in thofe nam'd; or double, as in *Blew-Bottle*, *Jacea*, and more of that rank. Next, the *Couch*, and *Plait* together in the fame *Flower*, as in *Marigolds*, *Daifies*, and all others of an agreeing form: where the firft apparent *Fould* or *Compofture* of the *Leaves* is in *Couch*; but the *Leaves* being erect, each likewife may be feen to lie in a double *Plait* within it felf. Then the *Rowl*, as in the *Flowers* of *Ladies-Bower*, the broad top of each *Leaf* being by a double *Rowl* foulded up inwardly. Next, the *Spire*, which is the beginning of a *Rowl*; and may be feen in the *Flowers* of *Mallows*, and others. Laftly, the *Plait* and *Spire* together, where the *Part* analogous to the *Foliation*, is of one piece, the *Plaits* being here laid, and fo carried on by *Spiral Lines* to the top of the *Flower*, as is in divers, and I think, in *Convolvulus Doronici folio*, more elegantly feen. Thefe and other *Foulds*, See in the *Figures* belonging to the *Second* Part of the *Fourth* Book. The reafon of all which varieties, a comparative confideration of the feveral *Parts* of the *Flower* may fuggeft. I'le only mention, That no *Flower*, that I find, hath a *Back-Rowl*, as hath the green *Leaf*. For two Reafons; becaufe its *Leaves* have not their *Fibres* ftanding out much on their backfide, as the green *Leaves* have; and becaufe of its *Attire*, which it ever embofomes, and cannot fo well do it by a *Back-Rowl*.

7. §. The ufual *Protections* of *Flowers* by the Precedents are exprefs'd, *fc. Green Leaves* and *Empalements*. Some have another more peculiar, that is a *double Veil*; as the *Spring-Crocus*. For having no *Empalement*, and ftarting up early out of the *Mould*, even before its *Green Leaves*, and that upon the firft opening of the Spring; left it fhould thus be quite ftarved, 'tis born fwath'd up in a double *Blanket*, or with a pair of *Sheets* upon its *Back*.

8. §. The *Leaves* of divers *Flowers* at their *Bafis* have an *hairy Tuft*; by which *Tufts* the Concave of the *Empalement* is filled up; That, being very choice and tender, they may thus be kept in a gentle and conftant Warmth, as moft convenient for them.

9. §. The *Leaves* of the *Flower*, though they are not hairy all over, yet in fome particular parts they are often fet with a fine Downy *Velvet*; that, being by their fhape and pofture in thofe parts contiguous to their delicate and tender *Attire*, they may thus give it a more foft and warmer touch. Thus in the *Flower* of *Ladies Bower*, thofe parts of its *Leaves* which rowl inward, and lie contiguous to the *Attire*, are Downy; whereas the other *Parts* are fmooth or bald: So the *Flowers* of *Peafe*, *Spanifh Broom*, *Toad-Flax*, and many others, where contiguous to their *Attires*, are deck'd with the like *Hairy Velvet*.

10. §. As upon the Green *Leaves*, so upon the *Flowers* are *Globulets* somtimes seen; as upon the backside of that of *Enula*. On none more plainly than that kind of *Blattaria* with the white *Flower*; where they are all transparent, and growing both on the *Stalk* and *Leaves* of the *Flower*, each shewing likewise its *Peduncle* whereon it is erected.

11. §. The use of the *Flower*, or the *Foliation* whereof we now speak, (that is, as to its private service) is for the protection of the *Attire*; This, as its under, and the *Empalement* as its upper Garments. As likewise of the *Fruit*: The necessity of which Service, in some Cases, by the different situation of the *Flower* and *Fruit*, with respect to each other, is evident; *Apples*, *Pears*, and several other *Fruits*, standing behind or under the *Flower*; but *Cherries*, *Aprecots*, and divers others, within it. For these, being of a very tender and pulpous Body, and withal putting forth with the colder part of the Spring; could not weather it out against the Variations and Extremities of the Air, (as those of a more solid *Parenchyma* can) except lodged up within their *Flowers*.

12. §. And as the *Flower* is serviceable to the safety of the *Fruit*, so is it to its growth; *sc.* in its Infancy, or *Embryo*-estate; for which purpose, as there is a *Flower*, so that *Flower* is greater or less, according as the nature of the *Fruit* to which it belongs, and the plenty of the *Sap* by which the *Fruit* is fed, doth require. Thus, where the young *Fruit* is of a solider Substance and the ascent of the *Sap* less copious, were there here no *Flower* to promote the said ascent thereof into the *Fruit* (in the manner as is effected by the Green *Leaves*) it must needs pine and die, or prove less kindly. On the contrary, should the *Flower* be over-large, it would not only promote the ascent of the *Sap* up to the *Fruit*, but being as yet over-proportionate to it, would likewise it self exhaust the same *Sap*, as fast as ascendent; like a greedy Nurse, that prepares the Meat for her Child, and then eats it up her self. Thus we see *Apples* and *Pears*, with a *Flower* of a moderate *Size*; like their *Body*, of a middle Constitution, and their *Sap*, of a middle quantity: But *Quinces*, being more solid, besides that they have as great a *Flower*, the *Impalers* of their *Flower* also thrive so far as to become handsom *Leaves*; continuing also after the *Flower* is fallen, firm and verdent a great while; so long, till the *Fruit* be able to provide for it self. On the other hand, *Plums* being more tender and Sappy than *Appels* and *Pears*, besides that their *Empalers* are much alike, their *Flower* is less. and *Goosberries* and *Currans*, which are still more Pulpy, and the course of the *Sap* towards them more free, have yet a *Flower* far less. And *Grapes*, whose *Sap* is still of quicker Ascent, have scarce any *Flower* at all; only some small resemblance thereof, serving just upon the setting of the *Fruit*, and no longer.

13. §. THE ATTIRE, I find to be of two kinds, *Seminiforme*, and *Florid*. That which I call *Seminiforme*, is made up of two general *Parts*, *Chives* and *Semets*, one upon each *Chive*. These *Semets* (as I take leave to call them) have the appearance, especially in many *Flowers*, of so many little *Seeds*: but are quite another kind of *Body*. For, upon enquiry, we find, that these *Semets*, though they seem to

be

be solid, and for some time after their first formation, are entire; yet are they really hollow; and their side, or sides, which were at first entire, at length crack asunder: And that moreover the *Concave* of each *Semet* is not a meer vacuity, but fill'd up with a number of minute Particles, in form of a *Powder*. Which, though common to all *Semets*, yet in some, and particularly those of a *Tulip* or a *Lilly*, being larger, is more distinctly observable.

Tab. 4.f.12.

14. §. These *Semets* are somtimes fastned so, as to stand erect above their *Chive*, as those of *Larks-heel*. Somtimes, and I think usually, so as to hang a little down by the midle, in the manner and figure of a *Kidney*; as in *Mallows*. Their *Cleft* or *Crack* is sometimes single, but for the most part double: At these *Clefts* it is that they disburse their *Powders*; which as they start out, and stand betwixt the two Lips of each *Cleft*, have some resemblance to the common Sculpture of a *Pomegranate* with its *Seeds* looking out at the *Cleft* of its *Rind*. This must be observ'd when the *Clefts* are recently made, which usually is before the expansion of the *Flower*.

f. 12. -a.

15. §. The Particles of these *Powders*, though like those of Meal or other Dust, they appear not easily to have any relugar shape; yet upon strict observation, especially with the assistance of an indifferent *Glass*, it doth appear, That they are a *Congeries*, usually, of so many perfect *Globes* or *Globulets*; Sometimes of other *Figures*, but always regular. That which obscures their *Figure* is their being so small: In *Dogs-Mercury*, *Borage*, and very many more *Plants*, they are extreamly so. In *Mallows*, and some others, more fairly visible.

16. §. Some of these *Powders*, are *yellow*, as in *Dogs-Mercury*, *Goats-Rue*, &c. and some of other Colours: But most of them I think are *white*; and those of *yellow Henbane* very elegant; the disburs'd *Powers* whereof, to the naked eye, are *white as Snow*; but each *Globulet*, through a Glass, transparent as Crystal; which is not a fallacy from the *Glass*, but what we see in all transparent *Bodies* whatsoever, lying in a *Powder* or small Particles together. The Parts of this *Attire*, see in *Tab.* 4. But especially, in the *Figures* belonging to the *Second* 𝔓art of the *Fourth* 𝔅ook.

17. §. The *Florid Attire*, is commonly known by the blind and rude Name of *Thrums*; as in the *Flowers* of *Marigold*, *Tansie*, &c. How inadequate its imposition is, observation will determine. For the several *Thrums* or rather *Suits*, whereof the *Attire* is made up, however else they may differ in various *Flowers*, in this agree, that they are ever consistent of more than one, sometimes of Two, and for the most part of Three *Pieces* (for which I call them *Suits*) and each *Piece* of a different, but agreeable and comely form.

Tab.4.f.13.a.

18. §. The *outer Part* of every *Suit*, is its *Floret*: whose *Body* or *Tube* is divided at the top (like that of the *Cowslip*) into five distinct *Leaves*. So that a *Floret*, is the Epitome of a *Flower*: and is all the *Flower* that many *Plants*, as *Mugwort*, *Tansie*, and others, have. What the Learned Sir *Thomas Brown* observeth of the number *Five*, as to the *Leaves* of the *Flower*, is still more universally holding in these of the *Floret*.

f. 13. b.

Treat. of the Quincunx.

19. §. Upon the Expansion of the *Floret*; the next *Part* of the *Suit* is from within its *Tube* brought to sight; which we may (with respect to that within it) call the *Sheath*. For this also, like the *Floret*,

f. 13. c.

is

is a *Concave Body*; in its shape very well resembling the Fistulous Pouches of *Wake-Robin*, or of *Dragon*.

20. §. The *Sheath*, after some time, dividing at the top, from within its Concave the Third and innermost part of the *Suit*, &c. the *Blade* advanceth and displayes it self. This Part is not hollow, as the other two, but solid; yet at its Point, is commonly, divided into two halves. *Tab.*4.*f.*13.d

21. §. About the said Point especially, there appears, *Globulets*, which are of the same nature with those of a *Semet*, though not so copious. So that all *Flowers* have their *Powders* or *Globulets*. The whole *Attire* may in *After Per*, *Blewbottle*, &c. where the *Suits* are large, be plainly observed without a *Glass*. The *Parts* of this *Attire*, See in *Tab.* 4. But especially in the *Figures* belonging to the *Second* **Part** of the *Fourth* **Book**.

22. §. The use of the *Attire*, how contemptibly soever we may look upon it, is certainly great. And though for our own use we value the *Leaves* of the *Flower*, or the *Foliation*, most; yet of all the three *Parts*, this in some respects is the choycest, as for whose sake and service the other two are made. The use hereof, as to *Ornament* and *Distinction*, is unquestionable; but is not all. As for Distinction, though, by the help of *Glasses*, we may make it to extend far; yet in a passant view, which is all we usually make, we cannot so well. As for Ornament, and particularly in reference to the *Semets*, we may ask, If for that meerly these were meant, then why should they be so made as to break open, or to contain any thing within them? Since their Beauty would be as good if they were not hollow; and is better before they crack and burst open, than afterwards.

23. §. Other uses hereof therefore we must acknowledge, and may observe. One is, for food; for Ornament and Distinction to us, and for *Food* to other *Animals*. I will not say, but that it may serve even to these for Distinction too, that they may be able to know one *Plant* from another, and in their flight or progress settle where they like best: and that therefore the varieties of these small parts are many, and well observed by them, which we take no notice of. Yet the finding out of Food is but in order to enjoy it: Which, that it is provided for a vast number of little *Animals* in the *Attires* of all *Flowers*, observation perswades us to believe. For why else are they evermore here found? Go from one *Flower* to another, great and small, you shall meet with none untaken up with these Guests. In some, and particularly the *Sun-Flower*, where the parts of the *Attire*, and the *Animals* for which they provide, are larger, the matter is more visible. We must not think, that *God Almighty* hath left any of the whole Family of his Creatures unprovided for; but as the Great Master, some where or other carveth out to all; and that for a great number of these little Folk, He hath stored up their peculiar provisions in the *Attires* of *Flowers*; each *Flower* thus becoming their Lodging and their Dining-Room, both in one.

24. §. Wherein the particular parts of the *Attire* may be more distinctly serviceable, this to one *Animal*, and that to another, I cannot say: Or to the same *Animal*, as a *Bee*, whether this for the *Honey*, another for their *Bread*, a third for the *Wax*: Or whether all only suck
from

from hence some *Juice*; or some may not also carry some of the *Parts*, as of the *Globulets*, wholly away.

25. §. Or lastly, what may be the Primary and Private Use of the *Attire* (for even this abovesaid, though great, yet is but Secondary) I now determine not.

CHAP. VI.

Of the FRUIT.

HE general composition of all *Fruits* is one, that is, their *Essential* and truly *Vital Parts*, are in all the same, and but the continuation of those which in the other *Parts* of a *Plant*, we have already observed. Yet because by the different *Constitutions* and *Tinctures* of these *Parts*, divers considerably different *Fruits* result; I shall therefore take aparticular view of the more known and principal of them, *sc.* *Apples*, *Pears*, *Plums*, *Nuts* and *Berries*.

2. §. AN APPLE, if cut traverse, appears constitued of four distinct *Parts*, the *Pilling*, the *Parenchyma*, *Branchery*, and *Coare*. The *Pilling* is only the spreading and dilatation of the *Skin*, or utmost part of the *Barque* in the *Branch*. The *Parenchyma*, when full ripe, is a tender delicate Meat. Yet as the *Pilling* is but the Continuation of the utmost part of the *Barque*; so is this, but the continuance and ampliation, or (as I may call it) the swelth and superbience of the *Inner Part* thereof; which upon observation of a young and Infant-*Apple* especially, is evident. Thus we see the *Pith*, which is often tough; in many *Roots*, as *Parsneps*, *Turneps*, &c. is tender and edible. So here, the *Parenchyma*, though originally no more than the *Barque*, yet the copiousness and purity of its *Sap* being likewise effectual to the largness and fineness of its growth, it thus becomes a soft and tender meat. The *Branchery* is nothing else but the Ramifications of the *Lignous Body* throughout all the parts of the *Parenchyma*; the greater *Branches* being likewise by the *Inosculations* of the less (as in the *Leaf*) united together. The main *Branches* are usually Twenty : Ten are spred and distributed through the *Parenchyma*, most of them enarching themselves towards the *Cork* or *Stool* of the *Flower* : The other Ten, running from the *Stalk* in a directer Line, at last meet the former at the said *Cork*, and are there osculated with them. Of these latter, five are originated from one; which running along the Center of the *Stalk*, and part of the *Parenchyma* of the *Fruit*, is therein at last divided. To these the Coats of the *Kernels* are fastned. So that whereas most of these *Branches* were originally extended even beyond the *Fruit*, and inserted into the *Flower* for the due growth
there-

thereof; the *Fruit* afterwards growing to some head, and so intercepting and preying upon the *Aliment* of the *Flower*, starves that and therefrom supersedes the service of the said *Branches* to it self, fifteen for its *Parenchyma*, and five for its *Seed*. The *Coar* is originated from the *Pith*; for the *Sap* finding room enough in the *Parenchyma*, through which to dispence it self all abroad, quits the *Pith*, which thereby hardens into a *Coar*. Thus we see the *Insertions*, although originate from the *Cortical Body*, yet their *Parts* being, by the *Inosculations* of the *Lignous*, so much compress'd and made to co-incide together, they become a *Body* very compact and dense. And in the *Barque* the same thing is effected by *Arefaction* only, or a meer *voydance* of the *Sap*; the *Inner Part* whereof, though soft and sappy, yet its superficial *Rind* is often so hard and smooth, that it may be fairly writ upon. The *Parts* of an *Apple*, See in the *Figures* belonging to the *Third* Part of the *Fourth* Book.

3. §. IN A PEAR there are five distinct *Parts*, the *Pilling*, the *Parenchyma*, *Branchery*, *Calculary*, and *Acetary*. The three former are here and in an *Apple* much alike; saving that here the *Inner* or *Seed-Branches* ordinarily stand double. The *Calculary* (most observable in rough-tasted, or *Choak-Pears*) is a *Congeries* of little stony *Knots*. They are many of them dispersed throughout the whole *Parenchyma*: But lying more continuous and compact together towards the Center of the *Pear*, surround the *Acetary* there, in a somewhat Globular Form. About the *Stalk* they stand more distant; but towards the *Cork* or *Stool* of the *Flower*, they still grow closer, and there at last gather (almost) into the firmitude of a *Plum-stone* it self. Within this lies the *Acetary*; 'tis allways sour, and by the bounding of the *Calculary* of a *Globular Figure*. 'Tis a simple *Body*, having neither any of the *Lignous* branched in it, nor any *Calculous Knots*. It is of the same substantial nature with the outer *Parenchyma*; but whether it be absolutely one with it, or be derived immediately from the *Pith*, my Enquiries yet made, determine not.

Tab.4.f.14.

4. §. The Original of the *Calculary* I seem to have neglected. But hereof we may here best say, that whereas all the other *Parts* are *Essential* and truly *Vital*; the *Calculary* is not: but that the several *Knots* whereof it consists, are only so many meer *Concretions* or *Precipitations* out of the *Sap*; as in *Urines*, *Wines*, and other *Liquors*, we often see. And that the *Precipitation* is made by the mixture and re-action of the *Tinctures* of the *Lignous* and *Cortical Bodies* upon each other: Even as all *Vegetable Nutrition* or *Fixation of Parts* is also made by the joynt efficiency of the two same *Tinctures*, as hath been said. Hence we find, that as the *Acetary* hath no *Branches* of the *Lignous Body*, so neither hath it any *Knots*. Hence likewise it is, that we have so different and contrary a tast in the *Parenchyma* beyond the *Calculary*, from that in the *Acetary*: For whereas this is sour, that, wherein the said *Precipitations* are made, is sweet; being much alike effect to what we find in mixing of *Corals*, &c. with *Vinegar* or other *acid Liquors*. The *Parts* of a *Pear*, See in *Tab.* 4. But especially in the *Figures* belonging to the *Third* Part of the *Fourth* Book.

5. §. IN A PLUM (to which the *Cherry, Apricot, Peach, Walnut*, &c. ought to be referr'd) there are four diſtinct *Parts*, the *Pilling*, the *Parenchyma, Branchery* and *Stone*. The *Pilling* and *Parenchyma* are, as to their Original, with thoſe of an *Apple* or *Pear*, both alike. As likewiſe the *Brunchery*; but differently ramified. In *Plums* (I ſuppoſe all) there are five main *Out-Branches*, which run along the Surface of the *Stone* from the *Baſis* to the point thereof, four of them by *Tab. 4. f. 15.* one *Ridge*, and one by the other oppoſite to it. In an *Apricot* there is the ſame number, but the ſingle *Branch* runs not upon the *Surface*, but through the *Body* of the *Stone*. There are likewiſe two or three ſmaller *Branches*, which run in like manner under the other *Ridge* for ſome ſpace, and then advancing into the *Parenchyma*, therein diſperſe themſelves: Theſe latter ſort in *Peaches* are numerous throughout.

6. §. But notwithſtanding the different diſpoſition of the *Branches* of the *Fruits* aforeſaid; yet is there one *Branch* diſpos'd in one and the ſame manner in them all. The entrance hereof into the *Stone* is at its *Baſis*; from whence running through its Body, and ſtill inclining *Tab. 4. f. 15.* or arching it ſelf towards its *Concave*, is at laſt, about its *Cone*, thereinto emergent, where the *Coats* of the *Seed* are appendent to it. Of the *Seed-Branch* 'tis therefore obſervable that after its entrance into the *Fruit*, 'tis always prolonged therein to a conſiderable length; as is ſeen not only in *Apples*, &c. where the *Seed* ſtands a good diſtance from the *Stalk*; but in *Plums* likewiſe, where it ſtands very near it; in that here the *Seed-Branch*, as is ſaid, never ſtrikes through the *Stone* into the *Coats* of the *Seed* directly, but runs through a *Chanel* cut in the *Stone*, till it iſſues, near the *Cone*, into the *Concave* thereof.

7. §. The *Stone* though it ſeem a ſimple Body, yet it is compounded of different ones. The Inner *Part* thereof, as it is by far the thinneſt, ſo is it the moſt *denſe, white, ſmooth* and *ſimple*. The Original is from the *Pith*; difficult, but curious to obſerve: For the *Seed-Branch*, not ſtriking directly and immediately quite through the *Baſis* of the *Stone*, but in the manner as is above deſcribed, carries a conſiderable *Part* of the *Pith*, now gather'd round about it, as its *Parenchyma*, along with it ſelf; which upon its entrance into the concave of the *Stone* about its farther end, is there in part ſpread all over it, as the *Lining* thereof. The outer and very much thicker *Part*, conſiſteth partly of the like *Precipitations* or concrete Particles, as in a *Pear*; being gathered here much more cloſely, not only to a *Contiguity*, but a *Coalition* into one entire *Stone*; as we ſee in *Pears* themſelves, eſpecially towards the *Cork*, they gather into the like Stonineſs; or as a *Stone, Mineral*, or *Animal*, is oftentimes the product of accumulated *Gravel*. But as the *Parenchyma* is mixed with the *Concretions* in the *Calculary*, ſo is it alſo, though not viſibly, with theſe in the *Stone*, the *ground* of the *Stone* being indeed a perfect *Parenchyma*; but by the ſaid *Concretions* ſo far alter'd, as to become dry, hard and undiſtinguiſhable from them. All which Particulars, are obſervable only in the ſeveral degrees of *Growth* in the young *Fruit*. And are repreſented in *Tab.* 4. But eſpecially by the ſeveral *Figures* belonging to the *Third* and *Fourth* 𝔓𝔞𝔯𝔱𝔰 of the *Fourth* 𝔅𝔬𝔬𝔨.

8. §.

8. §. IN A NUT (to which an *Akern* is analogous) there are three general *Parts*, the *Cap*, *Shell*, and *Pith*. The *Cap* is constituted of a *Pilling* and *Parenchyma*, derived from the *Barque*; and *Ramulets* from the *Lignous Body* of the *Branch*. The *Shell* likewise is not one simple Body, but compounded. The Superficial *Part* thereof is originated from the *Pilling* or *Skin* of the *Cap*, from the inside whereof it is, in a Duplicature, produc'd and spred over the *Shell*. Which, if you look at the *Basis* of the *Shell*, is farther evident: for that being continuous with the *Parenchyma* of the *Cap*, without the interposure of the *Skin*, the said superficial *Part* is there wanting. The thicker and inner *Part* of the *shell* consisteth of the same *Parenchyma* as that of the *Cap*, with a *Congeries* of *Precipitations* filled up, as in a *Stone*. And as the *Lignous Body* is branched in a *Stone*, so, with some difference, in a *shell*. The *outer Branches* or *Ramulets* are numerous, each issuing out of the *Parenchyma* of the *Cap*, and entring the *shell* at the *Circumference* of its *Basis*, and so running betwixt its superficial and inner *Parts* towards the *Cone*, round about. The *Inner* or *Seed-Branch* is single, entring in, as do the other, at the *Basis* of the *shell*, but at the *Center* thereof: from whence it runs, not through the *Shell*, as in *Plums* through the *Stone*; but through the *Pith*, as far as the *Cone*, where the *Coats* of the *Seed* hang appendent to it. The *Pith* whether derived from the same part both in name and nature in the *Branch* and *Stalk*; or from the *Cortical Body*, I yet determine not. The *Parts* of a *Nut*, See in the *Figures* belonging to the *Third* Part of the *Fourth* Book.

9. §. A BERRY, as a *Gooseberry* (to which *Corinths*, *Grapes*, *Hips*, &c. are to be referr'd) consisteth, besides the *Seed*, of the three general *Parts*, *Pilling*, *Parenchyma* and *Branchery*. The *Pilling* is originated as in the foregoing *Fruits*. The *Parenchyma* is double, as likewise in some other *Berries*. The *outer* is commonly, together with the *Pilling*, call'd the *Skin*, and is that part we spit out, being of a sour tast. Now as the *Pilling* is originated from the *outer*, so this from the *inner Part* of the *Barque*; and accordingly the *Pores* thereof may be observed plainly of a like shape with those both of the *Cortical Body* and *Pith*. The *Inner* or *Pulp* is of a sweet taste, and is the *Part* we eat: It is of a Substance so laxe and tender, as it would seem to be only a thicker or jellied *Juice*; although this likewise be a true *Parenchyma*, something like that of an *Orange* or *Limon*, with its *Pores* all fill'd up with *Liquor*. The *Branchery* is likewise double: The *Exterior* runs betwixt the *Pilling* and *Outer Parenchyma* in arched *Lines*, from the *Stalk* to the *Stool* of the *Flower*. These *outer Branches*, though of various number at the *Stalk*, yet at the *Cork* are usually ten principal ones; five for the five *Leaves* of the *Flower*, and five for the *Attire*. The *Inner main Branches* are two, diametrically opposite to each other, and at the *Cork* with the other inosculated. From these two are branched other smaller, every one having a *Seed* appendent to it, whose *Coats* it entreth by a double *Filament*, one at the *Basis*, the other at the *Cone*. They are all very white and turgent, and by a slaunt cut, may be observ'd concave; thus representing themselves analogous to so many true *spermatick Vessels*. The

Parts of a *Gooseberry*, See in the *Figures* belonging to the *Third Part* of the *Fourth Book*.

10. §. The Uses of *Fruits* are for *Man*, (sometimes also other *Animals*, as are *Akerns* and *Haws*) and for the *Seed*. For *Man*, they are so variously desirable, that till our *Orchards* and *Store-Chambers*, *Confectioners-Stoves* and *Apothecaries-Shops*, our *Ladies Closets*, their *Tables* or *Hands* are empty of them, I shall not need to enquire for what. If it be asked, how the *Fruit* becomes, generally above all the other *Parts*, so pleasant a Meat? It is partly from the *Sap*, the grosser portion thereof being deposited in the *Leaves*, and so the purer hereunto reserved. Partly from the *Globular Figure* of the *Fruit*. For the *Sap* being thus in a greater quantity herein, and in all Parts equally diffus'd, the *Concoction* hereof, as in a *Vessel*, is with greatest advantage favoured and promoted. Wherefore all *Fruits*, which we eat raw, how small soever, are of a *Globular* Form, or thereunto approaching; and the nearer, the delicater; amongst *Apples*, the *Pipin*; amongst *Pears*, the *Burgundian*; and amongst all *Fruits*, the *Grape*; and amongst *Grapes*, the roundest, are of all, the most dainty.

11. §. The visible cause of this *Globular Figure*, is the *Flower*; or the Inosculation of all the main *Branches* at the *Stool* of the *Flower*; and upon the fall of the *Flower*, the obtuseness, and with *Wind* and *Sun*, as it were the *seaing* of their several ends: For thus the *Sap* entering the *Fruit*, being not able to effect, either a *Disunion*, or a *shooting* forth of the said *Branches*, and so to carry on their *Growth* in length; they must of necessity be enarch'd, and with the *Parenchyma* more and more expand themselves. Whereas were they disposed and qualified otherwise, than as is said; instead of forming a *Fruit* within bounds, they would run out into all extravagance, and even into another little *Tree* or *Leafy Growth*.

12. §. To the *Seed*, the *Fruit* is serviceable; First, in order to its being supply'd with a due and most convenient *Sap*, the greater part thereof, and that which is less elaborated, being, in its passage towards the *Seed*, thereinto received; the *Fruit* doing the same office to the *Seed*, which the *Leaves* do to the *Fruit*; the *Sap* in the *Fruit* being, in a laxe comparison, as the *Wine*; and that for the *Seed*, a small part of the highest Spirit rectified from it.

13. §. So likewise for its Protection, in order to the prosperous carrying on and perfecting of its generation, and security being perfected. Which protection it gives not only to the Seminal *Sap* and *Seed* it self, but ever also to its *Seed-Branch*. Thus we see an *Apple*, besides that it is it self of ample compass, for the sake of its *Seed*, hath likewise its *Coar*; as if it were not sufficient, that the Walls of their Room are so very thick, unless also *wainscoated*. In a *Pear* again, where the *Parenchyma* is of less compass than that of an *Apple*, to what protection this affords, that of the *Calculary* is super-added. But in a *Plum*, where the *Parenchyma* is exceeding tender, and in a *Peach*, which hangs late, and till Autumn Frosts approach, we have not only the Rubbish of a *Calculary*, but stout Stone-Walls. Within which also, not only the *Seed* it self, but the *Seed-Branch* is evermore immur'd. Lastly, in a *Nut*, where the *Shell* being not surrounded with a *Parenchyma*, that protection is wanting without, 'tis answer'd by an ample

Pith

Pith within it; and the *Seed-Branch* likewise included, not meerly in the *Body* of the *Shell*, as in a *Plum*, but within the *Pith* it self. So necessary is this design, that what the Hen by Incubation or Hovering, is to the Egg or Chick; that the whole *Fruit*, by comprehension, is to the *Seed*.

CHAP. VII.

Of the SEED, *in its State of Generation.*

AS the Original, so the Ultimate end and Perfection of *Vegetation* is the *Seed*. How it is the former, and in its state apt for *Vegetation*, hath already been seen. How the latter, and in its state of Generation, we shall now lastly enquire. In doing which, what in the other state, was either not distinctly existent, or not so apparent, or not so intelligible, will occur.

2. §. The two general *Parts* of the *Seed* are its *Covers* and *Body*. The *Covers* in this estate are usually *Four*. The outmost, we may call the *Case*. 'Tis of a very various form; sometimes a *Pouch*, as in *Nasturtium*, *Cochlearia*; a *Cod*, as in all *Pulse*, *Galega*; sometimes not entire, but parted, or otherwise open, as in *Sorrel*, *Knotgrass*; with many other forms: I think alwaies more heterogeneous to that of the *Seed*, by which it differs from the proper *Coats*. To this the *Caps* of *Nuts*, and the *Parenchyma's* of other *Fruits* are analogous.

3. §. The two next are properly the *Coats*. In a *Bean* especially, and the like; from whence, to avoyd Confusion, the denomination may run common to the responding *Covers* of other *Seeds*. The Colour of the outer, is of all degrees, from White to the Blackness of *Jett*. It's Figure sometimes Kidney'd, as in *Alcea*, *Behen*, *Poppy*; Triangular, as in *Polygonatum*, *Sorrel*; Spherically triangular, in *Mentha*, *Melissa*; Circular, in *Leucoium*, *Amaranthus*; Globular, in *Napus*, *Asperula*; Oval, in *Speculum Veneris*, *Tithymalus*; half Globe, in *Coriander*; that which we take for *one* single round *Seed*, being a Conjugation of *two*; half Oval, in *Anise*, *Fennel*; Hastal, in *Lactuca*; Cylindrical, as, if I mistake not, in *Jacobæa*; Pyramidal, in *Geranium Althææ fol.* with many other differences. But the Perfection of one or two of the said Figures lieth in the *Case*. So that, as all *Lines* and *Proportions* are in the *Leaf* and *Flower*; so all Regular *Solids* in the *Seed*; or rather in its *Covers*.

4. §. 'Tis sometimes glistering, as in *Speculun Veneris*; Rough-cast, in *Catanance*; Studded, in *Behen*, *Balttaria*; Favous, in *Papaver*, *Antirrhinum*, *Lepidum annuum*, *Alcea Vesicaria*, *Hyoscyamus*, and many more, before the *Seeds* have lain long by; Pounced, in *Phalangium Cretæ*, *Lithospermum*; Ramified, in *Pentaphyllum fragiferum Erectum majus*,

resembling

resembling the *Fibers* of the *Ears* of the *Heart*; some just *Quinquenerval*, as in *Anisum*, and many more, the *Lignous Body* being in five main *Fibers* branched therein. The *Figures*, and *Surface*, of These, and other *Seeds*, See in the Tables belonging to the *Fourth* 𝔓𝔞𝔯𝔱 of the *Fourth* 𝔅𝔬𝔬𝔨.

5. §. The *Covers* of not only *Quince-Seeds*, and those of *Psyllium* (more usually taken notice of) but those also of *Horminum, Nasturtium, Eruca, Camelina, Ocymum,* and divers others, have a *Mucilage*. Which, though it be not visible when the *Seeds* are throughly dry; yet lying a while in some warm Liquor, or only on the Tongue, it swells more or less, and upon them all fairly shews it self, On that of *Ocymum* it appears grayish; on the other, transparent; and on that of *Nasturtium Hortense* very large; even emulous of the inner Pulp surrounding a *Gooseberry-Seed*. The putting of *Clary-seed* into the Eye, may have been brought into use from this *Mucilage*, by which alone it may become Medicinal. And thus far of the *Superficies*.

6. §. The *nature* of the outer *Coat* is also various, *Membranous, Cartilaginous* and *Stony*; the like *Precipitations* being sometimes made herein, as in a *Stone* or *shell*; as in that of the *Seeds* of *Carthamum, Lithospermum* and others. The Designment hereof, being either with respect to the *Seed* in its state of Generation; as where the *Case* is either wanting, or at least insufficient of it self, there for its due protection and warmth. Or, in its state of *Vegetation*, for the better Fermenting of its *Tinctures* and *Sap*; the Fermentations of some *Seeds* not well proceeding, unless they lie in their *Stony Casks* in the Ground, like Bottled Liquors in Sand.

7. §. All *seeds* have their outer *Covers* open; either by a particular *Foramen*, as in *Beans*, and other *Pulse*, as is said; or by the breaking off of the *Seed* from its *Peduncle* or *Stool*, as in those in *Cucumber, Cichory*; or by the entering and passage of a *Branch* or *Branches*, not only into the Concave thereof near the Cone, but also through the Cone it self; as in *Shells* and *stones*.

8. §. For the sake of this *aperture* it is, that *Akerns, Nuts, Beans, Cucumbers,* and most other *seeds*, are in their formation so placed, that the *Radicle* still standeth next to it; That So, upon *Vegetation*, it may have a free and ready passage into the Mould.

9. §. The Original of the outer *Coat*, though from Parts of the same substantial nature, yet is differently made. In a *Plum*, the *Seed-Branch* which runns, as is described, through the *Stone*, is not naked, but, as is said, invested with a thin *Parenchyma*, which it carries from the *Stalk* along with it; and which, by the *Ramification* of the said *Branch* within the *Stone*, is, in part, dilated into a *Coat*. That of a *Bean* is from the *Parenchyma* of the *Cod*; the superficial part of which *Parenchyma*, upon the large *peduncle* of the *Bean* becoming a thin *Cuticle*, and upon the *Bean* it self a *Cartilaginous Coat*.

10. §. The Original of the inner *Coat* of the *Bean* is likewise from the inner part of the said *Parenchyma*; which first is spred into a long *Cake*, or that which with the *Seed-Branch* maketh the *Penduncle* of the *Bean*; under which *Cake*, there is usually a black part or spot; by the length of which, the inner part of the *Cake* is next inserted into the outer *Coat*, and spred all over the Concave thereof, and so becomes the inner.

11. §.

11 §. Of this Inner *Coat* it is very obfervable, That allthough when the *Seed* is grown old and dry, 'tis fhrunk up, and in moft *Seeds*, fo far, as fcarcely to be difcern'd ; yet in its firft and juvenile Conftitution, it is a very Spongy and Sappy body ; and is then likewife (as the *Womb* in a Pregnant *Animal*) in proportion, very thick and bulky. In a *Bean*, even as one of the *Lobes* it felf: And in a *Plum* or *Apricot*, I think I may fafely fay, half an hundred times thicker than afterwards, when it is dried and fhrunk up, and can fcarcely be diftinguifhed from the upper *Coat*. Upon which Accounts it is, in this eftate a true and fair *Parenchyma*. The Delineation hereof, See in the *Figures* belonging to the *Fourth* Part of the *Fourth* Book.

12. §. In this Inner *Coat* in a *Bean*, the *Lignous Body* or *Seed-Branch* is diftributed: Sometimes, as in *French-Beans*, throughout the whole *Coat*, as it is in a *Leaf*. In the Great *Garden-Bean*, upon its firft entrance, it is bipartite, and fo in fmall *Branches* runs along the *Circumference* of the *Coat*, all meeting and making a kind of *Reticulation* againft the Belly of the *Bean*. In the fame manner the main *Branches* in the outer *Coat* of a *Kernel*, circling themfelves on both hands from the place of their firft entrance, at laft meet, and mutually inofculate ; as the *Veins* in the *Kidneys* of a Man or any *Quadrupede* ; Or the *Carotick Arteries* in the *Braine*.

13. §. So that all the *Parts* of a *Vegetable*, the *Root*, *Trunk*, *Branch*, *Leaf*, *Flower*, *Fruit* and *Seed*, are ftill made up of *Two* Subftantially different Bodies.

14. §. And as every *Part* hath *Two*, fo the whole *Vegetable* taken together, is a compofition of *Two* only, and no more: All properly *Woody* Parts, *Strings* and *Fibers*, are *One Body*: All fimple *Barques*, *Piths*, *Parenchyma's* and *Pulps*, and as to their fubftantial Nature, *Pills* and *Skins* likewife, all but *One Body*: the feveral *Parts* of a *Vegetable* all differing from each other, only by the various *Proportions* and *Mixtures*, and variated *Pores* and *ftructure* of thefe *Two Bodies*. What from thefe two general Obfervations might reafonably be inferr'd, I fhall not now mention.

15. §. The Fourth or Innermoft *Cover* we may call the *Secondine*. The fight of which, by cutting off the *Coats* of an *Infant-Bean*, at the Cone thereof, in very thin Slices, and with great Caution, may be obtain'd. While unbroken, 'tis tranfparent ; being torn and taken off, it gathers up into the likenefs of a *Jelly*, or that we call the *Tredle* of an *Egg*, when rear-boyl'd. This *Membrance* in larger or elder *Beans*, is not to be found diftinct. But (as far as our Enquiries yet difcover) it may in moft other *Seeds*, even full grown, be diftinctly feen ; as in thofe of *Cucumber, Colocynthis, Burdock, Carthamum, Gromwel, Endive, Mallows*, &c. 'Tis ufually fo very thin, as in the above-nam'd, as *Tab.* 4. *f.* 16. very difficultly to be difcover'd. But in fome *Kernels*, as of *Apricots*, 'tis very thick ; and moft remarquably fuch, in fome other *Seeds*. That all thefe have the Analogy of one and the fame *Cover*, which I call the *Secondine*, is moft probably argu'd from their alike Natures ; being all of them plain fimple *Membranes*, with not the leaft *Fibre* of the *Lignous Body* or *Seed Branch*, vifibly diftributed in them : As alfo from their Texture, which is in all of them more clofe. See this *Part* in *Tab.* 4. As alfo amongft the *Figures* belonging to the *Fourth* Part of the *Fourth* Book.

16. §.

16. §. The *Concave* of this *Membrane* is filled with a most transparent *Liquor*, out of which the *Seed* is formed; as in cutting a *petite* and *Infant-Bean*, may be seen; and yet better in a young *Walnut*. In *Beans* I have observed it to turn, upon boyling, into a tender white *Coagulum*.

17. §. Through this *Membrane*, the *Lignous Body* or *Seed-Branches* distributed in the inner *Coat*, at last shoot downright two slender *Fibres*, like two *Navel-strings*, one into each *Lobe* of the *Bean*. *Tab. 4. f. 18.* The places where the said *Fibres* shoot into the *Lobes*, are near the *Basis* of the *Radicle*; and by their *Blackishness* well enough remark'd: but the *Fibers* themselves are so very small, as scarcely to be discern'd. Yet in a *Lupine*, of the larger kind, both the *places* where the *Navel-Fibres* shoot into the *Lobes* (which here from the *Basis* of the *Radicle* is more remote) and the *Fibres* themselves, are fairly visible. For the *Seed-Branch*, upon its entrance into the *Coat* of the *Lupine*, is presently divided into two *main Branches*, and those two into other less; whereof some underly, others aloft, run along the *Coat*, and towards its other end meet and are inosculated: where about, two opposite, shallow, round, and most minute *Cavities*, answerable to two *Specks* of *Tab. 4. f. 17.* a *Cartilaginous* gloss, one in either *Lobe*, may be observed; which *Specks* are the ends of the said *Navel-Fibres*, upon the ripening of the *Seed* there broken off. These *Fibres* from the *Superficies* of each *Lobe*, descend a little way directly down: presently, each is divided into two *Branches*, one distributed into the *Lobes*, the other into the *Ra-* *Tab. 4. f. 18.* *dicle* and *Plume*, in the manner as in the *First Chapter* is described. And thus far the History. I shall now only with a brief account of the *Generation* of the *Seed*, as hereupon dependent, conclude this Discourse.

An Account of the Generation of the Seed.

18. §. LET US say then, that the *Sap* having in the *Root*, *Trunk* and *Leaves*, passed divers *Concoctions* and *Separations*, in the manner as they are said to be perform'd therein; 'tis now at last, in some good maturity, advanced towards the *Seed*.

19. §. The more copious and cruder part hereof is again separated by a free reception into the *Fruit*, or other *Part* analogous to it: being either sufficiently ample to contain it, or at least laxe enough for its transpiration, and so its due discharge. The more Essential part is into the *Seed-Branch* or *Branches* entertian'd. Which, because they are evermore of a very considerable length, and of a Constitution very fine, the said *Sap* thus becomes in its Current therein as in the *Spermatick Vessels*, still more mature.

20. § In this mature estate, from the *Seed-Branch* into the *Coats* of the *Seed*, as into the *Womb*, 'tis next delivered up. The meaner part hereof again, to the *Outer*, as *Aliment* good enough, is supplied. The finer part is transmitted to the *Inner*; which being, as is said, a *Parenechymous* and more spatious *Body*, the *Sap* therefore is not herein, as in the *Outer*, a meer *Aliment*; but in order to its being, by *Fermentation*, farther prepared.

21. §. Yet the Outer *Coat*, being on the contrary hard and dense; for that reason, as it admitteth not the Fermentation of the *Sap* so well within it self; so doth it the more promote and favour it in the Inner; being Bounds both to it and its *Sap*; and also quickneth the *process* of the whole Work in the formation of the *Seed*.

22 §.

22. §. Nor doth the Outer *Coat*, for the fame reafon, more promote, than declare the purity of the *Sap* now contained in the Inner: For being more hard and denfe, and fo not perfpirable, muft needs fuppofe the Parts of the *Sap* encompaffed by it, fince thus uncapable of any evacuation, to be therefore all fo choice, as not to need it.

23. §. The *Sap* being thus prepared in the Inner *Coat*, as a *Liquor* now apt to be the *Subftratum* of the future *Seed-Embrio*; by frefh fupplies, is thence difcharg'd. Yet that it may not be over-copious; which, becaufe of the laxity of the Inner *Coat*, from whence it iffues, it might eafily be: therefore, as the faid Inner *Coat* is bounded without, by the upper *Coat*; fo by the *Secundine*, is it bounded within. Through which *Secundine* the *Sap* being filtr'd, or, as it were, tranfpiring; the depofiture hereof, anfwerable to the *Colliquamentum* in an Egg, or to the *Semen Mulibre*, into its Concave at laft is made.

24. §. The other part of the pureft *Sap* embofom'd in the *Ramulets* of the *Seed-Branch*, runs a Circle, or fome progrefs therein; and fo becomes, as the *Semen Mafculinum*, yet more elaborte.

25. §. Wherein alfo, left its Current fhould be too copious or precipitant, by their *co-arcture* and *divarication* where they are inofculated, it is retarded; the nobleft portion only obtaining a pafs.

26. §. With this pureft *Sap*, the faid *Ramulets* being fupplied, from thence at laft, the *Navel-Fibres* fhoot (as the primitive *Artery* into the *Colliquamentum*) through the *Secundine* into the aforefaid *Liquor* depofited therein.

27. §. Into which *Liquor*, being now fhot, and its own proper *Sap* or *Tinctures* mixed therewith, it *ftrikes* it thus into a *Coagulum*; or of a *Liquor*, it becomes a *Body confiftent* and truly *Parenchymous*. And the fupply of the faid *Liquor* ftill continu'd, and the fhooting of the *Navel-Fibres*, as is above defcribed, ftill carried on, the faid *Coagulation* or *Fixation* is therewith likewife.

28. §. And in the Interim of the *Coagulation*, a gentle *Fermentation* being alfo made, the faid *Parenchyma* or *Coagulum* becometh fuch, not of any Texture indifferently, but is thus raifed (as we fee Bread in Baking) into a *Congeries* of *Bladders*: For fuch is the *Parenchyma* of the whole *Seed*.

FINIS.

THE
ANATOMY
OF
ROOTS;

Presented to the ROYAL SOCIETY at several times, in the Years, 1672 & 1673.

With an Account of the

VEGETATION OF ROOTS,

Grounded chiefly hereupon.

The SECOND BOOK.

By *NEHEMJAH GREW* M.D. Fellow of the *Royal Society*, and of the *College* of *Physicians*.

𝔗𝔥𝔢 𝔖𝔢𝔠𝔬𝔫𝔡 𝔈𝔡𝔦𝔱𝔦𝔬𝔫.

LONDON,

Printed by *W. Rawlins*, 1682.

TO THE
Right Honourable
WILLIAM
Lord Vi-Count *BROUNCKER*
THE
PRESIDENT
AND TO THE
Council and Fellows
OF THE
ROYAL SOCIETY.

MY LORD,

IF the Dedication of *Books* were not in use; yet here, I think, I might have been a Precedent. The promotion of *Phytological Science* is one Part of *Your* Work; and 'tis *You* have called me to the management of this Part; for some time, have intrusted me herein; and by *Your* most favourable and candid acceptance of what I have performed thus far, have encouraged me hereunto: I therefore present but *Your* Own, into *Your* Hands.

The great Honour and Advantage of *Your* Fellowship, I first obtained, by Mediation of Dr. *Wilkins*, the late most Reverend *Bishop* of *Chester*. Whom I cannot name, without saying thus much of him, That He was a Per-

The Epistle Dedicatory.

son of that eminent and happy Worth, which, as it was too good, to fear envy; so is it too great, to need an Elogie.

With Him, it was, *You* were pleased to commit to Me, the further prosecution of this *Work*; the Beginnings whereof, were by *Your Order* formerly made publique. Had I consulted my own Abilities altogether, I should scarcely have ventured upon it; seeing very little, for which I could think well of my self, saving, That I had learned, upon good grounds, to think of *You* with greatest Honour. But I also considered, That to insist hereon too much, might be a reflection upon *Your* Judgments, who had thought fit to make choice of Me. And, That *You* were not more the Patrons of Wit, than of Industry; and of All, who shall endeavour to find out, or to confirm the Truth of Things. Withal, I looked upon *Nature*, as a Treasure so infinitely full; that as all Men together, cannot exhaust it; so no Man, but may find out somewhat therein, if he be resolved to Try.

In compliance therefore with *Your* Commands, I have hereunto devoted a very considerable part of my Time. These, adding force to my own Desires, of being somewhat instrumental to the Improvement of Medicinal, and other wholesom Knowledge: if peradventure, as we increase herein, we may become better, and more happy. As to which Improvement, though I could not hope; yet, I would not dispair. I have already prepared the Soil, and made some Plantation: what remaineth behind, and the Vintage of the whole, will depend much upon the continued Influence of *Your* Beams: for how unpromising soever the Stock may be; yet the Fruit cannot but be somewhat matured, upon which *You* are pleased to shine. I am also confident, that the same Nobilty and Goodness, which accept the endeavours, will likewise pardon the faults, of,

September 1.
1673.

My Lord,
Your Lordships most humbly
and most sincerly
devoted Servant

NEHEMJAH GREW.

THE

THE CONTENTS.

The FIRST PART.

CHAP. I.

OF the Original of Roots, §. 1, 2, 3. Of their Figures, 4, to 8. Of their Motions, 9, to 15. And of their Ages, 16, to the end.

CHAP. II.

OF the Skin. Its external Accidents, and Original, §. 1, 2. Compounding Parts. Whereof the one Parenchymous, 3. The other Lignous, 4, to the end.

CHAP. III.

OF the Barque. Its Original and external Accidents, §. 1. Size, 2. Compounding Parts: Whereof the one Parenchymous, 3. The Bladders of the Parenchyma, 4, 5, 6. The Diametral Portions, 7, to 11. The other Part, Lignous, consisting of long Pipes or Vessels, 12, to 17. Of several Kinds, 18, to 23. In different Proportion, 24, 25. And in different and elegant Position. 26, to the end.

CHAP. IV.

OF that Part of the Root next within the Bark; in Trees and Shrubby Plants, called the Wood. Hereof the Parenchyma, §. 1, 2, 3, & 7. The Lignous Portion: of which, the Sap-Vessels, 4. The Aer-Vessels, 5, 6. The Position of the Former, 8, 9. Of the Latter, 10, 11, 12. Their Proportion, 13, 14, 15. The Latter, sometimes a little tapering. 16. Their Texture, 17, to 22. Content, 23.

CHAP.

The Contents.

CHAP. V.

OF the Pith. *Found in the upper part of most Roots*, §. 1. *Its size and shape*, 2. *Sap-Vessels*, 3. *Original*, 4, 5. *Bladders*, 6, *Fibres and Texture*, 7, *to* 11. *That of the Insertions and Barque the same*, 12. *Hence, the Original of the Aer-Vessels conjectured*, 13. *What the whole Body of a Root, concluded*, 14, 15. *The Contents of the Pith*, 16.

The SECOND PART.

THeology, *the Beginning and End of Philosophy*, §. 1, *to* 6. *The Divine Wisdom seen in the Growth of Plants*, 7. *If we observe*,

How the Ground is Prepared, 8, *to* 14.

How the Sap is Imbibed, and Distributed to the several Parts of the Root, 15, *to* 28.

How the several Parts are Nourished and Formed, 29, *to* 35.

How the several Parts receive their respective Situation, 36, *to* 40.

How Roots receive their different Size and Shape, 41, *to* 47.

How Roots receive their different Motions, 48, *to* 53.

How Roots are differently Aged, 54, 55, 56.

How the Liquors and other Contents of the several Parts are made 57 *to* 63.

How the Odors of Roots are made, 64.

How their Colours, 65, *to* 67.

How their Tasts, 68, *to the end.*

THE ANATOMY OF ROOTS;

PROSECUTED

With the bare EYE,

AND WITH THE

MICROSCOPE.

PART I.

CHAP. I.

Of the ORIGINAL, FIGURES, MOTIONS, and AGES of ROOTS.

BEING TO speak of *Roots*; it is requisite, for our better understanding of what follows, that some things, as to their *Original, Figures, Motions* and *Ages,* be premised.

1. §. *Roots,* taken altogether, have a Three-fold Original. Either from the *Radicle*; as all *Roots* which come of the *Seed*: or from the *Trunk* or *Caulis*, above ground; as in *Strawberry, Chamæmile,* and many other *Creepers*: or from the *Trunk* or *Caulis*, after it is sunk under ground; as in *Primrose, Bistort,* and many others; and presently shall be shewed how.

2. §. In the Growth of a *Bud,* and of a *Trunk-Root,* there is this observable difference; That the former, carries along with it, some portion of every *Part* in the *Trunk* or *Stalk*; whereof it is a *Compendium.* The latter, always shoots forth, by making a Rupture in the *Barque,* which it leaves behind, and proceeds only from the inner part of the *Stalk.*

3. §.

3. §. As alſo, That in a *Bud*, the *Lignous Part* is ſpread abroad, ſo as to encompaſs a *Pith*. Whereas in a *Trunk-Root*, it makes a ſolid Thred ſtanding in the Center. Which is the Cauſe of its deſcending into the Ground: as is already, in the *Firſt* 𝕭𝖔𝖔𝖐, and ſhall in This be further ſhewed.

4. §. ROOTS are generally diſtinguiſhed, as to their *Figures*, in being more Entrie, as is that of *Liquiriſh*; or Parted, as of St. *Johnswort*. Parted or Forked, either at the Bottom, as moſt *Roots*; or at the Top, as *Dandelyon*, and ſome others. A thing very odd, and unintelligible, without the knowledge of the *Motions* of *Roots*; whereof preſently.

5. §. Parted, again, are either Ramified, as that of *Cumfry*; or Manifold, as of *Crowfoot*: both are Parted; but the former, by the ſubdiviſion of greater *Branches*, into leſſer; theſe, when divers *Strings*, have all their diſtinct original from one *Head*. Some are Straight, as a *Radiſh*; others Crooked, as *Biſtort*. Smooth, as *Bugloſs*; or Stringy all round about, as *Columbine*. And to *Carnations*, this ſeems to be peculiar, That ſometimes many of the *Strings* run parallell with the *Wood* of the great *Root*, through the *Barque*, or betwixt the *Wood* and the *Barque*.

6. §. Again, ſome are Thick, as *Rhubarb*; Slender, as the *Vine*. Long, as *Fenil*; Short, as a *Turnep*: which are diſtinct from Great and Little; in that theſe, are ſo called with reſpect to ſeveral *Roots*; thoſe, with reſpect to the ſeveral Dimenſions of one. Short, are Stubbed, as *Iris tuberoſa*; or Round, as *Dracontium*. Round are Tuberous, or Simply Knobbed, as *Rape-Crowfoot*; Bulbous, that is Scaled, as ſome Lilys; or Shell'd, as an *Onion*. Where note, That all Bulbous *Roots*, are, as it were, Hermaphrodites, or *Root* and *Trunk* both together: for the *Strings* only, are abſolute *Roots*; the *Bulb*, actually containing thoſe *Parts*, which ſpringing up, make the *Leaves* or *Body*; and is, as it were, a Great *Bud* under ground.

7. §. *Roots*, again, are Even or Uneven; Even, are Cylindrical, as *Eryngo*; or Pyramidal, as *Borage*. Growing ſmaller Downwards, as do moſt; or Upwards, as *Skirrets*. Uneven, are Pitted, as *Potato's*, where the *Eyes* or *Buds* of the future *Trunks* lie inward; or Knotted, as *Jeruſalem-Artichoke*; where they ſtand out. Theſe Differences, are alſo Compounded: ſo ſome *Roots* are both Entire and Smooth, as *Peony*; others Entire, but Stringy, as *Clary*: that is, neither Ramifi'd, nor yet Bruſhy, or divided at the Top into ſeverall ſmall *Strings*; but a Single *Root* ſurrounded with many Hairy *Threds*. Some both Plain in ſome parts, and Knobbed in others, as *Filipendula*, *Lilium non bulboſum*, and others.

8. §. Some alſo have two or more *Roots*; and thoſe of one Kind: of which, ſome are diſtinctly faſtend to the bottom of the *Stalk*, as in *Dogſtones*; ſome ſtand one under another, ſo as only the uppermoſt is faſten'd to the *Stalk*, as in *Dragon*, *Crocus*, and others. And there are ſome, which have not only two *Roots*, at the ſame time; but thoſe alſo of two diſtinct Kinds, as in *Biſtort*; one of them, a ſlender ſtrait Cylindrick and horizontall *Root*; the other large and crooked, and bred of the Deſcending *Trunk*; as in ſpeaking next of the *Motions* of *Roots*, will be underſtood, how. All which, with other Differences

by

by Thofe that undertake the *Defcriptions* of *Plants*, are accurately to be Noted. But the Differences, above mentioned, will ferve for our prefent Purpofe.

9. §. THE MOTIONS of *Roots* are alfo divers. Sometimes Level, as are thofe of *Hops, Ammi, Cinquefoyle*; and all fuch as properly *Creep*. Sometimes Perpendicular, as that of *Parfnep*: Which is different from Straightnefs; for fome Straight *Roots*, are Level. Both of them are either Shallow or Deep: fome run Level, and near the *Turf*, as *Woodbind, Wild Anenomy*; others lower, as *Dogs-Grafs*. Some ftrike down, but a little way, as *Stramonium*; others grow deep, as *Horfe-Radifh*: Which is different from being Long; for many long *Roots*, are Level, as *Hops*.

11. §. Some again Defcend, as *Tulips*, and other Bulbous *Roots*, which differs from growing only Downwards; in that here, the *Head* of the *Root* is Immoveable; but in Defcending, the whole *Root* obteineth different Places, running deeper, time after time, into the *Earth*. Some alfo Afcend, fometimes, and in fome part, appearing above ground, as *Turneps*.

11. §. Thefe *Motions* are alfo Compounded; both in refpect of the feveral Parts of the *Root*, and of feveral Times. So the main *Root* of *Primrofe*, •is Level; the Strings are Perpendicular. The *Roots* of moft *Seedlings* grow Downward and Upward, or fhoot out in length at both Ends, at the fame time. Thofe of *Biftort, Iris*, and fome others, grow, in part, both Downward and Upward at feveral times: Whence it is, that *Biftort* is Crooked, with fome refemblance to an S, according to its *Name*; And that fome Parts of *Iris-Root* appear oftentimes above the ground.

12. §. There is alfo another *Motion*, in fome *Roots*, not heeded; and that is *Contortion*: whereby, without being moved out of their Place, they are Writhed or Twifted; as a piece of Cloath is, when the Water is wrung out of it; as in *Carduus, Sonchus*, and others: whether always I cannot fay. This *Motion* cannot be noted, without ftripping off the *Barque*; whereby the *Veffels* may be feen, fometimes, to make two or three Circumvolutions. This *Motion* feems to be governed by the winding of the *Stalk*; and therefore to begin at the Head, and terminate at the Poynt or lower end of the *Root*, which is immoveable.

13. §. BUT ABOVE all the *Motions* of *Roots*, not obferved, the moft remarkable is that of DESCENT. Which, although it hath been noted, by fome *Botanicks*, of *Bulbous Roots*; yet of thefe only: Whereas it is the Property, of a great many more; and thofe, of very different Kinds; probably, of the far greater number of *Perennial Roots* of *Herbs*; as of *Arum, Rape-Crowfoot, Valerian, Brownwort, Bearffoot, Tanfy, Lychnis, Sampier, Primrofe, Ammi, Avens, Wood-forrel, Iris*, and others. Of all which *Plants*, it is very obfervable, That their *Root* is annually *renewed*, or *repaired*, out of the *Trunk* or *Stalk* it felf. That is to fay, The *Bafis* of the *Stalk* continually, and by infenfible Degrees, defcending below the furface of the *Earth*, and hiding it felf therein; is thus, both in Nature, Place, and Office changed into a true *Root*. Which *Root*, by the continuance of the faid

O 2 *Motion*

Motion of the *Stalk*, also *Descends*; and so, according to the durableness of its Substance, becomes a shorter or longer *Root*; the Elder or Lower Portion thereof, Rotting off, by the same Degrees with the Generation of the Upper, out of the *Stalk*. So in *Brownwort*, the *Basis* of the *Stalk* sinking down by degrees, till it lies under Ground, becomes the upper part of the *Root*; and continuing still to sink, the next year, becomes the lower Part; and the next after that, rots away; a new Addition being still yearly made out of the *Stalk*, as the elder Parts yearly rot away. So in *Dragon, Crocus*, and the like, where the *Root* is double; the *Basis* of the *Stalk*, this year; the next, becomes the Upper-*Root*; after that, the Lower-*Root*; and at the length dies and is consum'd.

Tab. 5. *f.* 6, & 7.

14. §. The Demonstration hereof, is taken, more evidently, from some *Roots*, than from others; as from the Level and Knobed *Roots* of *Wood-sorrel, Primrose*, &c. For the *Leaves* of those *Plants* rotting off successively, and the *Bases* of those *Leaves* gradually descending into the Ground; each *Basis* is thus nourished with a more copious *Sap*, and so swelled into so many thick *Knots*. It may likewise be gather'd in some, from the like Position of the *Vessels* or Woody Parts, in the *Root*, as in the *Trunk*; as in *Bares-foot*, As also, from the *Root* of the *Iris Tuberosa*: where, although the *Leaves* fall off close to the Surface of the *Stalk*; yet after that is sunk down, and swell'd into a *Root*, the *Seats* of the perished *Leaves*, and the Ends of the *Vessels* belonging to them, are not obscurely visible; whereby the *Root* is wrought, as it were, with several *Seames* and *Prickt-Lines*; the *Seams* shewing the setting on of the *Leaves*; and the *Pricks*, the Terminations or broken Ends of the *Vessels*: which ends, are still more apparent, upon the stripping off the *Barque*. I considered likewise, That as among *Animals*, there are many, which are not Bred of *Eggs*, immediately; but are Transformed, one *Animal* into another: So, it is more than probable, That among *Plants*, there are not a few Instances of the like *Transformations*; whereof, this is one.

Tab. 5. *f.* 1, & 2.

Tab. 5. *f.* 4.

15. §. The *Cause* of this *Descent*, so far as it is dependent on the Inward Conformation of the *Root*, I shall shew in the following 𝔓art. But the Immediate Visible one, are the *String-Roots*, which this kind of *Trunks* frequently put forth: which, descending themselves directly into the Ground, like so many *Ropes*, lug the *Trunk* after them. Hence the *Tuberous-Roots* of *Iris* upon the rotting or fading away of the *String-Roots* hanging at them, sometimes a little Re-ascend. Hence also the *Shape* of some *Roots* is Inverted: For whereas most are parted downwards, into several *Legs*; some are parted upwards into divers *Necks*, as *Dandelyon*, and others. For these *Roots* sending forth at the top several *Trunk-Buds*, the said *Buds* successively put forth new, and cast their old *Leaves*; and continually also making their Descent, are at length formed into so many *Necks*, of three, four, five, or more Inches long, under Ground.

Tab. 5. *f.* 5.

16. §. HENCE ALso we understand, in what particular way, some *Roots* become *Perennial*. Some are wholly so, as those of *Trees, Shrubs*, and divers other woody *Plants*. Others, in part, or by a new *Progenies* of *Roots*, from the old Head or Body, in the room of those that die yearly, or after a certain Time: as of *Lilium non bulbo-*
sum,

sum, Jerusalem Artichoke, Potato, Dog-stones, Monks-hood, little Celandine, and others. In which *Plants*, one or more of their *Roots* are firm, the other spongy and superannuated; and partly, by the ravine of the *Trunk*, and other younger *Roots*, reduced to a Consumption and Death.

17. §. With these, *Tulips*, and other *Bulbous-Roots* consort: For the several *Rindes* & *shells*, whereof chiefly, the *Bulb* consists, successively perish and shrink up into so many thin and dry *Skins*: betwixt which, and in their Centre, other *Leaves* and *Shells*, being successively formed, the *Bulb* is thus perpetuated. In the same manner the *String-Roots* also succeed one another annually. So that at the end of divers Years, although it be still looked upon as the same *Individual Root*, yet it is, in truth, Another, as to every particle thereof.

18. §. Lastly, many other *Roots* are perpetuated by the aforesaid Descent of the *Trunk*; out of which, it is still annually Repaired, as by the gradual perishing of its lower parts, it is Diminished; as hath been said. Whence also we see the reason of the Rugged and Blunt extremities of these, and some other *Roots*, as of that *Plant* superstitiously called *Devils-bit*: because the end of it seems to be bitten off. Yet doth it not appear so originally; but the Lower part thereof rotting off, as the Upper descends; the living remainder, becometh stumped, or seemeth Bitten. Thus far of the *Original, Shapes, Motions*, and *Ages* of *Roots*.

Tab. 5. f. 3.

CHAP. II.

Of the SKIN.

NEXT proceed to the several *Parts* whereof a *Root* is Compounded. The outer *Part* of all is the *Skin*; which is common to all *Roots*. 'Tis diversly Coloured: Whiter in *Skirrets*; Yellow, in *Dock*; Red, in *Potato*; Brown, in *Lovage*; Black, in *Bugloss*. Its Surface, sometimes Smooth, as in *Horsradish*; Rough, as in *Scorzonera*. And the *Skins* of the several *Shells* of a *Tulip-Root*, taken up fresh, look as if they were perforated with a great many small holes. 'Tis of various Size; very Thin, in *Parsnep*; somewhat Thick, in *Bugloss*; very Thick in *Iris*. Sometimes it is Opacous, as in *Thistle*; and sometimes Transparent, as in *Madder*.

2. §. Every *Root* hath successively two kinds of *Skins*: the one, Coëtaneous with the other *Parts*; and hath its original from that which involveth the *Parts* of the *Seed* it self. The other, Postnate, succeeding in the room of the former, as the *Root* ageth; and is orinated from the *Bark*. So in *Dandelyon*, the old *skin*, looked upon about the beginning of *May*, seems to have been one of those several

Rings

Rings, which the precedent year compofed the *Cortical Body* of the *Root*: but by the Generation of a new *Ring*, next the *Wood*, is now thruft off and fhrunk up into a *Skin*. So alfo in the *Roots* of *Buglofs* and *Horfe-Radifh*, as far as the *Bladders* in the former, and the *Veffels* in the latter are Radiated; the *Cortical Body* feems either annually or oftener, to fhrink up into another new *Skin*, as, the old ones fall off. And fometimes, perhaps, as in *Afparagus*, the whole body of the Perpendicular *Roots*, except the woody *Fibre* in the Centre, becomes the fecond *Skin*. So that the wearing away of the old *Skin*, fucceeds the derivation of the new one; as in Defcending *Roots*, the Confumption of the Lower *Parts*, doth the Generation of the Upper. Becaufe the *Barque* fwells, and grows fometimes fafter than the *Skin* can fall off, or give way to it: therefore are the *Roots* of many *Herbs*, Barquebound, as well as the *Trunks* of *Trees*.

Tab. 14, 15.

Tab. 10.

3. §. This *Skin* is ufually, if not always, compounded of two Kinds of Bodies: which alfo is probable of the Coëtaneous. The one, *Parenchymous*, and frequently conftructed of exceeding little *Cells* or *Bladders*; which in fome *Roots*, as of *Afparagus*, cut traverfe, and viewed through a *Microfcope*, are plainly vifible. Thefe *Bladders* are of different Sizes; in *Buglos*, larger; in *Afparagus* lefs; and fometimes they coincide and difappear. But in thefe, and all other *Roots*, even where thefe *Bladders* appear not, the *Parenchyma* of the *Skin*, is of the fame Subftantial *Nature*, with that other more vivid and bulky one of the *Bark*: As is manifeft, from its being thence Originated; and alike Conformed, as fhall be feen; and not only adjacent to it, as a *Glove* is to the *Hand*; but continuous therewith, as the parts of a piece of flefh, are one with another.

Tab. 10.
Tab. 14.

4. §. OF THIS *Parenchymous Body*, the *Skin* confifteth chiefly, but not wholly; there being many *Lignous Veffels* which are Tubulary, mixed therewith: which, though hardly by the *Microfcope*, yet otherwife, is demonftrable. For in tearing the *Skin*, you fhall do it more eafily by the length, than bredth; becaufe, by the firft way, the continuity only of the *Parenchyma*, is diffolved; but by the latter, both of this, and of the *Veffels*, thefe being pofited by the length of the *Root*: So that, as by the fmalnefs of the *Bladders* of the *Parenchyma*, the *Skin* is Denfe; fo by thefe *Veffels*, is it Tough.

5. §. Again, if you cut a *Root* traverfe, and let it lie by for fome time, all the parts, where there are no *Veffels*, fhrink below the furface of the cut-end; but where-ever Thefe are pofited, there is no fhrinking; which oftentimes, evidently appears alfo in the *Skin*: becaufe the faid *Veffels*, though, as the *Bladders*, they may coincide; yet they cannot vifibly fhorten or fhrink up in length; no more than a *Straw*, whofe fides may yet be eafily crufhed together.

6. §. Further, the *Root* being cut traverfe, if, near the cut-end, you very gently prefs the fide of the *Root* with the edge of your Nail, the *Sap* will thereupon arife fometimes from the *Skin*; in the fame manner, as from any other part of the *Root*, where the like *Veffels* are pofited. And although the *Sap* may likewife be expreffed from the *Pith*, and other *Parts* where fometimes, there are none of thefe *Veffels*; yet not without a folution of there continuity; which here doth not follow; as appears, from the difappearing of the *Sap*, together with the intermiffion

termiſſion of the preſſure; the ſaid *Veſſels* then dilating themſelves by a Motion of *Reſtitution,* and ſo ſucking up the *Sap* again.

7. §. Hereunto may be added the Teſtimony of ſight; the very *Veſſels* themſelves, in many *Roots,* coming under an apparent view, and ſtanding in the utmoſt ſurface of the *Root* all round about, as in that of Liquiriſh, Columbine, Scorzonera, and others. Which *Experiments,* I have here, once for all, more particularly ſet down; becauſe I ſhall have occaſion, hereafter, to refer to them.

CHAP. III.

Of the BARQUE.

EXT WITHIN the *Skin* lieth the *Barque.* 'Tis ſometimes Yellow, as in *Dock*; Red, in *Biſtort*; but uſually, and in *Seed-Roots,* I think, always White. It is derived from the *ſeed* it ſelf; being but the extenſion or prolongation of the *Parenchyma* of the *Radicle*; One of the three *Organical Parts* of the *Seed,* deſcribed in the *Firſt* Chapter of the *Firſt* Book.

2. § It is variouſly Sized; ſometimes very Thin, as in *Jeruſalem Artichoke, Goats-beard,* and in moſt *Trees*; where it alſo retains the Name of a *Barque* or *Rind.* Sometimes 'tis more Thick, and maketh up the far greateſt protion of the *Root,* as in the String-*Roots* of *Aſparagus,* in *Dandelion,* and others. The thinneſt and the thickeſt are all analogous, and obtain the ſame general *Uſes.* The degrees of its Size, amongſt all *Roots,* may be well reckoned about Twenty, and ſeen in the following examples, *ſc. Beet, Dropwort, Jeruſalem Artichoke,* Tab. 7, 8, 9. *Orpine, Valerian, Goats-beard, Nettle, Brownwort, Columbine, Celandine, Aſparagus, Horſe-Radiſh, Peony, Bryony, Eryngo, Borage, Lovage, Dandelion, Parſnep, Carrot, &c.* In the Root of *Beet,* ſcarce exceeding a good thick *Skin*: but in a *Carrot,* half the Semidiameter of the *Root,* or above half an Inch over in ſome places: and that of *Dandelion,* ſometimes, in proportion with the *woody Part,* twice as thick: the reſt of Several intermediate Degrees: And to moſt *Roots,* this is common, To have their *Barque* proportionably thicker, at the bottom than at the top.

3. §. IT IS Compounded of two *Bodies.* The one *Parenchymous*; Continuous throughout; yet ſomewhat Pliable without a ſolution of its Continuity. Exceeding Porous; as appeareth from its ſo much ſhrinking up, in drying. The *Pores* hereof are extended much alike both by the length and bredth of the *Root*; therefore it ſhrinketh up, by both thoſe Dimenſions, more equally. And they are very Dilative; as is alſo manifeſt from its reſtorableneſs to its former bulk again, upon

its infusion in Water: that is to say, *It is a most curious and exquisitely fine wrought Sponge.* Thus much the Eye and Reason may discover.

4. §. The *Microscope* confirms the truth hereof, and more precisely shews, That these *Pores* are all, in a manner, Spherical, in most *Plants*; and this *Part*, an Infinite Mass of little *Cells* or *Bladders*. The sides of none of them, are Visibly pervious from one into another; but each is bounded within it self. So that the *Parenchyma* of the *Barque*, is much the same thing, as to its Conformation, which the Froth of *Beer* or *Eggs* is, as a fluid, or a piece of fine *Manchet*, as a fixed Body. The Sides also of these *Bladders* are as transparent, as those of Water; or the Bodies of some *Insects*.

Tab. 10, *& sequent.*

6. §. But their Size is usually much smaller; and their Posture more Regular than those in *Bread* or *Water*. In all *Roots* they are so small, as scarcely, without the *Microscope*, to be discerned: yet are they of different Size, both in the same, and in divers *Roots*; the varieties whereof, amongst all *Roots*, may be reduced to about Ten or Twelve according to the *Standard*, in *Tab.* 11. Some of those in *Dandelion*, being of the Smallest; and in *Buglofs*, of the Greatest. They are posited, for the most part, at an Equal Height; and piled evenly one over another: So that, oftentimes, they visibly run in Ranks or trains, both by the length and breadth of the *Roots*, as in the Root of *Buglofs*, or of *Dandelion*, split through the middle, may be seen. Although they are usually Spherical, yet sometimes, and in some places, they are more oblonge, as in the outward part of the *Barque* of *Buglofs*. These *Bladders*, are sometimes best seen, after the *Root*, being cut traverse, hath layn by a while, to dry.

Tab. 13, 14.

Tab. 14.

6. §. They are the Receptacles of *Liquor*; which is ever Lucid; and I think, always more Thin or Watery. They are, in all *Seed-Roots*, filled herewith; and usually, in those also which are well grown, as of *Borage*, *Radish*, &c.

7. §. THIS *Parenchymous Part*, in many *Roots*, is of one Uniform *Contexture*; as in *Asparagus*, *Horf-Radish*, *Peony*, *Potato*, and others. In many others, it is, as it were, of a Diversified *Woof*; the *Bladders* being, though every where Regular, yet either in Shape, Size, or Situation, different in some Parts hereof, from what they are, in other intermediate ones. For these Parts, are like so many White *Rays*, streaming, by the Diameter of the *Root*, from the inward Edge toward the Circumference of the *Barque*; as in Lovage, Melilot, Parsnep, &c. cut transverfly, is apparent. They are, though not in direct Lines, continued also by the length of the *Root*; so that they are, as it were, so many *Membrances*, by which the other Parts of the *Barque*, are disterminated.

Tab. 8, 9.

8. §. The Continuation of these Diametral *Rays*, or *Portions*, is divers: sometimes, but half through the *Barque*, or somewhat more, or less, as in *Melilot*. And it is probable, that to the *Roots* of all or most *Trefoyls*, and also of the *Leguminous* Kind, this is proper, To have their Diametral *Rays* come short of the Circumference. Sometimes, they run quite through to the very *Skin*, as in *Lovage*. And I think, in the *Roots* of all Umbelliferous *Plants*: In which therefore, the *Skin* feems to have a closer Communion with the Diametral *Rays*, and to be originated especially therefrom. They usually stand at an Equal Distance in the same *Root*: But with respect to divers *Roots*, their Distance

Tab. 9.

Tab. 8.

ſtance varies; ſo leſs, in *Parſnep*, greater in *Bugloſs*. They are commonly *Tab.* 7, 8.
Rectilinear, as in *Lovage*; but ſometimes winding to and fro, as in
a *Carrot*.   *Tab.* 8.

9. §. They are not always of one Size: in a *Carrot* near the Inner Edge of the *Barque*, exceeding Slender, and ſcarcely diſcernable; in others, Thicker, as in the Three greater ones of *Melilot*, and in common *Chervil*. Both by their Diſtance, and Size, they are alſo leſs or more Numerous; ſome, only as they are nearer; ſome, as ſmaller; others, as both. And 'tis proper, I think, to the *Intybous* kind, either to have none, or but a few. Sometimes they are of the ſame Thickneſs quite through the *Barque* from edge to edge, as in *Marſh-Mallow*. And ſometimes are conſiderably ſpread or dilated as they aproach the *Skin*, wherewith they are joyned, and whereinto they more viſibly run, as in *Parſley*, or the ſmaller part of the *Root* of *Lovage*. And in ſome *Roots*, as of *Scorzonera*, at ſome times of the year, when leſs ſucculent, almoſt the whole *Parenchyma* ſeems to be of the Nature of the Diametral *Rays*, in other *Roots*. The *Bladders* of theſe Diametral *Portions*, are ſometimes, greater than thoſe of the other *Parenchymous Parts*, as in *Parſley*; and I think ſometimes leſs. Yet as there, ſo here, variouſly ſized; to about ſix or eight Degrees; and thoſe of *Parſley* about the third, fourth, and fifth. Their Figure is Sometimes more oblong; and their direction or reſpect more towards the Center of the *Root*.   *Tab.* 8, 9.   *Tab.* 7.   *Tab.* 8.

11. §. As the other *Parenchymous Parts* of the *Barque*, are the Receptacles of *Liquor*; ſo theſe, (where they are) of *Aer*. This is argued, From their being more White, and not Tranſparent, as ſuch *Roots* and *Parts* uſe to be, which are more copiouſly and equally filled up with *Liquor*: as the *Pith* of *Elder*, which, in the old *Stalks*, is White; was once, and by being well ſoaked, will become, again Tranſparent. And from their being more dry and voyd of *Liquor*; whereupon their *Bladders*, which cannot be Vacuities, muſt be filled with more or leſs *Aer*, mixed with the *Sap* or the *Vaporous* parts thereof. This is more obſervable in thoſe Diametral *Portions*, which terminate upon, and run into the *Skin*.

12. §. THE BARQUE is not only of a divers *Woof*, but as is ſaid, of a Compounded Subſtance; there being a certain number of *Lignous Veſſels*, fewer or more, in ſome place or other, mixed with the *Parenchymous Part* above deſcribed; and ſome way or other, are demonſtrable in all *Roots* As by the Toughneſs of the *Barque*, when pulled by the length. By the viſible Continuation of the ſaid *Veſſels* through the length of the *Barque*, in the reſemblance of ſmall *Threds*. *Tab.* 6. And by the riſing up of the *Sap* in the traverſe cut of the *Root*, in ſuch places of the *Barque*, where theſe *Threds* terminate: as the exiſtence of the ſame *Veſſels* in the *Skin*, was proved in the Precedent 𝕮𝖍𝖆𝖕𝖙𝖊𝖗.

13. Theſe Tubulary *Threds*, run not through the *Barque* in direct lines; but are frequently Braced together in the form of Net-Work; The *Parenchymous Parts* every where filling up the ſpaces betwixt the Braced *Threds*; as in *Burnet*, *Scorzonera*, &c. the *Barque* being paired or ſtriped off, is apparent.   *Tab.* 6.

14. §.

14. §. They seem, at first, where they are Braced, to be Inosculated; so as to be pervious one into another. But a more accurate view, especially assisted by a *Microscope*, discovers the contrary. Neither are they woun'd any way one about another, as Threds are in a Rope: nor Implicated, as in ravled Yarn, or the Knots of a Net: but only contiguous or simply Tangent, as the several Chords in the Braces of a Drum: being thus joyned together by the *Parenchymous Parts*, as in speaking of the *Pith*, will be understood how. Yet do not always the same *Threds* belong and keep entire to one *Brace*; but are frequently parted into lesser *Threds*; which are transposed from *Brace* to *Brace*. Nor do they always, in whole or in part, presently after their contingence, mutually fall off again; but, oftentimes, run along collaterally joyned together for some space.

Tab. 6. 15. §. These *Braces* are of various number in divers *Roots*; more frequent in *Jerusalem Artichoke*, less in *Scorzonera*, more rare in *Cumfry*. The *Threds* likewise are variously Divaricated; sometimes more, where the *Braces* are frequent, as in *Jerusalem Artichoke*; and sometimes less, where the *Braces* are rare, as in *Scorzonera, Dandelion*: And in all *Roots*, more frequent towards the Inner Verge of the *Bark*.

16. §. By what is said, it is partly implied, That these *Threds*, are not Single *Vessels*; but a *Cluster* of them, Twenty, Thirty, or more or fewer of them together. Yet as the *Threds* are not Inosculated in the *Braces*; so neither are the *Vessels*, in the *Threds*. Nor yet Twisted; but only stand collateral together; as the several Single *Threds* of the *Silkworm*, do in Sleave-Silk. Neither are these *Vessels* pyramidal, so far as the *Glass* will discover; or, from probable Reason, may be conjectured. Nor Ramified, so as to be successively propagated one from another, after the manner of the *Veins* in *Animals*: but Cylindrical, and Distinctly continued, throughout the length of the *Root*; as the several *Fibres* in a *Tendon* or *Nerve*.

17. §. THESE VESSELS are either themselves of divers kinds, or serve, at least, to constitute divers Kinds, in divers *Roots*: of the different Natures whereof, although there may be other ways whereby to judge; yet so far as by Inspection, we may do it, chiefly, by the Diversity of those *Liquors*, which they severally contain. Sometimes they yield a *Lympha*; and that Thin, as they do in a *Parsnep*; especially those that make a *Ring*, at the inward extremity of the *Bark*. See the Root it self. That this Clear *Sap* ascendeth only from these *Vessels*, is certain. Because no *Liquor* will do the like, from any *Parenchymous Part*, as *Chap*. 2. hath been said. And because it is of a different nature from the *Sap* contained in the *Bladders* of the *Parenchyma*; although of the same *Colour*, yet sensibly more *Sweet*.

18. § Sometimes they yield a Thick and Mucilaginous *Lympha*, as in *Cumfry*, as appeareth by its tenacity. From the Mucilaginous Content of these *Vessels* it is, I suppose, that the *Sap* contained in the *Bladders* is rendred of the like nature, so far as it approaches hereto, which sometimes is more, as in *Marsh-mallow*; and sometimes but little as in *Borage*: For in pressing out the *Liquor* of this *Plant*, and then heating it over an indifferent fire; the far greater part hereof remaineth thin; only some certain strings and little bits of a gellied substance are mixed herewith; which as it seems, were originally the proper *Liquor* of these *Muciducts*.

19. §.

19. §. Oftentimes these Succiferous *Vessels* yield a *Milky* or White *Sap*; and sometimes Yellow, and of other colours as in *Sonchus*, and most *Cichoraceous Plants*; in *Angelica*, and most *Umbelliferous*; in *Burdock*, and divers *Thistles*, to which that is 'akin: in *Scorzonera*, Common *Bells*, and many other *Plants*, not commonly taken notice of to be *milky*. The *Milky Saps* of all which, although they differ in Colour, Thickness, and other Qualities; yet agree, in being more *Oyly* than any of the *Lymphous Saps*. It being the mixture of the *Oyly* parts with some other Limpid *Liquor*, but of a different Nature, which causeth them to be of a *Milky*, or other *Opacous* Colour, in the same manner as common *Oyl*, and a strong *Liquamen* of Tartar, shaked in a Bottle together, presently mix into a White *Liquor*. And although they will, for the greatest part, separate again; yet some of their parts, without any Boiling, or so much as the least Digestion with Heat, by Agitation only, or standing together for some time, incorporate in the form of a Thin *Milky-Sope*, which will also dissolve in *Water*. I suppose, therefore, That it is the *Volatile Salt*, chiefly, of these *Plants*, which being mixed with their *Oyl*, renders this *Liquor* of a White or other Opacous Colour.

20. §. Sometimes the *Oyl* will separate and discover it self: for if you cut a *Fenil-Root* traverse, after it hath layn some days out of the Ground; the same *Vessels*, which, in a fresh *Root*, yields *Milk*; will now, yield Oyl: the watery parts of the *Milk*, which in the drying of the *Root* are more evaporable, being spent.

21. §. All *Gums* and *Balsams* are likewise to be reputed the proper *Contents* of these *Vessels*: for These and *Milks*, are very near akin. So the *Milk* of *Fenil*, upon standing, turns to a Clear *Balsam*; of *Scorzonera*, *Dandelion*, and others, to a *Gum*. In the dryed *Root* of *Angelica*, &c. being split, the *Milk*, according to the Continuation of these *Vessels*, appeareth, as Blood clodders in the *Veins*, condensed to an hard and shining *Rosin*. And the *Root* of *Helenium* cut *Tab. 9.* transversely, presently yields a curious *Balsame* of a Citrine Colour, and sometimes of the Colour of *Balsame* of *Sulphur*. I call it a *Balsame*; because it will not dissolve in Water. Yet not a *Terebinth*; because, nothing near so viscid or tenaceous as that is. But the *Root* of Common *Wormwood*, bleeds, from large *Vessels*, a true *Terebinth*, or a *Tab. 10 E. Balsame* with all the defining properties of a *Terebinth*; although that word be commonly used only for the Liquors of some *Trees*.

22. §. There is yet another kind of *Sap-Vessels*, which may be called *Vapour-Vessels*; as in *Docks*, at least some of them. For by the *Sap-Vessels* it is, that the *Barques* of *Roots* do Bleed. Of which, some Bleed quick and plentifully, as the *Umbelliferous* and the *Cichoraceous* Kinds. Some, very slowly and scarce visibly, as all or most *Trefoyls*, and of the *Leguminous* Kind. And some seem not to Bleed, as the *Dock*. Yet that this *Root*, hath also *Vessels* distinct from those that carry *Aer*; doth partly appear, from the different Colour they produce where they stand; as will better be understood anon, in speaking of the Causes of the *Colours* of *Roots*. As also from the Toughness of the *Barque*, in pulling it by the length; neither the *Parenchyma*, nor the *Aer-Vessels*, being of themselves *Tough*. But because the *Succus* or *Sap* they carry, seems to be a kind of Dewy *Vapour*, therefore, they may not improperly be called *Roriferous* or *Vapour-Vessels*.

23. §. THE *Sap-Veſſels*, are not only of divers Kinds, in divers *Roots*, but in the ſame. Whether in all, I doubt: but in ſome it is certain they are: For if you cut a *Fenil-Root* traverſe, both *Milk* and *Limpid Sap*, will preſently aſcend, and, upon accurate inſpection, appear thereupon dinſtinctly. So the *Roots*, both of *Trachelium* and *Enula*, Bleed both a *Lympha*, and a Citrine *Balſame:* and *Wormwood*, both a *Lympha*, and a *Terebinth*, at the ſame time. So alſo the *Root* of *Dandelion* being cut in *November*, ſeems to bleed both a *Milk* and a *Lympha*; the latter being drowned by the former at another time when it is more copious. Whether all *Roots* have *Lymphæducts*, is doubtful; but 'tis moſt probable, that they have, more or fewer; ſtanding, for the moſt part, in a *Ring*, at the Inner Verge of the *Barque:* the *Sap* whereof, I ſuppoſe, is ſo far of common Nature in all *Roots*, as to be Clear, and leſs Oily.

Tab. 9.

24. §. THE Quantity of theſe *Veſſels* is very different: In *Borage*, *Peony*, *Biſtort*, but few; in *Aſparagus*, fewer: in *Parſnep*, *Celandine*, many; in *Fenil*, *Marſh-mallow*, many more: and betwixt theſe extreams, there are many Degrees, as by comparing the *Roots* of *Horſe-Radiſh*, *Turnep*, *Briony*, *Skirrets*, *Parſley*, *Goats-Beard*, and as many more as you pleaſe, may be ſeen. Amongſt the ſeveral Sorts of *Docks*, they ſeem in *Patience*, to be the feweſt; in *Red-Dock*, the moſt numerous. There are two ways of judging of their Number; Either as their Extremeties are viſible upon the traverſe cut of the *Barque*; or as the *Barque* is diverſly Brittle or Tough; being ſo, from the various Number of theſe *Veſſels* therein, as in the *Second Chapter* hath been ſaid.

Tab. 7,8,9.

25. §. The Quantity of the aſcending *Sap*, is a doubtful argument, whether of the Number, or Size of theſe *Veſſels*. For it is common to moſt *Milky-Roots*, for the *Milk* to aſcend more copiouſly: yet in ſome of them, the *Veſſels* ſeem, in proportion with the *Parenchymous Part*, not to be ſo numerous, as in ſome other *Roots*, where the aſcending *Sap* is leſs; as by comparing the *Lacteals* of *Dandelion*, and the *Lymphæducts* of *Fenil* together, may appear: ſo that it ſhould ſeem, that the bore of the *Lacteal Veſſels*, is greater than that of the *Lymphæducts*.

26. §. THE Situation of theſe *Veſſels*, as they appear, even to the naked Eye, in the tranſverſe Section, is Various and Elegant. Sometimes they are poſited only at the Inner Edg of the *Barque*, where they make a Ring, as in *Aſparagus*. In which place and poſition, they ſtand in moſt, if not in all, *Roots*, how variouſlly ſoever they are poſited alſo otherwiſe. The Common *Crow-Foot* with numerous *Roots*, hath a *Ring* of *Sap-Veſſels* next the *Skin*. So the *Barque* of *Monks-Hood*, is encompaſſed with a tranſparent *Ring* of *Sap-Veſſels*. The *Ring* is either more Entire, as in *Eryngo*, *Brown-Wort*, *Valerian*, *Hop*, *Madder*, &c. Or it is a Prick'd *Ring*, as in *Buttyr-Bur*. Sometimes they are chiefly poſtur'd in a Prick-Ring, towards the outward part of the *Barque*, as in *Peony*: and ſome *Roots* are pricked all over the *Barque*, as of *Melilot*. In others, they ſtand not ſo much in Pricks, as Portions or *Colums*, as in *Cumfry-*

Tab. 7,8,9, & 10.

27. §.

27. §. In others, again, they all ftand in more continued Lines, either Rays or Diametral, as in *Borage*; or Peripherial, as in *Celandine*. The *Vafcular* Rays are not equally extended in all *Roots*: in *Parfnep*, towards the Circumference of the *Barque*; in *Buglofs*, about half way. In all *Docks*, and *Sorrels*, the *Rays* are extended through about ¼ of the thicknefs of the *Barque*, towards the Circumference, whereabout, divers of them are always *arched* in, two and two together. In all or many *Trefoyls*, and of the *Leguminous* Kind, they are extended through no more than ⅓d of the *Barque*. In the *Umbelliferous*, they are Ralled in betwixt the Diametral *Portions* of the *Parenchyma*. In *Borage*, the Rays are more Continuous; in a *Carrot*, more Pricked. Here alfo the Pricks ftand in Even Lines; in *Lovage*, they are Divaricated. Of which, and thofe of fome other *Roots*, it is alfo Obfervable, That they are not all meer Pricks, but moft of them fmall, yet real Circles; which, after the *Milk* hath been frequently licked off, and ceafeth to afcend, are vifible, even without a *Glafs*. And note, that in obferving all *Milk-Veffels*, the *Milk* is to be taken off, not with the Finger but the Tongue; fo often, till it rifeth no more, or but little. And fome *Roots* may alfo be foaked in Water; whereby the Pofition of the *Milk-Veffels*, will be vifible by the darker Colour of the *Barque*, where they ftand. *Tab.* 7, 8, 9. *Tab.* 8.

28. §. The Rays fometimes, run more Parallel, and keep feveral, as in *Monkfhood*; and fometimes, towards the Circumference of the *Barque*, they are occurrent; as not only in *Docks*, but other *Plants*: In *Eryngo*, in a termination more Circular; and in *Bryony*, angular, or in the form of a *Glory*, as alfo in *Horfradifh*, through a *Microfcope*. The Peripherial Lines are in fome, more entire Circles, as in *Dandelion*; in others, made up of fhorter Chords, as in *Potato*, *Cumfry*, and the fmaller part of the *Root* of *Monks-hood*. In fome, the Pricks are fo exceeding fmall, and ftand fo clofe, that, to the bare eye, they feem to be continous *Rings*, which yet, through the *Microfcope*, appear diftinct, as in *Marfh-mallow* and *Liquirifh*. *Tab.* 7, 8, 9. *Tab.* 15. *Tab.* 12.

29. §. Sometimes Columns and Chords are compounded, as in *Burnet*; Pricks and Chords, in *Potato*; Rays and Rings, in *Monkfhood*; where the Ring is Single. In *Fenil*, there is a double or treple order both of Rays and Rings, the *Lymphæducts* ftanding in Rays and the *Lacteals* in Rings. And in *Marfh-mallow*, the *Veffels* are fo pofited as to make both thofe kinds of Lines at once.

30. §. In *Celandine*, they feem all, to the bare eye, to ftand in numerous Rings lying even one within another. As alfo in *Dandelion*; in which yet, being viewed through a *Microfcope*, there is an appearance of very many fmall Rays; which ftreaming from the Inner Verge of the *Barque*, crofs three or four of the fmaller Rings, and are there terminated. Whence it fhould feem that *Lymphatick* Rays and *Milky* Rings, are in that *Root*, fo far mixed together. Only the *Lympha*, being confounded with the *Milk*, cannot be difcerned. And where the *Milky-Veffels* are evacuated, or at fuch Seafons, wherein they are lefs full, divers *Milky Roots* will yield a clear *Liquor* at the Inner Verge of the *Barque*, where, at other times, they feem to yield only *Milk*. And this is the Defcription of the *Barque*. *Tab.* 13.

CHAP.

CHAP. IV.

Of the WOOD.

HAT Portion of the *Root* which ftandeth next within the *Barque*, and in *Trees*, and *Shrubby Plants*, is the *Wood*; is alfo compounded of Two Subftantially different Bodies, *Parenchymous* and *Lignous*. The *Parenchymous*, is of the fame Subftantial Nature with that of the *Barque*. And is originated from it; being not only adjacent to it, but all round about continuous therewith; even as that, is with the *Skin*; the *Parenchyma* of the *Barque*, being diftributed, from time to time, partly outward into the *Skin*, and partly inward, into the *Wood*.

2. §. The Pofition of the feveral parts hereof, is different. For the moft part it hath a Diametral Continuation, in feveral Portions, running betwixt as many more of the *Lignous*, from the Circumference towards the Center of the *Root*: all together, conftituting that, which in the *Second Chapter* of the *Firſt Book*, I call the *Inſertment*. In the *Roots* of many *Herbs*, thefe Diametral or Inferted Portions are more obfervable, as in *Cumfry*; which leadeth to the notice of them in all others, both of *Herbs* and *Trees*. Sometimes part of this *Parenchymous* Body is difpofed into *Rings*, as in *Fenil*. The Number and Size of which Rings differ: In *Fenil*, when the *Root* is grown large, they are in fome places broader, but fewer; in *Beet* they are narrower, but more. The Diametral Portions are here, in like manner, much varied; in *Cumfry*, *Celandine*, larger; in *Beet*, *Bugloſs*, meaner; in *Borage*, *Parfnep*, more, and fmaller; and in moft *Woody-Roots*, ftreaming betwixt the *Pith* and the *Barque*, as fo many fmall Rays. Their Continuation is alfo different; in fome *Roots*, to the Centre, as in *Columbine*; in others not, as in *Parfnep*. And fometimes different in the fame *Root*, as in the *Vine*.

Tab. 9.

Tab. 8, 9.

Tab. 17.

3. §. The Contexture of thefe *Parenchymous* Portions is fometimes Uniform, as in *Bugloſs*, *Peony*; and fometimes alfo, as it is in the *Barque*, different; in part, more fappy, and tranfparent; in part, more white, dry, and aery, as in *Carrot*, *Lovage*, *Scorzonera*, and others; which yet cannot be obferved without a wary view. But their general Texture is the fame being all made up of many fmall *Bladders*. Which are here of different Sizes, like thofe of the *Barque*, but for the moft part fmaller. Their Shape likewife, is ufually Round; but fometimes Oblong and Oval, as in *Borage*; or Oblong and Square, as in the *Vine*.

Tab. 17.

4. §. The *Lignous Part*, if not always, yet ufually, is alfo Compounded of Two Kinds of *Bodies*, *fcil. Succiferous* or *Lignous* and *Aer-Veſſels*. The *Lignous* as far as difcernable, are of the fame Conformation and Nature with thofe of the *Barque*, and in the tranfverfe cut

of the *Root*, do oftentimes, as those, emit a *Liquour*. They are also Braced; and many of them run in distinct *Threds* or Portions, collaterally together.

5. §. The *Aer Vessels* I so call, because they contain no *Liquor*, but an *Aery Vapour*. They are, more or less, visible in all *Roots*. They may be distinguished, to the bare Eye, from the *Parenchymous Parts*, by their Whiter Surface; and their standing more prominent, wheras those shrink below the transverse level of the *Root*, upon drying. They are frequently Conjugated divers of them together, sometimes fewer, and for the most part single, as in *Asparagus*; sometimes many, as in *Horf-Radish*. And their Conjugations are also Braced, as the Threds of the *Succiferous Vessels*. But they are no where Inosculated: nor Twisted one about another; but only Tangent or Collateral. Neither are they Ramified, the greater into less; but are all distinctly continued, as the *Nerves* in *Animals*, from one end of the *Root* to the other. *Tab.* 10, & 15.

6. §. Their Braces, as those of the *Succiferous Vessels*, are also of various number: in *Jerusalem Artichoke*, *Cumfry*, *Scorzonera*, more rare; in *Borage*, *Burnet*, more frequent; as by stripping off the *Barque* of such *Roots*, where it is easily separable, may be seen. And they often vary in the same *Root*; so in *Borage*, *Scorzonera*, &c. they are more frequent in the Centre, and next the *Barque*, than in the Intermediate space, as by splitting those *Roots* down the middle doth appear. They also vary from those of the *Succiferous Vessels*; those being usually more frequent, as in *Jerusalem Artichoke*, than these of the *Aerial*. *Tab.* 6. *Tab.* 6.

7. §. Betwixt these Braced *Aer-Vessels*, and the rest, which make the true *Wood*, run the *Parenchymous Parts* above described; as they do betwixt the *Succiferous* in the *Barque*: and so make up two Pieces of *Net Work*, wherof one is the filling up of the other. *Tab.* 6.

8. §. The Position of both these Kinds of *Vessels*, is Various. The *Succiferous* or *Lignous*, are sometimes posited in diametral lines or portions; as in the *Vine*, and most *Trees*. Sometimes, oppositely to the *Aerial*, as in *Beet*; each Ring herein being double, and made both of *Sap*-and *Aer-Vessels*. *Tab.* 17. *Tab.* 8.

9. §. In *Nettle* the Position is very peculiar, from what it is in the *Roots* of other *Herbs*; being curiously mixed; the *Succiferous* running cross the *Aerial*, in several, *viz*. Five, Six, Seven, or more *Rings*. In *Bryony* the several Conjugations of the *Aerial*, are distinctly surrounded with the *Succiferous*. In *Patience*, the *Succiferous* are disposed, besides Rays, into many small Rings, of different Sizes, sprinkled up and down, and not, as in other *Roots* having one common Centre; within divers whereof, the *Aer-Vessels* are included: especially within those which are drawn, not into *Rings*, but, as it were, into little stragling *Hedges*. *Tab.* 8. *Tab.* 7.

10. §. That also of the *Aer-Vessels*, is Various and Elegant: especially in the upper part of the *Root*. In *Ammi*, *Lilium-non-bulbosom*, they make a Ring. In these, a Prick'd-Ring; in *Peony*, a Ring of Rays; in *Valerian*, a Ring of Pricks and Rays. In others, they make not Rings, but longer Rays, extended either towards the Centre, as in *Scorzenera*; or meeting in it, as in *Columbine*. In the Common *Dock*, they stand more in single Rays: in the other *Species* of *Docks*, both in Rays, and collateral Conjugations between. *Tab.* 7, 8, 9.

II. §.

Tab. 8, 9.

11. §. In *Beet*, they stand in several Rings; and every Ring, made of Rays. In *Cumfry*, the Rays and Rings are separate; those stand without, these next the Centre. In *Dandelion*, they stand altogether, and make a little Rope, in the Center it self. In *Geranium*, and others of that Kindred, they make a little Thred, in the same place. And in *Skirret*, they stand in two Threds, near the Centre.

Tab. 8, 9.

12. §. In *Celandine*, they stand in almost parallel Lines. In *Monks-hood*, of a wedged Figure; divided in the smaller part of the *Root*, into Three little Wedges, with their poynts meeting exactly in the Centre. In *Cinquefoyle*, and *Strawberry*, they are also postur'd in three Conjugations, triangularly. In the young *Roots* of *Oak*, they stand neither in Radiated, nor otherwise strait, but Winding Lines. And in *Borage* the position, of many of them, is Spiral. As likewise, sometimes, in *Mercury*, or *Lapathum unctuosum*. In *Horse-Radish*, they stand more confused neither in Rings nor in Rays; yet their several Conjugations, are radiated: with very many other differences.

Tab. 15.

13. §. The Quantity of these *Vessels*, as to the space they take up in the *Root*, is to be computed Two ways, By their Number, and Size. Their Number may, in some *Roots*, and in some measure, be judged of, by the bare Eye; having, frequently, a whiter surface than the other *Parts*. As also their Size; the Bore of these *Vessels* being greater than that of the *Lignous* in all *Roots*; especially in some. For if you take the *Roots* of *Vine, Fenil, Dandelion, Plum-tree, Elder, Willow,* &c. and lay them by, for some time, to dry; and then, having cut off a very thin Slice of each, transversely; if you hold up those Slices before your Eye, so as the Light may be trajected through the said *Vessels*, they hereby become visible, as notably different, both in Number and Size.

14. §. But undeceitful and accurate Observation of both their Number, and Size, must be made by the *Microscope*; and so they will appear to be much more various. In *Bistort, Skirret*, they are very few; in *Beet*, very many: betwixt which extreams there are all Degrees; as in *Orpine, Venus Looking-Glass, Scorzonera,* Great *Celandine, Peony, Borage, Fenil,* &c. may be seen. So their Size, in some is extream small, as in *Strawberry, Bistort, Valerian*; in others very great, as in *Asparagus, Bugloss, Vine.* They are also of several Sizes in one and the same Numerical *Root*; but in some, are less varied, as in *Lilium non bulbosum, Asparagus, Bugloss*; in others, more, as in *Bryony, Lovage.* Amongst all *Roots*, they vary by about Twenty Degrees; as by comparing the *Roots* of *Vine, Thorn-Apple, Bryony, Lovage, Fenil,* Wild *Carrot, Saxifrage, Parsley, Peony, Hore-hound, Cinquefoyl, Strawberry,* &c. together, may be seen. Some of those in the *Vine*, being of the greatest Size; appearing through a good *Glass*, at least one Third of an Inch in Diametre: those in *Strawberry*, and that Kind, of the smallest; most of them appearing, in the same *Glass*, no bigger, than to admit the poynt of a small Pin, according to the Standard, in *Tab.* 12. See also the Figures of so many of them as are drawn.

Tab. 10. to the 17.

Book II.   of Roots.   73

15. §. In some *Roots*, they are Small, and Few; as in *Jerusalem Artichoke*; in others Small, but Many, as in *Horse-Radish*: in *Bugloss*, they are Great, but Few; in the *Vine*, Great and Many. So that the proportion, which those of a *Vine*, their Number and Size being taken together, bear to those of *Jerusalem Artichoke*, may be, at least, as Fifty, to One. Of the smallest Kinds, as those of *Cinquefoyl*, *Jerusalem Artichoke*, and the like; It is to be noted, That they are scarce ever visible in the fresh Slices of these *Roots*; but after they have layn by a while, at last, by a good *Glass*, Clear Light, and steddy View, are discernable. *Tab.* 11, 14, 15, 17.

16. §. In some *Roots*, the greater of these *Vessels* stand in or next the Centre, as in *Taraxacum*, or *Dandelion*; in others next the circumference, as in *Horse-Radish*. Sometimes each of them is from one end of the *Root* to the other, of a more equal Size, or more Cylindrical, as in *Marsh-mallow*; but usually, they widen, more or less, from the Top, to the Bottom of the *Root*, as in *Thorn-Apple*: about the Top of which, they are, for the most part, but of the Sixth, Seventh, and Eighth, Magnitude; some of the Fifth, but none of the Third; but about the Bottome, they are most of the Third, and Fifth: whence it is manifest, That some of them are, in the manner of *Veins*, somewhat Pyramidal. Yet is it observable, That their ampliation proceedeth not towards, but from their Original, as in *Nerves*. *Tab.* 13, 15.

17. §. Of these *Vessels* Seignior *Malpighi* hath observed; *Componuntur* (saith he) *expositæ fistulæ Zona tenui & pellucida, velut argentei coloris lamina, parum lata; quæ, spiraliter locata, & extremis lateribus unita, Tubum, interius & exterius aliquantulum asperum, efficit.*

18. §. To whose Observation I further add, That the *Spiral Zone*, or *Lamina*, as he calls it, is not ever one Single *Piece*; but consisteth of Two or More round and true *Fibres*, although standing collaterally together, yet perfectly distinct. Neither are these Single *Fibres* themselves *flat*, like a *Zone*; but of a *round* forme, like a most fine *Thred*. According as fewer or more of these *Fibres* happen to break off, from their Spiral location, together; the *Zone* is narrower, or broader: usually, Narrower in the *Trunk*, and Broader in the *Root*.

19. §. Of these *Fibres* I also Observe, That they are not *Inosculated* side to side, but are *Knit* together by other smaller *Fibres*; those being, as it were, the *Warp*, and these the *Woof* of the *Aer-Vessels*. Yet I think the several *Fibres* are not interwoven just as in a *Web*; but by a kind of Stitch, as the several *Plates* or *Bredths* of a Floor-Mat. A clear and elegant sight of these *Fibres*, and of their *Interweftage*, by splitting a *Vine-Root*, or a piece of *Oak*, may, with a good *Glass* in the sides of their Greater *Aer-Vessels*, be obtained; having much of the resemblance of *Close Needle-work*.

20. §. The Spiration of the *Fibres* of these *Vessels*, may more easily be observed in the *Trunk*, than in the *Root*. And better in younger *Plants*, than other. And not so well by Cutting as by Splitting, or by Tearing off some small Piece, through which they *run*: their Conformation being, by this means, not spoiled. Yet this way, the *Vessels* are seen, chiefly, *Unresolved*.

21. §. But in the Leaves and Tender *Stalks* of all such *Plants*, as shew, upon breaking, a kind of *Doune* or *Wool*; they may be seen *Resolved* and Drawn out, and that sometimes even to the naked Eye,

an Inch or two Inches in length. This *Wool* being nothing elſe, but a certain number of *Fibres* Reſolved from their Spiral poſition in theſe *Veſſels*, and Drawn out in Length; and ſo cluſtred together, as ſo many *Threds* or little *Ropes*: appearing thus more or leſs, in the *Leaves* and ſome other *Parts* of moſt *Plants*; but more remarquably in ſome, as in the *Vine*, *Scabious*, and others. As alſo in the Scales of a *Squill*. In which laſt, for example, they are ſo eaſily ſeparable, as further to ſhew, what before was obſerved; *viz.* That the *Plate* or *Zone*, into which the *Aer-Veſſels* are uſually Reſolved, is not one Single Piece, or meer *Plate*; but made up of ſeveral Round *Fibres*, all ſtanding and running parallel, and ſo knit together by other ſmaller ones, tranſverſly, in the form of a *Zone*. For if you break or cut a *Leaf* or *Shell* of a freſh *Squill*, till you come to the *Aer-Veſſels*, and having ſoftly drawn them out, for about an Inch or more (to the naked *Eye*) in length, you then ſingle out one or two of them from the reſt, and rowl them, as they hang at the *Shell*, eight or nine times round, each *Veſſel* will appear, through a *Glaſs*, to conſiſt of 8, 10, or 12 ſmall *Fibres*; which, in the Unreſolved *Veſſel*, run parallel; but by this means, are all ſeparated one from another. See the *Figures* belonging to the *Third* and *Fourth* Books.

22. §. The Proceſs of their Spiration, is not, ſo far as I have obſerved, accidental, but conſtantly the ſame; *ſcil.* In the *Root*, by *South*, from *Weſt* to *Eaſt*: But in the *Trunk*, contrarily, by *South*, from *Eaſt* to *Weſt*.

23. §. The *Content* of theſe *Veſſels*, is, as hath already been intimated, more *Aery*. The Arguments for which, are, That upon a tranſverſe Cut of the *Root*, the *Sap* aſcendeth not there, where Theſe ſtand. Being alſo viewed through a *Microſcope*, they are never obſerved to be filled with *Liquor*. Beſides a *Root* cut and immerſed in Water, till the Water is in ſome part got into theſe *Veſſels*, and then the *Root* taken out and cruſhed; the other *Parts* will yield *Liquor*, but Theſe, only *Bubbles*: which *Bubbles* are made, by ſome ſmall quantity of *Liquor* mixed with the *Aer*, before contained in the ſaid *Veſſels*. To which, other Arguments will ariſe out of thoſe Things that follow in the *Second* Part. As alſo for this *Content*, its not being a pure or ſimple, but *Vaporous Aer*. Whether theſe *Veſſels* may not, in ſome *Vegetables*, and at ſome times, contain *Liquor*, is doubtful. (*a*) Thus far of the *Lignous Part*.

(*a*) See Book 3.

CHAP. V.

Of the PITH.

ITHIN the *Lignous Part* lyeth the *Pith*. This *Part* is not common to all *Roots*, for some have none, as *Nicotian*, *Srtamonium*, and others. Yet many which have none, or but little, throughout all their lower parts, have one fair enough about their tops, as *Mallow*, *Bourage*, *Dandelion*, and the like. See the *Roots*. And in many others there are *Parenchymous Parts*, of the same substantial nature with the *Pith*, distributed betwixt the several Rings of *Vessels*, and every where visible, from the top to the bottom, as in *Beet*, *Fenil*, &c. *Tab.* 6. *Tab.* 8.

2. §. The Size of the *Pith* is varied by many Degrees, easily reckoned an Hundred; in *Fenil*, *Dandelion*, *Asparagus*, but small; in *Horse-Radish*, *Valerian*, *Bistort*, great. The Shape hereof, in the lower parts of most *Roots*, is Pyramidal; but at the tops, Various, according to the different Distribution of the *Vessels*, as in *Carrot*, Hyperbolick, in *Parsley*, Oval; as appeareth, in cutting the *Roots* lenghtways. *Tab.* 6.

3. §. The *Pith*, for the most part, especially in *Trees*, is a *Simple* Body: but sometimes, it is, as the *Barque*, compounded; some certain number of *Succiferous Vessels* being mixed herewith; as in *Jerusalem Artichoke*, *Horse-Radish*, &c. upon a traverse cut, by a strict view, may be discerned. Their Position is sometimes Confused, as in a *Carrot*; and sometimes Regular, as in *Parsley*; appearing, by the traverse cut, in Rings, and in cutting by the length, in Arches. And sometimes the *Pith* is hollow; as in the *Level-Roots* of *Bishops-Weed*: these *Roots* being made out of the *Stalk*, as in the *First* 𝕮𝖍𝖆𝖕𝖙𝖊𝖗 hath been shewed, how. *Tab.* 6, & 8. *Tab.* 6. §. 13,14,15.

4. §. As all the other Parts of the *Root*, are originated from the *Seed*; so, sometimes, is the *Pith* it self. But sometimes, it hath its more immediate Derivation from the *Barque*. Hence it is, that many *Roots*, which have no *Pith* in their lower parts, have one at their top, as *Columbine*, *Lovage*, &c. For the *Parenchymous Parts* of the *Barque* being, by degrees, distributed into Diametral *Portions*, running betwixt those of the *Lignous Body*, and at length, meeting and uniting in the Centre, they thus constitute the *Pith*. In the same manner, at the top of some *Roots*, the *Pith* is either made or augmented, out of the *Parenchymous Rings* above described; these being gradually distributed to, and embodied in the Centre; as in *Fenil*, and some other *Roots*, their lower and upper parts compared together, may be seen. Even as in *Animals*, one *Part*, as the *Dura Mater*, is the original of divers others. *Tab.* 4.

5. §. From hence, it also appears, That the *Pith* is of the same Substantial Nature with the *Parenchyma* of the *Barque*, and with the Diametral *Portions*; and that therefore they are all one body, differing in no Essential Property, but only in their Shape and Place. The same is also evident from the Continuity of the *Pith* with the Diametral *Portions*, as of These, with the said *Parenchyma*. And from their Contexture, which, by a *Microscope*, appeareth to be of one and the same general kind, in all *Plants*, both in the *Parenchyma* of the *Barque*, in the *Insertment* or Diametral *Portions*, and in the *Pith*, all being made up of *Bladders*.

Tab.9, & 15.   6. §. The *Bladders* of the *Pith*, are of very different Sizes; seldom less, than in the *Barque*, as in *Asparagus*; usually much bigger, as in *Horse-Radish*. They may be well reckoned to about fifteen or twenty degrees; those in *Jerusalem Artichoke*, of the largest; in *Valerian*, *Horse-Radish*, of the meaner; in *Bistort*, *Peony*, of the smallest. Their Position is rarely varied, as it is oftentimes, in the *Barque*; but more uniform, and in the transverse Cut, equally respective to all parts of the *Root*: yet being piled evenly, one over another, in the long cut, they seem to run, in Direct Trains, by the length of the *Root*. Their Shape also is, usually more orbicular; but sometimes,

Tab. 11.   somewhat angular, in the larger kinds, as in *Jerusalem Artichoke*.

7. §. THUS FAR the Contexture of the *Pith* is well discoverable in the *Root*. In the *Trunk*, farther, and more easily. Whereof therefore, in the next *Book*, I shall give a more particular *Description* and *Draught*. Yet since I am speaking of it, I shall not wholly omit here to observe, That the *Sides*, by which the aforesaid *Bladders* of the *Pith* are circumscribed, are not meer *Paper-skins*, or rude *Membranes*; but so many several Ranks or Piles of exceeding small *Fibrous Threds*; lying, for the most part, evenly one over another, from the bottom to the top of every *Bladder*; and running cross, as the *Threds* in the Weavers *Warp*, from one *Bladder* to another. Which is to say, That the *Pith* is nothing else but a *Rete mirabile*, or an Infinite Number of *Fibres* exquisitely small, and admirably Complicated together: as by cutting the *Pith* with a *Razor*, and so viewing it with a good *Glass*, may be seen. See the *Figures* belonging to the *Third Book*.

8. §. All *Plants* exhibit this Spectable, not alike distinctly; those best, with the largest *Bladders*. Nor the same *Pith*, in any condition; but best, when dry: Because then, the *Sap* being voided, the spaces betwixt the *Fibrous Threds*, and so the *Threds* themselves, are more distinctly discernable. Yet is it not to be dryed, after Cutting; Because its several parts, will thereupon coincide and become deformed. But to be chosen, while the *Plant* is yet growing; at which time, it may be often found dry, yet undeformed; as in the *Trunks* of *Common Thistle*, *Jerusalem Artichoke*, &c.

9. §. Neither are these *Threds*, so far as I can observe, Single *Fibres*; but usually, consist of several together. Nor are they *simply* Collateral, but by the weftage of other *Fibres*, in their natural Estate, knit together; much after the same manner as the *Spiral Fibres* of the

Aer-

Aer-Vessels. This Connexion I have no where so well seen, as in the White *Bottoms* of the *Bladders* of a *Bulrush,* being cut traverse; wherein they have the appearance, of very Fine and close *Needle-work.*

10. §. The *Fibres* by which the said *Threds* are knit together, I think are all Single: and are seldom and scarcely visible, except by obliquely Tearing the *Pith*; by which means, they will appear through the Glass, broken off, sometimes, a quarter or half an Inch, or an Inch in Length; and as small as one Single *Thred* of a *Spiders Webb.* In a *Bulrush,* they are sometimes discernable in cutting by the Length. These *Fibres,* and the *Threds,* they knit together, for the most part, are so pellucid, and closely situate, that they frequently seem to make One entire Body, as a piece of *Ice* or a film of *Water* it self: or even as *Animal Skins* sometimes shew, which yet are known to be *Fibrous.*

11. §. The Situation of these *Threds,* is contrary to that of the *Vessels,* as those by the Length, so these, chiefly, by the Bredth of the *Root,* or horizontally, from one edge of the *Pith* to the other. They are continued circularly; whereby, as oft as they keep within the compass of the several *Bladders,* the said *Bladders* are Round: But where they winde out of one *Bladder,* into another, they mutually Intersect a *Chord* of their several *Circles*; by which means, the *Bladders* become Angular.

12. §. The Contexture, likewise, both of the *Parenchymous Part* of the *Barque,* and of the Diametral *Portions* inserted betwixt the *Lignous*; is the same with this of the *Pith,* now described; that is, *Fibrous.* Whence we understand, How the several *Braces* and *Threds* of the *Vessels* are made: For the *Vessels* running by the length of the *Root,* as the *Warp*; by the *Parenchymous Fibres* running cross or horizontally, as the *Woof*: they are thus *knit* and as it were *stitched* up together. Yet their *weftage* seemeth not to be *simple,* as in Cloath; but that many of the *Parenchymous Fibres* are *wraped* round about each *Vessel*; and, in the same manner, are continued from one *Vessel* to another; thereby knitting them altogether, more closely, into one *Tubulary Thred*; and those *Threds,* again, into one *Brace*: much after the manner of the *Needle work* called *Back-Stitch* or that used in Quilting of Balls. Some obscure sight hereof, may be taken in a *Thred* of *Cambrick,* through a *Microscope.* But it is most visible, in the *Leaves* and *Flowers* of some *Plants.* The Delineation of these Things I shall therefore omit, till we come hereafter to speak of the other *Parts.*

13. §. From what hath been said, it may be conjectured; That the *Aer Vessels* successively appearing in the *Barque,* are formed, not out of any *Fluid Matter,* as are the original ones: But of the *Parenchymous Fibres*; sc. by changing them from a *Spherical* to a *Tubulary Forme.*

14. §. From the precedents, it is also manifest, That all the *Parenchymous Parts* of a *Root,* are *Fibrous.*

15. §. And lastly, That the whole Body of a *Root,* consisteth of *Vessels* and *Fibres.* And, That these *Fibres* themselves, are *Tubulous,*

or so many more *Vessels*, is most probable: There only wanteth a greater perfection of *Microscopes* to determine.

16. §. The *Contents* of the *Pith* are, sometimes *Liquor*, and sometimes a *Vaporous-Aer*. The *Liquor* is always Diaphanous, as that of the *Parenchymous* Part of the *Barque*; and in nature, not much differing from it. The *Aer* is sometimes less, and sometimes more *Vaporous*, than that of the *Barque*. By this *Aer* I mean, that which is contained in the *Bladders*. Within the Concaves of the *Fibres* which compose the *Bladders*, I suppose, there is another different Sort of *Aer*. So that as in the *Bladders* is contained a more *Aqueous*; and in the *Vessels*, a more *Essential Liquor*: So sometimes, in the same *Bladders*, is contained a more *Vaporous*; and in the *Fibres*, a more Simple and *Essential Aer*.

An Account of the

VEGETATION

O F

ROOTS

Grounded chiefly upon the foregoing

ANATOMY.

PART II.

TO *Philosophize*, is, To render the *Causes* and *Ends* of Things. No man, therefore, that denieth *God* can do this, Truly. For the taking away of the *first Cause*, maketh all things *Contingent*. Now, of that which is *Contingent*, although there may be an *Event*; yet there can be no *Reason* or *End*: so that Men should then study, That, which *is not*. So the *Causes* of Things, if they are Contingent, they cannot be *Constant*. For that which is the *Cause* of This, now; if it be so *Contingently*, it may not be the *Cause* hereafter: and no *Physical* Proposition, grounded upon the *Constancy* and Certainty of Things, could have any foundation. He, therefore, that philosophiseth, and denieth *God*, playeth a childish Game. *Theology the Beginning and End of Philosophy.*

2. §. Wherefore *Nature*, and the *Causes* and *Reasons* of Things, duly contemplated, naturally lead us unto *God*; and is one way of securing our Veneration of Him: giving us, not only a general Demonstration of his *Being*; but a particular one, of most of the several *Qualifications* thereof. For all *Goodness*, *Righteousness*, *Proportion*, *Order*, *Truth*, or whatever else is Excellent and Amiable in the *Creatures*; it is the Demonstration of the like in *God*. For it is impossible, that *God*

should

should ever make any thing, not like Himself, in some degree or other. These Things, and the very Notions which we have of them, are *Conceptions* issuing from the *Womb* of the *Divine Nature*.

3. §. By the same means, we have a greater assurance of the Excellency of his *Sacred Word*. That He, who hath *Done* all things so transcendently well; must needs *Speak* as well, as he hath *Done*. That He, who in so admirable a manner, hath *made* Man; cannot but know best, What his true *Principles* and *Faculties* are; and what *Actions* are most agreeable thereunto: and, that having adorned him with such *Beauteous* and *Lovely* ones; it is impossible, He should ever put him upon the Exercise of those *Faculties*, in any way *Deformed* and *Unlovely*. That He should do all things, so well *Himself*; and yet require his *Creatures*, to do otherwise, is unconceivable.

4. §. And as we may come, hereby, to rectifie our Apprehension of His *Laws*; so also, of His *Misteries*. For there are many Things, of the *Manner* of whose Existence, we have no certain Knowledge. Yet, of their *Existence*, we are as sure, as our *Senses* can make us. But, we may as well deny, what *God* hath Made, *To be*; as, what he hath Spoken, *To be true*, because we understand not *how*. And the knowledge of *Things* being gradually attained, we have occasion to reflect, That some *Things*, we can now well conceive, which we once thought unintelligible. I know, therefore, what I *understand* not; but, I know not, what is *unintelligible*: what I know not now, I may hereafter; or if not I, another; or if *no Man*, or other *Creature*, it is sufficient, That *God fully understandeth Himself*. It is not, therefore, the *Knowledge* of *Nature*, but they are the *wanton phansies* of Mens minds, that dispose them, either to Forget *God*, or to Think unduly of Him.

5. §. Nor have we reason to fear going *too far*, in the Study of *Nature*; more, than the *entring* into it: Because, the higher we rise in the true Knowledg and due Contemplation of *This*; the nearer we come to the *Divine Author* hereof. Or to think, that there is any Contradiction, when *Philosophy* teaches that to be done by *Nature*; which *Religion*, and the *Sacred Scriptures*, teach us to be done by *God*: no more, than to say, That the *Ballance* of a *Watch* is moved by the next *Wheel*; is to deny that *Wheel*, and the rest, to be moved by the *Spring*; and that both the *Spring*, and all the other *Parts*, are caused to move together by the *Maker* of them. So *God* may be truly the *Cause* of *This Effect*, although a Thousand other *Causes* should be supposed to intervene: For all *Nature* is as one Great *Engine*, made by, and held in His Hand. And as it is the Watch-makers *Art*, that the *Hand* moves regularly, from hour to hour, although he put not his Finger still to it: So is it the Demonstration of *Divine Wisdome*, that the Parts of *Nature* are so harmoniously contrived and set together; as to conspire to all kind of Natural Motions and Effects, without the Extraordinary and Immediate Influence of the *Author* of it.

6. §. Therefore, as the *Original Being* of all Things, is the most proper Demonstration of *Gods Power*: So the *successive Generations*, and *Operations* of Things are the most proper Demonstration of his *Wisdom*. For if we should suppose, that *God* did now make, or do any Thing, by any Thing; then, no *Effect* would be produced by a *Natural Cause*: and consequently, He would still be upon the Work of *Creation*: which yet *Sacred Scripture* assureth us, He *resteth from*. And we might expect
the

the Formation of a *Child*, in an *Egg*, as well as in a *Womb*; or of a *Chicken*, out of a *Stone*, as an *Egg*: And all Sorts of *Animals*, as well as *Plants*, might propagate their *Species*, without Coition: and the like. For *Infinite Power*, needeth not make any difference in the Things it undertakes to manage. But in that, these Things are not only *made*, but *so made*, that is, according to such certain Natural Laws, as to produce their *Natural Effects*; here is the Sensible and Illustrious Evidence of his *Wisdom*. Wherefore as the Wisdom of Government, is not seen, by the King his interposing Himself in every Case; but in the contrivance of the *Laws*, and Constitution of *Ministers* in such sort, that it shall be as effectually determin'd, as if he did so indeed: So the more complicated and vastly Numerous, we allow the *Natural Causes* of Things to be; the more duely we conceive of that *Wisdom*, which thus disposeth of them all, to their several *Effects*: All Things being thus, as *Ministers* in the Hands of *God*, conspiring together a Thousand *Ways*, towards a Thousand *Effects* and *Ends*, at one time; and that with the same certainty, as if he did prepose to each, the same Omnipotent *Fiat*, which he used at the Creation of the World.

7. §. THIS *Universal Monarchy*, as it is eminently Visible in all other Particular *Oeconomies*; so is it, no less, in that of *Vegetables*. Infinite Occurrences, and secret Intrigues, 'tis made up of; of which we cannot skill, but by the help of manifold *Means*; and those, in the foregoing *Idea*, have been lately proposed. Wherein, although some *Experiments* have been briefly touch'd: yet that which I have hitherto chiefly prosecuted, hath been the *Anatomical* Part; and that not throughly neither. Notwithstanding, so far as Observations already made will conduct us, I shall endeavour to go. And if, for the better clearing of the way, I have intermixed some Conjectures; I think they are not meerly such, but for which I have layd down some Grounds, and of which, the *Series* also of the following *Discourse*, may be some further proof. *The Divine Wisdom seen in the Growth of Plants; if we observe,*

8. §. LET US say then, that the *Root* of a *Plant* being lodged in some Soil, for its more convenient growth; 'tis necessary the Soil should be duly prepared for it. The *Rain*, therefore, falling and soaking into the Soyl, somewhat diluteth the Dissoluble *Principles* therein contained; and renders them more easily communicable to the Root: Being as a *Menstruum*, which extracteth those *Principles*, from the other greater and useless part of the Soil. *First, How the Ground is prepared.*

9. §. And the warm *Sun*, joyned with the diluting *Rain*, by both, as it were a *Digestion* of the Soil, or a gentile *Fermentation* amongst its several Parts, will follow: whereby the Dissoluble Parts therein, will rot and mellow: that is, those *Principles* which as yet remained more fixed, will now be further *resolved and unlocked*, and more copiously and equally spread themselves through the Body of the Soil.

10. §. These *Principles*, being with the growth of *Plants* continually exhausted, and needing a repair; the successions, therefore, of Wet, Wind, and other Weather, beat down and rot the *Leaves* and other *Parts* of *Plants*. Whereby these (as *Weeds* which are wont to be buried under ground) become a *natural Manure*, and Re-impreg-

nate the Soil: Being thus, in part, out of their own Resolved *Principles*, annually Compounded again.

11. §. Many of these *Principles*, upon their *Resolution*, being by the *Sun* more attenuated and volatilized; continually ascend into the *Aer*, and are mixed therewith. Where, although they lose not their *Vegatable Nature*, yet being amongst other purer *Principles*; themselves also, depositing their Earthy feculencies, become more subtile, simple and Essential Bodies.

12. §. And the *Aer* being of an *Elastick* or *Springy* Nature, pressing, more or less, upon all Bodies; it thereby forceth and insinuateth it self into the Soil, through all its permeable Pores. Upon its own entrance, it carries also many of the said *Vegetable* and *Essential Principles* along with it; which, together with the rest, are spread all over the Body of the Soil. By which means, though a less Vehement, yet more Subtil *Fermentation*, and with the least advantage of warmth, continuable, will be effected.

13. §. The *Principles* being thus farther *resolved* and subtilized, would presently exhale away, if the *Rain*, again, did not prevent. Which, therefore, falling upon and soaking through the Ground, is as a fresh *Menstruum*, saturate or impregnate with many of them. And as it still sinketh lower, it carries them along with it self, from the Superficial, to the Deeper parts of the Ground: thus, not only maturing those parts also, which, otherwise, would be more lean and cold; but therein likewise, laying up and securing a *Store*, more gradually and thriftily to be bestowed upon the Upper parts again, as they need.

14. §. And *Autumn* having laid up the *Store*, *Winter* following thereupon, doth, as it were, lock the doors upon it. In which time, some warmer Intervals, serve further and gradually to mature the stored *Principles*, without hazard of their being Exhaled. And the *Spring* returning, sets the doors open again, with warmer and more constant *Sun*, with gentle and frequent *Rain*, fully resolves the said *Principles*; and so furnisheth a plentiful Diet, for all kinds of *Vegetables*: being a *Composition* of *Water* chiefly, wherein are resolved, some portions of *Earth*, *Salt*, *Acid*, *Oyl*, *Spirit*, and *Aer*; or other Bodies of Affinity herewith.

Then, How the Sap is imbib'd, and distributed to the several Parts.
(a) P.1.c.3. §. 3.
(b) §. 11, 12

15. §. THE ROOT standing in the *Ground* thus prepared, and being always surrounded with a *Barque*, which consisteth chiefly of a *Parenchymous* and *spongy* Body; (a) it will thus, as *Sponges* do, naturally suck up the watry parts of the *Soil* impregnate with the said *Principles*. Which *Principles* notwithstanding, being in proportion with the watry parts, but few, and also more *Essential*; (b) therefore in this *Parenchymous Part*, are they never much discovered, either by *Colour*, *Taste*, or *Smell*. As it is probable, that some distilled *Waters*, which discover nothing, to Sense, of the *Plants* from which they are distilled, may yet, in part, retain their *Faculties*. And it is known, that many Bodies; as *Crocus Metallorum*, convey many of their parts into the *Menstruum*, without any sensible alteration thereof. So *Frost* and *Snow* have neither *Taste* nor *Smell*; yet from their *Figures*, 'tis evident, that there are divers kinds of *Saline Principles* incorporated with them; or at least, such *Principles* as are common to *them* and divers kinds of *Salts*.

16. §.

16. §. The entrance of this *Impregnate Water* or *Sap* is not without difference, but by the Regulation of the intervening *Skin*; being thereby *strained* and rendred more pure : the *Skin*, according to the thickness (*a*) or closeness thereof, becoming sometimes only as a *brown paper*, sometimes as a *Cotton*, and sometimes as a *Bag of Leather* to the transient *Sap*, as the nature of it doth require. By which it is also *moderated*, lest the *Barque*, being spongy, should suck it up too fast, and so the *Root* should be, as it were, surcharged by a *Plethora*. And divers of the *Succiferous Vessels* being mixed herewith (*b*) and lying next the Soil, usually more or less *mortified*, and so their *Principles* somewhat *resolved*; the *Sap* is hereby better *specified*, and further *tinctured*; such parts of the *Sap* best entring, as are most agreeable to those *Principles*; which the *Sap* also carries off, in some part, as it passeth into the *Barque*.

(*a*) P.1.c.2. §.1.

(*b*) P.1.c.2. §.4.

17. §. The *Sap* thus *strained*, though it be *pure*, and consisteth of *Essential* parts; yet being *compounded* of *heterogeneous* ones; and received into the *Parenchyma* of the *Barque* a laxe and spongy Body, they will now easily and mildly *ferment*. Whereby they will be yet further prepared, and so more easily insinuate themselves into all the *Bladders* of the said *Parenchyma*; swelling and dilating it as far as the *Continuity* of its parts will bear. Whereupon, partly from the continued entrance of fresh *Sap*, and partly by a Motion or Pressure of *Restitution* in the swollen and Tensed *Bladders* of the *Parenchyma*, the *Sap* is forced thence into the other parts of the *Root*.

18. §. And because the *Parenchyma* is in no place openly and Visibly Pervious, but is every where composed of an Infinite Number of small *Bladders* (*c*); the *Sap*, therefore, is not only *fermented* therein, and fitted for Separation; but, as it passeth through it, is every part of it, *strained* an Hundred times over, from *Bladder* to *Bladder*.

(*c*) P.1.c.3. §.4.

19. §. The *Sap* thus *fermented*, and *strained*, is distributed to the other *Organical Parts*, according as the several *Principles* of This, are agreeable to those whereof the said *Organical Parts* consist. As the *Sap* therefore passeth from *Bladder* to *Bladder*, such *Principles* as are agreeable to those of the *Fibres* of the said *Bladders*, will adhere to, and insinuate themselves into the Body of the *Fibres*; *sc.* *Watry* chiefly, next *Acid*, then *spirituous*, *Earthy*, *Aery*, and *Oleous*. (*d*)

(*d*) Idea, §. 50, 52.

20. §. And the *Sap* by its continual appulse and *percolation*, as it leaveth some parts upon the said *Fibres*; so as it is *squeezed* betwixt them from *Bladder* to *Bladder*, it licks and carries off some others from them, in some *union* together with it; and so is *Impregnate* herewith: as *Water*, by passing through a *Mineral Vein*, becomes *tinctured* with that *Mineral*.

21. §. The *Sap* thus *Impregnate* with some *united* Principles of the *Parenchymous Fibres*, passeth on to the *Lignous Vessels*, whereinto their correspondent *Principles* also enter; *sc.* *Watry*, *Saline*, *Oleous* and *Earthy* chiefly. (*e*) And because the *Parenchymous Principles* mixed with them, are in some degree *united*, and so more ready to *fix*; some of these therefore will likewise enter into the said *Vessels*. Whereupon, the *Alkali oleosum* of the one, and the *Acidum spirituosum* of the other, meeting together; These, with the other *Principles*, all *concentre*, and of divers *fluids*, become one *fixed* Body, and are gradually *agglutinated* to the *Vessels*; that is, The *Vessels* are now *nourished*.

(*e*) Idea, §. 51, 52.

22. §. The supply of the *Sap* still continued, the *Principles* thereof will not only enter into the *Body* of these *Parts*, but also their *Concaves*. And the *Parenchymous Fibres* being *wrapped* about the *Vessels*, (*a*) as often as the said *Fibres* are more *turgid* with their own contained *Fluid*, they will thereby be somewhat *shortned*, or contract in length; and so must needs *bind* upon the *Vessels*, and thereby, as it were, *squeeze* some part of the *Fluid*, contained both within themselves and the *Vessels*, back again into the *Bladders*.

(*a*) P.1.c.5. §. 12.

23. §. And the *Sap* herein, being thus *tinctured* with some of the *united Principles* of the *Vessels*, divers of them will now also insinuate themselves into the *Parenchymous Fibres*, and be incorporated with them: Whereby, the said *Fibres*, which before were only *relaxed* and *dilated*, are now also *nourished*, and not till now. Some portion of the *united Principles* both of the *Parenchymous* and *Lignous Parts*, being necessary to the true *nutrition* of Each: As the Confusion and joynt assistance of both the *Arterious* and *Nervous Fluids*, is to the nourishment or coagulation of the *Parts* in *Animals*.

24. §. Some portion of the *Sap* thus doubly *tinctured*, is at the same time transmitted to, and enters the *Body* of the *Aer-Vessels*; consisting chiefly of *Water*, *Aer*, and *Acid*; and, in like manner, as in the other *Parts* is herein *agglutinated*. And the appulse and pressure of the *Sap* still continued, some portion hereof is also trajected into the *Concaves* of the said *Vessels*; existing therein as a most *Compounded Fluid*; partaking, more or less, both of the *Principles* and *Tinctures* of the other *Organical Parts*, and of the *Aer-Vessels* themselves; being as it were, a *Mixed Resolution* from them all.

25. §. And the *Parenchymous Fibres* being *wrapped* about These, as about the other *Vessels*, (*b*) and, in like manner, *binding* upon them; they thus frequently *squeeze* part of the said contained *Fluid* out again: As necessary, though not to the immediate Nourishment of the *Parts*, yet the due Qualification of the *Sap*; being a Constant *Aerial Ferment*, successively stored up within the *Aer-Vessels*, and thence transfused to the *Sap*, in the other *Organical Parts*.

(*b*) P.1.c.5. §. 12.

26. §. And that there may be a better Transition of the *Sap* thus *tinctured*, to the several *Organical Parts*; therefore, none of them are close set and compact within themselves, severally: For so, they would be inaccessible to the *Sap*, and their inward Portions, wanting a due supply of *Aliment*, would be starved. But the *Vessels*, both of *Aer* and *Sap*, being every where divided into *Braced* Portions, and other *Parenchymous* Portions, filling up the spaces every where betwixt them (*c*); there is therefore a free and copious communication of the *Sap*, (and so of all the *Tinctures* successively transfused into it) from *Part* to *Part*, and to every Portion of every *Part*: The *Parenchymous* Portions, running betwixt the *Braces*, as the smaller *Vessels* do throughout the *Viscera*, in *Animals*. Whereby, none of them want that Matter, which is necessary either for their *Nutrition*, or for the good Estate of their *Contents*, or for the due period of their Growth.

(*c*) P.1.c.3. §. 13. & c.4. §. 4,5,7.

27. §. For the better *Tempering* of the several parts of the *Sap*, serve the *Diametral* Portions of the *Parenchymous Body* which run sometimes directly through the *Barque*, as in *Lovage*, *Parsley*, &c. is described and figur'd (*d*) Which being, all or most of them, continued betwixt both the *Succiferous* and the *Aer-Vessels*, from the Circumference

(*d*) P.1.c.3. §. 7, 8.

to

to the Centre; they hereby carry off a more Copious and *Aerial Ferment* from the One, and communicate it unto the Other. For as the *Sap* enters the *Barque*, the more *liquid* part, still passeth into the *succulent* Portions thereof; the more *Aery*, is separated into those White and Dryer *Diametral* ones; and in its passage betwixt the *Portions* of the *Aer-Vessels*, is all along communicated to them. Yet is it not a pure or *simple Aer*, but such as carries a *Tincture* with it, from the *Succiferous Vessels*. And therefore it is observable, That when the *Diametral* Portions are more distant, the *Sap-Vessels* run not in a Straight Line betwixt them, but are Reciprocally so inclined, as to touch upon them; as in *Lovage* is visible: Thereby communicating their *Tincture* to the *Aer*, as it passeth by them, through the said *Diametral* Portions.

28. §. By the continual appulse of fresh *Sap*, some, both of the *aery*, and of all the other parts thereof are transmitted into the *Pith*; where, finding more room, it will yet more kindly be *digested*. Especially having the advantage herein of some degree of Warmth; being herein remoter from the *Soil*, and, as it were, *Tunn'd* up within the *Wood*, or the *Mass* of surrounding *Vessels*. So that the *Pith*, is a *Repository* of better *Aliment* gradually supplied to those *Succiferous Vessels*, which are frequently scattered up and down therein, and which ascend into the *Trunk*. (*a*) But where no *succiferous Vessels* are mixed, herewith, it usually becomes Dryer, and is replenished with a more *Aerial* and *Warmer Sap*; whereby the growth of the *Caulis* is promoted, as by an *Hot Bed* set just under it. And in many *Plants* with divers knobbed *Roots*, the younger are more succulent, serving chiefly to feed the *Stalk*: the Elder are spongy and fill'd with *Aer*, for the fermenting of the *Sap*, and more early growth of the *Stalk*: as in little *Celandine*, *Dogstones* and all of that Kindred. And thus all the *Parts* have a fit *Aliment* provided for their *Nourishment*

(*a*) P.I.c.5. §. 3.

29. §. IN THIS *Nourishment*, the *Principles* of the *Sap* are, as is said, *concentred* and *locked* up one within another: (*b*) Whence it is, that the *Organical Parts*, being cleansed of their *Contents*, have none of them any *Taste* or *Smell*, as in the *Piths* of *Plants*, *Paper* and *Linen Cloth* is evident. (*c*) Because till by *Digestion*, violent *Destillation*, or some other way, they are resolved, they cannot act upon the *Organs* of those *Senses*. For the same reason, they are never *tinctured*, excepting by their *Contents*: and although, to the bare Eye, they frequently shew *White*, yet viewed through a *Microscope*, they all appear *transparent*. In like manner, as the *Serum* of *Blood*, *Whites* of *Eggs*, *Tendons*, *Hairs* and *Horns* themselves are *transparent*, and without much *Smell* or *Taste*, their *Principles* being, in all of them, more or less *concentred*: But when ever these *Principles*, are forcibly *resolved*, they are ever variously invested with all those *Qualities*.

How the several Parts are Nourish'd and Form'd.
(*b*) §. 21.
(*c*) Idea, §. 49, 51.

30. §. And as from the *Concentration* of the *Principles*, in every *Organical Part*, the said *Parts* do thus far, all agree: So, from the *Predominion* of the *Principles* of each *Part*, the rest are controuled, not only to a *Concentration*, but an *Assimilation* also; whereby, the *Specifick Differences*, of the several *Organical Parts*, are preserved. Hence the *succiferous Vessels* are always *Tough* and very Pliable; for so are all *Barques*, wherein these *Vessels* abound; so is a Handful of *Flax*, which is nothing else but a heap of the *succiferous Vessels* in the *Barque* of

that

that *Plant*. For besides *Water*, and *Earth*, an *Alkaline Salt* and *Oyl* are, as is said, the predominant *Principles* of these *Vessels*. (*a*) It is then the *Oyl*, chiefly, by which these *Vessels* are *Tough*: for being of a tenacious Nature, by taking hold of other *Principles*, it marries them together; and the *Alkaline Salt* and *Earth*, concentred with it, addeth to it more *strength*. Hence the *Caput Mortuum* of most Bodies, especially those that abound with *Oyl* and a *Sal Alkali*, is *brittle* and *friable*; those *Principles*, which were the *Ligaments* of the rest, being forced away from them. From the same Cause, the *Parenchymous Parts* of a *Root*, even in their Natural State, are *brittle* and *friable*; *sc.* Because their *Earthy*, and especially *Oleous* and *Saline Principles* are, as is said,(*b*) so very few. Therefore all *Piths* and more *simple Parenchyma's*, break *short*, so *Corn*, and the *Roots* of *Potato's*, and divers other *Plants*, being dryed, will easily be rub'd to *Meal*; and many *Apples*, after Frosts, eat *mealy*; the *Parenchymous Parts* of all which, are not only by *Analogy*, but in Substance or Essence, the self same Body. (*c*)

(*a*) §. 21.

(*b*) §. 19.

(*c*) Lib. 1. c.7. §. 14.

31. §. And as the *Consistence* of the several *Organical Parts*, is dependent on their *Principles*; so are their *Figures*. And first, the *Succiferous Vessels*, from their *Alkaline Salt*, (*d*) grow in Length. For by that Dimension, chiefly, This *Salt* always *shoots*: And being a less moveable *Principle* than the rest, and so apt more speedily to *fix* or *shoot*: It thus overrules them to its own *Figure*. And even as the Shape of a *Button* dependeth on the *Mould*, the *Silk* and other Materials wrought upon it, being always conformable thereunto: so here; the *Salt* is, as it were, the *Mould*; about which, the other more passive *Principles* gathering themselves, they all consort and fashion to it. Hence also the same *Sap-Vessels* are not *pyramidal*, as the *Veins* of *Animals*; but of an equal bore, from end to end; the *shootings* of the said *Salt*, being also figured more agreeably to that *Dimension*. And as by the *Saline Principle*, these *Vessels* are *Long*; so by the *Oleous*, (*e*) they are every where *Round*, or properly *Cylindrical*; without some joynt Efficacy of which *Principle*, the said *Vessels* would be *Flat*, or some way Edged and *Angular*, as all *saline shoots*, of themselves, are; as those of *Alum*, *Vitriol*, *Sal Ammoniac*, *Sea Salt*, *Nitre*, &c. And because the *Spirituous* and more *Fluid* part of the *Principles*, is least of all apt to *fix*, while therefore, the other parts fix round about, This will remain moveable in the *Centre*; from whence every *Vessel* is formed, not into a *solid*, but *hollow Cylinder*; that is, becomes a *Tube*.

(*d*) P. 2. §. 21.

(*e*) Ib.

32. §. The *Lactiferous Vessels* are *tubulary*, as the *Lymphæducts*, but of a somewhat wider *Concave* or *Bore*. For being their *Principles* are less *Earthy* and *Oleous*, and also more loosely *Concentred*; as from their easie corruption or *Resolution* by the *Aer*, it appears they are: they are therefore more tender, and so more easily dilative, and yielding to the said *Spirituous* part in the Centre. And by this means, obtaining a wider *Bore*, they are more adapted to the free motion of the *Milky Content*: which being an *Oleous* and Thicker *Liquor*, than that in the *Lympheducts*; and having no advantage of *pulsation*, as the *Blood* hath in *Animals*; might sometimes be apt to stagnate, if the *Vessels*, through which it moves, were not somewhat wider.

33. §.

33. §. As the *Saline Principle* is the *Mould* of the *Succiferous*, so is the *Aerial* of the *Aer-Veſſels*. (*a*) Now the Particles of *Aer* ſtrictly ſo called, at leaſt of that part of it concerned in the Generation of the *Aer-Veſſels*, I ſuppoſe, are crooked: and that by compoſition of many of thoſe crooked ones together, ſome of them become Spiral, or of ſome other winding *Figure*: and that thereupon dependeth the *Elaſtick* Property of the *Aer*, or its being capable of *Rarefaction* and *Condenſation* by force. Wherefore, the ſaid *crooked* Particles of the *Aer*, firſt *ſhooting* and *ſetting* together, as the *Mould*, the other *Principle* cling and *fix* conformably round about them. So that, as by force of the *Saline Principles*, the reſt of them are made to *ſhoot* out in *Long continued Fibres*; ſo by force of the *Aerial*, thoſe *Fibers* are ſtill diſpoſed into *Spiral Lines*, thus making up the *Aer-Veſſels*. And according as there are fewer of theſe *Aerial* Particles, in proportion to the *Saline*, the *Concave* of the *Aer-Veſſels* is variouſly wider, or the *Fibres* continue their *ſhooting* by wider *Rings*; as thoſe that come nearer to a *right Line*, and ſo are more complient to the *Figure* and ſhooting of the *Saline* parts. And whereas the *Lympheducts*, ſhooting out only in length, are never ſenſibly *amplified* beyond their original ſize: Theſe, on the contrary, always, more or leſs, enlarge their *Diameter*; becauſe their *Fibres*, being diſpoſed into *Spiral Lines*, muſt needs therefore, as they continue their growth, be ſtill dilated into greater and greater *Rings*. And being at the bottom of the *Root* more remote from the *Aer*, and ſo having ſomewhat fewer Particles purely *Aerial*, there ingredient to them, then at the top; they fall more under the government of the *Saline*, and ſo come nearer to a *right Line*, that is into greater *Circles*; and ſo the *Aer-Veſſels*, made up of thoſe *Circles*, are there generally wider. (*b*)

(*a*) P.1.c.4. §. 23. & P.2. §. 24.

(*b*) P.1.c.4. §. 16.

34. §. By mediation of their *Principles*, the *Parenchymous Parts* likewiſe of a *Root* have their proper *Contexture*. For from their *Acid Salt* they are *Fibrous*; from their *Oyl*, the *Fibres* are *Round*, and in all parts even within themſelves; and from their *Spirit*, it is moſt probable, that they are alſo *hollow*. But becauſe the *Spirit* is, here, more copious than the *Aer*; and the *Saline Principle* an *Acid*, (*c*) and ſo, more under the government of the *Spirit*, than is an *Alkali*; therefore are not the ſaid *Fibres* continued in *ſtraight Lines*, as the *Sap-Veſſels*; or by one *uniform* motion, into *ſpiral* lines, as the *Fibres* in the *Aerial*; but *winding*, in a circular manner, to and fro a *thouſand ways*, agreeable to the like *motions* of the *Spirit*, that moſt *active*, and here moſt *predominant Principle*. And the *Spirituous Parts* being, as is ſaid, here more copious and redundant, they will not only ſuffice to fill up the Concaves of the *Fibres*, but will alſo gather together into innumerable little ſpaces, without them: whence the *Fibres* cannot wind cloſe together, as Thred, in a *Bottom* of Yarn; but are forced to keep at ſome diſtance, one parcel from another, and ſo are diſpoſed, as Bread is in baking, into *Bladders*. (*d*)

(*c*) §. 19.

(*d*) P.1.c.3. §. 4.

35. §. And the *under Fibres* being ſet firſt, as the *Warp*, the *ſpirituous* parts next adjacent, will incline alſo to *fix*, and ſo govern an *over-work* of *Fibres*, wrapping, as the *Woof*, in ſtill ſmaller Circles round the other: whereby they are all knit together. (*e*) For the ſame reaſon, the *Lympheducts*, being firſt formed, the *Parenchymous Fibres* ſet and wrap about Theſe alſo. (*f*) And the *Aer-Veſſels* being formed

(*e*) P.1.c.5. §. 9.

(*f*) P.1.c.5. §. 12.

formed in the Center, the *succiferous* run along those likewise (as *volatile Salts shoot* along the sides of a Glass, or *Frost* upon a Window) and so are, as it were, Incrustate about them in a *Ring*.

How the several Parts come to be Situate or Dispos'd.

36. §. SOME OF THE more Ætherial and Subtile parts of the *Aer*, as they stream through the *Root*, it should seem, by a certain *Magnitisme*, do gradually dispose the *Aer-Vessels*, where there are any store of them, into *Rays*. This Attraction (as I take leave to call it) or *Magnetick power* betwixt the *Aer* and these *Vessels*, may be argued, From the nature of the *Principles* common to them both: From the *Electral* nature of divers other Bodies; the *Load-stone* being not the only one which is attractive: And from other Effects, both before (*a*) and hereafter mentioned. Wherefore in the inferiour parts of the *Root*, they are less Regular; (*b*) because more remote from the *Aer*. And in the upper parts of many *Roots*, as *Cumfery, Borage, Parsnep*, where those that are next the Centre are confused, or differently disposed; those next the *Barque*, and so nearer the *Aer*, are postured more Regularly, and usually into *Rays*. For the same reason it may be; that even the *Sap-Vessels* in the *Barque*, as often as the *Aer Vessels* are more numerous, are usually disposed into *Rays*, as following the direction of the *Aer-Vessels*. And that the *Parenchyma* of the *Barque*, is disposed into *Diametral Portions*: and that where the *Aer Vessels* are fewer or smaller, these *Portions* are likewise smaller or none; as in *Chervil, Asparagus, Dandelion, Orpine, Bistort, Horse-Radish, Potato's*, &c.

(*a*) Lib. 1. c.2. §.25. & c.4. Append. §. 2, 3.
(*b*) P.1.c.4. §. 10. Tab. 7, 8, 9.

Tab. 7, 8, 9, &c.

37. §. The said Ætherial parts of the *Aer*, have a Power over the *Aer-Vessels* not only thus to Dispose them; but also to Sollicite and spread them abroad from the Center towards the Circumference of the *Root*. By which means, those *Roots* which have no *Pith* in their lower parts, obtain one in their upper. (*c*) And the same *Pith*, which in the lower part, is ratably, small, in the upper, is more or less enlarged. (*d*)

(*c*) P.1. c.5. §. 1.
(*d*) Ib. §. 4.

38. §. The *Spreading* of these *Vessels* is varied, not only according to the *Force* the *Aer* hath upon them, but also their own greater or less *Aptitude* to yield thereto. As often therefore, as they are Slenderer, they will also be more Pliable and recessive from the Centre, towards the Circumference. Hence, in such *Roots* where they are small, they stand more distant; as in *Turnep, Jerusalem Artichoke, Potato's*, and others; and so their *Braces* are fewer: and in the same *Root*, where they are smaller, their distance is greater. Besides, in these smaller *Aer-Vessels*, the *Rings* being less, and the *Spiral Fibres* whereof they are made, continuing to *shoot*; the said *Rings* therefore, must needs be so many more, as they are smaller; and so take up more space by the length of the *Root*; and so, not being capable of being crowded in a *right line*, every *Vessel* will be forced to recede to a *crooked* or *bowed* one.

Tab. 2, & 6.

39. §. The *Sap Vessels*, being by the *Parenchymous Fibres* knit to *these*, will likewise comply with Their motion, and spread abroad with them. Yet being still smaller (*e*) and more pliable than the *Aer-Vessels*, and so more yielding to the intercurrent *Fibres* of the *Parenchyma*, their braced *Threds* will, sometimes, be much more divaricated, than those of the *Aer-Vessels*; as in *Jerusalem Artichoke*. And because the *Succiferous Vessels*,

(*e*) P.1. c.3. §. 16.

Tab. 6.

Veſſels, although they are joyned to the *Aerial* by the *Parenchymous Fibres*, (a) yet are not continuous with them; neither fall under the like *Attractive Power* of the *Aer*, as the *Aerial* do; the *Aerial* therefore, upon their ſpreading, do not always carry all the *Succiferous* along with them; but often, if not always, leave many of them behind them ſprinkled up and down the *Pith*; as in *Parſley, Carrot, Jeruſalem Artichoke, Turnep*, &c. may be ſeen. (a) P.1. c.5. §. 12.

Tab. 6.

40. §. The *ſpreading* of the *Aer-Veſſels* ſtill continued, ſeveral of them, at length, break forth beyond the circumference of the *Root*; and ſo are diſtributed, either in the lower parts, into *Branches* and *Strings*; or at the top, into *Leaves*. And leſt they ſhould all ſpread themſelves into *Leaves*, and none be left for the *Caulis*; as where they are very ſmall, or the *Sap-Veſſels* to bound them, are but few, they might; therefore divers of them are, oftentimes, more frequently *braced* in the Centre; for which reaſon, they cannot ſo eaſily ſeparate and ſpread themſelves from thence, but run more inwardly up into the *Caulis*, as in *Borage*.

Tab. 6.

41. §. FROM THE various *Sizes, Proportions*, and *Diſpoſitions* of the Parts, *Roots* are variouſly *ſized, ſhaped, moved* and *aged*. Thoſe which, by their Annual Growth, are large; have fewer, both *Aerial*, and *Sap-Veſſels*, and a more copious *Parenchyma*. So that the *Aer-Veſſels*, or rather, the *Aery Ferment* contained in them, *volatilizing* only a ſmaller portion of the *Sap*; the ſaid *Sap* is leſs capable of advancement into the *Trunk*; and ſo muſt needs remain and *fix* more copiouſly in the *Root*, which is thereby more augmented. And where the *Sap-Veſſels* alone, are but few, the *Root* is yet, ratably, ſomewhat large: but where they are numerous, it is never ſo, as to its Annual Growth, in any proportion to their Number: Becauſe their *Tincture*, which is *Alkaline*, will go farther in ſetting the *Parenchymous Parts*: than the *Tincture* of *Theſe*, which is *Acidulate*, will go, in ſetting Them. (b)

How the whole *Root* is differently ſized and ſhaped.

(b) P. 2. §. 31.

42. §. When the *Aer-Veſſels* are more pliable and ſequent to the *Attraction* of the *Aer*, and ſo ſpread themſelves, and the *Succiferous* together with them, more abroad; in the manner as hath been ſaid; the *Root* alſo will grow more in *Breadth*; the nutrition of the *Parenchymous Parts*, to which the *Veſſels* are adjacent, being thus, by the ſame dimenſion, more augmented; as in *Turnep, Jeruſalem Artichoke*, &c. But where theſe are not ſpread abroad, the *Root* is but ſlender; as in *Aſparagus, Dandelion*, &c.

Tab. 2, & 7.

Tab. 7, & 8.

43. §. If the *Aer-Veſſels* be contracted into, or near the Centre, and are ſomewhat Large or Numerous; and the *Succiferous*, alſo more copiouſly mixed with, or ſurrounding them; the *Root* grows very Long; as do thoſe of *Fenil, Vine, Liquiriſh*, &c. For the *Aer-Veſſels* containing a more copious *Ferment*, it will well digeſt and mature the *Sap*: Yet the *Succiferous* being over proportioned to them; the *Sap* will not therefore, be ſo far volatilized, as to aſcend chiefly into the *Trunk*; but only to ſubſerve a fuller Growth of their *Veſſels*: which being more numerous, and ſo more ſturdy, and leſs ſequent to the expanſive motion of the *Aerial*; this their own Growth, and conſequently, that of all the other *Parts*, cannot be ſo much in Breadth, as Length.

Tab. 2, & 17.

44. §.

44. §. Where the same *Aerial Vessels* are Fewer, or more Contracted, or sheathed in a Thicker and Closer *Barque*; the *Root* is smooth, and less *Ramified*, as in *Asparagus, Peony, Dandelion*. But where more Numerous, sheathed in a Thinner *Barque*, Smaller, or more Dilated; the *Root* is more *Ramified*, or more *Stringy*, as in *Columbine, Clary, Beet, Nicotian*. For being, as is said, by these means, more sequent to the Attraction of the *Aer*; approaching still nearer the circumference of the *Barque*, they at last strike through it, into the *Earth*. And the *Parenchymous Fibres* being wrapped about them, and the *Succiferous Vessels* knit to them by those *Fibres*; (a) therefore they never break forth naked, but always invested with some quantity of these *Parts* as their *Barque*: where by, whatever *Constitutive Part* is in the *main Body* of the *Root*, the same is also in every *Branch* or *String*.

Tab. 7, 8.

Tab. 2, & 7.

(a) P. 1. c. 5. §. 12.

45. §. From the same Expansion and Pliability of the *Aer-Vessels*, the *Root* oftentimes putteth forth *Root-Buds*; which gradually shoot up and become so many *Trunks*. In the Formation of which *Buds*, they are pliable and recessive all kinds of ways; being not only invited Outward, toward the Circumference of the *Root*, as in *Root-strings*, but also spread more Abroad every way, so as to make a *Root-Bud*: Where as in the said *Root-strings*; they are always more Contracted. Which, in respect of the Disposition of the *Parts*, is the principal difference betwixt the *Root* and the *Trunk*, as hath been said. (b) Hence, those *Roots*, chiefly, have *Root-Buds*, which have the smallest *Aer-Vessels*; (c) these, as is said, being the most pliable and Expansive.

Tab. 6.

(b) P. 1. c. 1. §. 2, 3. *Tab.* 11. (e) P. 1. c. 4. §. 15.

46. §. But because the expansiveness of the *Vessels*, dependeth also, in part, upon the Fewness of their *Braces*; therefore the said *Buds* shoot forth differently, in divers *Roots*. Where the *Braces* are fewer, the *Buds* shoot forth beyond the Circumference of the *Root*, as in *Jerusalem Artichoke*; where more close, as in *Potato's*, the *Buds* lie a little absconded beneath it; the *Aer-Vessels* being here, by their *Braces*, somewhat checked and curbed in, while the *Barque* continueth to swell into a fuller Growth.

Tab. 6.

47. §. If the *Aer-Vessels* are all along more equally sized, the *Root* is so also, or *Cylindrical*; as are those of *Eryngo, Horse-Radish, Marshmallow, Liquirish*, &c. But if unequal, growing still *wider* towards the *bottom* of the *Root*; then the *Root* is unequal also: But groweth, as is observable, quite contrarily to the *Aer-Vessels*; not Greater, as They do; but still smaller, or *pyramidally*; as in *Fenil, Borage, Nettle, Patience, Thorn-Apple*, &c. is apparent. For the *Aer-Vessels* peing considerably wider about the *bottome* of these *Roots*; they there contein a more Copious *Ferment*: Whereby the *Sap* is there also more volatilized, and plentifully advanced to the Upper Parts. Withal, thus receiving into themselves, and so trasmitting to the upper Parts, a more plentiful *Vapour*, they hereby rob the *Parenchymous Parts* of their Aliment, and so stint them in their Growth.

How Roots are diffently Mov'd. *Tab.* 8.

48. §. FROM THE different *Proportions* and *Situation* of the *Parts*, the *Motions* of *Roots* are also various. For where the *Are-Vessels* are spread abroad and invested with a thinner *Barque*; the *Root* runs or lies *Level*, as in the *level-Roots* of *Primrose, Bishops-weed, Anemone*, &c.

may

may be feen. So that thefe *Roots*, as by the *Perpendicular Strings*, which fhoot from them into the *Earth*, and wherein the *Aer-Veffels* are contracted into their Center, they are Plucked down (*a*): So by the *Aer-Veffels*, which ftand nearer the *Aer*, and more under its Attractive Power (*b*) they are invited *upwards*; whereby they have neither *afcent* nor *defcent*, but keep *level*, betwixt both.

(*a*) P.I.c.1, §.15.
(*b*) P. 2. §. 36.

49. §. But if thefe *Veffels* are Contracted, ftanding either in, or near the Centre, and are invefted with a *Barque* proportionably Thick; the *Root* ftriketh down *perpendicularly*, as doth that of *Dandelion, Buglofs, Parfnep*, &c. And therefore the faid *Veffels*, although they are fpread abroad in the *level* Roots, yet in the *perpendicular* ones of the fame *Plant*, they are always contracted; as by comparing the *Level* and *Down-right Roots* of *Ammi, Primrofe, Jerufalem Artichoke, Cowflip*, and others, is manifeft.

Tab. 7, 8.

50. §. If the *Aer-Veffels* are Contracted, and Environed with a greater number of *Succiferous*, the *Root* grows *deep*; that is, *perpendicular* and *long*. (*c*) *Perpendicular*, from the Contraction of the *Aer-Veffels*; (*d*) and *long*, from the Predominion of the *Succiferous*, which in their growth, are extended only by that Dimenfion, as in *Liquirifh, Eryngo*, &c.

(*c*) P.I.c.1. §.9.
(*d*) P. 2. §. 49.

51. §. If the *Succiferous* are over proportioned to the *Parenchymous Parts*, but under to the *Aer-Veffels*; the *Root* is *perpendicular* ftill, but groweth *fhallow*: The *Succiferous* being fturdy enough to keep it *perpendicular*; But the *Aer-Veffels* having a predominion to keep it from growing *deep*; as in *Stramonium, Nicotian, Beet*, &c.

Tab. 7.

52. §. If, on the contrary, the *Parenchymous Parts* are predominant to the *Aer Veffels*; and that, both in the *Root* and *Trunk*; then the whole *Root* changeth place, or *defcends*. (*e*) For the faid *Aer-Veffels*, having neither in the *Trunck*, nor in the *Root*, a fufficient Power to Draw it *upwards*; it therefore gradually yields to the Motion of its *String-Roots*; which, as they ftrike into the Soil, Pluck it down after them. And becaufe the old *Strings* annually rot off, and new ones fucceffively fhoot down into the *Ground*, it therefore annually ftill defcendeth lower; as in *Tulip, Lily*, &c. may be obferved.

(*e*) P.I. c.1. §. 10. Lib.1. c.2. §.25, and 4. Append. §. 10. P. 2. §. 36.

53. §. Where the *Aer-Veffels* are much *fpread* abroad, and alfo numerous, the *Root* oftentimes, as to its feveral parts, *defcends* and *afcends* both at once. So *Radifhes* and *Turneps*, at the fame time, in which their nether parts *defcend*; their upper, (where the faid *Veffels* are more loofely braced, and fpread more abroad than in the lower parts) do *afcend*, or make their Growth *upward*. Hence alfo, the upper part of moft young *Roots* from *Seed*, afcends: Becaufe the firft *Leaves*, being proportionably large, and ftanding in a free *Aer*, the *Aer-Veffels* therein, have a dominion over the young *Root*; and fo themfelves yielding to the folicitation of the *Aer*, *upwards*; they draw the *Root*, in part, after them.

Tab. 2.

54. §. BY THE *Situation* and *Proportions* of the *Parts*, the *Age* of the *Root* is alfo varied. For if the *Sap-Veffels* have the greateft Proportion, the *Root*, is *Perennial*, and that to the fartheft extent, as in *Trees* and *Shrubs*. Becaufe thefe *Veffels* containing a more copious *Oyl*; (*f*) and their feveral *Principles* being more clofely *Concentred*, they are lefs fubject to a *Refolution*, that is, a Corruption or Mortification by the *Aer*.

How *Roots* are differently Aged.

(*f*) P. 2. §. 21.

55. §.

55. §. If the *Parenchymous Parts* have much the greatest, the *Root* seldom liveth beyond Two Years; but afterwards perisheth either in whole, or in part; as do divers *bulbous, tuberous,* and other *Roots*; whether they are more Porous and Succulent, or more Close and Dry. If Porous, all the *Liquid Principles* standing herein more abundant, either by a stronger Fermentation, or otherwise, Resolve the *fixed ones* of the *Organical Parts*; whence the whole *Root*, rots; as in *Potato's*. So also *Parsneps,* and some other *Roots*, which, in a hard and barren *Soil*, will live several years, in another more rank, will quickly rot. If the *Parenchyma* be Close, then the *Aer*, chiefly, entring in and filling it up, thus mortifies the *Root*; not by Rotting the *Parts*, but over Drying them; as in *Satyrion*, *Rape-Crowfoot*, *Monkshood*, &c. (*a*)

(*a*) P.1.c.1. §. 13,16. & P.2.§. 28.

56. §. But if the *Aer-Vessels* have the greatest Proportion, and especially if they are more *large*, and withall, are *spread* more abroad: the *Root* is *Annual*, as in *Thorn-Apple*, *Nicotian*, *Carduus Ben.* &c. And of the same Kindred, if any, those are *Annual*, which have the most *Aer-Vessels*. So *Endive* and *Sonchus*, which have store of *Aer-Vessels*, are both *Annual*: whereas *Cichory*, in which they are fewer, is a *Perennial Root*. For hereby a more copious *Aer* being Transfused into all the other *Parts*; (*b*) they are thus, by degrees, hardned, and become sticky; and so impervious to the *Sap*, which ought to have a free and universal Transition from *Part* to *Part*. As *Bones*, by *Precipitations* from the *Blood*, at length, cease to grow. Or the same more abundant *Aer*, so far *volatilizeth* the *Liquors* in the *Root*, that they are wholly advanced into the *Trunk*, and so the *Root* is starved. Whence also the *Aer-Vessels* of the *Trunk*; where they are numerous, and over proportioned to the *Bulk* of the *Root*, as in *Corn*; they so far promote the advance of the *Sap*, as to exhaust the *Root*, sucking it into a Consumption and Death.

Tab. 9.

(*b*) P. 2. §. 25, 26.

How the Contents of the several Parts are made.
(*c*) Idea, §. 54.

57. §. FROM THE *Principles* of the *Parts*, their *Contents* and the several *Qualities* hereof are also various; (*c*) the *Fluid* of each *Organical Part*, being made, chiefly, by *Filtration* through the sides thereof; such of the *Principles* in the *Sap*, being admitted into, and transmitted through them, as are aptest thereunto. In the like manner, as when *Oyl* and *Water*, being poured upon a Paper, the *Water* passeth through, the *Oyl* sticks: or as the *Chyle* is strained through the *Coats* of the *Guts*, into the *Lacteal Vessels*: or as *Water* in *Purgations*, is strained through the *Glands* of the same *Guts*, from the *Mesenterical*.

58. §. The *Principles* therefore of the *Parenchymous Fibres* being *spirituous, acid,* and *aerial,* they will also admit the like into them; excluding those chiefly which are *Alkaline* and *Oleous*. (*d*) And as by the *Conjugation* of such *Principles* in the *Fibres*, the like are capable of admittance into their Body: so the *Proportion* and *Union* of the same *Principles,* regulates the *transmission* hereof into their *Concave*. Wherefore, the predominant *Principles* of the *Fibres* being chiefly *acid*, next *spirituous*, and *aery*, the more *aery* ones will be *transmitted*. For if more of *them* should *fix* they must do so by similitude and adhesion: But where there are fewer similary parts to adhere to, fewer must adhere. The *Fibres* therefore contain so many parts of *Aer*, as to *admit many*

(*d*) P. 2. §. 19.

many more into their Body; but not to *fix* them; which therefore must needs, upon admiffion, pafs through into their *Concave*; where, together with fome other more *fpirituous* parts, they make an *Ætherial Fluid*. And becaufe fome *aqueous* or *vaporous* parts will alfo ftrain through with them; hence it is, that as more and more of thefe enter, they by degrees ftill thruft out the *aery* ones; which quitting the more *fucculent Fibres* of the *Parenchyma*, are forced to betake themfelves to the *dryer* ones, *fcil.* all thofe, whereof the *Diametral Portions* do confift. For the fame reafon the *Aery* parts being gradually excluded the *fucculent Fibres* of the *Barque*; they are forced to recede and tranfmigrate into thofe of the *Pith*. And the *Fibres* of the *Pith* themfelves being filled, aud the *Aery* parts ftill forced into them; they at length alfo ftrain through the *Fibres* into the *Bladders*: whence it comes to pafs, that while the *Barque* is *fucculent*, the *Pith* is often times filled with *Aer*.

59. §. The *Lymphæducts* being more *earthy*, *Salinous*; *oleous*, and *aqueous*, will both *admit* and copioufly *fix* the like *Principles*, as their proper Aliment. The *Water* being more *perfluent* than the reft, will therefore ftrain, with a lighter *Tincture* of them, into their *Concave*. Efpecially the *Oleous* parts of thefe being *rampant*, and lefs apt to *fix* and *feize* the *aqueous*, upon their entrance, than the *faline*. *(a)* P. 2, §. 21.

60. §. The *Lactiferous*, appearing to be made, chiefly, by the Conftipation of the *Parenchymous Parts* all round about their Sides; the *Liquor* conteined in thofe *Parts*, although it may eafily enough be transfufed into the Hollow of thefe *Veffels*; yet feems it not, with equal facility, to be refunded thence: So that the *thinner* and more *aqueous* Portion only, paffing off; the remainder, is, as it were, an *Oleous Elixyr*, or extract, in the form of a *Milk*.

61. §. The *Fluid Ferment* contained in the *Aer-Veffels*, is alfo in part, dependent on the *Principles* of thofe *Veffels*, being in their percolation *tinctured* therewith. But becaufe the percolation is not made *through* the *Body* of the *Fibres* whereof the *Veffels* are compofed, but only *betwixt* them; therefore the tranfient *Principles* more promifcuofly, yet with an over porportion of dryer Particles, pafs into the *Concaves* of thefe *Veffels*, and fo are herein all immerfed in a Body of *Aer*. *(b)* The *Fibres* themfelves, in the mean time, as thofe of the *Parenchyma*, admitting and containing a more *Aery* and *Ætherial Fluid*. *(b)* P. 2. §. 24.

62. §. The *Contents* are varied, not only by the *Nature*, but alfo the *Proportion* and *Situation* of the *Parts*, whereby the faid *Contents* are with different Facility and Quantity, communicated one to another. Hence it is, partly, that a *Vine*, or that *Corn*, hath fo little *Oyl*: *fc.* Becaufe their *Aer-Veffels*, in proportion with the other *Parts*, are fo Great and Numerous: in *Corn*, the *Stalk* being alfo very hollow, and fo becoming as it were, one Great *Aer-Veffel*. For the *Oily* parts of the *Sap*, are fo exceedingly *attenuated (c)* by the *Aery Ferment* contained in thefe *Veffels*; that they are, for the moft part, fo far *immerfed* in the *Spirit*, or mixed therewith, as not, by being collected in any confiderable Body, to be diftinguifhable from it. And the affinity that is betwixt *Spirits* and *Oils*, efpecially *Effential*, is manifeft: Both are very inflammable; Both will burn all away; The *Odors*, which we call the *Spirits* of *Plants*, are lodged in their *effential Oyl*; Both, being duly *(c)* P. 2. §. 25, & 56.

duly *Rectified*, will mix as easily together, as *Water* and *Wine*. So that, although *Oyl*, by the separation of its *earthy* and *Saline* parts, which give it its sensibly *oleous* Body, may not be so far *attenuated*, as to *produce a Spirit*; yet that it may so far be attenuated, and so be mixed therewith, as *not to be discerned from it*, as in the forementioned *Plants*, will be granted.

(a) Tab. 9. & 16.

63. §. Hence it is, that the *Lactiferous* standing more remote from the *Aer-Vessels*, and the *Succiferous* interposing; (a) the *Liquor*, therefore, contained in them, is not so much under the government of the *Aerial Ferment*, and is thence, partly, more *Oily*. For the same reason, all *Roots* which are *Milky*, so far as I have observed, have an under-proportion of *Aer-Vessels*; these being either Fewer or Smaller.

How the Odours of Plants are made.

64. §. FROM what hath been said, we may receive some information, likewise, of the *Odours*, *Colours*, and *Tastes* of *Plants*. And for *Odours*, I suppose, That the chief Matter of them, is the *Aerial Ferment* contained in the *Aer-Vessels*. Not but that the *other Parts* do also yield their smell; but that *these* yield the *strongest* and the *best*, and *immediately* perceptible in *fresh*, *undryed* and *unbruised Plants*. For the *Aer* entring into, and passing through the *Root*, and carrying a *Tincture*, from the several *Organical* and *Contained* Parts, along with it, and at last entring also the *Concaves* of the *Aer-Vessels*; it there exists the most *Compounded* and *Volatile Fluid*, of all others in the *Plant*, and so the fittest matter of *Odour*: and such an *Odour*, as answers to that of all the *Odorous* parts of the Plant. (b) Wherefore the *Organical Parts*, being well clensed of their Contents, smell not at all; Because the *Principles* hereof are, as hath been said, so far *fixed* and *concentred* together. Hence also the *Contained Parts* themselves, or any other Bodies, as their *Principles* are any way more *fixed*, they are less *Odorous*: So is *Rosin*, less than *Turpentine*, and *Pitch*, than *Tar*; and many the self same Bodies, when they are *coagulated*, less than when they are *melted*. So also *Musk*, which is not so liquid as *Civet*, is not so strong; nor *Ambergreece*, as *Musk*: For although it hath a more excellent smell, than *Musk* hath, yet yieldeth it not so easily; since it is a more *fixed* Body, and requireth some *Art* to be opened. Hence also the *Leaves* of many *Plants* lose their *Odour* upon *rubbing*: Because the *Aer-Vessels* being thereby broken, all their *contained odorous Fluid* vanisheth at once: which before, was only strained gradually through the *Skin*. Yet the *fixed* Parts themselves, upon drying, are so far altered by the *Sun* and *Aer*, as to become resoluble, and volatile, and thence odorous.

(b) P. 2. §. 24.

How their Colours.

(c) P.1. c.2. §. 2, 4.

65. §. SO ALSO of their *Colours*. As whence the *Colours* of the *Skins* are varied. For divers of the *Sap-Vessels*, together with the *Parenchymous Parts* successively falling off from the *Barque* into the Skin (c) by their proximity to the *Earth* and *Aer*, their *Sulphureous* or *Oleous Principle* is more or less *resolved*, and so produceth divers *Colours*. So those *Roots* which turn *purple* any where within, have usually a *blacker Skin*; the one of those two *Colours* being, by a resolution and corruption of parts, easily convertible into the other, as in *Cumfry*, *Thistle*, &c, So the *Milk* of *Scorzonera*, contained in the *Vessels* of the *Barque*, upon drying, turneth into a *brown Co-*

lour: Wherefore the *Skin*, in which there are divers of those *Vessels*, is of the *same*. So both the *Milk* and *Skin* of *Lovage* is of a brownish *yellow*. But *Parsnep* hath a clearer *Sap* in all its *Vessels*, and a whiter *Skin*. So *Potato's*, being cut traverse, after some time out of ground, have divers *red* specks up and down where the *Vessels* stand, and their *Skin* is accordingly *red*.

66. §. The reason, I say of these *Colours*, is the *resolution* or *reseration* of the *Principles* of the several *Parts*, chiefly, by the *Aer*, and a *lighter mixture* of them consequent thereupon: whereby the *Sulphureous* or *Oyly Parts*, which were before *concentred*, are now more or less *rampant*, discovering themselves in divers *Colours*, according as they are diversly mixed with the other *Principles*. Hence these *Colours* are observable, according to the nature of the *Parts wherein* they *are*, or whereunto they are *adjacent*: So where the *Lympheducts* doe run, there is a *Red*, or some other *Sulphureous Colour*; the *Oleous Principles* being, as is said, *(a)* more copious in these *Vessels*; as in the *Bark* of *Peony*, the inward parts of *Potato's*, &c. may be seen. But the *Parenchymous Parts*, where more remote from the said *Vessels*, they are usually *White*, or but *Yellow*: the *Sulphureous Principle* of these *Parts*, being, as hath been said, but sparing. *(b)* The same is seen in those *Roots* which shew both *Red* and *Yellow*: those *Parts*, principally, where the *Succiferous Vessels* run, being *Red*; but those *Parts*, where only the *Aer-Vessels* are mixed with the *Parenchymous*, being *Yellow*; as in *Patience*. So likewise the *pithy* part of a *Carrot*, where the *Aer-Vessels* have very few *Succiferous* mixed with them, is *Yellow*; but the *Barque*, where the *Succiferous* are very numerous, is *Red*. For the same reason, many *Roots*, which are *Whiter* in their upper parts, are *Purple* or *Reddish* in their inferiour, as *Avens*, *Strawberry*, &c. Because those lower parts, having lain longer *(c)* under ground (these being *descending Roots*) their *Principles* are, thereby, somewhat more *resolved*, and so the *Oleous*, ramp and spread all over the rest in that *Colour*.

(a) P. 2. §. 21.

(b) P. 2. §. 20.

(c) P. 1. §. 13.

67. §. And that the *Resolution* of the *Sulphureous* and other *Principles* is partly effected by the *Aer*, appears, In that, where the *Aer* hath a free access to the *Succiferous Vessels*, the *Colours* are there, chiefly produced, or are more conspicuous. So in *Potato's*, where the *Succiferous Vessels* are either next to the external *Aer*, as in the *Skin*; or contiguous with the *Aer-Vessels*, as in the *Ring* within the *Barque*; there, they produce a *Red*: but where more remote from both, as in the middle of the *Barque*, and Centre of the *Root*, there they produce none. Hence also it is, that the *Leaves* and *Flowers* of some *Plants*, as *Bloodwort*, *Wood-sorrel*, *Radish*, *Jacea*, &c. although *Green* or *White* in the greatest portion of their *Parenchymous Part*; yet where the *Succiferous* and *Aer-Vessels* run together, they are of *Red*, *Blue*, and other *Colours*; the *Oleous parts* of the *one*, being *unlocked* and *opened*, by the *aery* of the *other*.

68. §. AND LASTLY, of their *Tasts*. Most *Roots* which are *acres* or *bitting*, have a very copious *Parenchyma* in proportion with the *Succiferous Vessels*, as of *Arum*, *Dragon*, and others: Because the *Saline* and other *Principles* are not so much *hot*, by any sufficient quantity of *Sulphureous*, from those *Vessels*, in which the *Sulphur*, as is said, is more abundant; *(a)* but rendred rather *pungent*, from

How their *Tastes*.

(a) P. 2. §. 21.

some

some *Spirit* and *Aer*. But divers *Umbelliferous Roots*, especially which abound with *Lactiferous Vessels*, are *hot*; as *Fenil, Lovage, Angelica,* &c. Yet is it not their *Oyl* alone that makes them *hot*, but the combination thereof with the *Saline* Parts: as is manifest, from the nature of the *Seed* of these *Plants*; wherein, as the *Oyl* is most copious; So being held to a Candle till they burn, constantly *spit*; which cometh to pass, by the eruption of the *saline* Parts: and is the very same effect, with that which followeth upon burning of *Serum* or *Blood*. And therefore, as these *seeds* are more *hot*, they also *spit* the more; So those of *Cumine*, which, though fulsom, yet are not so *hot*, *spit* less; *Fenil* and *Dill*, which are *hotter*, more; there being a greater quantity of *volatile Salt* contained herein. Hence all *Essential Oyls* are *hot*, the *Spirit* and *volatile salt*, being incorporated herewith. And some of them will *shoot*, and crystallize as *Salts* do, as that of *Anise*; which argues a mixture of a considerable quantity of *volatile Salt*. As also doth the Nature of these *Oyls*, in being amicable to the *Stomach, Carminative*, and sometimes *Anodyne*; *scil.* as they kill some *fetid*, or *corrosive Acid*: for *volatile salts* themselves will have the like operation in some cases as these *Oyls*.

69. §. Many *Lactiferous Roots*, as *Taraxacum* and others of that kind, are not so much *hot*, as *bitter*. For although by the *Lactiferous Vessels* they are very *Oyly*; yet those *Vessels* being posited in *Rings*, and not in *Rays*, and having no *Diametral Portions* running through their *Barque* to the *Aer-Vessels*; the *Acido-Aerial Parts* do hereby, although not *mortifie*, yet so far *refract* the *saline*, lightly *binding up* the *Oleous* therewith, as to produce a *bitter Taste*. So, many *sweet* Bodies, upon burning, become *bitter*; the *Acid* Parts, now becoming *rampant*, and more copiously mixed with the *Oleous*.

ab. 13.

70. The *Roots*, or other Parts, of many *Umbelliferous Plants*, have a *sweetish Taste*, as both the *sweet*, and *Common Chervil*; both the *Garden*, and *wild Carrot*; *Parsnep, Fenil,* &c. the *Saline Principles* being *concentred* in the *Oyly*, and both of a moderate quantity with respect to the rest. For by the *Oyly*, the *Saline* is rendred more *smooth* and amicable; and both being moderate, they are not therefore *hot*, as in some other *Umbelliferous Roots*; but by the predominion of the other *Principles*, made *mild*. Hence it is, that *Sugar* it self is *sweet*, *scil.* because it is an *Oleous Salt*; as is manifest, from its being highly *inflammable*; its easie *dissolution* by a moderate, *Fire*, without the addition of *Water*; and in that, being melted with *Turpentine*, and other *Oily* Bodies, it will *mix* together with them. So also the *Acid Parts* of *Vinegar*, being *concentred* in the *Salino-sulphureous* of *Lead*, produce a *Sugar*. Hence *Barley*, which upon *Distillation* or *Decoction* yeildeth only an *acid*; being turned into *Mault*, becomes *sweet*. Because, being *steeped, couched*, and so *fermented*, the *oleous* parts are thereby *unlocked*, and becoming *rampant*, over the other *Principles*, altogether produce that *Taste*. And the *Bile* it self, which, next to *Water* and *Earth*, consisteth most of *oily* parts, and of many both *saline* and *acid* is a *bitter-sweet*. Wherein, as some of the *Saline* and *Acid* parts, smoothed by the *Oleous*, produce a *Sweet*: So, some of the *Oleous*, impregnated with the *Saline*, and the *Acid*, doe hereby produce a *Bitter*.

THE

THE
ANATOMY
OF
TRUNKS,

With an Account of their

VEGETATION.

Grounded thereupon.

The Figures hereunto belonging, Presented to the ROYAL SOCIETY in the Years, 1673 & 1674.

The THIRD BOOK.

By *NEHEMJAH GREW* M.D. Fellow of the *Royal Society*, and of the *College* of *Physicians*.

𝔗𝔥𝔢 𝔖𝔢𝔠𝔬𝔫𝔡 𝔈𝔡𝔦𝔱𝔦𝔬𝔫.

LONDON,

Printed by *W. Rawlins*, 1682.

T

TO THE

Right Honourable

WILLIAM

Lord Vi-Count *BROUNCKER*

THE

PRESIDENT;

AND TO THE

Council and Fellows

OF THE

ROYAL SOCIETY.

MY LORD,

THE *Commands* I received from *Your Lordship*, and the *Royal Society*, To profecute the *Subject* treated of in the Two former *Books*; have produced This which follows. And I humbly fubmit the fame to *Your Lordships* Judgment:

The Epistle Dedicatory.

ment: which must needs be Candid and Benign, because it is Great. I have only this to say,

———— Ἐς Τροίαν πειρώμενοι ἦνθον Ἀχαιοί;

Your Lordship will not disapprove the Enterprise, although it falls short of perfection. It being the result of *Your Lordships* manifold *Virtues* and *Abilities,* That *You* know how far to Encourage the meanest Attempts; as well as rightly to Value and Assist the greatest Performances.

I am,

My Lord,

Your Lordships

most humble

and

most obsequious

Servant

NEHEMJAH GREW.

London,
August 20.
1675.

THE CONTENTS.

The FIRST PART.

CHAP. I.

A Description of several Stalks or Trunks, as they appear to the Naked Eye.

OF the Stalk of Maze, §. 1, 2. Of Dandelion, 3, to 6. Of Borage, 7, to 10. Of Colewort, 11, to 16. Of Holyoak, 17, to 20. Of Wild Cucumer, 21, to 23. Of Scorzonera, 24, to 26. Of Burdock, 27, to 29. Of Endive, 30, 31. Of Vine, 32, to 35. Of Sumach, 36, to 38. Cautions to be had in observing the Parts, 39. Some Particulars better observed in cutting by the length, 40, 41.

CHAP. II.

Of the Barque, as it appears through a good Microscope.

First, a General Description of the several Parts of the Barque, 2, to 9. Next, a Particular Description of the Barques of 8 several Trunks; sc. Of Holly, Hazel, Barbery, Apple, Pear, Plum, Elm, Ash; The Vessels of all whose Barques are Lymphæducts: and those of two kinds, 10, to 13. Of 3 more, sc. Wallnut, Fig, and Pine: the Vessels of the Barques of the Two first, being Lymphæducts and Lactiferous, Of the next, Lymphæducts and Resiniferous, 14, to 20. Of 3 more, sc. Oak, Common Sumach, and Common Wormwood; the Vessels of whose Barques are of 3 Kinds, 21, to 29. Some further Observations and Conjectures of the Sap-Vessels, 30, to 37.

CHAP. III.

Of the Wood.

WHat in all Trunks, §. 1. A Description of its Parts, in the several Trunks aforesaid. Of the Parenchymous Part, or Insertions, 2, to 9. Of the true Wood, 10, to 15. Of the Aer-Vessels, 16, to 26. Some further Observations and Conjectures of their Form, 27. Texture, 28, to 32. Nature, 33, 34. And Original, 35.

CHAP.

CHAP. IV.

Of the Pith.

A *Description of the Pith, in General,* §. 1. *In the several Trunks or Brances aforesaid. As of the Size,* 2, 3. *Vessels,* 4. *Parenchyma and Bladders,* 5, *to* 9. *Apertures or Rupturss,* 10. *Some further Observations of the Pith. And of all the Pithy and Parenchymous Parts. And thence of the True Texture of a Plant,* 11, *to* 15.

The SECOND PART.

CHAP. I.

Of the Motion and Course of the Sap.

CHAP. II.

Of the Motion and Course of the Aer.

CHAP. III.

Of the Structure of the Parts.

CHAP. IV.

Of the Generation of Liquors.

CHAP. V.

Of the Figuration of Trunks.

CHAP. VI.

Of the Motions of Trunks.

CHAP. VII.

Of the Nature of Trunks, as variously fitted for Mechanical Use.

THE ANATOMY OF TRUNKS;

PROSECUTED

With the bare EYE,

And with the

MICROSCOPE.

PART I.

CHAP. I.

The Descriptions of several Trunks, *as they appear to the bare Eye.*

TO *the* end we may clearly understand, what the *Trunk*, *Stalk*, or *Branch* of a *Plant*, is; I shall by these *Figures* here before us, Describe the several *Parts*, whereof it is compounded.

1. §. And for examples sake, I shall in the first place, Describe the *Trunks* of some *Plants*, as being cut tranversly, and accurately observ'd, they appear to the naked Eye. And some others, as by the length. Which having done, I shall next proceed to a more particular Description of divers other *Trunks* and *Branches* as they appear through a good *Microscope*. In both shewing, not only what their several *Parts* are, as generally belonging to a *Branch*; but also, by a Comparative Prospect, in what respects they are *specifically* distinguished one from another, in the several *Sorts* of *Branches*.

2. §.

Tab. 18.

2. §. I SHALL begin where the Work of *Nature* appears less *Diversify'd*: as in the *Stalk* of *Maze* or *Indian Wheat*. In which, although there are the same *Parenchymous* and *Lignous Parts*, as in all other *Plants*; yet is there neither *Barque*, nor *Pith*; the *Vessels* being dispersed and mixed with the *Parenchyma*, from the Circumference to the Centre of the *stalk*: Saving, that in and next the *Skin*, there seems to be no *Aer-Vessels*. Every where else, they run up, like fine *Threds*, through the length of the *Stalk*: Each *Thred* being also surrounded with *Sap Vessels*; which in a Slice cut transversly, appear in very small and dark colour'd *Rings*. The like structure may also be seen in the *Sugar-Cane*, and some other *Plants*.

Tab. 18.

3. §. LET the next *Trunk* be that of *Taraxacum*, or *Dandelion*. In a *slice* whereof, being cut transversly, is seen next the *skin*, first, a simple, white, and close *Parenchyma* or *Barque*; made up of *Vesicles*; but such as are exceeding small; and hardly visible without a *Glass*.

4. §. Within This, stand *Milk-Vessels* in seven or eight distinct *Colums*, of different size: each *Colum* being also made up of seven or eight *Arched Lines*. Betwixt these *Colums*, run as many *Diametral Portions*, derived from the *Barque*, into or towards the *Pith*.

5. §. Next within These, stand the *Aer-Vessels*. Which are likewise divided, by the said *Diametral Portions*, into divers *Arched Lines*. The *size* of these *Vessels*, as well as their *number*, is small.

6. §. Within These, stands the *Pith*, consisting of very small *Vesicles* or *Bladders*, as the *Barque*. 'Tis very small, the *Diameter* hereof, being scarce one *fifth*, of that of the *Pith* of *Borage*. But the *Barque* of *Borage* is not half so thick as this of *Dandelion*.

7. §. FOR a Third *stalk*, we may take that of *Borage*; wherein there is some further Variety. For in a *slice* hereof, cut transversly, there appears, first a Tough, yet Thin and Transparent *Skin*. Within this *Skin*, and *Continuous* therewith, there is also a Thin *Ring* of *Sap-*

Tab. 18.

Vessels: which, without being crushed in the least, do yeild a *Lympha*.

8. §. Next standeth the *Parenchyma* of the *Barque*. Which is made up of a great number of very small *Vesicles* or *Bladders*. Upon the inner *Verge* of this *Parenchyma*, standeth another *Ring* of *Sap-Vessels*: which also yield a *Lympha*; and that different, as is probable, from the *Lympha* in the utmost *Ring*. Hitherto goes the *Barque*.

9. §. Adjacent to the *Ring* of *Sap-Vessels*, on the inner *Verge* of *Barque*, stand the *Aer-Vessels* on the outer *Verge* of the *Pith*. Not in a *Ring*; but in several *Parcels*; some *Parcels* or *Conjugations*, in the figure of little *Specks*; others, in little *Arched Lines*, almost like an V *Consonant*. And being viewed in a good *Glass*, there appears to be within the compass of every larger *Speck* or *Parcel*, about 20 or 30 *Aer-Vessels* and within the smallest, about 8 or 10.

10. §. The *Pith*, in a well grown *stalk* of this *Plant*; is always hollow. But originally, it is *entire*. It is likewise wholly made up of a great number of *Vesicles*: of which, through a *Glass*, some appear *Pentangular*, others *Sexangular*, and *Septangular*. Most of them are larger than those of the *Barque*; so as to be plainly visible to a naked Eye.

11. §. A FOURTH *Trunk*, shall be that of *Colewort*, which seems likewise, to have at least, two Sorts of *Lymphæducts*. For being cut transversly, as the former, we may observe, next the *Skin*, a

very

very close *Parenchyma*, of a darkish Green. Wherewith are mixed some few *Sap Veffels*, which give it that *Colour*.

12. §. Within This, ftands a *fcalloped Parenchymous Ring*, or a *Ring* of many fhort and flender white *Arches*. Which all round about the *Barque*, meeting together, run in fo many white *Diametral Portions*, or extream fmall *Rays*, into the *Pith*. *Tab.* 18.

13. §. Betwixt thefe white *Rays*, and next of all to the faid white *Arches*, ftand as many fmall *Parcels* of *Sap-Veffels*, like fo many little *Half-Ovals*. Within each of which, is included a white *Parenchyma*.

14. §. On the inner *Verge* of the *Barque*, ftands another Sort of *Sap-Veffels*, in one flender and entire *Ring*. And fo far goes the *Barque*.

15. §. Next within this *Ring* ftand the *Aer-Veffels*, in feveral *Parcels*, diametrically oppofite to the faid white *Parenchymous Parcels* next without the *Sap-Ring*.

16. §. Laft of all, and more within the *Pith*, ftand the fame kind of *Sap-Veffels*, as thofe of the *Half-Ovals*. Both thefe, by fmall lines, run one into another; thus, on both fides, hemming in the *Aer-Veffels*, and fo making altogether, fo many little *Pyramids*.

17. §. LET a *Fifth* be that of *Holyoake*. In which, the Curiofity of *Nature*, is ftill more copious: prefenting us, as it is feems, with Three forts of *Lymphæducts*; Of which, two yield a *Thin*; the Third, a *Thick Lympha*. For being cut, as before, next to the *Skin*, ftands the *Barque*; fomewhat clofe, and, in proportion, *Thick*.

18. §. Towards the inner *Verge* hereof, ftand one fort of *Sap-Veffels*, poftur'd in fhort *Rays*. Thefe *Veffels* yield a *Mucilage*. And on the inner *Verge* of the *Barque*, ftands a Thin *Ring* of other *Sap-Veffels*, which yield a *thinner Liquor*. *Tab.* 18.

19. §. Next within the *Barque* ftand the *Aer-Veffels*, poftur'd likewife in fhort *Rays*, diametrically oppofite to thofe in the *Barque*. In every *Ray*, there are about twelve or fixteen *Veffels*.

20. §. Laftly, and more within the *Pith*, there ftand other *Sap-Veffels*, all in very Thin or Slender *Arched-Lines*; thus hemming in the ral *Parcels* of *Aer-Veffels*.

21. §. FOR a Sixth, I will take that of Wild *Cucumer*: Wherein is alfo found a *Mucilaginous Lympha*. For firft of all, next to the *Skin*, there is a *Ring* of *Sap-Veffels*. Which *Ring* is alfo radiated, the *Rays*, all poynting towards, and moft of them terminating on, the *Skin*. *Tab.* 18.

22. §. Next of all, there is a thick, and fimple *Parenchymous Ring*. On the inner *Verge* whereof, there are other *Sap-Veffels* ftanding in *Parcels*, alfo in a *Ring*. So far goes the *Barque*.

23. §. Next within, ftand the *Aer-Veffels*, in as many *Parcels*, contiguous to thofe of the *Sap-Veffels* aforefaid. To which likewife are adjoyned as many more *Parcels* of *Sap-Veffels* within the *Pith*, oppofite to the faid *Sap-Veffels* within the *Barque*.

24. §. FOR a Seventh, we may, choofe that of *Scorzonera*. In which, the *Veffels* are both *Lymphæducts*, and *Lactiferous*. All of them, with the *Aer-Veffels*, in a radiated pofture. For firft next the outer Edge of the *Barque*, ftand the *Lactiferous*, in little *Specks*. Next to thefe, on the inner Edge of the *Barque*, ftand the *Lymphæducts*, in the fame form. *Tab.* 18.

25. §.

25. §. Hereunto adjacent, on the outer Edg of the *Pith*, stand the *Aer-Veſſels*, some in *Specks*, and some in extream short *Lines*; hardly distinguished, without a very nice Inspection.

26. §. Within Theſe, are placed other *Lymphæducts*, oppoſite to those in the *Barque*. And within these *Lymphæducts*, ſtill in the ſame radiated Line, run more of the *Milk-Veſſels*.

27. §. AN EIGHTH, may be that of *Burdock*; Wherein firſt, there are a Sort of *Lymphæducts*, which ſtand in *Arched Parcels*, round the *Trunk*, adjacent to the *Skin*.

Tab. 18.

28. §. Within these, about the middle of the *Barque*, run the *Milk-Veſſels*, in the form of ſmall round *Specks*.

29. §. Next to these on the inner Edg of the *Barque* are placed other *Lymphæducts*. Which, together with more of the ſame in the *Pith*, and the *Aer-Veſſels* betwixt them, ſtand all in Radiated Lines, of ſeveral Lengths, and all ſharpning towards the Centre.

30. §. LET the *Ninth*, be that of *Endive*: In which there is alſo much curious Work. Next to the *Skin*, there is, firſt, a thick and ſimple *Parenchyma*. Then there is a kind of *Undulated Ring* of *Milk-Veſſels*. Within which ſtand a Sort of *Lymphæducts*, in ſeveral *Parcels*; some, in *Arched Half-Ovals*; others, in ſhort ſlender *Rays*. Betwixt these *Parcels*, many of the *Milk-Veſſels* likewiſe ſtand.

Tab. 18.

31. §. Next there is an *undulated Ring* of other *Lymphæducts*, parting as in moſt *Trunks*, betwixt the *Barque* and the *Pith*. Within which, are the *Aer-Veſſels*. And within Theſe, more *Sap-Veſſels*. Both of them, in ſmall *Specks*, anſwerable, or oppoſite to the *Rays* in the *Barque*.

32. §. I SHALL give alſo one or two Examples of *Trees*, or *Arborescent Plants*; the *Vine* and *Common Sumach*. In a Slice of the former cut tranſverſly, next the *Skin*, there is a Thin *Barque*. In the inner part whereof, adjacent to the *Wood*, ſtand the *Lymphæducts* in ſeveral *Half-Oval Parcels*, oppoſite to ſo many Radiated Pieces of the *Wood*.

Tab. 18.

33. §. The *Wood* is divided into the ſaid Pieces, by as many *Parenchymous Rays*, inſerted from the *Barque*, and ſo continuous therewith.

34. §. Within these Radiated Pieces of *Wood*, ſtand the *Aer-Veſſels*; the largeſt of which, eſpecially if held up againſt the light, are plainly viſible to the bare Eye.

35. §. Within the hollow of the *Wood*, ſtands the *Pith*; in the young Growths always large. In the utmoſt *Verge* whereof, adjacent to the *Wood*, ſtand a few more *Sap-Veſſels* of the ſame Sort with thoſe in the *Barque*.

36. §. IN A like Slice of *Common Sumach*, contiguous to the hairy *Skin*, there is a *Ring* of *Lymphæducts*. Next to this a Simple *Parenchyma*. Then ſeveral *Arched Parcels* of *Lymphæducts*. Within theſe, a *Ring* of *Milk-Veſſels*. And then a *Ring* of other *Lymphæducts*. Thus far the *Barque*.

Tab. 18.

37. §. Within the *Barque*, ſtands the *Wood*, divided into ſeveral *Portions*, by the Diametral *Inſertions* divided from the *Barque*. In the Body of the *Wood*, ſtand the *Aer-Veſſels*, very much ſmaller than in the *Vine*.

38. §.

38. §. The hollow of the *Wood* is filled up with the *Pith*. In the Circumference of which, stands a *Ring* of *Lymphædctus*, of the same sort with those next to the *Wood* without.

39. §. All the *Parts* of these *Trunks*, may, as I have now described them, be observed without a *Microscope*: excepting the *Bladders* and number of *Aer-Vessels*. Yet Three things are hereunto necessary; *viz.* a good *Eye*, a clear *Light*, and a *Rasor*, or very keen *Knife*, wherewith to cut them with a smooth surface, and so, as not to Dislocate the *Parts*.

40. §. UPON *Inspection* also by the length, there are some particulars, common, more or less, to most *Plants*, yet better observable in some, than in others. As first, the *Reticulation* of the *Vessels*, (formerly described) not only in the *Wood*, but in the *Barque*: which is evident in a young *Branch* of *Corin*, upon the very Surface thereof, when some of the *Vessels* begin to be cast off into the *Skin*. And so, by stripping off the *Skin*, upon the Surface of the *Wood*. *Tab.* 19.

41. §. In cutting by the length, as well as transversly, the young *Fibres*, which grow within the *Wood* in the Edg of the *Pith*, are also seen. As likewise the manner of the Derivation of the *Parts* of the *Bud* from the *Branch* or *Stalk*; as in *Sonchus*. There are also many Varieties in the *Pith*, such as those hereafter mentioned (*a*) which fall under observation only in cutting by the length. (*a*) *Chap.* 4.

CHAP. II.

Of the Barque, *as it appears through a good Microscope.*

NOW proceed to a more particular *Description* of several *Trunks* and *Branches*, as they appear through good *Glasses*.

1. §. Now the *Trunk*, or *Branch* of every *Tree*, hath Three General *Parts* to be described; *sc.* the *Barque*, the *Wood*, and the *Pith*. That likewise of every *Herbaceous Plant*, hath either the same Three *Parts*; or else Three *Parts* Analogous; *sc.* the *Cortical*, the *Lignous*, and the *Pithy Parts*.

2. §. The *Barque* consisteth of two *Parts*, *sc.* the outmost or *Skin*, and the *Main Body*. The *Skin* is generally composed, in part, of very small *Vesicles* or *Bladders*, cluster'd together. That is, originally it is so; but as the *Plant* grows, the *Skin* dries, and the said *Bladders*, do very much shrink up and disappear.

3. §. Amongst these *Bladders* of the *Skin*, there are usually intermixed a sort of *Lignous Fibres*, or *Vessels*, which run through the length of the *Skin*; as in *Mallow*, *Nettle*, *Borage*, *Thistle*, and most *Herbs*. Which is argued not only from the Toughness of the *Skin* by means of the said *Vessels*; but in some *Plants*, may be plainly seen, as in *Teasle*. In which, *Tab.* 20.

the several *Fibres*, which run by the length of the *Stalk*, are also conjoyned by other smaller ones, which stand transversly.

4. §. Whether they are *Aer-Vessels*, or *Sap-Vessels*, is dubious. For, on the one hand, because they emit no *Sap*, or *bleed* not, and also stand adjacent to the *Aer*; 'tis probable that they are *Aer-Vessels*. On the other hand, they may be *Sap-Vessels*; notwithstanding that they *bleed* not: Because the non-emission of *Sap* is not an infallible and concluding argument of an *Aer-Vessel*. For there are some *Plants* which *bleed* not. Which yet are furnished with *Sap-Vessels*, as certainly as any others which *bleed*. (*a*)

(*a*) B. 2. P. I. c. 3. §. 22.

5. §. The *Skin* of the *Trunk* is sometimes visibly porous. But no where more, than in the better sort of walking *Canes*; where the *Pores* are so big, as to be visible even to the naked Eye: like to those, which are observable in several parts of the Ball of the *Hand*, and upon the ends of the *Fingers* and *Toes*.

Tab. 20.

6. §. THE *Main Body* of the *Barque* consisteth likewise of two *Parts*, sc. *Parenchyma*, and *Vessels*. The *Parenchyma* is made up of an innumerable company of small *Bladders* cluster'd together. Differing in nothing from those aforesaid in the *Skin*; saving, that they are much larger; and generally rounder.

Tab. 22. & Sequent.

7. §. This *Parenchyma* of the *Barque* is the same, as to its *Substance*, both in the *Root* and *Trunk*. Yet as to the *Texture* of its *Parts*, in the *one*, and in the *other*, there is This observable difference, *viz.* That in the *Barque* of the *Root*, cut transversly, the said *Parenchyma* (as hath been shew'd) is usually, more or less, disposed into *Diametral Rays*; running through the *Barque*, after the same manner, as do the the *Hour-Lines* through the Margin of the *Dial-plate* of a *Clock* or *Watch*: as in *Marsh-Mallow*, *Lovage*, *Melilot*, and others. Whereas here in the *Barque* of the *Trunk*, the said *Parenchyma* is rarely thus disposed into *Dimetral Rays*: Nor when it is, are those *Rays* continued to the Circumference of the *Barque*; as in the *Barque* of the *Root* they frequently are. So in *Rhus* or *Sumach*, although part of the *Parenchyma* be dispos'd into *Diametral Rays*: yet are those *Rays* extended not half way through the *Barque*. So also in *Fig-tree*, *Worm-wood*, *Thistle*, and others. What is further observable in the *Texture* of the *Parenchyma*, I shall shew in the description of the *Pith*.

Tab. 7, 8, 9.

Tab. 22. & Seq.

Tab. 31, 34, &c.

8. §. THE *Vessels* of the *Barque*, are, as I shall also shew, diversifyed many ways. But there are some Things, wherein, in all Sorts of *Plants*, they agree. *First*, in standing, most numerously, in or near, the inner Margin of the *Barque*. *Secondly*, in being always, and only *Sap-Vessels*. I have viewed so many, that at least, I can securely affirm thus much, That if there be any Heteroclital *Plants*, wherein they are found otherwise, there is not One, in Five Hundred. *Thirdly*, in being always *Conjugated* or *Braced* together in the form of *Net-work*. Although the Number and Distances of the *Braces*, are very different: as I have already shewed in the *Anatomy* of *Roots*.

Chap. 3. Tab. 6.

9. §. THE Properties, whereby the said *Vessels* of the *Barque* are specificated and distinguished one from another, both in the same *Plant*, and in the several Species of *Plants*, are very many. Which Properties, are not Accidental, but such as shew the Constant and Universal Design of Nature. All which shall be demonstrated by the *Description* of several *Quarters* of the *Slices*, of so many Kinds of

Branches

Branches, cut Transversly: and by the several *Figures* which represent them.

10. §. FIRST then, for the *Eleven* first *Quarters*, the *Vessels* of the *Barque* are only of *Two Kinds*. And these, in the first *Eight*, seem to be *Roriferous* (described also in the *Anatomy* of *Roots*) (a) and those which are common to most, if not to all *Plants*, sc. the *Lymphæducts*. Yet in all the *Eight*, they are, in respect both of their *Proportion*, and *Position*, very different. So in *Hazel* and *Ash* they are but few. In *Holly* and *Barberry* more. In *Apple*, *Pear*, *Plum*, *Elm*, still more numerous. And of those three *Fruits*, in an *Apple*, or *Plum*, more than in a *Pear*. *Tab.* 22. &c. seq. (a) P. 1: c. 3. §. 22.

11. §. Again, as their *Proportion*, so likewise their *Position* is divers. For in *Holly*, the *Lymphæducts* or inner *Vessels* next to the *Wood*, stand in *Rays*. Yet so numerous and close together, as to make one Entire *Ring*. In *Hazel*, they stand more in Oblong *Parcels*. In *Barberry*, they stand likewise in *Parcels*, but they are so many *Half-Ovals*. The utmost *Vessels* or *Roriferous* of all Three, make a *Ring*. *Tab.* 22, 23, 24.

12. §. Again, in *Apple*, *Pear*, and *Plum*, the *Lymphæducts* are *Radiated*. The *Roriferous* are neither *Radiated*, nor make an *entire Ring*; but stand in *Peripherial Parcels*. Much after the same manner, they also stand in *Elm*. In *Ash*, the *Vessels* make **Two** *Rings*; but neither of them *Radiated*: the inmost *Ring* or *Lymphæducts*, consisting of *Arched Parcels*, and the utmost or *Roriferous Vessels*, of *Round* ones. And whereas in all the foregoing, the *Lymphæducts* are still contiguous to the *Wood*; and the *Roriferous* more or less, distant from the *Skin*: here, on the contrary, the former are distant from the *Wood*, and the latter contiguous to the *Skin*. *Tab.* 25, to 28. *Tab.* 29.

13. §. And that these *Vessels* in each *Barque* of the said Eight *Branches*, are of **Two** distinct Kinds, seems evident, as from some other reasons, so from hence; In that their *Positions* are altogether Heterogeneous: Yet in both Constant, Regular and Uniform. I say, there seems to be no Reason, why the self *same Kind* or Species of *Vessels*, should have a different, yea a contrary *Position* in one and the same *Plant*; and that Contrariety, not *Accidental*, but *Regular* and *Constant*.

14. §. FOR the Three next *Quarters* sc. the *Ninth*, *Tenth* and *Eleventh*, the *Vessels* of the *Barque* are also different in *Number*, *Position*, *Size* and *Kind*. In *Pine*, which is the *Eleventh*, they are fewer. In *Walnut* the *Ninth*, more. In *Fig*, the *Tenth*, most numerous. *Tab:* 30, 31, 32.

15. §. So for their *Position*. In *Pine*, the inmost make a *Radiated Ring*. The utmost stand stragling up and down, without any certain order. In *Wallnut* the inmost make also a *Radiated Ring*; The utmost make a Double *Ring*; not *Radiated*, but of *Round Parcels*. In *Fig*, the inmost make also a *Radiated Ring*. But the utmost make a *Double* and sometimes *Treble Ring*, not of *Radiated*, nor *Round*, but *Arched Parcels*.

16. §. Thirdly, they are also different in *Kind*. Those, I think, of the two former, *Wallnut* and *Fig*, are thus different: those certainly, of the *Fig*, are so; being *Lymphæducts* and *Lacteals*. The *Lymphæducts* make the inmost *Radiated Ring*. The outmost which make the other *Rings* in *Arched Parcels*, are the *Lactifers*.

17. §. That they are distinct *Kinds* of *Vessels*, is evident for two Reasons. *First*, from their *Position* in the *Barque*; which is altogether

ther different, as hath been said. *Secondly*, from the most apparent Diversity of the *Liquors* or *Saps*, which they contain, and which, upon cutting the *Branch* transversly, do distinctly *Bleed* from them. Which is one way, whereby we do distinguish the *Vessels* of *Animals* themselves. As in the *Liver*, it were hard to say, which is a *Blood-Vessel*, and which is a *Bile-Vessel*, where they are very small, if it were not for the *Contents* of them both.

18. §. Those in the *Barque* of *Pine*, are likewise of Two *Kinds*. The inmost are *Lymphæducts*, as in the two former. The utmost are not *Milk-Vessels*, but *Gum-Vessels*, or *Resiniferous*; which stand stragling, and singly, about the midle of the *Barque*. Out of these *Vessels* all the clear *Turpentine*, that drops from the *Tree*, doth issue.

19. §. Few, but very great. So that besides the difference of their *Number* and *Position*, and of the *Liquors* which they contain, and *Bleed*; there is yet a Fourth, and that is, their *Size*. Most of these *Turpentine Vessels*, being of so wide a *bore*, as to be apparent to the naked Eye: and, through a good *Glass*, above $\frac{1}{3}$d of an Inch in Diametre. Whereas that of the *Lymphæducts*, can hardly be discovered by the best *Microscope*.

20. §. The same *Turpentine-Vessels* of *Pine*, are likewise remarkably bigger, not only than the *Lymphæducts*, but many times, than the *Milk-Vessels* themselves: as those of the *Fig*, which, in comparison, are exceeding small; every *Arch*, not being a *single Vessel*, but a *Parcel* or *Cluster* of *Vessels*; Whereas one single *Gum-Vessel* in *Pine*, is sometimes as big as two whole *Arched Clusters*, that is, as some Scores of the *Milk Vessels* in a *Fig-tree*. And the said *Gum-Vessels* of *Pine*, being compared with the *Lymphæducts* of the same Tree, one *Gum-Vessel*, by a moderate estimate, may be reckoned *three* or *four hundred* times *wider* than a *Lymphæduct*. The like prodigious difference may be observed in the *Size* of the several *Kinds* of *Vessels* of many other *Plants*.

21. §. THE *Three* next *Quarters* of *Branches*, are of *Oak*, *Common Sumach*, and *Common Wormwood*. In the *Barque-Vessels* whereof, there is observable some father Variety. For in all or in most of the above named, there are only Two *Kinds* of *Vessels* in the *Barque*. But in Each of these, there are, at least, Three *Kinds*.

Tab. 33.
22. §. And first, in that of *Oak* there are *Lymphæducts, Roriferous*, and a Sort of *Resiniferous*. The inmost or *Lymphæducts*, make a *Radiated Ring*, contiguous to the *Wood*. The utmost or the *Roriferous* make also a *Ring*, but not *Radiated*. Those which are a sort of *Rosin-Vessels*, stand in Round *Parcels*; the greater *Parcels* betwixt the Two *Rings* of *Roriferous* and *Lymphæducts*; and the lesser, betwixt the *Roriferous* and the *Skin*.

23. §. That these last are different *Vessels* from both the other, seems evident, from the difference of their *Position*, as aforesaid. And that they are a sort of *Resiniferous*, is argued from hence; In that, not only *Galls* are very full of *Rosin*, but that the *Barque* of *Oak* it self is also somewhat *Resinous*. For the conveyance of whose *Resinous* parts, it is most unlikely that any other *Vessels* should subserve, but a peculiar *Kind*; which may therefore be properly called *Resiniferous*.

Tab. 34.
24. §. The next is a *Branch* of *Common Sumach*. In the *Barque* whereof, there are likewise Three *Kinds* of *Vessels*. First of all, there is a thick *Radiated Ring* of *Lymphæducts*; standing on the inner *Margin*

gin of the *Barque*, contiguous with the *Wood*. These *Vessels* exhibit their *Lympha* very apparently. A second kind of *Vessels*, sc. *Roriferous*, are situate towards the outer *Margin* of the *Barque*, and are composed into distinct *Arched Parcels*, all standing in a *Ring*.

25. §. Betwixt these Two *Kinds* stand the *Milk-Vessels*. Every single *Milk-Vessel* being *empaled* or hemmed in with an *Arch* of *Roriferous*. The *Milk-Vessels* are extraordinary large, almost as the *Gum-Vessels* of *Pine*; so as distinctly to be observed without a *Microscope*; after they are evacuated of their *Milk*; and without difficulty will admit a *Virginal Wyer*; being two or three hundred times as big as a *Lymphæduct*. Besides these Three sorts of *Vessels*, there is also a *Ring*, adjacent to the *Skin*; which seems to be another sort of *Roriferous*.

26. §. The Last, is a *Branch* of *Common Wormwood*. In the *Barque* whereof, there are likewise Three *Kinds* of *Vessels*. First of all, there is a thin *Radiated Ring* of *Lymphæducts*, contiguous with the *Wood* or on the inner *Margin* of the *Barque*. Yet the *Ring* is not entire, but made up of several *Parcels*; which are intercepted by as many *Parenchymous* inserted into the *Pith*. *Tab. 35.*

27. §. A Second Sort of *Vessels*, which seem to be *Roriferous*, are situate about the middle of the *Barque*: and are composed into *Arched Parcels*, which likewise stand all even in a *Ring*.

28. §. Beyond these *Arches*, and towards the outer *Margin* of the *Barque*, stand a Third Sort of *Vessels*. Different from the *Milk-Vessels* in *Sumach*, both as to their *Situation*, *Size* and *Content*. For in *Sumach*, the *Milk-Vessels* stand within the *Arched Lymphæducts*: whereas these in *Wormwood*, stand without them. Likewise, being the *Vessels* of an *Herb*, they are far less; sc. about the compass or width of a small *Wheat-straw*. Their *Content*, is not a *Milk*, but a *liquid*, most *Oleous* and *viscid Gum*. Or which, for its pleasant *Flavour* may be called an *Aromatick Balsom*. For it perfectly giveth whatever is in the *Smell* and *Taste* of *Wormwood*: being the *Essence* of the whole *Plant*, which nature treasureth up in these *Vessels*. So that they are, in all respects, analogous to the *Turpentine Vessels* in *Pine*. There are divers other *Herbs* and *Trees*, which in the like *Vessels*, contain a *Turpentine*, or rather *Aromatical Balsom*; as *Angelica*, *Helenium* and others; the *Vessels* being so very large, that they may be easily traced with a knife, in cutting by the length of a *Branch* or *Staltk*.

29. §. Whether in some *Plants*, there are not more *Sorts* of *Vessels*, in the *Barque*, than have been now mentioned, I cannot say: Though we have not much reason to doubt of it. Because we see, there is so great variety in the *Viscera* of *Animals*. For what the *Viscera* are in *Animals*; the *Vessels* themselves are in *Plants*.

30. §. CONCERNING the *Form* and *Texture* of the *Lymphæducts*, there are some things, which though they are best observed in the *Wood*, yet in regard I am now describing the said *Vessels*, I shall here therefore add. I have already said, and shewed, in the former *Books*, That the *Lignous* and *Towy Parts* of all *Plants*, are *Tubulary*. And that the *Lympha* is conveyed, by the length of a *Plant*, through an innumerable company of small *Tubes* or *Pipes*.

31. §. The *Question* may be yet further put: If the *Towy Parts* of the *Barque* are made of *Tubes*, What are these *Tubes* themselves made up of? *I answer*, That these *Tubes* or *Lymphæducts*, are not only

themselves

Tab. 40.

themselves *Organical*; but their very *Sides* also, seem to be composed of other *Parts*, which are *Organical*, *sc.* of *Lignous* or *Towy Fibres*. Which *Fibres*, standing close or contiguous in a round *Figure*, they make one *Tubulary Body*, which I call the *Lymphæduct* of a *Plant*. And it is probable, That these *Fibres* themselves, are also *Tubulary*. That is, that a *Lymphæduct*, is a small *Tube*, made up or composed of other, yet much smaller *Tubes*, set round together in a *Cylindrick Figure*. As if we should imagine a company of *Straws*, which are so many small *Pipes*, to be joyned and set round together, so as to make another greater *Pipe*, answerable to a *hollow Cane*. The *Cane*, I say, is as the *Lymphæduct*; and the *Straws* are as the *Fibres* whereof it is composed. By which also appears, the admirable smallness of these *Fibres*. For there are some *Lymphæducts*, which may be reckoned fifty times smaller than a *Horse-Hair*. Allowing therefore but Twenty of the aforesaid *Fibres* to make a *Thred* so big as one *Lymphæduct*; then one of the said *Fibres*, must be a Thousand times smaller than a *Horse-Hair*. That these *Fibres*, whereof the *Lymphæducts* are made, are themselves made up of other *Fibres*, is not altogether improbable.

32. §. These *Fibres*, although parallel; yet are they not *coalescent*, but only contiguous; being contained together in a *Tubulary Figure*, by the Westage of the *Cortical Fibres*, as in *Chapter* the *Fourth* will better be understood.

33. §. The first notice I took of the *Composition* and *Texture* of these *Vessels*, so far as the best *Glasses* yet known, will admit; was in a very *white* and *clear* piece of *Ash-wood* torn, with some care, by the length of the *Tree*, and objected to a proper *Light*. They seem also sometimes discernable in some other *clear Woods*, as in very *white Fir*, &c. And having formerly demonstrated, that the *Lignous Part* of a *Plant*, is annually made or augmented out of the inner part of the *Barque*, wherein the *Lymphæducts* always stand: we may reasonably suppose the same *Lymphæducts* to have the like Conformation in the *Barque*, as in the *Wood*.

34. §. And I am the rather induced to believe, that I am not mistaken in this Description, upon these two Considerations. *First*, that herein the Analogy betwixt the *Vessels* of an *Animal* and a *Plant*, is the more clear and proper. For as the *Sanguineous Vessels* in an *Animal* are composed of a number of *Fibres*, set round, in a *Tubulary Figure*, together: so are these *Lymphæducts* of a *Plant*. *Secondly*, in that herein, there is a more genuine respondence betwixt these, and the other *Vessels* of a *Plant* it self; *sc.* the *Aer-Vessels*; which are made up of a certain number of *Round Fibres*, standing collaterally, or side to side as I have already observed in the *Anatomy* of *Roots*. So that it is the less strange, that the *Lymphæducts* should be made up of *Fibres*, since the *Aer-Vessels* are evidently so made. Only with this difference, that whereas in the *Aer-Vessels*, the *Fibres* are postured or continued *Spirally*: here, in the *Lymphæducts*, they stand and are continued only in *straight Lines*.

Tab. 20.

35. §. THE STRUCTURE of the *Lactiferous* and *Gum-Vessels*, which have a very ample *Bore*, is more apparent. And, by the best *Glasses* I have yet used, they seem to be made, chiefly, by the Constipation of the *Bladders* of the *Barque*. That is to say, That they are so many

Chanels,

Book III.   of Trunks.   113

Chanels, not made or bounded by any walls or sides proper to themselves, as a *Quil* thrust into a *Cork*, and as the *Aer-Vessels* are in the *Wood*: but only by the *Bladders* of the *Parenchyma*; which are so postured and crouded up together, as to leave certain *Cilyndrick Spaces*, which are continued by the length of the *Barque*.

36. §. One difference betwixt the *Vessels* or *Chanels* now describ'd, and the Tubulary *Hollows* and other *Apertures* in the *Pith*, is this; That these never exist originally with the *Pith*; but are so many *Ruptures* supervening to it in its Growth. Caused, partly, by the Stretch or Tenter it suffers from the Dilatation of the *Wood*: (*a*) and partly, the drying, and so the Shrinking up of its *Bladders*, and of the *Fibres* whereof they are composed. Whereas the said *Vessels* in the *Barque*, are many of them originally formed therewith. And those which are *post-nate*, not made by any *Rupture*, but only such a Disposition of the *Parenchymous Fibres*, and Constipation of the *Bladders*, as is thereunto convenient.

(*a*) B. I. c. 3. §. 22, & 23.

37. §. In paring the *Barque* of a *Branch* of Pine, Sumach, &c. they appear, neither *parallel*, nor any where *Inosculated*: but run, with some little obliquities, distinct one from another, through the length of the *Branch*: and so, we may believe, through the length of the Tree.

CHAP. III.

Of the WOOD.

THE next general *Part* of a *Branch*, is the *Wood*; which lyeth betwixt the *Barque* and the *Pith*. And this likewise evermore consisteth of Two General *Parts*, *sc.* of a *Parenchymous Part*, and that more properly called *Lignous*. The *Parenchymous Part* of the *Wood*, though much diversifyed, yet in the *Trunks* of all *Trees* whatsoever, hath this property, To be disposed into many *Rays*, or *Diametral Insertions*, running betwixt so many *Lignous Portions*, from the *Barque* to the *Pith*: as in any of the *Quarters* here before us may appear.

2. §. But these *Insertions* are much diversifyed, according to the several *Sorts* of *Plants*. So in *Barberry, Ash, Pine, Worm-wood,* they are less numerous. In *Elm, Wallnut, Fig, Sumach,* they are more. And in *Holly, Pear, Plum, Apple, Oak, Hazel,* are most numerous.

Tab. 22, to 35.

3. §. The same *Insertions*, in *Barberry, Wormwood,* and some in *Oak,* are very Thick. In *Pine, Fig, Ash,* of a middle *Size*. In *Pear, Holly,* and most of them in *Oak,* are exceeding Small. Again, in *Barberry, Elm, Ash, Sumach, Fig,* they are of an Equal *Size*. In *Holly, Hazel, Pear, Plum, Oak,* they are very Unequal: some of those in *Holly,* being *Four* or *Five* times thicker than the rest; in *Plum, Six* or *Seven* times; and in *Oak, Ten* times at least.

Ibid.

X   4. §.

4. §. In some *Plants*, they are Equidistant; in others, not: in some, the Great ones are Equidistant; in others, the Lesser; in others, both; in some, neither. Which *Varieties* are not accidental; but constant to the *Species* in which they are severally found.

5. §. They are not always visibly continued from the Circumference to the Centre of the *Wood*: but in some *Branches*, as of *Sumach*; and in most *Trunks* of many years growth, declining, in some places, under or over, from a Level, are thereby, upon a Transverse Section, in part cut away.

Tab. 34.

6. §. They have yet one more Diversity, which is, That in divers of the aforesaid *Branches*, they run not only through the *Wood*; but also shoot out beyond it, into some *Part* of the *Barque*, as in *Elm*, *Sumach*, *Wormwood*, &c. Whereas in *Pine*, and some of the rest they either keep not distinct from the other parts of the *Parenchyma* of the *Barque*; or are so small, as not to be distinguished there form.

Tab. 28, 34, 35. Tab. 32.

7. §. The *Texture* likewise of these *Insertions* is somewhat various. For in *Wormwood*, and most *Herbs*, they are manifestly composed of small *Bladders*: differing in nothing from those of the *Barque* or *Pith*, saving, in their being much less. Yet in *Herbs*, they are much larger than they are in *Trees*. And in many *Trees*, as *Apple*, *Pear*, *Plum*, *Pine*, &c. they are either quite lost, or so squeezed and pressed together by the hard *Wood* standing on both sides, as to be almost undiscernable.

Tab. 35.

Tab. 36, 37.

8. §. So that although the *Parenchyma* of the *Barque* or *Pith*, and the *Insertions* in the *Wood*, are of the same *Specifick* Nature or Substance: yet there is this difference betwixt them; That the *Fibres* of the former, are so Netted together, as to leave several round Vacuities; or to make a great many little *Bladders*, whereas, in the latter, they are usually so far crowded up, as to run (as when a Net is stretched out) like a *Skein* of *Parallel Threds*.

9. §. Of these *Insertions* in the *Wood*, it is further observable, That they do not only run betwixt the *Lignous Portions*; but that many of their *Fibres* are likewise all along distributed to the several *Fibres*, of which the *Lignous Portions* consist, and are interwoven with them; both together thus making a piece of *Linsy-Woolsy Work*, or like many other *Manufactures* in which the *Warp* and the *Woof* are of different Sorts of *Stuff*: as in the end of the *Fourth Chapter* is further explained.

Tab. 40.

10. §. THE WOOD is likewise compounded of Two Sorts of Bodies; That which is strictly *Woody*; and the *Aer-Vessels* mixed herewith. The true *Wood* is nothing else but a mass of antiquated *Lymphæducts*, viz. those which were originally placed on the inner Margin of the *Barque*. For in that place, there grows, every year, a new *Ring* of *Lymphæducts*. Which losing its original softness by degrees, at the latter end of the year, is turned into a dry and hard *Ring* of perfect *Wood*.

11. §. So that every year, the *Barque* of a *Tree* is divided into Two Parts, and distributed two *contrary* ways. The outer Part falleth off towards the *Skin*; and at length becomes the *Skin* it self. In like manner, as hath been observed of the *Skin* of the *Root*. Or as the *Cuticula* in *Animals*, is but the efflorescence of the *Cutis*. I say, that the elder *Skin* of a *Tree*, is not originally made a *Skin*; but was once, some of the *midle* part of the *Barque* it self, which is annually cast off, and

and dryed into a *Skin:* even as the very *Skin* of an *Adder,* upon the gradual generation of a new one underneath, in time, becomes a *Slough.* The inmost portion of the *Barque*, is annually distributed and added to the *Wood* : the *Parenchymous Part* thereof making a new addition to the *Insertions* within the *Wood* ; and the *Lymphæducts* a new addition to the *Lignous pieces* betwixt which the *Insertions* stand. So that a *Ring* of *Lymphæducts* in the *Barque* this year, will be a *Ring* of *Wood* the next ; and so another *Ring* of *Lymphæducts*, and of *Wood,* successively, from year to year. So the *Table,* for an *Apple-Branch,* sheweth a quarter of a *Slice* of a *Branch* cut transversly, of Three years growth : That of *Barbery,* of Two ; That of *Sumach,* of One only ; That of *Elm,* of Five. *Tab.* 25. 24, 34, 28.

12. §. Hereby two things may be the better noted. First, the difference betwixt the degrees of the *annual* growths of several *Trees* : three years growth in an *Oak,* being as thick as five in an *Elm.* Secondly, the difference betwixt the *Annual* growths of the *same Tree* ; being not of a constant proportion, but varying in thickness, as it should seem, according to the *season* of the year : whereby it may appear, what *season,* or kind of year, doth most of all favour, the latitudinal growth, or the *thickening* of any *Tree.* *Tab.* 33, 28.

13. §. The *Lymphæducts* thus antiquated or turned into *Wood,* do rarely, if ever, *Bleed :* but only transmit a kind of *Dewy* or *Vaporous Sap.* And some of them, as in the *Heart* of some *Trees,* it is probable, That they transmit not any *Sap,* either in the form of a *Liquor,* or a *Vapour :* and so being gradually deprived of their *Watery Parts,* become the *Heart.*

14. §. There is this further variety in the *Wood* ; represented in *Walnut, Fig* and *Oak.* That some certain parcels hereof, make either several small and white *Rings,* as in *Oak* ; or else divers white and crooked *Parcels,* transverse to the *Insertions,* as in *Wallnut* and *Fig.* For it seemeth, that, at least, in many *Trees,* some portion of all the *Kinds* of *Vessels* in the *Barque,* are not only annually distributed to the *Wood,* but do likewise therein retain the same, or somewhat like *Position,* which they originally had in the *Barque.* So that as all those bigger and darker Portions of the *Wood,* were originally, the *Radiated Lymphæducts* of the *Barque* : so the little white *Circles,* or *Parcels* of *Circles,* in the same *Wood,* were originally another sort of *Sap-Vessels* in the Barque, *sc.* those which have a circular *Position* therein. *Tab.* 30, 31, 33.

15. §. In the *Branches* of *Fir, Pine,* and others of the same *Kindred,* there are some few *Turpentine-Vessels* scattered up and down the *Wood* ; and represented by the larger Black Spots. Which *Vessels* are *eadem numero,* the self same, which did once appertain to the *Barque* ; and do even here also in the *Wood,* contain and yield a liquid *Turpentine.* Only, being pinched up by the *Wood,* they are become much smaller *Pipes.* *Tab.* 32.

16. §. THE *Aer-Vessels,* with the *Insertions,* and true *Wood,* altogether make up That, which is commonly called, The *Wood* of a *Tree.* The *Aer-Vessels* I so call, not in that they never contain any *Liquor* ; but, because all the principal time of the growth of a *Plant,* when the *Vessels* of the *Barque* are filled with *Liquor,* these are filled only with a *Vegetable Aer.*

17. §. In almost all *Plants,* not one in some hundreds excepted, this is proper to the *Aer-Vessels* ; To have a much more ample *Bore* or *Cavity,*

vity, than any other in the *Wood*. In the *Wood*, I say; for in the *Barque*, there are many *Sap-Vessels* bigger than the biggest *Aer-Vessels* that be.

18. §. The Varities hereof are very many; in respect both of their *Number*, *Size*, and *Position*; being, as to these, the same, in no two *Sorts* of *Plants* whatsoever. First in respect of their *Number*. So in *Hazel*, *Apple*, *Pear*, they are very numerous; but in different degrees: and are represented in the *Figures* already referred to, by all the black spots in the *Wood*. In *Holly*, *Plum*, *Barberry* somewhat numerous. In *Oak*, *Ash*, *Walnut* fewer. In *Pine*, and others of that *Kindred*, very few; *sc.* fewer than in any other kind of *Plant*.

Tab. 23, 25, 26. 22, 24, 27. 29, 30, 33. 32.

19. §. Secondly, in respect of their *Size*; which from the first or greatest, to the least, may be computed easily to about Twenty Degrees. Thus, many of those in *Elm*, *Ash*, *Wallnut*, *Fig*, *Oak*, are very large. In *Barberry*, *Plum*, not so large. In *Hazel*, *Sumach*, smaller. In *Holly*, *Pear*, of a still smaller *Size*. So that many of those in *Elm*, or *Oak*, are Twenty times bigger, than those in *Holly* or *Pear*.

Tab. 28, 33. 22, 26.

20. §. In an ordinary joynted *Cane*, they are so wide, that if you take one a yard, or a yard and ½ long, and putting one end into a Basin of Water, you blow strongly at the other; your Breath will immediately pass, through the *Aer-Vessels*, the length of the *Cane*, so as to raise up the Water into a great many Bubbles.

21. §. And as they have a different Size in divers Kinds of *Plants*; so likewise, according to the place where they stand, in the self same. So in *Holly*, *Hazel*, *Apple*, their *Size* is more equal throughout the bredth of the *Tree*. But in *Barberry*, *Elm*, *Oak*, *Ash*, very different: Not fortuitously, but always much after the same manner. For in all the last named *Branches*, the *Aer-Vessels* that stand in the inner *margin* of each annual *Ring*, are all vastly bigger, than any of those that stand in the outer part of the *Ring*.

Tab. 22, 23, 25. 24, 28, 29. 33.

22. §. Thirdly, these *Aer-Vessels* are also different in their *Situation*. So in *Apple*, *Wallnut*, *Fig*, they are spread all abroad in every annual *Ring*; not being posited in any one certain *Line*. In others, they keep more within the compass of some *Line* or *Lines*; either *Diametral*, or *Peripherial*. So in *Holly* they are *Radiated*, or run in even *Diametral Lines* betwixt the *Pith* and the *Barque*. So also are some of of them in *Hazel*; and some few in *Wallnut*.

Tab. 25, 30, 31.

Tab. 22. 23, 32.

23. §. Whether they stand *Irregularly*, or are *Radiated*, it is to be noted, That Nature, for the most part, so disposeth of them, that many of them may still stand very near the *Insertions*. So in *Apple*, she will rather decline making an even *Line*; or in *Holly*, will rather break that *Line* into *Parcels*, than that the *Aer-Vessels* shall stand remote from the *Insertions*. To what end this is done, shall be said hereafter.

24. §. Again, in *Ash*, the *Aer-Vessels* are none of them *Radiated*, but most of them stand in *Circles* on the inner *Margin* of every annual *Ring*. Which *Circle* is sometimes very thick, as in *Ash* and *Barberry*. In others but thin, the *Vessels* standing, for the most part, single throughout the *Circles*; as in *Elm*. Sometimes again, they both make a *Circle*, and are also spread abroad; as in *Pear* and *Plum*.

Tab. 24. 29. 28. 26, 27.

25. §. Those likewise which are spread abroad, are sometimes *Regularly* posited. So in *Barberry*, besides those larger, that make the *Circle*, there are other smaller ones, that stand, in oblique *Lines*, athwart

Tab. 24.

athwart one another; almoſt like a *Bend,* or ſometimes, an entire or broken *Saltyr* in an *Eſcutcheon.* In *Oak,* they make rather certain *Tab.* 33. Columns, in the poſture of the *Pale.* And in *Elm,* they make, as it 38. were, many *croſs Parcels,* in the poſture of the *Feſs.*

26. §. This great difference in the *Size* and *Poſition* of the *Aer-Veſſels,* in the ſame individual *Plant,* is one ground, for which, I think it probable, That there are *divers Kinds* of *Aer-Veſſels,* as well as of *Sap-Veſſels.* Even as in *Animals,* there are divers *Kinds* of *Organs* for *Spiration,* and the ſeparation of *Aer: Fiſhes* having their *Branchiæ; Land-Animals* their *Lungs;* and thoſe in *Frogs, &c.* being of a ſomewhat peculiar *Kind.*

27. §. THE *Form* and *Texture* of theſe *Veſſels,* and the various ways whereby they may be beſt obſerved, I have already deſcribed and ſhewed in my *Anatomy of Roots.* As to their *Form,* one thing *P.* 1. *c.* 4. remarqued was this; That they are never *Ramified,* but diſtinctly continued from one end of a *Plant,* ſmall or great, to the other: as the *Nerves* are in *Animals.* A further and eaſie proof whereof, may be made, only by holding up a piece of an ordinary *Cane,* about ½ a foot long, cut very ſmooth at both ends, againſt a full light: whereupon, if you keep it in a ſtraight *Line* betwixt the *Light,* and the caſt of your *Eye,* and then look ſteadily, you may ſee quite through it, that is, through the *Aer-Veſſels,* which run ſtraight along the *Cane* from end to end.

28. §. As to their *Texture;* whereas, oftentimes, the *Aer-Veſſels* appear to be *unroaved* in the form of a very ſmall *Plate,* it is to be noted, That it is not only of different bredth, in divers *Plants,* and uſually much broader in the *Root,* than in the *Trunk:* but alſo, that in the *Trunk,* many times, the ſaid *Veſſels* are unroaved or reſolved, not in the form of a *Plate,* but of a *Round-Thred.* The Cauſes of which Diverſity, are principally Three; *viz.* The *Weſtage* of the *Fibres* of which the *Aer-Veſſels* conſiſt; The deference betwixt the ſaid *Fibres,* or betwixt the *Warp* and the *Woof;* And the different *Kinds* of *Woof.*

29. §. By the *Weſtage* of the *Fibres,* it is, That the *Veſſels,* oftentimes, *unroave* in the form of a *Plate.* As if we ſhould imagine a piece of fine narrow *Ribband,* to be woun'd ſpirally, and Edg to Edg, round about a Stick; and ſo, the Stick being drawn out, the *Ribband* to be left in the *Figure* of a *Tube,* anſwerable to an *Aer-Veſſel.* For that which, upon the *unroaving* of the *Veſſel,* ſeems to be a *Plate,* *Tab.* 39. or one ſingle *Piece,* is, as it were, a *Natural Ribband,* conſiſting of ſeveral Pieces, that is, a certain number of *Threds* or *Round Fibres,* ſtanding parallel, as the *Threds* do in an *Artificial Ribband.* And as in a *Ribband,* ſo here, the *Fibres* which make the *Warp,* and which are Spirally continu'd; although they run parallel, yet are not coalleſcent; but conteined together, by other Tranſverſe *Fibres* in the place of a *Woof.*

30. §. And as the ſaid *Fibres* are tranſverſly continued, thereby making a *Warp* and *Woof:* So are they (as in divers woven *Manufactures*) of very different *Bulk;* thoſe of the Former, being much bigger, and therefore much ſtronger, than thoſe of the Latter. By which means, as *Cloth* or *Silk* will often Tear one way, and not another; ſo here, while the *Warp* or thoſe *Fibres* which are Spirally continued,

tinued, are usually *unroaved* without breaking; those smaller ones, by which they are *stitched* or *woven* together, easily tear in sunder all the way.

31. And because the *Fibres* of the *Woof*, are themselves also of different *Bulk*; therefore it is, That where they are more sturdy, as usually in the *Root*, they require a greater quantity of *Warp*, that is, a broader *Plate*, to overmatch them. Whereas, where they are more extream small, as in the *Trunk* and *Leaves*, one *Thred* of the *Warp*, that is, one *Spiral Fibre*, will be strong enough of it self, and so, sometimes, be singly *unroav'd*.

32. §. From the extream Tenuity of these *Fibres*, it is, That they are very rarely discern'd, and not without the greatest difficulty. As also, from their great Tenderness; whereby not enduring to be drawn out, they all break off close to the Sides of the *Spiral ones*. In the *Pith*, the like Transverse *Fibres* are a little more visible: which first conducted Me to the notice of them here also.

33. §. All the *Fibres* of the *Aer-Vessels*, both the *Warp* and the *Woof*, are of the same Substantial Nature with the *Pith* and the other *Parenchymous Parts* of a *Plant*. From whence it is, That whereas the *Towy Parts* of a *Plant*, whereof all *Linen Manufactures* are made, are very *Strong* and *Tough*; these, as is abovesaid, are extream *Tender* and *Brittle*, like those of the *Pith* and all the *Pithy Parts*. To which therefore, the *Aer-Vessels* are to be referr'd. And the *Content* of both, is oftentimes the same.

B. 1. c. 7. §. 13, 14.

34. §. From whence, we have a further proof of what I have formerly asserted, which is, That in all *Plants*, there are *Two* Substantially different *Parts*, and no more than *Two*, *viz*. the *Pithy*, and the *Towy* or *Lignous Parts*.

35. §. From hence also we have some ground to conjecture, That so many of the *Aer-Vessels*, at least, which are not formed with the seed, but *post-nate*, are originated from the *Parenchymous Parts*; which seem by some alteration in the *Quality*, *Position* and *Texture* of the *Fibres*, to be Transformed into *Aer-Vessels*, as *Caterpillars* are into *Flies*. And as the *Pith* it self, by the Rupture and Shrinking up of several *Rows* of *Bladders*, doth oftentimes become Tubulary: So is it also probable, that in the other *Parenchymous Parts*, one single *Row* or *File* of *Bladders* evenly and perpendicularly piled; may sometimes, by the shrinking up of their Horizontal *Fibres*, all regularly break one into another and so make one *continned Cavity*; or a *Tube*, whose *Diametre* is the same with that of the *Bladders*, whereof it is composed. All which, will appear more probable, and what hath been said, be yet better understood, when we come, in the next *Chapter*, to the Description of the *Pith*.

CHAP. IV.

Of the PITH.

HE Third General *Part* of a *Branch* is the *Pith*. Which though it have a different name from the *Parenchyma* in the *Barque*, and the *Insertions* in the *Wood*; yet, as to its *Substance*, it is the very same with them both. Whereof there is a double evidence, *sc.* their *Continuity*, and the sameness of their *Texture*. Their *Texture* shall be shewed presently. As to their *Continuity*, it is to be noted, That as the *Skin* is continuous with the *Parenchyma* of the *Barque*; and this *Parenchyma* likewise, with the *Insertions* in the *Wood*; so these *Insertions* again, running through the *Wood*, are also continuous with the *Pith*. So that the *Skin*, *Parenchyma*, *Insertions*, and *Pith*, are all One entire piece of Work; being only filled up, in divers manners, with the *Vessels*.

2. §. The *Size* of the *Pith* is various, being not the same in any two *Branches* here represented. In *Wormwood, Sumach, Fig, Barbery*, 'tis very large; *sc.* betwixt 5, and 7 *Inches Diametre*, as it appears through the *Microscope*. In *Pine, Ash, Holly, Walnut*, not so large; from 3 *Inches Diameter* to 4. In *Oak, Apple, Pear, Hazel*, lesser, scarce from 2, to 3. In *Damascene*, not above an *Inch* and half. And in *Elm*, scarce an *Inch Diameter*. Note also, that of all *Plants*, both *Herbs*, and *Shrubs*, have generally the largest *Piths*, in proportion with the other *Parts* of the same *Branch*, as in *Sumach, Fig, Barbery*, is manifest. *Tab.* 24, 31, 34, 35, 22, 29, 30, 32. 23, 25, 26, 35. 27. 28.

3. §. It is also worth the noting, That wheras, in most *Plants*, the *Barque* and *Wood* do both grow thicker every year: the *Pith*, on the contrary, groweth more slender; So that in a *Branch* of one years growth, it is apparently more ample, than in one of two; and in a *Branch* of two, than in one of three; and so on.

4. §. The *Pith*, for the most part, if not always, in the *Branch*, as well as the *Root*, is furnished with a certain number of *Sap-Vessels*. They are here usually so postur'd, as to make a *Ring* on the *Margin* of the *Pith*. Where they are more numerous, or large, they are more evident; as in *Walnut, Fig, Pine*, and others. They are also of divers *Kinds*, answerable to those in the *Barque*; as in *Walnut, Lymphæducts*; in *Fig, Lacteals*; in *Pine, Resiniferous*. *Tab.* 30, 31, 32.

5. §. The *Parenchyma* of the *Pith* is composed of *Bladders*. Which are the very same with those in the *Barque*, and oftentimes in the *Insertions* within the *Wood*. Only these in the *Pith*, are of the largest *Size*; those in the *Barque*, of a lesser; and those of the *Insertions* least of all: for which reason they are less obvious than in the *Pith*.

6. §. The *Bladders* of the *Pith*, though always comparatively Great; yet are of very different *Sizes*. Being easily distinguished, even as to their *Horizontal Area*, to *Twenty* Degrees. Those of *Fig, Barbery*, and some others, are somewhat large. And of many *Herbs*, as *Tab.* 24, 31.

of *Thistle*, *Borage*, and others, three times as big again; appearing in the *Microscope*, like to the largest *Cells* of an *Hony-comb*. Those of *Plum*, *Worm-wood*, *Sumach*, less. Of *Elm*, *Apple*, *Pear*, lesser. Of *Holly* and *Oak*, still less. So that the *Bladders* of the *Pith* in *Borage* or *Common Thistle*, are of that *Size*, as to contain, within the compass only of their *Horizontal Area*, about twenty *Bladders* of the *Pith* of *Oak*. Wherefore one whole *Bladder* in *Thistle*, is, at least an hundred times bigger, than another in *Oak*.

Tab. 39.

Tab. 32.

7. §. Of the *Size* of these *Bladders* of the *Pith*, 'tis also to be noted, That it doth not at all follow the *Size* of the *Pith* it self; but is still varied, according as Nature designeth the *Pith* for various use. Thus, whereas the *Pith* of *Sumach*, is Larger than that of *Barberry*; it might be thought, that the *Bladders*, whereof it is composed, should be likewise Larger: Yet are they Three times as Small again in *Sumach*, as they are in *Barberry*. So the *Pith* of *Plum*, is far Less, than that of *Pear*; yet the *Bladders* of the former are Four or Five times as big, as those of the latter. So the *Pith* of *Hazel* is almost Three times as Little again, as that of *Holly*; yet the *Bladders* in *Hazel*, are Ten times bigger, than in *Holly*.

Tab. 24, 34. 26, 27. 22, 23.

8. §. The *Shape* of the *Bladders* hath also some Variety. For although, for the most part, they are more round; yet oftentimes they are angular: as in *Reed-grass*, a *Water-plant*; where they are *Cubical*; and in *Borage*, *Thistle*, and many others, where they are *pentangular*, *sexangular* and *septangular*.

9. §. Of the *Texture* of the *Bladders*, 'tis also to be noted, that many times, the *Sides* of the greater *Bladders* are composed of lesser; as is often seen in those of *Borage*, *Bulrush*, and some other *Plants*. In the same manner, as the *Sap-Vessels*, are but greater *Fibres* made up of lesser.

10. §. The *Pith*, though always originally composed of *Bladders*, and so *One Entire Piece*; yet in process, as the *Plant* grows up, it hath divers openings or *Ruptures* made in it: oftentimes very regularly, and always for good use, and with constancy observed in the same *Species* of *Plants*. In *Sharp-poynted Dock*, many of the *Pores* are considerably prolonged by the length, like small *Pipes*. In *Walnut* it shrinketh up into transverse *Filmes* or *Membranes*; as likewise sometimes in *Spanish-Broom*. Sometimes the *Pith* is hollow or *Tubulary*: either throughot the *Trunk*, as in *Thistle*, *Endive*, *Scorzonera*, *Marsh-Mallow*: or so, as to remain entire at every joynt; as in *Sonchus*, *Nettle*, *Teasle*; in which it is divided as it were into several *Stories*: and divers other ways.

Tab. 19.

11. §. I SHALL conclude this discourse with a further illustration of the *Texture* of the *Pith*, and of the whole *Plant*, as consequent thereupon. I say therefore, (and have given some account hereof in the *Anatomy* of *Roots*) That as the *Vessels* of a *Plant*, sc. the *Aer-Vessels* and the *Lymphæducts* are made up of *Fibres*; according to what I have in this Discourse above said; so the *Pith* of a *Plant*, or the *Bladders* whereof the *Pith* consists are likewise made up of *Fibres*. Which is true also of the *Parenchyma* of the *Barque*. And also of the *Insertions* in the *Wood*. Yea, and of the *Fruit*, and all other *Parenchymous Parts* of a *Plant*. I say, that the very *Pulp* of an *Apple*, *Pear*, *Cucumber*, *Plum*, or any other *Fruit*, is nothing else but *a Ball of most extream small transparent Threds or Fibres, all wrapped and stich'd*

ſtitch'd up (though in divers manners) together. And even all thoſe *Parts* of a *Plant*, which are neither formed into viſible *Tubes*, nor into *Bladders*, are yet made up of *Fibers*. Which, though it be difficul to obſerve, in any of thoſe *Parts* which are cloſer wrought and principally in the *Inſertions* of ſome *Trees*: yet in the *Pith*, eſpecially of ſome *Plants*, which conſiſteth of more open *work*, they are more viſible. Which introduceth the obſervation of them in all other *Parenchymous Parts*. So in the *Pith* of a *Bulruſh* of the *Common Thiſtle*, and ſome other *Plants*; not only the *Threds* of which the *Bladders*; but alſo the ſingle *Fibres*, of which the *Threds* are compoſed; may ſometimes with the help of a good *Glaſs*, be diſtinctly ſeen. Yet one of theſe *Fibres*, may reaſonably be computed to be a Thouſand times ſmaller than an *Horſe-Hair*. *Tab.* 38.

12. §. The *Fibroſity* of the *Parenchyma* is alſo viſible in ſome *Woods*, in which, it is apparently mixed with the *Lignous Parts*, not only by *Inſertions*, but *per minimas Partes organicas*. That is to ſay, The *Parenchymous Fibres*, like ſmaller *Threds*, are either *wraped round about* both the *Lignous* and the *Aer-Veſſels*, or at leaſt *interwoven* with them, and with every *Fiber* of every *Veſſel*: as in very white *Aſh* or *Fir-Wood*, with an advantagious poſture and light, may be obſerved. *Tab.* 39.

13. §. WHENCE it follows, that the whole *Subſtance*, or all the *Parts* of a *Plant*, ſo far as *Organical*, they alſo conſiſt of *Fibres*. Of all which *Fibres* thoſe of the *Lymphæducts*, run only by the *Length* of the *Plant*: thoſe of the *Pith*, *Inſertions*, and *Parenchyma* of the *Barque*, run by the *breadth* or horizontally: thoſe of the *Aer-Veſſels*, fetch their Circuit by the *Breadth*, and continue it by the *Length*.

14. §. By which means, the ſaid *Parenchymous Fibres*, in fetching their *horizontal Circles*, do thus *weave*, and make up the *Bladders* of the *Pith*, in *Open-Work*. And the ſame *Fibres* being thence continued; they alſo *weave* and make up the *Inſertions*, but in *Cloſe-Work*. Betwixt which *Inſertions*, the *Veſſels* being likewiſe tranſverſly interjected, ſome of the ſame *Fibres wrap* themſelves alſo about theſe; thus *tying* many of them together, and ſo making thoſe ſeveral *Conjugations* and *Braces* of the *Veſſels*, which I have formerly deſcribed. And as ſome of theſe *Horizontal Fibres* are *wraped* about the *Veſſels*; ſo alſo about the *Fibres*, whereof the *Veſſels* are compoſed. By which means it is, that all the *Fibres* of the *Veſſels* are *Tacked* or *Stitched* up cloſe together into One Coherent Piece. Much after the ſame manner, as the *Perpendicular Splinters* or *Twigs* of a *Basket*, are, by thoſe that run in and out *Horizontally*. And the ſame *Horizontal Fibres*, being ſtill further produced into the *Barque*; they there compoſe the ſame *work* over again (only not ſo *open*) as in the *Pith*. *Tab.* 40.

15. §. SO THAT the moſt unfeigned and proper reſemblance we can at preſent, make of the whole *Body* of a *Plant*, is, To a piece of *fine Bone-Lace*, when the Women are working it upon the *Cuſhion*, For the *Pith*, *Inſertions*, and *Parenchyma* of the *Barque*, are all extream Fine and Perfect *Lace-Work*: the *Fibres* of the *Pith* running *Horizontally*, as do the *Threds* in a Piece of *Lace*; and bounding the ſeveral *Bladders* of the *Pith* and *Barque*, as the *Threds* do the ſeveral *Holes* of the *Lace*; and making up the *Inſertions* without *Bladders*, or with very ſmall ones, as the ſame *Threds* likewiſe do the *cloſe* Parts of

Y the

the *Lace*, which they call the *Cloth-Work*. And laftly, both the *Lignous* and *Aer-Veffels*, ftand all *Perpendicular*, and fo crofs to the *Horizontal Fibres* of all the faid *Parenchymous Parts*; even as in a Piece of *Lace* upon the *Cuſhion*, the *Pins* do to the *Threds*. The *Pins* being alfo conceived to be *Tubular*, and prolonged to any length; and the fame *Lace-Work* to be wrought many Thoufands of times over and over again, to any thickneſs or hight, according to the hight of any *Plant*. And this is the true *Texture* of a *Plant*: and the *general compoſure*, not only of a *Branch*, but of all other *Parts* from the *Seed* to the *Seed*.

An Account of the
VEGETATION
OF
TRUNKS
Grounded upon the foregoing
ANATOMY.

PART II.

AVING before given the *Anatomy* of *Trunks*; I shall next proceed to see, what *Use* may be made thereof; and principally, to explicate the manner of their *Vegetation*. In doing which, that former *Method*, which I used in shewing the manner of the *Growth* of *Roots*, I shall not exactly follow. For so, in regard the *Organical Parts* of the *Root* and *Trunk* are the same, and consequently their *Nutrition* and *Conformation* are effected in the same way; I should hereby be obliged to a nauseous and unprofitable repetition of many things already said. The Explication therefore of all those Particulars, which more especially belong to the *Trunk*, or are more Apparent therein, and not spoken of, or not so fully, in the former *Books*, will be my present Task. The chief *Heads* whereof, shall be these Seven following, *viz.*

FIRST, the *Motion* and *Course* of the *Sap*.
SECONDLY, The *Motion* and *Course* of the *Aer*.
THIRDLY, The *Structure* of the *Parts*.
FOURTHLY, The *Generation* of *Liquors*.
FIFTHLY, The *Figuration* of *Trunks*.
SIXTHLY, The *Motion* of *Trunks*.
SEVENTHLY, And lastly the *Nature* of *Trunks* as variously fitted for *Mechanical Use*.

CHAP. I.

Of the Motion *and* Course *of the* Sap.

FIRST, as to the *Course* of the *Sap*, there are Three *Parts* in which it *moveth*; *sc.* the *Pith*, the *Wood*, and the *Barque* First the *Pith*; in which the *Sap* moveth the *First year*, and *only the First year*. Or, it is *Proprium quarto modo*, to the *Pith* of every *Annual Growth*, and to the *Pith* of such a *Growth* only, To be succulent. That is, whether of a *Sprout* from a *Seed*, or of a *Sucker* from a *Root*, or of a *Cyon* from a *Branch*; The *Pith* is always found the *First year* full of *Sap*. But the *Second year*, the same individual *Pith*, always becomes *dry*, and so it continues ever after.

2. §. One cause whereof is, that the *Lymphæducts* in the *Barque*, being the first year adjacent to the *Pith*; they do all that time, transfuse part of their *Sap* into it, and so keep it always *Succulent*. But the same *Lymphæducts*, the year following, are turned into *Wood*; and the *Vessels* which are then generated, and carry the *Sap*, stand beyond them, in the *Barque*. So that the *Sap* being now more remote from the *Pith*, and intercepted by the new *Wood*, it cannot be transfused, with that sufficient force and plenty as before, into the *Pith*; which therefore, from the first year, always continues dry.

3. §. THE SECOND *Part* in which the *Sap moves, sub forma liquoris*, is the *Wood*. Which yet, it doth not in all *Plants*, but only in some; and visibly, in very few; as in the *Vine*: In a *Vine*, I say, the *Sap* doth *visibly ascend* by the *Wood*. And this it doth, not only the first year, but every year, so long as the *Vine* continues to grow. But although this *ascent*, in or through the *Wood*, be every year; yet it is only in the *Spring*, for about the space of a Month; *sc.* in *March* and *April*.

4. §. There are many other *Trees*, besides the *Vine*, wherein, about the same time of the year, the *Sap ascendeth*, though not so copiously, yet chiefly, in the *Wood*. For if we take a *Branch* of two or three years growth, suppose of *Sallow*, and having first cut the same transversely; if the *Barque* be then also transversely, and with some force, pressed with the back of the knife, near the newly cut end; the *Sap* will very plainly rise up out of the utmost *Ring* of *Wood*. And if it be pressed in the same manner, or a little more strongly, about an Inch lower, the *Sap* will ascend out of every *Ring* of *Wood* to the Center. Yet at the same time, which is to be noted, there *ariseth* no *Sap* at all out of the *Barque*.

5. §. Whence appears the Error of that so Common Opinion, That *the Sap always riseth betwixt the Wood and the Barque*. The contrary whereunto is most true, That it never doth. For the greater part of the year, it riseth in the *Barque*, *sc.* in the inner *Margin* adjacent

cent to the *Wood*, and in *Spring*, in or through the *Wood* it self, and there only.

6. §. THE THIRD *Part* in which the *Sap* ascends, is the *Barque*, as was above hinted, and may be observ'd in almost any *Branch*, if cut cross, in the *late Spring* and in *Summer*; either as the *Sap* issueth spontaneously, or upon pressing, as aforesaid. So that when the *Sap* ceaseth to ascend, *sub forma liquoris*, by the *Wood*, then it begins to ascend by the *Barque*.

7. §. Besides the difference of *Time*, the *Organical Parts* likewise, in which these two *Saps* ascend, are divers. For in the *Barque*, it ascendeth *visibly*, only in the *Succiferous*, whereas in the *Wood*, it ascendeth only by the *Aer-Vessels*.

8. §. FROM what hath been said, we may understand, what is meant by the *Bleeding* of *Plants*. If we take it generally, it properly enough expresses, *The eruption of the Sap out of any Vessels*. And so, almost all *Plants*, in *Summer* time, do *Bleed*, that is, from *Sap-Vessels*, either in the *Barque*, or in the *Margin* of the *Pith*: the *Saps* they *Bleed*, having either a *Sower*, *Sweet*, *Hot*, *Bitter*, or other *Tast*. At which time, the *Vessels* also, in the *Barque* of a *Vine-Branch*, do *Bleed* a *Sower Sap*.

9. §. But that which is *vulgarly* called *Bleeding*, as in a *Vine*, is quite another thing; both as to the *Liquor* which issueth, and the *Place* where it issues: that is to say, it is neither a *Sweet*, nor *Sower*, but *Tastless Sap*; issuing, not from any *Vessels* in the *Barque*, but from the *Aer-Vessels* in the *Wood*. So that there is as much difference betwixt *Bleeding* in a *Vine*, or the *Rising* of the *Sap* in any other *Tree*, in *March*, and in *July*; as there is betwixt *Salivation* and an *Hæmorrhage*; or betwixt the *Course* of the *Chyle* in the *Lactiferous Vessels*, and the *Circulation* of the *Blood* in the *Arteries* and *Veins*.

10. §. NOW the *Cause* from whence it comes to pass, that the *early Spring-Sap* of a *Vine*, and other *Trees*, ascendeth by the *Wood*, is, In that the *Generation* of the young *Sap-Vessels* in the *Barque*, by which the *Sap* ascendeth all the *Summer*; is, in the beginning of *Spring*, but newly attempted. So that the *Sap* having not yet these *Vessels* to receive it, it therefore (*pro hac vice*) runs up the *Aer-Vessels* in the *Wood*. But so soon as the said *Vessels* in the *Barque* begin to be considerably encreased, the *Sap*, declining the *Aer-Vessels*, betakes it self to *These*, as its most proper *Receptacles*.

11. §. THE CAUSE also, why the *Vessels* of almost all *Plants*, upon cutting, do yield *Sap*, or *Bleed*; is the *Pressure* which the *Parenchyma* makes upon them. For the *Pith* and other *Parenchymous Parts* of a *Plant*, upon the reception of *Liquor*, have always a *Conatus* to *dilate* themselves. As is manifest from *sponges*, which are a Substance of the same Nature, and have a somewhat like structure. As also from *Cork*, which is but the *Parenchyma* or *Barque* of a *Tree*. I say therefore, that the *Parenchyma* being fill'd and swell'd with *Sap*, hath thereby a continual *Conatus* to *dilate* it self; and in the same degree, to press together or contract the *Vessels* which it surroundeth. And the said *Vessels* being cut, their actual *Contraction* and the *Eruption* of the *Sap*, do both immediately follow.

12. §. IT may be also noted, That the *Trunk* or *Branch* of any *Plant* being cut, it always *bleeds* at both ends, or upwards and downwards, alike

alike freely. Which, as well as divers other *Experiments* plainly shews, That in the *Sap-Vessels* of a *Plant*, there are no *Valves*.

13. §. FROM what we have now above, and elsewhere formerly said, we may also understand the *manner* of the *Ascent* of the *Sap*. As to which, I say, *First*, That considering to what heigth and plenty, the *Sap* sometimes ascends; it is not intelligible, how it should thus ascend, by virtue of any one *Part* of a *Plant*, alone; that is neither by virtue of the *Parenchyma*, nor by virtue of the *Vessels*, alone. Not by the *Parenchyma* alone. For this, as it hath the Nature of a *Sponge* or *Filtre*, to suck up the *Sap*; so likewise, to suck it up but to a certain heigth, as perhaps, about an Inch, or two, and no more.

14. §. Nor by the *Vessels* alone, for the same reason. For allthough we see, that small *Glass-Pipes* immersed in Water, will give it an ascent for some Inches; yet there is a certain *period*, according to the *bore* of the *Pipe*, beyond which it will not rise. We must therefore joyn the *Vessels* and the *Parenchyma* both together in this Service; which we may conceive performed by them in the manner following.

Tab. 39.

15. §. Let A B be the *Vessel* of a *Plant*. Let C E D F be the *Bladders* of the *Parenchyma*, wherewith, as with so many little *Cisterns*, it is surrounded. I say then, that the *Sap*, in the *Pipe* B A, would, of it self, rise but a few Inches; as suppose, from D to L. But the *Bladders* D P, which surround it, being swelled up and turgid with *Sap*, do hereby press upon it; and so not only a little contract its bore, but also transfuse or strain some *Portion* of their *Sap* thereinto: by both which means, the *Sap* will be forced to rise higher therein. And the said *Pipe* or *Vessel* being all along surrounded by the like *Bladders*; the *Sap* therein, is still forced higher and higher: the *Bladders* of the *Parenchyma* being, as is said, so many *Cisterns* of *Liquor*, which transfuse their repeated Supplies throughout the length of the *Pipe*. So that by the supply and pressure of the *Cisterns* or *Bladders* F D, the *Sap* riseth to L; by the *Bladders* Q L, it rises to M; by the *Bladders* N M, it rises to I; by the *Bladders* O I, it rises to K; by the *Bladders* P K, it rises to E; and so to the top of the *Tree*. And thus far of the *Motion* of the *Sap*.

CHAP. II.

Of the Motion *and* Courſe *of the* Aer.

THE NEXT enquiry to be made, is, into the *Motion* and *Courſe* of the *Aer*. Where this queſtion will firſt of all be asked ; *ſc.* Which way the *Aer* firſt *enters* the *Plant* ; whether at the *Trunk*, *Leaves*, and other *Parts* above ground; or at the *Root* ? I anſwer, *That it enters in part, at them all*. For the *Reception*, as well as *Extramiſſion* whereof, the *Pores* are ſo very large, in the *Trunks* of ſome *Plants*, as in the better ſort of thick walking *Canes*, that they are viſible, to a good Eye, without a *Glaſs* ; but with a *Glaſs*, the *Cane* ſeems, as if it were ſtuck top full of holes with great *Pins* : being ſo large, as very well to reſemble the *Pores* of the *Skin* in the end of the *Fingers* and *Ball* of the Hand. *Tab.* 19.

2. §. In the *Leaves* of *Pine*, they are likewiſe through a *Glaſs*, a very Elegant Show ; ſtanding all moſt exactly, in *rank* and *file*, throughout the length of the *Leaves*. The *Figure* whereof ſhall be given hereafter, when we come to the *Anatomy* of the *Leaf*.

3. §. But although the *Aer* enters, in part, at the *Trunk* and other *Parts*, eſpecially in ſome *Plants* ; yet its *chief* entrance, is at the *Root*. Even as ſome Parts of *Aer*, may continually paſs into the *Body* and *Blood*, by the *Habit*, or *Pores* of the *Skin* ; but the chief entrance hereof, is at the *Mouth*. And what the *Mouth* is, to an *Animal* ; that the *Root* is to a *Plant*.

4. §. Again, if the chief entrance of the *Aer*, were at the *Trunk* ; then, before it could be mixed with the *Sap* in the *Root*, it muſt deſcend ; and ſo move not only contrary to its own Nature, but likewiſe in a contrary Courſe to the *Sap*, throughout the *Plant*. Whereas, by its *Reception* at the *Root*, and ſo its *Tranſition* from thence ; it hath a more natural and eaſie motion of Aſcent. For while the *Sap* aſcends, that the *Aer*, in the ſame *Plant*, ſhould continually deſcend, cannot reaſonably be ſuppoſed.

5. §. The ſame is further argued, From the fewneſs and ſmallneſs of the *Diametral Portions* in the *Trunk* in compariſon with thoſe in the *Root*. In which *Nature* hath plainly deſigned the ſame, for the *Separation* of the *Aer* from the *Sap*, after they are both together received thereinto. So that the *Reception* and *Courſe* of the *Aer*, is made on this manner following.

6. §. THE *Aer* being a *Springy* Body, it inſinuates into all the *Holes* and *Cranies* of the *Earth* ; and ſo is plentifully mixed therewith. Whereupon, as the *Sap* enters the *Root*, more or leſs *Aer* ſtill intrudes it ſelf together with it. The *Liquid* Portion of the *Sap*, ſwells and fills up the *Succulent Parts* of the *Barque*. The *Aery* Part, is, as was ſaid, ſeparated from the *Liquid*, into the *Diametral* Portions. Which

running

running from the *Barque* towards the Centre of the *Root*, and so passing along betwixt the *Aer-Vessels*; do hereby convey the *Aery* Part of the *Sap* from the *Barque*, into the same.

7. §. Being thus received into the *Aer-Vessels*, and the *Reception* thereof, by the same means continued; it is by them advanced into the *Trunk*. In which advance, it is again, more or less, disbursed into all the *Parts* of the *Trunk*, as it goes. *Partly*, inwards to the *Pith*. From whence, the *Pith* is always, at length, filled with *Aer*. *Partly*, into the *Insertions*; by which it is conveyed outward into the *Barque*. Wherein, it is in some part, transfused through the *Sap*: and so the rest, with part of the *Sap*, remitted, in *perspirations*, back again into the *Aer*.

8. §. So that, whereas the *Diametral Portions* in the *Root*, do serve to convey the *Aer* from the *Sap* in the *Barque*, into the *Aer-Vessels*, in the *Wood*: on the contrary, the *Insertions* here in the *Trunk*, serve to convey the *Aer* from the *Aer-Vessels* in the *Wood*, into the *Sap*, in the *Barque*. Wherefore, as the *Aer-Vessels* advance the *Aer*, or the *Aery Part* of the *Sap*, and so convey it by the *length* of the *Trunk*; so the *Insertions* filter it, and convey it by the *breadth*.

9. §. AND that the *Insertions* have this *Office* or *Subservience* unto both *Kinds* of *Vessels*; doth yet further appear, if we consider, That the *Aer-Vessels* are always so postured, as to touch upon the said *Insertions*, or at least to stand very near them. For either they are large, and so do frequently touch upon them on both sides; as in *Elm*, *Ash*, *Wallnut*, &c. Or if they are small; then they either run along in even lines collateral and oftentimes contiguous with the said *Insertions*, as in *Holly*: or at least, are reciprocally, some on one side, and some on another, inclined to them; as in *Apple*. By all which means, the *Aer* is more readily conveyed from the *Vessels* into the *Insertions*.

Tab. 28.
29.
22.

25.

10. §. A further evidence hereof is this, That generally, the bigger and the more numerous the *Aer-Vessels* be; the bigger, or at least, the more numerous also are the *Insertions*: Especially, if the comparison be made (as in all other cases it ought to be, as well as here) betwixt the several *species* of the same *Kind*. So *Corin*, which hath small *Aer-Vessels*, hath also very small *Insertions*. But the *Vine*, hath both very large: and so for others.

Tab. 17.

11. Wherefore, the *Insertions* minister betwixt the *Aer-Vessels*, and the *Succiferous*; in the same manner, as the *Vesiculæ* of the *Lungs*, do betwixt the *Bronchiæ* and the *Arteries*. That is to say, as in an *Animal*, the *Bronchiæ* deposite the *Aer* into the *Vesiculæ* of the *Lungs*; which administer it to the *Arteries*: so in a *Plant*, the *Aer-Vessels* deposit the *Aer* into the *Insertions*, that is into the *Vesiculæ* of the *Insertions*; by which it is gradually *filtred* off into the *Barque* and the *Sap-Vessels* therein.

CHAP. III.

Of the Structure *of the* Parts.

A THIRD enquiry, is into the *Generation* and *Structure* of *Parts*. The manner whereof I have already endeavoured to *explicate* (*a*) from the *A-* (*a*) *Lib. 2. natomy* of the *Root*, throughout all particulars. *P. 2.* Some whereof I shall yet further clear.

1. §. As *First*, the *Union* of the *Barque* to the *Body* of the *Tree*, Contrary to the common Opinion, *That they are not continuous*; but that the *Barque* only surrounds the *Body*, as a *Scabbard* does a *Sword*, or a *Glove* the *Hand*. As also seemeth to be proved, by the easy *slipping* of the *Barque* of *Willow*, and most other *Trees*, when full of *Sap*, from the *Wood*.

2. §. But, notwithstanding this, they are as truly continuous, as the *Skin* of the *Body* is with the *Flesh*: *sc.* by means of the *Parenchyma*; which is one entire *Body*, running from the *Barque* into the *Wood*, *Tab.* 19. and so uniting both together; as in a *Branch* of *Vine* or *Corin-Tree*, when the *Barque* is stripped off, is apparent; the Spaces between the several *Parts* of the *Wood*, being filled up with the *Parenchymous*, inserted from the *Barque*.

3. §. Now the reason why the *Barque* nevertheless slips so easily from the *Wood*, is plain, *viz*. Because most of the young *Vessels* and *Parenchymous Parts*, are there every year successively formed; that is, betwixt the *Wood* and *Barque*: where the said *Parts* newly formed, are as tender, as the tenderest *Vessels* in *Animals*. And we may imagine, how easie it were at once to tear or break a thousand *Vessels* or *Fibres* of an *Embrio*, of a *Womb* or *Egg*.

4. §. THE same *Vessels* of the *Barque* being always *braced*, and gradually falling off, together with the *Parenchyma*, into the utmost *Rind*: Hence it is, that the *Barques* of many *Trees*, are as it were, lat- *Tab.* 19. ticed with several *Cracks* of divers *Sizes*, and sometimes in the *Figure* of *Rombs*: the said *Fissures* representing the *Position* and *Tract* of the *Vessels* in their *Braces*. Hence also it is, that the *Barque* of some *Trees*, as of *Corin*, *Cherry*, &c. falleth off in *Rings*, *sc.* because the *Sap-Vessels* are posited in the same manner in the *Barque*.

5. §. The *Sap-Vessels*, as they are generated at the inner *Verge* of the *Barque*: so likewise, in a small quantity, at the utmost *Verge* of the *Pith*. These being not only fed with a more vigorous *Sap*, but with great caution, secured within the *Wood*, for the propagation of the succeeding *Buds*.

6. §. Hence also it is, that is, by the annual accretion of these *Vessels*, that the *Pith* is sometimes less in the *Trunk*, than in the *Branches*; *Tab.* 18. and less in the elder *Branches*, than in the younger; and sometimes 'tis almost wholly filled up. By which means, as the *Branches* carry every year a greater burthen; so they become still more sturdy the better to support it.

7. §. SOMETIMES also the *Pith* breaks and shrinks up, thus making the *Trunk* a *Pipe*. The cause whereof, is either the *Largeness* of its *Pores*, or the *Thinness* of the *Sides* of the said *Pores*; upon both

which

which accounts, the *Pith* doth more easily tear, and upon tearing shrink up, and so become hollow: as in *Cichory, Lampsana, Sonchus, Teasel, Brownwort*, and others; wherein the *Pores* of the *Pith* are *Large*, and the *Sides* of the *Pores, Thin*. Whereas, upon contrary accounts, the *Piths* of most *Trees*, remain perpetually entire.

8. §. THE Reason why *Plants* are made thus to become *hollow*, is *partly*, for the *ripening* of the *Fruit* or *Seed*; which is the better effected by a more plentiful supply of *Aer* continually received into their hollow *Trunks*. For by means of that *Aer*, part of the *Sap*, is dryed up, and the remaining part of it made warmer, and so sooner matured.

9. §. *Partly*, for the better determining the due *Age* of the *Plant*. Hence it is, that the greater part of *Annual Trunks*, are hollow: the *Aer* contained in that hollow, drying up the *sap*, and shrinking up the *Sap-Vessels* so far, as to hinder the free motion of the *Sap* therein; from whence the *Plant* must needs perish. So that as the *Content* of the *Aer-Vessels*, is a kind of *Vegetable Aer*, whose Office is to *Attenuate*, and *Ferment* the *Juyces* of *Plants*: so the *Content* of these *Cavities*, cometh nearer to a more common *Aer*, designed chiefly, so soon as it is convenient, to *dry* them up.

10. §. AGAIN, as to the *Aer-Vessels*, divers questions may be asked. As how it comes to pass, that they are generally less in the *Trunk* of the same *Plant*, than in the *Root*? The Cause whereof is, that here in the *Trunk* they are more under the power of the *Aer*; both that which entreth in at the *Trunk*, and that which of its own *Nature* ascendeth up into it from the *Root*. For the *Aer*, as we have elsewhere said, is the *Mould* of the *Aer-Vessels*; to whose crooked or at least, *Acid Parts*, the *Saline*, and other *Principles* concurring to their generation, do conform. To which they do best, the smaller they are: the *Fibres* of the larger *Aer-Vessels* making greater *Circles*, and so coming nearer to a *right Line*, answerable to the *Figure* of the Particles, not of the *Aerial*, but of the *Saline Principle*.

11. §. Wherefore as the *Aer-Vessels* may be observed still to be dilated or widened towards the lower parts of the *Root*; the *Aerial Principle* being there less predominant, and the *Saline* more: So towards the upper part of the *Trunk*, to be contracted or grow smaller; the *Aerial Principle* being here more predominant, and the *Saline* less.

12. §. FOR the same cause it may be observed, That the *Aer-Vessels* of the Second years Growth, and the several years succeeding, are usually nearer of one *Size*, than those of the *Second* and *First*; all being under a less power of the *Aer*, than the *First*. For the first year the *Pith* being full of *Liquor*, the *Aer-Vessels* themselves, are the only *Repositories* of the *Aer*. Whereas, after the first year, the *Pith* becoming dry, or another great *Repository* for the *Aer*; the *Aer-Vesseles* are henceforth filled with a moister or more *Vaporous* and *Saline Aer*, and so made to grow wider.

13. §. Hence the very *Size* of the *Pith*, hath much *influence* upon the *Aer-Vessels*, and the manner of *Nutrition*, and the *Generation* of *Liquors* in *Plants*.

14. §. BUT for the most part, the *Aer-Vessels* are somewhat, more or less, amplified in every new *Annual Ring*; or at least to a certain number of years. Probably, because in the elder *Branches*, the *Spiral Fibres*, of which the *Vessels* consist, are more bulky; and so make a

Vessel

Veſſel of a wider, as a more agreeable bore. Nature obtaining hereby, that the Quantity of *Aer*, ſhall always be anſwerable to the Growth of the *Plant*, or at leaſt, be ſufficient to maintain its *Vegetable Life* and *Vigour*.

15. §. And therefore, as is above hinted, it ſeems likely, That after a certain number of years, the *Aer-Veſſels* are no longer amplified, but ſtand at a ſtay, and perhaps may grow ſmaller, according as the *Tree* is leſs or more *Longæve*; and that after this period, it is ſome way or other in its Declining State.

16. §. LASTLY, from the *Content* and *Governing Principle* of the *Aer-Veſſels*, the *Time*, when they begin every year to be formed, or to appear, is always *later*; at leaſt with reſpect to the *ſeaſon* of the *Tree*. So that whereas the *Sap-Veſſels* begin to be formed in *Spring*: theſe, not till the latter end of *Summer*, or there about; at leaſt not till about that time to *appear*. That is, when the *Sap* begins to decreaſe, and to grow more *Aery*; and ſo more fit matter for the *Generation* of the ſaid *Aer-Veſſels*.

CHAP. IV.

Of the Generation *of* Liquors.

UPON the *Structure* and *Formation* of the *Parts*, dependeth the *Generation* of *Liquors*, as was lately intimated. The *manner* whereof I have formerly ſhewed, in diſcourſing of the *Root*. Yet ſome things I ſhall here further *explicate*. And *Firſt*, what we have formerly aſſerted, *ſc.* That the concurrence of two *ſpecifically* diſtinct *Fluids*, is as neceſſary to *Nutrition* in *Plants*, as in *Animals*. Which appears, as from divers other conſiderations, ſo from the very *Structure* of a *Plant*: where in all the *Organical Parts*, or the *Parenchyma* and the *Veſſels*, are every where mixed together *per minima*, that is, *per minimas partes organicas*, or *Fiber* with *Fiber* of ſeveral *Kinds*. Every ſmall part of a *Tree*, or of the *Barque* of a *Tree*, being as I may ſay, a ſort of *Linſy-Woolſey*. So that there is not the leaſt part of the *Sap*, which is not impregnate with divers *Eſſential Tinctures*, as it is continually *filtred* from the *Fibres* of one *Kind*, to thoſe of another; ſtanding every where *woun'd* and *ſtitch'd* up together for the ſame purpoſe.

2. §. FROM the ſpecial *Nature* and *Structure* of the *Parts*, the *Liquors* of *Plants* are likewiſe *ſpecified*. The *Veſſels* being the chief *Viſcera* of a *Plant*. For all *Liquors* in a *Plant*, are certainly made by that *Plant*. And ſince the *Plant* hath no *Viſcera* (ſo called) I would then know, what its ſeveral *Liquors* are made by? If in the *Parenchyma*, ſurely by that *Parenchyma*. If in the *Veſſels*, by the *Veſſels*. And if of divers *Kinds* by divers *Kinds* of *Veſſels*. So that what the *Viſcera* are in *Animals*, the *Veſſels* themſelves are in *Plants*. That is to ſay, as the *Viſcera* of an *Animal*, are but *Veſſels* conglomerated: ſo the *Veſſels* of a *Plant*, are *Viſcera* drawn out at length.

3. §. AGAIN, as the *specifying* of the *Sap* dependeth chiefly on the *special Nature* of the *Parts*: so partly, upon the *Structure* of the *Whole*. Whereby every *Part* is still better accomodated with its own *Juyce*. Thus the *Aer-Vessels* are necessary, not only and barely for a *supply* of *Aer*; but also by their *Number*, *Size*, and *Position* to *adjust* the quantity of that *Aer*, to the government of *Nutrition*, and the *Generation* of the *Specifick Liquors* of every *Plant*. Which is evident from hence, in that they do not follow the *size* of the *Plant*; but are great and many, in some small *Plants*; and small and few, in some others that are large. So *Vines*, and *Corn*, as we have formerly observed, have proportionably a great number of *Aer-Vessels*, and those very large. By which means the *Sap* is *attenuated* and less *Oyly*, and more copiously impregnated with a *Subtile*, *Volatile* and *Winy Spirit*.

4. §. For the same reason, the *Stalk* of *Maze* or of *Indian Wheat*, which when it is *Green* yieldeth a very sweet *Juyce*; and the *Canes*, whereof *Sugar* (which aboundeth with a *volatile* and *inflammable Spirit*) is made; these, I say, obtain the like over proportion of *Aer-Vessels*, to what we see in most other *Plants*. Hence also it is, that none of the said *Plants* have any considerable *Barque*; that so the attenuating and subtilizing *Aer*, may have a more easie and plentiful admission at the *Trunk* also. For which reason likewise the *Pores* of the *Skin* of some *Canes* are, as hath been said, remarkably wide.

5. §. Hence also it is observable, that of the same *Species* or *Kindred*, those *Plants* which have the most, and especially the largest *Aer-Vessels*; have also the greatest abundance either of a *sweet*, or of a *winy* Liquor. So in *Apple*; they are larger than in *Crab*; In *Warden*, larger than in *Qnince*; and in *Pear-Tree*, larger than in *Warden*. So also in *Corin*, larger than in *Gooseberrey*; and in *Vine*, larger than in *Corin*: and so in others.

6. §. AND as the *Aer-Vessels*, by their *Multitude* and *Largeness*, are accommodated to the better making of a *Winy Sap*: so by their *fewness* and *smallness*, of an *Oylie*. As is remarkably seen in *Fir*, and other *Resiniferous Trees*: these having, if not the *smallest*, yet the *fewest Aer-Vessels* of all other *Trees*.

7. §. IF it be asked, how a *Plant* comes to have any *Oyl* at all in any *Part*? Since we see, that the *Sap* by which the *Root* is fed, seemeth to be nothing else but *Water*: and that many *Plants* which yield a great deal of *stillatitious Oyl*, as *Mint*, *Rue*, and others, will yet grow in *Water*: I say, if it be enquired how this *Water*, is made *Wine* or *Oyl*? I answer, that there is no such matter. But that the *Oyl*, and all other *Vegetable Principles* are actually existent in, and mixed *per minima*, though in an extraordinary small proportion, with the *Water*. Even as we see the distilled *Waters* of *Anise Seeds*, *Penyroyal*, and the like to be impregnated with their own *Oyls*, which give the *Taste* and *Smell* to such *Waters*.

8. §. Wherefore, as a certain quantity of any *Salt* may be dissolved in *Water*; beyond which, it will not mix therewith, but remains under its own *Form*: So is there a certain proportion of *Oyl*, though far less, which may also be perfectly mixed with *Water*; and is certainly so, more or less, with all the *Water* in the world. But if that proportion, or degree of impregnation be once exceeded; the particles of *Oyl* do then, and not till then, gather into a body, and appear under their own *Form*.

9. §.

9. §. I say therefore, that all kinds of *Vegetable Principles*, are either in or together with the *Water*, with less difference first received into a *Plant*. But when they are once therein; they are then *separated*, that is to say, *filtred*, some from others, in very different *Proportions* and *Conjunctions* by the several *Parts*; the *Watery* by one *Part*, the *Aery* by another, the *Oyly* by another, and so the rest: and so every *Part* is the *Receptacle* of a *Liquor*, become peculiar, not by any *Transformation*, but only the *Percolation* of *Parts* out of the common *Mass* or *Stock* of *Sap*. And so all those parts of the *Sap*, which are *superflous* to any kind of *Plant*, are at the same time, discharged back by *Perspirations*, into the *Aer*.

10. §. AND, that *Nature*, in the various *Percolations* and *Separations* of the *Sap*, may still the better answer her end; hence, it is, that she carefully seeth, not only to the special *Nature* and *Proportion* of the *Organs*, by which she doth her work; but likewise to their very *Position*. Thus it is observable, That whereas the *Lymphæducts*, which carry a more *Watery Liquor*, are still placed on the inner *Verge* of the *Barque*, next to the *Aer-Vessels*: the *Lactiferous* and *Resiniferous Vessels* of *Plants*, to whose *Oylie Liquor* a mixture of much *Aer* is incongruous; do usually stand, neither on the inner, nor the outer verge of the *Barque*; but in the *midle*. By which means, they are at the greatest distance, and so most secure, from the *Aer*; either that which enters the *Barque* at the Circumference, or from the *Wood* and *Pith*.

11. §. AND because the *Resinous Liquors* of *Plants* are more *Oily*, than their *Milky*; their security therefore, from the approach of the *Aer*, is yet further contrived. In that in *Pine*, and other *Resinous Trees*, the *Diametral Insertions* are never found; or at least, not *visible*: which yet in other *Trees*, are *conspicuous*; being those *Parts*, whose office it is, to introduce the *Aer* from the *Aer-Vessels* into the *Barque*.

12. §. AGAIN, the *Milky Liquors* of *Plants* being thinner than the *Resinous*, and having a considerable quantity of *Water* mixed with their *Oyl*; hence it is, that in *Milky Plants*, as in *Rhus*, there are a greater number of *Lymphæducts*; and those standing nearer to the *Milky Vessels*, than they do in *Pine* and the like, to the *Resinous*. By which means they are better fitted to affuse their *Aqueous Parts* more plentifully to the said *Milky Liquor*.

13. §. FROM the *Mixture* of *Watery Parts* with the *Oylie*, it comes to pass, that whereas all *Lympha's, Mucilages*, and *Rosins* are transparent; the *Aquæ-oleous Liquors* of *Plants* are *Milky* or white, or otherwise *Opacous*. For the same thing is the cause of the whiteness of *Vegetable*, as of *Animal-Milk*: that is to say, a more copious mixture of *Watery* and *Oily Parts per minima*, or into one Body. For even the *Serous* and *Oylie Parts* of *Animal Milk*, when throughly separated one from the other, they become very transparent. So the *Stillatitious Oyl* of *Anise Seeds*, is most transparent and *limpid*, even as *Water* it self: yet there is a known sort of *White Anise-Seed Water*, as it is commonly called: that it is to say, wherein the *Oyl*, in distillation, ariseth and is mixed more plentifully with the *Water*. And the *Water*, wherein the stillatitious *Oyl* of any *Vegetable* is dissolved, becomes a perfect white *Milk*; as in this *Honourable and Learned Presence*, I have formerly had occasion to shew the Experiment. (a)

(a) See the *Discourse of Mixture*

14. §.

14. §. AND that the *Milky Liquors* of all *Vegetables* whatsoever, are more *Oylie* than their *Lympha's*, is most certain. For all those *Gums*, which dissolve either in *Oyl* or in *Water*, as *Galbanum*, and the like, are originally the *Milky Juyces* of *Plants*. And if you take the *Milk* of any *Plant*, as for instance, the *Milk* of common *Sumach*, or of any *Taste*, *Bitter*, *Astringent*, *Hot*, *Cold*, or any other whatsoever; and having well *dryed* it, and then fired it at a candle; it will thereupon burn with a very bright and durable flame, even like that of *Tar* or *Turpentine* it self.

15. §. FROM what hath been said, we may likewise gather the most genuine import of the word *Gum*, and the distinction thereof both from a *Rosin* and a *Mucilage*. First, a *Rosin*, is originally a *Turpentine*, or *Acidoleous Liquor*, having an exceeding small quantity of *Watery Parts* mixed therewith; and which, for that reason, will not be dissolved in *Water*, but only in *Oyl*. Of this kind are *Mastick*, *Benzoine*, *Taccamahacca*, and divers others, commonly, in our *Bils* to *Apothecaries*, called *Gums*. Yet in strict speaking they are all so many *Rosins*.

16. §. Secondly, a *Gum*, and every *Oylie Gum*, is originally a *Milky Liquor*, having a greater quantity of *Water* mixed with its *Oyly Parts*; and which for that reason, will be made to dissolve either in *Water* or *Oyl*. Of this kind are *Sagapen*, *Opopanax*, *Ammoniac*, and others.

17. §. The third sort of *Gum*, is that which is *Unoylie*, and which therefore dissolveth only in *Water*, as *Gum-arabick*, the *Gum* of *Cherry-Tree*, and others such like. This *Gum*, though commonly so called, yet is properly but a dryed *Mucilage*: being originally nothing else but the *Mucilaginous Lympha* issuing from the *Vessels* of the *Tree*. In like manner, as it doth from *Cumfry*, *Mallow*, and divers other *Plants*: and even from the *Cucumer*. The *Vessels* whereof, upon cutting cross, yield a *Lympha*, which is plainly *Mucilaginous*, and which being well dryed, at length becomes a kind of *Gum*, or rather a hardened *Mucilage*. In like manner, the *Gums* of *Plum-tree*, *Cherry-tree* and the like, are nothing else but *dryed Mucilages*. Or, if we will take the word in its widest sense, then all *Gums* are originally, either a *Terebinth*, or a *Milk*, or a *Mucilage*.

18. §. I have likewise made divers Observations of the *Tasts*, *Smells*, and *Colours* of *Plants*, and of their *Contents*, since those I last published: and that both for the finding out the true *Causes* of their *Generation*, and also the applying of them unto *Medical* and other *Uses*. Of which hereafter.

CHAP. V.

Of the Figuration *of* Trunks.

THE *Fifth Head*, shall be, of the *Figuration* of *Trunks*. Which also, as well as the *making* of *Liquors*, dependeth upon the *Structure* of the *Parts*. As *First*, almost all *Shrubs* (*cæteris paribus*) have a greater number of *Aer-Vessels*; and those of a smaller *Size*; and consequently much spread abroad, as most easily yielding to the *magnetick* Power of the *Aer*, according as we have more fully demonstrated, in speaking of the *Vegetation* of *Roots*: as in *Elder, Hazel, Fig, Sumach*, and the like. By which spreading, the said *Aer-Vessels* do sooner, and more easily strike into the *Barque*, and so produce collateral *Buds* and *Branches*, and that upon the first rising of the *Body* from the *Root*: that is, the *Plant* becomes a *Shrub*.

2. §. BUT if the said *Aer-Vessels* are very large, they will not yield so easily to shoot out collaterally; and so the *Trunk* grows up taller and more entire: as in *Oak*, *Wallnut, Elm*, &c. wherein they are exceeding large, is seen. Hence also the *Vine*, if supported, will grow to a prodigious length. And *Hops* and *Bryony*, are some of the tallest, amongst all *Annual Growths*: the *Aer-Vessels* of all which, are very large. Whereas *Borage*, and many other like *Plants*, although the *Pores* of their *Parenchyma*, are vastly wide, and filled with *Sap*; yet because their *Aer-Vessels* are small, they are therefore but *Dwarf-Plants*. Wherefore the tallness or advancement of a *Plant* or *Tree*, dependeth not upon the *Plenty* of *Sap*, how great soever, but on the *Largeness* of the *Aer-Vessels*.

3. §. AGAIN, as a *Plant* or *Tree* grows either *Shrubby*, or *Tall* and *Entire*, according to the *Size* of the said *Vessels*: so from their *Position*, doth it grow *Slender* or *Thick*. So, where they keep more within the compass of a *Ring*, as in *Elm*, and *Ash*, the *Tree*, in proportion, usually grows taller, and less thick. But where the said *Vessels* are spread more abroad, and especially are postured in *Rays*, as they are in *Oak*, the *Tree* grows very thick. Because the said *Vessels* thus standing all along nearer to the *Insertions*, there is a more ready and copious passage of the *Aer* out of the one into the other; and so the *Diametral* growth of the *Wood* is more promoted.

4. §. LASTLY, from the same general cause it is, That the *Trunks* of *Vegetables* are either *Round* or *Angular*. Those of all *Trees* are *Round*. Because the *Barque*, being here thicker, and the *Aer-Vessels* bound up with a greater quantity of *Wood*; the *Aer* hath not sufficient power to move them, and the *Barque* with them, into those various *Positions* or *Figurations*, as the *Trunks* of *Herbs* do yield to.

5. §. Yet the cause of the various *shapes* of the *Trunk*, is not the *Aer* alone; but partly, the *Principles* of the *Plants* themselves, in conjunction therewith; according to the predominion whereof, and chiefly of some certain kind of *Salt* or *Salts*, as I shall hereafter (a) more particularly explicate) the *Trunk* is *Square, Triangular, Pentangular*, or otherwise *Figured*. And thus much in general of the *Figuration* of *Trunks*. (a) B.4.P. 1. Ch. 6.

CHAP. VI.

Of the Motions *of Trunks.*

HE *Motions* alſo of *Trunks* are various. Principally *Four*; ſc. *Aſcending, Deſcending, Horizontal,* and *Spiral*. The cauſe of the *Aſcent* of a *Plant*, is a certain *Magnetick Correſpendence* betwixt the *Aer* and the *Aer-Veſſels* of a *Plant*; the *Motion* and *Tendency* whereof, the whole *Plant* follows. This I have aſſerted, and I think, clearly demonſtrated in my *Firſt* and *Second* Books of the *Anatomy* of *Plants*. I will here add this plain *Experiment*.

2. §. Take a Box of *Moulds*, with a hole bored in the bottom, wide enough to admit the *Stalk* of a *Plant*, and ſet it upon ſtilts half a yard or more above ground. Then lodg in the *Mould* ſome *Plant*, for Example a *Bean*, in ſuch ſort, that the *Root* of the *Bean* ſtanding in the *Moulds* may poynt upwards, the *Stalk* towards the ground. As the *Plant* grows, it will follow, that at length the *Stalk* will riſe upward, and the *Root*, on the contrary, arch it ſelf downward. Which evidently ſhews, That it is not ſufficient, that the *Root* hath *Earth* to ſhoot into, or that its *Motion* is only an *Appetite* of being therein lodged, which way ſoever that be: but that its nature is, though within the *Earth* already, yet to change its *Poſition*, and to *move Downwards*. And ſo likewiſe of the *Trunk*, that it riſes, when a *Seed* ſprouts, out of the Ground, not meerly becauſe it hath an *Appetite* of being in the open *Aer*; for in this Experiment it is ſo already; yet now makes a new *Motion* upwards.

3. §. BUT although the *Natural Motion* of the *Trunk* be to *Aſcend*; yet is it forced oftentimes to *Deſcend*. For the *Trunk-Roots* growing out of ſome *Plants* near the ground, and ſhrinking thereinto, like ſo many *Ropes*, do pluck the *Trunk* annually lower and lower into the ground together with them; as may be ſeen in *Scrophularia, Jacobæa,* and many other *Plants*.

4. §. IF theſe *Trunk-Roots* break out only about the *bottom* of the *Trunk*, as in the aforeſaid *Plants*, then the *Trunk* gradually *Deſcends* into the *Earth*, and is turned into a *Root*. But if it be very *ſlender*, and the *Trunk-Roots* break forth *all along* it, then it *Creeps* horizontally; the ſaid *Roots* tethering it, as it trails along, to the ground; as in *Strawberry, Cinquefoyl, Mint, Scordium,* &c.

5. §. AS to their *Spiral Motion*, it is to be noted; That the *Wood* of all *Convolvula's* or *Winders*, ſtands more cloſe and round together in or near the Center, thereby making a round, and ſlender *Trunk*. To the end, it may be more tractable, to the power of the external *Motor*, what ever that be: and alſo more ſecure from breaking by its winding *Motion*.

6. §.

6. §. Wherefore, *Convolvula's* do not wind by any peculiar Nature or *Genius*, which other *Trunks* have not; but becaufe their *Parts* are difpofed fo, as to render them more fequaceous to the external *Motor*. Even as the *Claspers* of a *Vine*, having the like *Structure*, have alfo a *Motion* of *Convolution*: whereas the *Branches* themfelves upon a contrary account, move in a *straight Line*.

7. §. The *Convolution* of *Plants*, hath been obferved only in thofe that Climb. But it feems probable, that many others do alfo *wind*; in which, the main *Stalk*, is as the *Axis* to the *Branches* round about. Of which number, I conceive, are all thofe whofe *Roots* are twifted; a *Motion* we obferved in fpeaking of the *Root*. Whether it be fo, or not the Experiment may eafily be made by tying a *Thred* upon any of the *Branches*; fetting down the refpect it then hath to any Quarter in the *Heavens*: for, if it fhall appear in two or three Months, to have changed its Situation towards fome other Quarter; it is a certain proof hereof. And that hereby the *Roots* of many *Plants* become twifted; the *Motion* beginning in the *Stalk*, and ending at the bottom of the *Root*, which ftands always fixed in the fame place. B. 2. P. 1. Ch. 1.

8. §. The *Convolution* of *Trunks*, is made not *one*, but *divers* ways; fome moving by *South* from *East* to *West*; and others from *West* to *East*. Wherefore it feemeth, that as the *Efficient Caufe* of *Convolution*, is not within the *Plant*, but external: fo alfo, that it is not *One*, but that there are *Two Great Efficients* of this *Motion*; fc. the *Sun* and the *Moon*. Some winding together with the *Sun*, in its *Diurnal Motion*, (or, if the the Earth moves, then, Inclining to the *Sun*) by *South* from *East* to *West*. And others winding with the *Moon*, in its *Monthly Motion*, from *West* to *East*.

9. §. This poffibly, may alfo be one *fenfible* way of diftinguifhing betwixt *Solar*, and *Lunar Plants*. Thus far, in general, of the *Motions* of *Trunks*.

CHAP. VII.

Of the Nature of Timber or Trunks, as they ferve for Mechanick Ufe.

THE laft thing I purpofed to fpeak of, is, Thofe feveral *Qualities* of *Timber* or of *Trunks*, by which they are fitted for *Mechanical Ufe*. As *Hardnefs*, *Softnefs*, *Faftnefs*, *Clevefomenefs*, *Toughnefs*, *Brittlenefs*, *Durablenefs*, or any of the fame *Qualities* compounded. The *Vifible Caufes* whereof are obfervable, *Partly*, in the *Structure* of the feveral *Parts*; fc. the *Infertions*, *Sap-Veffels* and *Aer-Veffels*; as to the *Number*, *Size*, or *Pofition* of any of them. And *partly*, in the *Nature* of the *Parts*; I mean fuch as is *manifeft to fenfe*. According to our clear and diftinct obferving of all which Caufes, we may underftand, Wherefore any *Wood* is made ufe of for any certain purpofe. And alfo, wherein fitly to apply it to further *Ufe*. In order to which, a compleat

compleat History of the *Mechanical Uses* of *Vegetables* would very much conduce. I shall for the present give some *Instances*.

2. §. AS *First*, some *Woods* are *soft*, as *Deal*, and *Sallow*. Yet from different Causes. *Deal*, from the great *Porosity* of the *Wood* it self, or the large *Pores* amongst the *Sap-Vessels*. But *Sallow*, from the great number of *Aer-Vessels* spread all over it. And therefore, though they are both *soft*, yet will not serve for the same purposes; *Sallow* being well wrought upon, which way soever you cut it: but *Deal*, especially the white *Deal*, if it be cut cross, it tears, and will never polish or work smooth.

3. §. Again, in *Sallow*, by the equal spreading of the *Aer-Vessels*, the *Softness* is equal or alike in all *Parts*. For which cause it maketh an excellent *Coal* for *Painters Scribets*. Because it doth not only make a *light Stroak*, but every where *certain*; and so doth not disturb the even *Motion* of the *Hand*. For the same cause, *shoomakers* also make use of it for their *Carving-boards*. Because being every where equally soft, it turns not the edge of their *Knives*, Which *Deal* would presently do; because though very soft in some places, yet in others 'its hard; that is to say, on the inner *Verge* of every annual *Ring* of *Wood*, where the old *Sap-Vessels* grow much more compact and close together.

4. §. AGAIN, some *Woods* are *soft*, but not *fast*; others are both, as *Linn*: its *Softness*, depending on the numerousness and equal spreading of the *Aer-Vessels*: its *Fastness*, on the closeness of the true *Wood*, and the shortness, and smallness of the *Insertions*. For which cause, it is of excellent use for many purposes; and particularly, for *small Sculpture*: such as may sometimes be seen for the Frames of *Looking-Glasses*, or of smaller *Pictures* in *Water-Colours*.

5. §. SOME *Woods*, again, are *fast*, and *hard*, as *Elm*. Its *hardness* depending upon the closeness of the *Wood*. Its *fastness*, *Partly*, upon the same cause; and *partly*, on the smalness of the *Insertions*; as also on the fewness of the *Aer-Vessels* in proportion with the *Wood*; and on the *thwart* and *cross Position* of many of them. Hence it is, that *Elm*, of all others, is the most *Cross-grain'd Timber*; that is, cleaveth so unevenly, to and fro, according to the *cross Position* of the said *Vessels*.

6. §. Hence also it cleaveth the *most Difficultly*. Even then, when it is without any *Knots*. For which reason it is always used, as best for the *Hub* of a great *Wheel*. As also for *Water-Pipes*, and for *Pumps*. Not because it is the most *durable Wood*; but because it will not *split* or *crack*, either in the *working*, or *afterwards*. For the very same reason, it is used for *Coffins*; that is, because, it will not *split* in working: not because it will endure longest under ground; for *Pales* are always made of *Oak*. So also the *Ladles* and *Soles* of a *Mill-wheel* are always made of *Elm*; as also the *Keel* of a *Boat*, &c. lest they should *split*: but the other *Parts* are made of *Oak*.

7. §. It may here also be noted, That the *Planks* commonly called *Groaning-Boards*, lately exposed, as a kind of *Prodigy*, to the view and hearing of many People, were of *Elm*. The *Aer-Vessels* of this *Wood*, being, though not more numerous, yet more ample, than in any other *Timber*. So that upon the application of the *Red-hot-Iron*, as was usual, and thereby the *Rarifaction* of the *Aer* and *Watery Parts* in the *Timber*; every *Vessel* became, as it were a little *Wind-Pipe* for

their

their *Expiration*. And as a great many Drops falling together in a shower of Rain; so a great many of these *Pipes* playing together, might make a kind of big or groaning noyse.

8. §. AS *Elm*, of all *Woods*, is one of the *fastest*; So, on the contrary, of all *hard Woods*, *Oak* is the most *Cleavesome*, or *splitteth* the most *easily*. The cause whereof is, partly, the *Largeness* of the *Insertions*; and partly, the *Diametral* or *Radiated Position* of most of the *Aer-Vessels*: upon both which accounts, wherever a *crack* is once begun, 'tis easily continued throughout the *Diameter* of the *Trunk*.

9. §. AGAIN, some *Woods* are hard, fast, and tough. So is *Ash*, and especially *Beech*. Hard and fast, from some of the same Causes, as *Elm*. Tough not from the *Structure*, but from the *Nature* of the *Parts*; whose *Principles* are united in a more exact proportion. Wherefore *London-Cars* have the *Rings* of their *Wheels* of *Beech*; because it *tears* more difficulty than even *Ash* it self. Whence also for *large Screws*, there is no *Wood* like it. But for *Small Screws*, of about an *Inch Diameter*, *Birch* is the best; as being, though not so *hard*, yet more *tough*.

10. §. THE more *Brittle* a *Wood* is, 'tis likewise usually the more durable. So *Oak*, which, with respect to its *hardness*, is not a *tough*, but very *brittle Wood*, is almost as *durable* as any. Whereas *Beech*, *Birch*, and the like, although very *tough*; yet for *Duration*, are of no service; for there are no *Woods* will rot sooner: and therefore, though strong enough, yet unfit to make any *Standing Parts* of *Building*, or of *Furniture*; especially in wet and moist places. Because, these *Woods*, having a less proportion of *Oyl*, than there is in *Oak*; they are apter to imbibe the moisture even of a *dank Aer*; by which moisture, they either *Rot*, or breed *Worms*, which destroy them.

11. §. HENCE it is, that what we call the *Heart* of *Timber*, as it is more *brittle*, so also more *durable*; *sc.* Because more *Oylie*. So that which is called the *Sap* of *Oak*, is much more *tough* than the *Heart*, although the *Heart* be more *durable*. That is to say, the older the *Wood* is, the *Watery Parts* are the more evapourated, whilst the *Oylie* still remaine, as a kind of *Tincture* or *Extract* in the *Wood*. Even as we see, that the older *Seeds* of any one *Kind*, are more *Oylie* than those that are green and young. So that the *Oylie* or *Rosinous Parts* of the *Sap*, are a kind of *Embalming* to the *Heart*, or older *Part* of a *Tree*, securing it from the destructive impressions of the *Aer*. For which Cause it is, that *Oak*, *Yew*, *Cocus*, *Guajacum*, &c. which are *Oylie Woods*, have always much *Heart*, whereas *Birch*, *Alder*, *Beech*, *Maple*, which are very *Unoylie*, have never any *Heart*.

12. §. FROM hence likewise we may understand the Cause of the *Toughness* of *Flax*: what we call *Flax*, being only the *Sap-Vessels*, or *Lignous Fibres* of the *Barque*. And generally, the *Barque* of any *Tree*, as of *Willow* (whereof are usually made a sort of *Ropes*) is very *tough*. The *Vessels* being here younger, and less *Oylie* than in the *Wood*. So likewise *Hemp*, is nothing else but the *Sap-Vessels* of the *Barque* of the *Plant* so called. And *Scotch-Cloath*, is only the *Housewifery* of the same *Parts* of the *Barque* of *Nettle*.

13. §. WHENCE it is very probable, that there are many other *Plants*, as well as the above named, whereof might be made good *Tow*. And of some, especially in some respects, better than of *Flax* it self. Because that even *Hemp*, although it will not make so

fine a *Staple*, as *Flax* (for all our fine *Hollands* are made of *Flax*) yet *Flax*, which is but of the same fineness as *Hemp*, will never, by all the Art yet known, be made so white as *Hemp* is made. The *Qualities* therefore of the best *Tow*, that can be in Nature, are that the *Staple* be *long, small, tough,* and *white*. So that if in the *Barque* of any *Plant*, we can find these *Qualities*, or any of them, to excell; we may be sure, it will be of better use, in some respects, for the making of *Cloath*, or other purpose, than *Flax* it self.

14. §. I WILL conclude with one *Instance* more, and that is as to *Grafting*. The good and happy success whereof, doth certainly depend upon the suitableness or respondence betwixt the several *Parts* of the *Stock* and *Cyon*; as the *Barque, Wood,* and *Pith*; and that both as to the *Number, Size,* and *Position* of the said *Parts*, and of their several *Pores* or *Vessels:* according to the degrees whereof, the *Conjunction* (*cæteris paribus*) will be more or less prosperous. So that of all such *Conjunctions* as are found to be apt and taking, and which some have learned not without long Practice and Experience; another, only by comparing the *Branches* of *Trees* together, may with little trouble, and in much less time, inform himself. By the same means, some *Conjunctions* which seem to be strange, as *Quince* and *Pear, White Thorn* and *Medlar,* &c. do yet, by the respondence of their *Parts*, as well as by *Experience*, appear to be good. And there is no doubt, but that many *Conjunctions* not yet tryed, or not known to have been so, may upon the same ground, be tryed with good success.

15. §. The chief Use of *Grafting* and *Inoculation*, is, That they *Accelerate* the growth of *Good Fruit*. The *Cause* whereof, is the *Knot*, which is always made in the *Conjunction*. By means of which, all the *Sap* is strained, and so ascendeth up into the *Graff* or *Bud*, both *Purer* and in less *Quantity*; and is therefore better and sooner concocted. Hence, the smaller the *Fruit* of any *Tree*, though it be not the best, yet the *Sap* being there, in *less Qantity*, is the *sooner ripe*. On the contrary, where the *Sap* ascendeth too freely, it doth not only *retard* the growth of the *Fruit*, but produceth *Barrenness*; as is seen in those luxuriant *Branches*, where it runs all up to *Leaves*. Hence also *Vines*, by *Bleeding*, become more *Fruitful:* that is, by the Effusion of *Part* of the *Sap*, there is a more easier *melioration* of that which remains. Even as *Phlebotomy* doth oftentimes produce a more healthful and better *Habit* of our own *Bodies*. To conclude, the *lessening* the *Quantity*, and thereby the *melioration* of the ascending *Sap*, by *Knots*, is *Natures* own contrivance; as is seen in *Sugar-Cane, Corn,* and other *Plants*.

THE
ANATOMY
OF
LEAVES, FLOWERS, FRUITS and SEEDS.

In Four Parts.

The FOURTH BOOK.

By *NEHEMJAH GREW* M.D. Fellow of the *ROYAL SOCIETY*, and of the *COLLEGE* of *PHYSICIANS*.

LONDON,

Printed by *W. Rawlins*, 1682.

THE CONTENTS

OF THE

First Part.

CHAP. I.
Of the Protections *and* Folds *of* Leaves.

CHAP. II.
Of those Things which appear upon the Surface *of the* Leaf.

CHAP. III.
Of the Figures *of* Leaves; *and the* Apparent Position *of the* Fibres.

CHAP. IV.
Of the Parts *and* Texture *of the* Leaf.

CHAP. V.
Of the Duration *of* Leaves, *and the* Time *of their* Generation.

CHAP. VI.
Of the Manner *of the* Generation *of the* Leaf. *Where also, that of the* Two General Parts *of a* Plant, *fc. the* Lignous *and* Parenchymous, *is further explain'd.*

To the Honourable
Robert Boyle Esq;

SIR,

AFTER I had finished the foregoing Books, *In which, I conceive, as far as* Glasses *will yet lead us, I have clearly De-scrib'd and Delineated the* Structure of a Plant; *and have endeavour'd, in some part, to Unfold the* Reason *and* Scope of Nature there-in: *I was willing to sit down, and leave what remained, to the* Improvements *of the* Present *and* Succeeding Ages.

But in Discourse upon this Subject, You *have been pleased frequently to insist, That I should by no means omit, to give likewise, some Examples of the* Mechanisme of Nature *in all the other* Parts. *The Performance whereof therefore, next to the* Obedience *I owe to the* Royal Society, *is to be looked upon, as a Due to the Authority which Your Judgment hath over me.*

This I have said, that, if what is herein done, shall prove acceptable unto Learned Men; *they may know, To whom they are oncemore to give their Thanks: After they have so often done it, upon (a better score) the Publishing of Your own Excellent Works. In which, there seems to be a Question, Whether Your Continual Endeavours, to enlarge the Bounds of* Natural Knowledge, *or Your Successes therein, have been the Greater. So that, whereas* Nobility *in some, doth only*

The Epistle Dedicatory.

only serve to lift them, like Jupiter's Satellits, *out of sight:* You, *by giving a greater Light, have drawn all Mens Eyes upon* You. *And whilest there are many, in all Ages, fond of Preheminency in the Conduct of Popular Affairs; who yet rarely hit the Mark they aim at; or aim at That they pretend:* You *have thought fit, rather to separate* Your Self, *to that more Innocent, and more Noble Sort of* Wisdom, *which lieth, not in the Arts of Conceiling, but in* Discovering, *the Truth of Things.*

That we may have many to imitate You *herein, cannot but be heartily wish'd by all, who regard the Honour of their own Country; as it is, with much Zeal, by*

Sir,

Your most obedient

Servant

NEHEMJAH GREW.

THE

THE ANATOMY OF LEAVES,

PROSECUTED

With the bare EYE,

And with the

MICROSCOPE

Read before the *Royal Society*, *Octob.* 26. 1676.

PART I.

CHAP. I.

Of the Protections *and* Folds *of* Leaves.

IN THE General *Anatomy* of *Plants*, I have assigned one whole *Chapter* (a) to the *Germen* and *Leaf.* Since then, I have occasionally made divers *Remarques* of the same; both with the Naked Eye, as there, and also with the *Microscope.* The *Principal* whereof, I shall here set down; without repeating any from thence; or obliging my self strictly to the *Order* there used.

(a) *Lib.* 1. *Ch.* 4.

2. §. That which in a *Germen*, first occurs to the *Eye*, is the *Protection* of the *Leaves*, or the various *Methods* which Nature takes to preserve them from the *Injuries* both of the *Ground*, and of the *Weather.* To the Instances formerly given, I shall add these that follow.

B b   3. §.

3. §. AND *First*, it is observable of the young *Buds* of *Ammi*, that left they should be bruised, or starved, upon their first *Eruption* from under the Ground; they are couched, as *Firn* is rowl'd, inward; each *Bud*, against the *Base* of the *Stalk* of the foregoing *Leaves*, and most exactly laid up within the *Membranes* thence produced: Just as the Child in the Womb, lies with his *Head* against his *Knees*; or as it is afterwards embraced with the *Armes* of the Nurse. And it is a general Rule of Nature, where the *Stalks* of the *Leaves* are so long, that they cannot lap one over another, and where no other special *Protection* is provided; for the bottoms of the *Stalks* to be produced into broad *Membranes*, as Blankets to the succeeding *Buds*; as in *Crowfoot, Dovesfoot, Claver, Cransbill, Strawberry, Yarrow*, and others. And sometimes instead of two *Skins* lapped one over another, there is one entire *Skin*, produced from the *Stalk*, in which as within a *Secundine*, the *Bud* is safely shrowded; and which, in its Growth, it gradually breaks open.

Tab. 41.

4. §. THE same is also observable in *Dock, Sorrel, Bistort*, and all other *Plants* of this Kindred; with this difference, That every *Veil* or *Secundine* is not here produced from the *Stalk* of the *Leaf*; but hath its *Original Distinct* from it. And whereas in the former, every *Bud* hath only one to it self: in these *Plants*, every lesser *Leaf*, together with its own proper *Veil*, is always inclosed, with the next greater *Leaf*, in another *Veil* common to them both; and both these with the next, in another; and so on to the greatest. These *Veils* are extream thin, and have very few *Vessels*; being so many meer transparent *Skins*. For which reason, there is always found a *Mucilage* or clear *Gelly*, between every *Leaf*, and its *Veil*, and between *Veil* and *Veil*. The one, thus preserving the other, (as do the *Humors* and *Membranes* of the Eye) from drying and shrinking up, and thereby from becoming useless for the *Protection* of the *Plant*.

5. §. THE *Orchis*, and other *Plants* of this kindred, because they *Spring* and *Flower* early, when the mornings are cold, have a double *Sheath*, or *Blanket* over all. The *Buds* of some *Herbs* (as of *Plantain*) having no *Hairs* growing on them, are covered with *Hairy Thrums*. And the *Nettle* hath *Bastard-Leaves*, or *Interfoyls* between *Leaf* and *Leaf*, for the preservation of its *Stings*.

Tab. 41.

6. §. ANOTHER Sort of *Protection* is seen in *Wild Clary, White Archangel*, and other *Plants* of a like Shape. In which, the greater *Leaves* do still cover and inclose the lesser, not by being lapped over them, as where the *Leaves* are more numerous, is usual; but by a *Double Fore-Curl* at the bottom of every two greater *Leaves*; by which the little *Under-Bud* is embraced, and so kept safe and warm.

Tab. 42.

7. §. THE *Leaves* of *Onions* are all *Pipes* one within another. These *Pipes* are every where entire, saving about the middle, where they have a small *Aperture*; common to all of them, even the most minute in the Centre: not being a forced *Crack*, but a *Door* originally formed, for the issuing of every lesser *Pipe*, out of a greater.

Tab. 41.

8. §. THE LAST I shall give, is that which is remarkable in *Common Sumach*. The *Buds* whereof, being exceeding tender, Nature appears sollicitous in a peculiar manner, for their preservation. For whereas in other *Plants*, they are well enough secured only by standing behind the *Stalks* of the elder *Leaves*: here they are lodged within the

very

Book IV. *of Leaves.* 147

very *Body* of the *Stalk*; as entirely, as a *Kernel* is within an *Apple*, or a *Fœtus* in the *Womb*. From whence it comes to pafs that the *Bafis* of every *Stalk* is extreamly fwelled, as going Great with a *Bud*.

9. §. UPON THE removal of thofe *Parts*, which are contrived for the *Protection*; the *Foulds* and *Compofture* of the *Leaves* do next appear: all which are moft aptly fuited both to the *Number* and *Shape* of the *Leaves*, and alfo their *Pofition* upon the *Branch*. In the *Firſt Book* (*a*) I have given Examples of thefe *Eight Sorts*, fc. the (*a*) Ch. 4. *Plain Lap*, the *Plicature*, the *Duplicature*, the *Multiplicature*, the *Single Roll*, the *Double Back-Roll*, the *Double Fore-Role*, and the *Treble-* Tab. 42. *Roll*. To which I fhall add *Four* or *Five* more.

10. §. And *Firſt*, in fome *Plants*, as *Ground-Ivy*, *St. Johns Wort*, and divers others, where the *Leaves* are fmall, pretty numerous, and grow by pairs, they have no *Fould*, but ftand Flat and Tangent, like a pair of *Battledores* clapt together.

11. §. They have the like *Poſture* in *Baum*; faving, that here the Edges of the *Leaves* are a little *curled* backward. Not *Rolled*, a *Curl* being but the beginning of a *Roll*. So the feveral *Labels* of a *Grounfel-Leaf* are all laid in a *Back-Curl*.

12. §. The *Leaves* of fome *Plants*, as *Horehound*, *White Lamium*, *Nettle*, and others, are likewife only Tangent, but are fet with a *Fore-Curle*. And the feveral *Labels* or *Scallops* of the *Leaf* of *Common Crowfoot*, are all *Curled* Inward. But thofe of *Hepatica aurea*, are compofed into *Double Fore-Rolls*.

13. §. THE *Leaves* of *Sage*, *Scabious*, *Red Lamium*, *Lychinis Sylveſtris*, and others, are neither couched one over another, as in the *Bow-Lap*; nor plated, as in the *Flat Lap*; but being loofely foulded, of every pair of *Leaves*, the half of one is reciprocally received between Tab. 42. the two halfs of another, and may therefore be called the *Cleep*. A *Poſition* very well fuited to the Smalnefs of their Number, and the Equality of their Size, not fo well agreeing with the *Bow-Lap*; and the fomewhat inward Poſture of the *Fibres*, not allowing the *Flat Lap*. Sometimes, as in *Syringa*, where the *Leaves* are broader, the *Cleep* is joyned with a *Fore-Curle*.

14. §. THE laft I fhall mention, is the *Plaite-Roll*, as in the *Lapathum Alpinum*, which fome call *Englifh Rhubarb*. The *Leaves* whereof are fo very large, and the *Fibres* fo prominent; that befides and under the two *Back-Rolls*, they are alfo laid in feveral *Plaits*, and under thofe *Plaits*, again with leffer ones, all moft exquifitely Tucked up between the faid *Fibres*: So, as neither to bruife the fame, nor yet to leave any Vacuity: whereby every *Leaf*, and the whole *Bud*, lie clofe and round within their *Veils*.

CHAP. II.

Of those things which appear upon the Surface of the Leaf.

THESE are *Globular Excrescences, Spots, Hairs, Thorns* and *Prickles:* of all which, except *Spots*, I have spoken in the *Appendix* to the *Chapter of Leaves* in the *First Book.*

2. §. Of the *Globulets*, it may here be further noted, That those which are white, and lie sometimes like a fine Powder upon the *Leaf,* were once transparent, as in *Bears-Eear*; their cleer *Liquor* beeing now evaporated to an *Extract* or *White Flowers.* This, if licked off, will give you the *Tast* of the more *Essential Content* of the *Plant*; different from that perceived in chewing the *Leaf.*

Tab. 43.

3. §. For the observing of them, it may also be noted, That although they often grow on both sides the *Leaf* alike; yet sometimes, as in *Ground-Ivy,* only or chiefly on the Back-Side. And that in many *Plants,* where the elder *Leaves* have none; on the young *Buds* they are very numerous; as in *Corin Tree, Sorrel,* and others.

4. §. AS for *spots,* the smaller ones are observable not only in *St. Johns-worts,* (in which *Plant* only they are commonly taken notice of) but also in *Rue, Ground-Ivy, Pympernel* or *Anagallis,* and divers other *Plants,* when held up against the *Light.* The original whereof seems to be, at least in some, from the *Globulets* above mentioned; that is, when they break and dry away. So the *Spots* of *Rue-Leaves,* which in the *Reflection* of *Light* look *black,* but upon the *Trajection* thereof are *transparent,* are so many little *Holes,* pounced half way through the thickness of the *Leaf,* and seem as made, by the breaking and drying away of as many *Globulets.* Whence also, as the *Globulets* are best seen in the younger *Leaves,* so these *Spots* in the elder.

Tab. 43.

5. §. BESIDES these, and some others (as those in *Ladies-Thistle*) which are Natural to the *Leaf*; there are also some *Spots,* or rather *Streaks,* which are *Adventitious*; as those in the Leaves of *Sonchus.* The Cause whereof, is a small flat *Insect,* of a grey *Colour,* and about ⅒th of an Inch long. Which neither ranging in bredth, nor striking deep into the *Leaf*; eats so much only as lies just before it, and so runs scudding along betwixt the *Skin* and the *Pulp* of the *Leaf*; leaving a whitish *Streak* behind it, where the *Skin* is now loose, as the measure of its Voyage.

Tab. 43.

6. §. THE Original and several kinds of *Thorns,* I have describ'd in the above said *Appendix.* I only add, that the very *Leaves* of some *Plants,* if they stand till the second year, are changed into so many *Thorns,* as in the *Furz.*

7. §.

7. §. They are of Use, not only for the *Protection* of the *Bud*; but likewise, for the support of the *Plant*; as is observable in those *Climbers*, which are neither strong enough to stand of themselves; nor yet, from their fragility, are capable of winding about another, without being torn all to pieces. For which end also, these *Thorns* grow not like *Buds*, erected; but poynt all downwards, like so many *Tenters* or *Hanging-hooks*: as in the *Bramble*, chiefly on the *Stalks*; and in *Clivers*, also on the *Leaves* themselves; whereby they catch at any Thing that stands next them; and so, although such slim and feeble *Plants*, yet easily climb to a very great hight. *Tab.* 43.

8. §. OF THE several *Figures* of *Hairs*, and their Use, I have also spoken. As to one Use, *sc*. the *Protection* they give to the *Leaf*, I shall here further note, That the design of *Nature*, is the more evident if we consider, That all *Leaves* are not alike *Hairy*, nor at all times, nor in every part: but differently, according to their *Age, Substance, Texture*, and *Foulding* up. Their *Age*; for there are many young *Buds* covered with a thick warm *Hair*, which afterwards dries up and disappears, as useless; as those of the *Vine, Golden Liverwort*, &c. Their *Substance*; so those *Buds* which are tenderest, and would sooner feel the cold, if naked, have the fullest *Hair*; as of *Thistle, Mullen, Burdock*, and others. Their *Structure*; therefore those *Leaves*, whose *Fibres* stand more prominent or above their *Surface*, lest the cold should nip them, are covered with greater Store of *Hair*; as in *Moth-Mullen, Garden-Clary*, and the like. And their *Fould*; it being observable, That those *Leaves* which are folded up inward, have little or no *Hair* on their inner, but only on their Back-Sides, which are open to the *Aer*; as is visible in *Corin, Warden, Golden Liverwort*, and others. B. 1. Ch. 4. *Tab.* 43.

9. §. Add hereto, That where there is Store of *Hair*, Nature is the less sollicitous for other *Covers*; and where there is not, she is more. So the *Leaves* of *Beans* and *Peasen*, of *Nettle, Plantain*, &c. not being *Hairy*, have each a *Surfoyl*, or else certain *Hairy Thrums*, to protect them. And those *Plants* which have neither, are such as have a *Hotter Juyce*, and so less subject to the impressions of *Cold*, as *Speerwort, Scurvygrass, Watercress, Fenil*, and most of the *Umbelliferous* Kind.

10. §. *Hair* is of use to preserve young *Buds*, not only, from the cold *Aer*, but also from too much *Wet*; which, if it were contiguous, especially in *Winter*, would often rot and destroy them. But being made to stand off in drops at the ends of the *Hair*, doth not hurt, but refresh them. Thus doth *Nature* make the meanest Things sometimes subserve to the best Ends.

CHAP. III.

Of the Figure *of the Leaf; and the Apparent* Position *of the Fibres.*

HAT which in the *Leaf* offers it self next to be obſerved, is its *Figure*. This is infinitely varied with the ſeveral Kinds of *Plants*: and there are ſome, which have *Leaves* (beſides the two firſt *Diſſimilar* ones) of Two Kinds or Two diſtinct *Figures*; as the *Bitter-Sweet*, the common *Little Bell*, *Valerian*, *Lady-Smocks*, and others. For the *Under Leaves* of *Bitter-Sweet*, are Entire; the Upper, with two *Lobes*: the Under *Leaves* of the *Little Bell*, like thoſe of *Pancy*; the Upper, like thoſe of *Carnation*, or of *Sweet-William*. And in ſome *Plants*, Nature affecteth a Kind of *Irregularity*; the *Leaves* whereof are of no one certain *Figure*; as in *Dragon*, *Peony*, *Biſhops-Weed*, &c.

2. §. BUT the *Leaves* of moſt *Plants*, have a Regular *Figure*; and this Regularity, both in Length and Circuit, always defineable. In *Length*; by the Proportion between the ſeveral *Leaves* upon one

Tab. 46.

ſtalk, or between the ſeveral *Lobes* upon one *Leaf*. So the *Leaves* of *Clematis Sylv. major*, which ſtand by Ternaries, ſhorten by equal *Proportions*, that is to ſay, if, the chief *Fiber* of each, be divided into equal *Parts*; their ſeveral Lengths are not as Ten, Eight, and Four; but as Ten, Eight, and Six. So the *Lobes* and *Fibers* of *Clematis Virginiana Hederæ folio*, of *Artenuiſa*, &c. ſhorten in like manner by equal Proportions. The ſame is obſervable in meaſuring, upon a *Gooſeberry-*

Tab. 46.

Leaf, from the Poynt of the firſt *Lobe*, to the firſt *Angle*; from thence, to the ſecond Poynt; from thence, to the ſecond Angle; and from thence to the third Poynt.

3. §. But in many, the Proportion is different. So in the *Leaves* of the *Leſſer Maple*; the ſhortning of the *ſmaller Lobes*, with reſpect to the middelmoſt; is not Equal, but Double to that of the middlemoſt, with reſpect to the Greater. For if their chief *Fibres* be divided into Equal Parts, they are as Eleven, Nine, and Five. On the contrary, in the *Leaves* of *Althæa fruticoſa Pentaphylloidea*, the middlemoſt *Lobes* ſhorten by a greater Proportion than the Leaſt; all three being as Ten, Fourteen, and Twenty.

4. §. WITH reſpect to the *Circumference*, the *Figure* of moſt *Leaves* is very Complex. Yet Two things are evident. Firſt, that all Regular *Leaves*, are defined or meaſured out by *Circles*; that is, by the *Arches* or *Segments* of ſeveral *Circles*, having either the ſame, or divers *Centers* and *Diameters*. Secondly, That the Length of the *Leaf*, or of the chief *Fiber* thereof, is the *ſtandard Meaſure* for the *Diameters* of theſe *Circles*: theſe being either its full Length, or certain equal parts ſubſtracted, or multiplied; as half its Length, or its Length and half, &c.

5. §.

5. §. TO make this appear, I shall give several Instances: of some, where both the Edges are of one Measure; and of others, where they are different. And of both kinds, where they are measured by fewer, and where by more *Circles*.

6. §. The *Leaf* of *Lagopus major fol. pennat.* is measured by One *Circle*, the same on both Edges, whose *Diametre* is Thrice the Length of the *Leaf*.

7. §. That of *Syderitis Salviæ fol.* by Two *Circles*: the *Diameter* of the Lower, being Twice the Length of the *Leaf*; of the upper, the Length and half. In both these the *Circles* are drawn Outward; that is, with their Centers some where upon the middlemost or chief *Fiber* of the *Leaf*. *Tab. 44.*

8. §. That of *Orange-Tree*, is also measured by Two *Circles*: but one of them repeated with *Opposite Centers*. That next the *Cone* of the *Leaf*, is drawn Inward; that is, with the *Center* no where upon the *Leaf*, but without it. The *Diameter* hereof is just the Length of the *Leaf*. The midle part of the Edge is measured by the same *Circle*, only drawn Outward. The lower *Circle* next the *Stalk*, is drawn Inward, as the upper; and its *Diameter* Three times the Length of the *Leaf*. *Tab. 44.*

9. §. The *Leaf* of the *Venetian Vetch*, is measured by Three *Circles*. That next the *Cone*, drawn Inward; the *Diameter* whereof, is Twice the Length of the *Leaf*; the next is drawn Outward; whereof the *Diameter*, is just the Length. The third or lowermost, is drawn also Outward; and its *Diameter*, half the Length. So that they all lessen by an Equal Proportion. *Tab. 44.*

10. §. The *Leaf* of *Great Laserwort*, is also measured by Three *Circles*; all drawn Outward, and one of them Repeated. The *Diameter* of that next the *Cone*, is Half the Length of the *Leaf*; of the next, Thrice the Length; of the Third, just the Length; the lowermost, is the same with the First. *Tab. 45.*

11. §. That of *Broad Leav'd Laserwort*, is also measured with Three *Circles*; and one of them repeated with *Opposite Centers*. The *Diameter* of the First, is Half the Length of the *Leaf*; of the Second, Twice the Length; of the Third, just the Length: all of them drawn Outward. That next the *Stalk*, is the same with the First; only drawn Inward. *Tab. 44.*

12. §. The *Figure* of the *Leaf* of the *Cornelian Cherry*, is exactly that of the foregoing, Inverted: the same measure there beginning at the *Base*, and ending at the *Cone*; which here begins at the *Cone*, and ends at the *Base*: as by comparing their Draughts together may be observ'd. *Tab. 44.*

13. §. IN ALL, the foregoing Examples, both the Edges of the *Leaves* have the same *Measure*. But they have oftentimes, different ones; as in these that follow.

14. §. The *Leaf* of *Althæa fruticosa*, is measured by Three *Circles*. The left Edge (as the *Leaf* lies with the backside upward) by One *Circle*, but Twice repeated. For the *Diameter* of the First, is the Length of the *Leaf*; the Second is the same, but drawn upon another Center; the Third also the same, but drawn Inward. The right Edg, is measur'd by Two *Circles*: the *Diameter* of the First, being the Length of the *Leaf*; of the Second, Half the Length. *Tab. 45.*

15. §.

15. §. That of *Black Poplar*, by Three; and each Edge by Three repeated. On the left, the *Diameter* of the Firſt, is the Length of the Leaf; of the Second, Half the length; of the Third, the Length and Half. The *Meaſure* of the right Edge, is that of the left, Inverted: the ſame *Meaſure* there beginning at the *Baſe*, and ending at the *Cone*; which here begins at the *Cone*, and ends at the *Baſe*.

Tab. 45.

16. §. That of *Doronicum*, is meaſured by Three *Circles*, whereof, one is repeated Once; and another Thrice. The right Edge by Two, and One repeated. For the *Diameter* of the Firſt or that next the *Cone*; is the Length of the *Leaf*; the next is the ſame, but drawn Outward; the *Diameter* of the Third, is Half the Length. The left Edge, by Three *Circles*; whereof One is repeated on the ſame Edge, and Two, the ſame, as on the other. For the *Diameter* of the firſt, is the Length of the *Leaf*; of the Second, Four times the Length; the Third, the ſame as the Firſt; and of the Fourth, Half the Length.

Tab. 45.

17. §. Laſtly, that of *Mountain Calamint* is meaſured by Four *Circles*. The left Edge, by Three *Circles*, of which, the lowermoſt is once repeated: the right Edge alſo by Two; whereof the nether is likewiſe once repeated.

Tab. 45.

18. §. It may ſeem, even from theſe Inſtances, no very unobvious Concluſion; That all *Crooked Lines, Spiral, Helick, Elliptick, Hyperbolick, Regular*, or *Irregular*; are made up of the *Arches* of *Circles*, having either the ſame, or divers *Centers* and *Diameters*. And, as otherwiſe, ſo from the *Contemplation* of *Plants*, men might firſt be invited to *Mathematical* Enquirys.

19. §. TOGETHER with the *Figure* of the *Leaf*, the *Poſition* of the *Fibers*, as it is apparent before Diſſection, is obſervable; eſpecially on the back of the *Leaf*. Whereof I ſhall add, to what I have ſaid in the *Firſt Book*, the following *Remarques*.

20. §. Firſt, that there are ſome *Leaves*, in which the firſt Collateral *Fibres* make *Right Angles* with the Great one in the midle: as the *Great-Maple*, the *Great Celandine, Chondrilla*, and the reſt, or many, of the *Intybous* Kind; with ſome few others. But that generally all the chief *Fibers* of a *Leaf*, make *Accute Angles* together: both where they ſtand collateral with the midle *Fiber*, as in *Strawberry*; and where they all part at the *Stalk*, as in *Mallow*.

21. §. Again, that of theſe, there are ſome few, any two of whoſe Defining *Fibres* making two *Rays* of equal Length, take in One Eighth Part of a *Circle*, as in *Mallow*; and in ſome one Tenth: but in moſt they take in either one Twelfth part, as in *Holy-Oak*; or one Sixth, as in *Sirynga*. So that where the *Fibres* ſtand Collateral with one in the the midle, if you ſuppoſe them to be drawn out at *Oppoſite Angles*; or where the chief *Fibers* part at the *Stalk*, you only take in the *Stalk*; you will thereby divide a *Circle* into Eight, Twelve, or Six equal Parts; as in *Sirynga*, the *Vine* and others. And ſo likewiſe, where there are ſeveral *Sprigs* upon one *Stem*, as in *Fenil, Hemlock*, and the like: as will beſt be underſtood by the *Figures*.

Tab. 46, & 47.

Tab. 46, 47.

CHAP.

CHAP. IV.

Of the Parts and Texture of the Leaf.

 I COME next to observe the several *Parts*, whereof the *Leaf* is composed: and first the *skin*. This being stript off the *Leaf*, although to the bare Eye it looks no otherwise than a *skin* of *Isinglass*: yet being viewed through a good *Glass*, with a clear and true Light, and in an advantagious Position; it appears to consist not only of *Organical Parts*, as do the *skins* of *Animals*; but these also Regularly mixed together; that is, of *Parenchymous* and *Lignous Fibres*, all very curiously interwoven as it were, into a piece of admirably fine white *Sarcenet*: as in *Flag*, *Tab.* 48. *Tulip*, and the like.

2. §. From hence, it is easy to conceive how the *skins* of all *Plants*, as well as those of *Animals*, are perspirable; *sc.* between the several *Fibers* of which they consist. But as the *skins* of *Animals*, especially in some *Parts*, are made with certain open *Pores* or *Orifices*, either for the Reception, or the Elimination of something for the benefit of the *Body*: so likewise the *skins*, of at least many *Plants*, are formed with several *Orifices* or *Pass-ports*, either for the better *Avolation* of *Superfluous Sap*, or the *Admission* of *Aer*.

3. §. THESE *Orifices* are not in all *Leaves* alike; but varied in *Bigness*, *Number*, *Shape*, and *Position*: Serving to the different *Nature* of the *Plant*, or *Leaf*; and giving the *Leaf*, as it were, a different *Grain*. *Princes Feather*, *i. e.* a Sort of *Sanicle*, they stand only on the Edges of the *Leaf*; but are very ample. In the *White Lily*, they are *Oval*, very white, and each surrounded with a slender white *Border*. They stand about a 6th or 8th part of an Inch distant, as they appear through *Tab.* 48. a good *Glass*, all over the *Leaf*, but not in any regular Order. These *Orifices* are the cause of the Greyish *Gloss* on the upper side the *Leaf*: for the Back side, in which there are none of them, is of a dark *Sea-Green*.

4. §. In the *Leaf* of *Pine*, they are also *Oval*, and about the same *Bigness* and *Number*, as in that of a *Lily*; yet without a Border. But their *Position* is very Elegant, standing all, most exactly, in *Rank* and *Tab.* 48. *File* from one end of the *Leaf* to the other.

5. §. NEXT TO the *skin*, lies the *Pulpy* part of the *Leaf*; which by the same latitude, as *Use* hath taught us in many other Words, I call the *Parenchyma*. This *Parenchyma* or *Pulp* of the *Leaf*, like the *Pith*, and all other *Parenchymous Parts* of a *Plant* is made up of incomparably small *Cylindrick Fibres*: and these *Fibres*, in most *Leaves*, woven and wound up into little *Bladders*.

6. §. The *Bladders* are here of several Sizes, as in the *Pith*: but generally more visible in the *Stalk*, than in the *Body* of the *Leaf*. Va- *Tab.* 49. ried, as in the *Pith*, so here, not according to the *Size*, but the *Nature* of the *Leaf*. So in *Common Dock*, and *Moth Mullein*, both Great

Leaves, they are Small; in *Wild Clary*, a Leſſer *Leaf*, they are very Large. In the *Body* of the *Leaf*, ſometimes the *Sides* of the greater *Bladders*, are made up of leſſer ones; as in *Borage*.

Tab. 50.

7. §. In ſome *Leaves*, theſe *Parenchymous Fibres* are all drawn cloſe up together. In the Former, they are as the *Threds* in the *Open-work* of *Bone-Lace*; in Theſe, as the ſame *Threds*, in the *Cloth-work*.

8. §. The *Pithy Part*, in the *Stalk*, and almoſt up to the Top of the chief *Fiber*, in many *Leaves*, is Tubular; even whilſt they are yet Young and Sappy: as in *Sweet Chervil, Hemlock, Endive, Cichory, Lampſana, Dandelion, Burdock, Daiſye, Scorzonera*, and others. And ſometimes the ſaid *Pithy Part* is opened into ſeveral little *Pipes*, like ſo many *Aer-Veſſels*, above ½ a Foot long; as in the *Common Dock* and the *Little Spurge*, by ſome called *Wart-Wort*.

9. §. THE *Strings* of the *Leaf*, or thoſe *Fibres* which are viſible to the bare Eye, are compoſed of *Veſſels* of the Two General Kinds, ſc, for *Sap*, and for *Aer*. They are joyntly diſtributed throughout the *Leaf*: Yet not ſo, as to run meerly parallel; as in *Animals*, every *Artery* hath its *Vein*: but the *Aer-Veſſels* are every where Incloſed, or as it were *ſheathed* in the *Sap-Veſſels*.

10. §. THEIR *Poſition* is various and regular, not only in the *Body* of the *Leaf*, as is above ſhewed; but likewiſe in the *Stalk*: of which alſo I have given ſeveral Inſtances in the *Firſt* 𝕭𝖔𝖔𝖐. I ſhall here note, and more particularly deſcribe, One or Two more. In the *Stalk* of a *Mallow-Leaf*, they ſtand in Six Oblong *Parcels* of equal Size, and in a *Ring* near the Circuit. Whereby the *Stalk* is ſtronger, the Growth hereof, before and behind, more equal, and ſo the poſture of the *Leaf* more erect.

Tab. 49.

11. §. In *Dandelyon*, they ſtand in *Five Parcels*: of which the Greater ſtands a little *behind* the Centre of the *Stalk*; figured into a very ſmall *Half-Moon* or *Semi-Tube*, whoſe *Diametre*, through a *Glaſs*, is not above ⅛th of an Inch. The other Four, are extream ſmall *Cylinders*. Altogether make an *Angle*, twice as big as that of a V Conſonant. Whereby, although the *Stalk* be ſtrong enough to ſupport the younger *Leaves*; yet thoſe which are grown longer, and ſo not only by their *Bulk*, but their farther Extenſion from the Center of *Gravity*, are become more weighty; commonly lie flat on the Ground.

Tab. 49.

12. §. In *Wild Clary*, they ſtand alſo in *Five Parcels*, the Greater ſtands not *behind*, but *before* the *Center*; making an *Arch*, whoſe *Chord* in a *Glaſs*, is above ½ an Inch long; and belongeth to a *Circle*, whoſe *Diameter* is an Inch and half. The other Four, are ſmall *Cylinders*, alſo different from thoſe in *Dandelion*; the two bigger, there ſtanding hindmoſt; but here, the two Leſs, and the two Bigger, within the two round *Ridges* of the *Stalk*.

Tab. 49.

13. §. From hence it is, that the *Leaves* of this *Plant* have not only a *Prone* or *Horizontal Poſture*, but alſo make that *Forceable Preſſure* on the Ground, which can by no means be imputed to their *Weight*. For the *Great Arched-Fibre* ſtanding before the *Centre* of the *Stalk*, and the two Longer Round ones being uppermoſt, in the *Ridges* of the *Stalk*; they put on the upper parts thereof to a more full and forward Growth, and ſo to bow the *Leaf* back-ward. And the *Fibrous Arch* being, though broad, yet almoſt flat, doth hereby the more eaſily yield to that *Motion*.

14. §.

Book IV. of Leaves. 155

14. §. In *Borage*, and *Moth-Mullen*, they stand also in *Five Parcels*. In the former, the largest maketh still a more bulky *Arch*, than that of *Clary*; being thicker, as broad, and of a lesser *Circle* or more bowed. But in *Mullein*, it maketh almost an entire Oval. *Tab.* 49.

15. §. By means of this *Figuration*, a sufficient number of *Vessels* for such large *Leaves*, are not only more conveniently Distributed into them; but also stand more safely in the *Stalk*. For were the *Arch* contracted into a *solid Cylinder*, it could not so presently be resolved into small *Fibers*. And were it laid into a flat *Plate*, or straight out, either the *Figure* of the *Stalk*, and so of the *Leaf*, must be altered; or else, the two ends of the *Plate*, would come too near the Circumference of the *Stalk*, and so be more liable to the *Impressions* of the *Weather*: as may be observed in cutting the *Stalk* transversly, and by the *Figures*.

16. §. IN the *Body* of the *Leaf*, besides the *Positions* of the *Fibrous strings* or *Threds*; above expessed, there is one *Thred*, bigger or less, which in all *Plants*, runs round the Edge of the *Leaf*, and hems in all the rest; but can hardly be well observed in any, without stripping off the *Skin* of the *Leaf*. When the *Fibres* of the *Leaf*, are bigger, or less tender, as in *Holly*, the *Skin* and the *Pulp* are sometimes found either rotted off, or eaten away with *Insects*; whereby, both the said surrounding *Fiber*, and the rest, are all very fairly visible. *Tab.* 50.

17. §. THE *Vessels* seem to be continu'd, in the *Leaf*, by being Ramified out of Greater into Less, as *Veins* or *Arteries* are in *Animals*. But if the *Skin* and *Pulp* of the *Leaf*, as suppose a *Borage-Leaf* be taken off, and the *Vessels* laid bare; by the help of a good *Glass*, it will appear; That they are all of the same Size, every where in the *Leaf*; and also continued throughout the same, all several and distinct *Pipes* one from another, as the *Threds* in a *Skein* of *Silk*. And that therefore the *Distribution* of the *Threds* which the *Vessels* compose, is not the Ramifying of Greater *Pipes* into Less; but the dividing a greater Cluster of *Pipes*, into several lesser Clusters, till at last they come to be single; as in the *Distribution* of the *Nerves*. *Tab.* 50.

18. §. The *Vessels* seem also to be Inosculated, not only side to side, but the ends of some into the Sides of others. But neither is this ever really done: the lesser *Threds*, being only so far diducted, as sometimes to stand at *Right-Angles* with the greater. So that they are Inosculated only End to End or Mouth to Mouth, after they are come at last to their final distribution. *Tab.* 50.

19. §. The *Aer-Vessels*, are not only, as is said, Existent in the *Leaves* of all *Plants*; but are herein also discoverable without the help of *Glasses*: For upon breaking the *Stalk* or chief *Fibers* of a *Leaf*; the likeness of a fine *Woolly* Substance, or rather of curious small *Cobwebs*, may be seen to hang at both the broken Ends. This is taken notice of, only in some few *Plants*, as in *Scabious*, where it is more visible. But may also be seen more or less, in most other *Plants*, if the *Leaves* be very tenderly broken: as I have noted near twenty years since; and thence conjectur'd them a Sort of *Vessel* common to *Plants*. Now this fine *Wool*, is really a *Skein* of *Aer-Vessels*, or rather of the *Fibers* of the *Aer-Vessels*, unroaved from their *Spiral Position*, and so drawn out in Length. As they appear thus unroaved and drawn out at Length, both to the bare Eye, and through a good *Microscope*, I have represented in two Exemples, the one a *Scabious Leaf*, the other that of a *Vine*. *Tab.* 51, & 52.

Cc 2 20. §.

20. §. THE *Westage* of the *Strings* and *Parenchymous Fibers* together, is here made in the same manner, as hath been described in the *Anatomy* of the *Root*, and *Trunk*: the former being in some Sort as the *Warp*, the latter as the *Woof* of the *Leaf*.

21. §. And one Example we have (it may be more than one) wherein Nature shews, though not a greater, yet a different *Art*; and that is the *Palm-Net*. For whereas in other *Plants*, the *Webb* is made betwixt the *Lignous-Strings* and the *Fibers* of the *Parenchyma*, only visible through a *Microscope*: here the said *Strings* themselves are Interwoven, and the *Westage* apparent to the bare Eye. Of these *Palm-Nets* or *Sacks*, there are several Sorts. One of them is composed in this manner. It hath a Fivefold *Series* of *Lignous Strings* or *Fibers*. The greatest whereof swell out above the rest; and like so many *Ribs*, are obliquely produced on both hands, so as to encompass the *Sack*. Along each of these *Ribs*, on the inside the *Sack*, runs a small *Whitish Line*; being a *Thread* of *Aer-Vessels* growing thereto. Betwixt these *Ribs* or larger *Strings*, there are others much less, Two or Three betwixt *Rib* and *Rib*, *Parallelly* interjected. On the inside, there is a Third *Series*, which is also obliquely produced; but transversly to the former. The Fourth and Fifth, consist of the smallest *Strings*; not only Transversly produced, but also Alternately, from the outside to the inside of the *Sack*, & *vice versa*. By these two last, all the rest are most elaborately woven into one entire and strong piece of Work.

CHAP. V.

Of the Duration *of Leaves, and the* Time *of their* Generation.

AN *Evergreen*, is one degree above a *Plant* which is simply *Perennial*: of This, only the *Trunk* and *Buds* live all the *Winter*; of That, also the Expanded *Leaves*. And an *Evergrow*, is a degree above an *Evergreen*: here, the *Buds* and young *Sprigs*, do only *live*; there, they *grow* and are put forth.

2. §. An *Evergreen*, is made such, either by the *Toughness* of the *Skin*, and *Closeness* or *Density* of the *Parenchyma*, whereby the *Leaf* is better able to endure Cold; as in *Holly*: or by the extream *Smalness* or *Fewness* of the *Aer-Vessels*, whereby the *Sap* is less dryed up, and so sufficient, even in *Winter*, for the Nourishment of the *Leaf*; as in *Box*, and *Yew*, as also *Fir*, and all *Resiniferous Plants*.

3. §. The perpetual Growth of a *Plant*, seemeth to depend chiefly on the Nature of the *Sap*. For all *Juyces* will not ferment alike, nor with the same degree of *Heat*. So that whereas many *Plants* require a greater *Heat*, as that of *Summer*, for the fermenting and distribution of their *Juyces*, and so their growth; the *Warmth* of *Spring* is sufficient for many others; and for some few, that of *Winter* it self.

4. §. AS TO the *Time* wherein the *Leaves* are formed; First, it is very probable, That in those *Plants* which have *Leaves* (besides the *Dissimilar*) of Two distinct *Figures*, as hath the Little Common

Bell,

Book IV. *of Leaves.* 157

Bell, and some others; the *Under-Leaves*, which differ in Shape from the rest, are all at first formed in the *Plume*, before it begins to *sprout*; and the rest afterwards; That is to say, that the former *Leaves*, are all formed (out of *Sap* from the *Trunk*) with the *Seed* it self, and so compose one Principal *Part* thereof, *sc.* the *Plume*: the latter, not till after the *Seed* is sow'n, and so the *Plume* supply'd with *Sap* immediately from the *Root*. Which *Sap*, it seems, is so far different from the former, as sometimes to produce a different Sort of *Leaves*.

5. §. SECONDLY, of the *Buds* of all *Trees*, and of *Perenni-Stalks*, it appears, That they consist of a great number of *Leaves*, all perfectly formed to the Centre; where, notwithstanding, they are sometimes, not half so big as a *Cheese-Mite*. So that all the *Leaves* which stand upon a *Branch* or *Cien* of one whole Years Growth, were actually existent in the *Bud*. It is also very observable, That although these *Buds* begin to be expanded not till *Spring*, yet are they entirely Formed, as to all their *Integral Parts*, in the *Autumn* foregoing. So that the whole Stock of *Leaves* which grow upon a *Tree*, or any Perennial *Stalk*, this year; were made, or actually in being, the last year. A greater *Heat*, more subtilized *Aer*, and better concocted *Juyce*, being requisite for their *Generation*, than for their bare *Expansion* and *Growth*.

6. §. LASTLY, of all *Annual Plants*, in which there are several Successive *Generations* of *Buds*, one under another in one year; although I have not made the Remarque, yet am apt to believe, That as the *Leaves* in every *Bud* are all formed together, as in other *Plants*: so likewise, that the Successive *Generations* of the *Under-Buds*, begin at certain stated *Terms*: as in some *Plants*, at every *New Moon*; in others, at the *Full Moon*; and in some perhaps; with both, or every Fourt'night.

CHAP. VI.

Of the Manner *of the* Generation *of the* Leaf. *Where also, that of the* Two General Parts *of a* Plant, *sc. the* Lignous *and* Parenchymous, *is further explain'd.*

THE *Visible Causes* of the *Figures* of *Leaves*, have B.1. Ch. 4. been formerly mentioned. It may here be further noted, That the greater *Fibers* of the *Leaf*, being never Braced in the *Stalk*; it is a good preparative for their better spreading in the *Leaf*. As also, that the same is much favour'd, by the extream smalness of the *Aer-Vessels* herein: whereby they are more easily divaricated, in the lesser *Fibers*, and so the *Leaf* dilated.

2. §. BUT these and the like are to be reckoned a secondary Order of *Causes*; which serve rather to carry on and improve, that which Nature hath once begun. And therefore, we must not only consider Idea, §.53. the visible *Mechanism* of the *Parts*; but also the *Principles* of which they are composed; wherewith, Nature seems to draw her first Strokes. 3. §.

3. §. Now of these, I have formerly, and as I conceive upon good ground, supposed, the chief Governing *Principle*, to be the *Saline*, whether *Alkaline*, *Acid*, or of any other Kind: being in some sort as the *Mold* of a *Button*, to which the other *Principles*, as its *Attire*, do all conform. Or the *Salts* are, as it were, the *Bones*; the other *Principles*, as the *Flesh* which covers them.

Lib. 2. P. 2. §. 31, &c.

4. §. A further Argument hereof may be deduced from the *Cuticular* and other *Concretions*, commonly called *Mothers*, in Distill'd *Waters*, *Vinegar*, and other *Liquors*. For in these *Concretions*, there is allways a tendence to *Vegetation*; and many of them are true *Vegetables* in their Kind; as shall hereafter be seen. Now the *Liquors*, in which these are generated, do always, wholly or in part, lose their *Taft* and *Smell*, and so become *Vapid*. The more sensible *Principles* therein having made their *Transit* from the *Fluid*, into the *Concrete Parts*. So, I have known, sometimes, *Vinegar* it self, to become by these *Concretions*, almost as *Tasteless* as Common *Water*. Whereby it seems evident, That of *Vegetable Principles*, there are some, more Masterly than others: and that of these, the *Saline* is the chief. The same is likewise argued, from the frequent Experiment of many good Husband-men; that most *Bodies* which abound with *Salt*, are the greatest *Nourishers* of *Plants*.

5. §. This *Saline Principle*, as is above hinted, is to be understood, a *Generick* Name, under which divers *Species* are comprehended; and of some whereof, it is always compounded, as in other *Bodies*, so in *Plants*. As shall be made to appear, by divers Experiments, when we come, hereafter, to speak of *Vegetable Salts*. Whereby we are conducted, yet further to enquire, What are the *Princinples* of this *Principle*?

6. §. NOW these seem to be Four; a *Nitrous*, an *Acid*, an *Alkaline*, and a *Marine*. The *Admixture* of the First, is argu'd from the Place, which Nature hath assigned for the *Generation* and *Growth* of most *Plants*, *fc*. neither in *Caverns* under Ground, as for *Minerals*; nor above it, as for *Animals*; but the Surface of the Earth, where this Sort of *Salt* is copiously bred. And doth therefore prove, not only a *Mixture*, but a good Proportion hereof with the other *Principles* of a *Plant*. Hence it is, that *Dew* or *Water* on *Windows* or Plain and Smooth *Tables*, by virtue of a *Nitro-Aerial Salt*, is often frozen into the resemblance of little *Shrubs*. And the like *Figure* I have often seen in a well filtred *Solution* of the *Salt* of any of our *Purging Waters*, as of *Epsom*, &c. being set to *shoot*. Produced, as I conceive, by the *Nitre*, which with the *Rain* or other *Waters*, is washed down from the *Surface* of the *Earth*, and so mixed with the *Mineral Salts*.

7. §. The other Three *Salts* are exhibited, by the several ways of Resolving the *Principles* of a *Plant*. Many *Plants*, even in their *Natural Estate*, do yield an *Acid Juyce*. And the *Juyces* of many more, by *Fermentation*, will become *Acid*. And most, by *Distillation* in a *Sand-Furnace*, yield an *Accid Liquor*.

8. §. By *Calcination*, all Sorts of *Plants*, yield more or less, both of a *Fixed* and a *Volatile Alkaly*: the former, in the *Ashes*; the latter, in the *Soot*. And, at least the generality, by *Fermentation* also, yield a *Volatile* one; or such a kind of *Salt*, which, whether we call an *Urinous*, or otherwise, hath the like *Odour* and *Taft* with that of *Urine*, *Harts-Horn*, *Soot*, and the like.

9. §.

9. §. The *Marine*, is obtained no other way, that I know of, but from a Solution of the *Alkaline*, upon its being expofed to the *Aer*. The procefs wherof, I fhall particularly fet down in a following *Difcourfe*. Of thefe *Salts*, mixed in a certain proportion, together, and alfo Impregnated with fome of the other *Active Principles* of a *Plant*, and not without an *Admixture* of fome *Parts* from the *Aer*; I fuppofe, that which I call the *Effential*, is produced: of which, I fhall alfo give an account in the fame *Difcourfe*.

10. §. ALL THE Four *Salts* above mentioned, feem in their *Order*, to have a fhare in the *Formation* of a *Leaf*, or other *Part* of a *Plant*: And firft of all, the *Marine*. For all *Generations* are made in fome *Fluid*: But in every *Fluid* there is a perpetual *Inteftive Motion* of *Parts*. So that the firft Intention of *Nature* is, That fome of thofe *Parts* be difpofed to *Reft*. Now of all the *Principles* of a *Plant*, there are none hereunto more difpofed, than their *Salts*; whofe *Particles*, being figu'rd with plain Sides, as often as they touch Side to Side, like two *Marbles* exquifitely polifhed, they will adhere together. And the *Particles* of *Marine Salt*, being *Cubick*; and fo, with refpect to their *Figure*, of greater *Bulk* than thofe of any other *Salt*; they will hereby, be moft and firft of all difpofed to *Reft*; and fo become, as it were, the Foundation of the following *Superftructure*.

11. THE Second Intention of Nature is, That the *Particles* be brought to *Reft*, in a certain *Pofition*, agreeable to the *Figure* of the *Parts* which are to be formed. And therefore in the next place, all thofe *Parts* of a *Plant* which are truly *Lignous*, by the *Marine Salt*, with the affiftance of the *Alkaline*, but efpecially of the *Nitrous*, are made to fhoot out in Length, or into an innumerable company of fmall *Cylindrick Fibres* : thefe *Salts* being, altogether, fturdy enough to refift thofe *Impulfes* which might incline them to conform to any other *Figure*. Tab. 53.

12. §. THE next Intention is, That thefe *Fibres*, at the fame time in which they are formed, may likewife receive fuch a *Pofture* as will beft anfwer the indented *Shape* of the *Leaf*. Which *Pofture*, although in the *Growth* of the *Leaf* it is much Govern'd by the *Aer-Veffels*; yet in the *Generation* hereof, feems to be firft determined by the forementioned *Salts*, according to their feveral *Angles*, whereby they are differently applicable one to another.

13. §. Now all the *Sides* of the *Marine Salt*, and the *Sides* and the *Ends* of the *Nitrous*, properly fo called, ftand at *Right Angles*. And it is very probable, from the *Figure* of the *Cryftalls* in *Spirit* of *Blood*, and fome other *Bodies*, that the *Particles* of the *Alkaline* are Square at one End, and Poynted at the other. And thofe of the *Acid*, at both; And that, withal, they are Shorter and more Slender. Tab. 53.

14. §. It fhould therefore feem, That where the *Alkaline Salt* is any way predominant, and that the *Particles* thereof are placed End to End; there the *Lignous Fibres* (as the larger ones in many *Leaves*) declining their parallel *Growth*, begin to fhoot out obliquely, or at *Angles* one with another, and thofe *Acute*. Tab. 53.

15. §. If the same *Salt* be predominant, and some of its *Particles* placed, with the Pointed End of one, to the Side of another, or the Square End of one, to the Poynted End of another; there the said *Fibres* begin to shoot at *Angles* less *Acute*.

Tab. 53.

16. §. But if either the *Marine* or *Nitrous Salt* is predominant; or some *Particles* of the *Alkaline*, are placed with the Square End of one, to the Side of another; there the *Fibres* begin to make, not *Acute*, but *Right Angles*; as do the greater *Fibres*, in some *Leaves*; and the smaller, in all.

Tab. 53.

17. §. IN the same manner, the *Fibre* in the Circumference of the *Leaf* is also governed; the *Particles* of the said *Salt*, being reduceable, not only to any *Angle*, but also to any *Circle*, or other *Crooked Line*, as they are variously applyed. For if the *major part* be applied End to End, and only every Third or Fourth applied End to Side, they produce a great *Circle*. But if the Poynted End of each, be set to the Side of another, they make a less. And if the Application be the same, but to the contrary Side, they thence begin a new *Circle* with the same *Diameter*, but with another *Center*, answerable to the intended *shape* of the *Leaf*.

Tab. 53.

18. §. AFTER the same manner, the *Aer-Vessels* may be formed by the *Particles* of the *Acid Salt*. Which, without being supposed to be crooked (as those of the *Aer*, at least the compounded ones, probably be) only by applying the lesser Side of one, to the greater Side of another, will also be reduced to any either *Circular* or *Spiral Line*. And so, likewise, for the production of the winding *Fibres*, which compose the *Bladders* of the *Pith* and other *Parenchymous Parts* of a *Plant*.

Tab. 53.

19. §. Thus doth *Nature* every where γεωμετρεῖν. For what She appears in Her *Works*, She must needs be also in their *Causes*.

THE

THE
ANATOMY
OF
FLOWERS,

PROSECUTED

With the bare EYE,

And with the

MICROSCOPE.

Read before the *Royal Society, Novemb.* 9. 1676.

The SECOND PART.

By *NEHEMJAH GREW* M.D. Fellow of the *ROYAL SOCIETY*, and of the *COLLEGE* of *PHYSICIANS*.

LONDON,

Printed by *W. Rawlins,* 1682.

THE CONTENTS OF THE Second Part.

CHAP. I.

Of the EMPALEMENT.

CHAP. II.

Of the FOLIATURE.

CHAP. III.

Of the ATTIRE SEMINIFORM.

CHAP. IV.

Of the FLORID ATTIRE.

CHAP. V.

Of the USE of the ATTIRE.

CHAP. VI.

Of the TIME of the Generation of the Flower.

The Appendix.

Being a METHOD proposed, for the ready finding, by the Leaf and Flower, to what Sort any Plant belongeth.

THE ANATOMY OF FLOWERS.

PART II.

CHAP. I.

Of the EMPALEMENT.

I NEXT proceed to the *Flower*. Where I intend not to repeat thofe things, which have been by Me already noted in the *First Book*. And the forego- ing Difcourfe of *Leaves*, will excufe me from di- vers particulars, common to *Thefe* and the *Flower*. I fhall here therefore remarque fome things not be- fore mentioned, or but *in tranfitu*, and fuch as are more particular to the *Flower*. *Ch.* 5.

2. §. And Firft, it may be noted; That where the *Leaves* of the *Flower* are few; thofe of the *Empalement* or *Green Border*, are either of the fame Number, or juft half as many, whether even, or odd. So in *Leucanthenum* and *Chickweed*, there are Five *Leaves*; in the former Five *Empalers*; in the latter, Ten. In Great *Celandine*, there are Four *Leaves*, and but Two *Empalers*; and fo in *Poppy*. The *Arith- metick* of *Nature* being every where fuitable to Her *Geometry*.

3. §. Of this *Part* of the *Flower* it is likewife obfervable, That it is rarely, if ever, entire or one piece, but parted into divers little *Leafy Pales*, efpecially in all *Flowers* with the *Florid Attire*, as of *Mari- gold, Daify* and the like; being fo numerous, as to make a *Double*, and often a *Treble, Quadruple* or *Quintuple Border*. Whereby they are apt- ly defigned, not only to *protect* the *Leaves* of the *Flower* in the *Bud*; and after their Expanfion, to keep them tite: but alfo, by receding, *Bredthways*, one from another, and fo making a greater *Circle*, gra- dually to give way for the full Growth and fafe fpreading of the *At- tire*. Which, in regard it confifts of *Parts* fo exquifitely tender, were

it pinched up too clofe, would be killed or fpoyled before it came to the *Birth*. As *Teeming Women*, gradually flaken their *Laces*; or as *Taylors* ufe to fplit their *Stomachers* into feveral *Lappets*, to fpread, as their *Belly* rifes.

4. §. Nor is the *Pofture* of the *Parts* in the *Empalement* lefs fuitable: not being filed one juft over another, but alternately. Whereby the *Pales* or *Pannciles* of every *Under-Order*, ferve to ftop up the gaps made by the *Recefs* of the *Upper*. And fo, notwithftanding they all make more roome, yet all confpire to keep the *Aer* out.

5. §. It is alfo worth the notice, That, for the fame purpofe, the Edges at leaft, of the feveral *Pales*, are neither *Fibrous*, nor *Pulpy*; but fo many extream fine tranfparent *Skins*, as in *Chamemile*. Whereby they clofe fo exactly one over another, that it is impoffible for any *Aer* to creep in, or any *steams* ufeful to the *Attire* or *seed*, over haftily to perfpire. As we ufe, when we have put a *Cork* into a *Bottle*, to tie a *Bladder* over it.

CHAP. II.

Of the FOLIATURE.

Tab. 54.
Ch. 5.

THE *Leaves* of the *Flower* are folded up in fuch Sort, as is moft agreeable to their own *Shape*, and that of their inclofed *Attire:* whereof I have given Inftances in the *Fift* Book. I fhall here add fome further *Remarques*.

2. §. The *Leaves* of the *Flower* of *Blattaria*, although of different *Size* and *Shape*; are *Tab.* 54. fo lapped one over another, as to make an Equilateral Pentangle.

Tab. 54.
3. §. The *Spiral Fold*, which is proper to the *Flower*, and never feen in the *Green Leaves*; as it is it felf immediately vifible on the *Surface*, fo by cutting off the top of the *Flower* before it is expanded, feems alfo to make a *Helix*; as in *Perwincle*, the larger *Convolvulus*, &c.

4. §. In fome *Flowers*, where the *Attire* is lofty or fpreading, as in *Holioak*, together with the *Spiral Fold*, the *Leaves* are all at the top tacked down a little; thereby making a blunter *Cone*, and fo a more ample *Pyramid* for the inclofed *Attire*.

5. §. In *Poppy*, although the *Leaves* are extraodinary broad, yet being but few, and inclofing a fmall *Attire*; they could not be well re-*Tab.* 54. duced to any regular *Fold*, without leaving fuch a *Vacuity*, as by being filled with *Aer*, might be prejudicial to the *Seed*. For which reafon, they are cramb'd up within the *Empalement* by hundreds of little *Wrinckles* or *Puckers*; as if Three or Four fine *Cambrick Handcherchifs* were thruft into ones *Pocket*.

6. §.

Book IV.   of Flowers.   165

6. §. In *Ladies-Bower*, the *Leaves* are neither laped one over another, as is moſt uſual, nor ſet Edge to Edge, as ſometimes, but Side to Side, anſwerable to their *Shape*, and the *Diſtribution* of their *Fibres*. Their broad *Tops* being alſo rowled up ſo as to make a *Cone*. In *Ladys-Looking-Glaſs*, they ſtand alſo Side to Side, but in a different manner: in the Former with the Sides ſtanding inward, but here, bearing outward.   *Tab.* 54.

7. §. In the *Marvel* of *Peru*, the *Fold* is likewiſe very peculiar. For, beſides the ſeveral *Plates*, about Six, whereby the *Flower* is gathered in the Midle; the *Top* of it is alſo gathered up by as many diſtinct *Plates*, underneath the former; and theſe *rowled* or *wreathed* up together ſo exactly, that the like could hardly be imitated by a very dextrous Hand.   *Tab.* 54.

8. §. OF the *Hairs* upon *Flowers* and their *Uſe* to the *Attire*, I have alſo ſpoken in the *Firſt* Book. I ſhall here add, That they are likewiſe of *Uſe* to the *Leaves* themſelves, that is, for their cloſer and faſter *Conjunction*. For of ſome *Flowers* it is obſervable, That they are all over *ſmooth*, ſaving on their Edges, which are border'd with *Fringes* of *Hair*; as of *Spaniſh Broome*, *Dulcamara*, and others: In which, the *Hairs* on the Edge of one *Leaf*, are ſo complicated, or at leaſt indented, with thoſe of another, that all the *Leaves* ſeem to be but one piece. *Nature* ſeeing it fit, by this meanss to tie them together, leſt they ſhould be expanded before it be due time.   *Ch.* 5.   *Tab.* 55.

9. §. Many *Flowers* inſtead of *Hairs*, are beſet round about, with a great Number of ſmall *Parts*, not ending in a *Poynt*, but having a *Head*. Sometimes oval, as in *Snap-Dragon*, like the *Horns* of a *Butterfly*, or a *Plummers Sodering-Iron*. But uſually *Globular*, as in *Deadly Nightſhade*, like ſo many little *Muſhrooms* ſprouting out of the *Flower*.

10. §. Out of theſe *Heads*, doth ſometimes iſſue a *Gummy* or *Balſamick Juyce*. From whence proceeds that *Clamminess* of ſome *Flowers*, whereby, being handled they ſtick to our *Fingers*, as do thoſe of *Blataria*, and of *Marigold*; and thoſe of *Colus Jovis*, where the ſaid *Heads* are ſo ſoft and ſucculent, that they reſemble ſo many little *Drops* of *Balſame*. The *Clamminess* which is felt upon freſh *Carduus*, may perhaps proceed from the like *Cauſe*.

11. §. THE *Number* of the *Leaves* of the *Flower* hath been noted by the Learned Sir *Thomas Brown*, to be uſually *Five*. And this Nature ſo far affecteth, that many times where the *Leaves* of the ſame *Flower* are of a different *Size*, yet they keep to this *Number*, as in *Blattaria*.   *Treat.* of the *Quinc.*   *Tab.* 54.

12. §. I alſo add, That even thoſe *Flowers*, which are not properly parted into *Leaves*, have yet their *Tops* uſually divided into *Five* great *Scallops*; as thoſe of *Toad-Flax*, *Snap-Dragon*, *Coded-Arſmart*, *Clary*, *Broom*, and others. And when the *Flower* hath more than *Five*, even many times *Five Leaves*; yet the *Top* of each *Leaf* is indented into *Five Parts*; as in *Scorzonera*, *Cichory*, and all the *Intybous Kind*, with many others.   *Tab.* 54.

13. §. From whence and other like *Inſtances*, it may ſeem, That there is ſome certain *Species* of *Salt* in Nature, and that in moſt *Plants*, of whoſe *Agency* there are ſtill ſome *Footſteps* or other in the *Flower*.

14. §.

14. §. The *Number* of the *Leaves*, as hath been said, is commonly *Five*. Yet some *Flowers* have fewer, and some more, and that with Constancy, in divers *Numbers*, from *One* to *One* and *Twenty*; perhaps in all, so far. The *Flower* of *Acanthus Syriacus*, is in a manner one single *Leaf*, that of *Monks-Rubarb*, *Three-Leav'd*; of *Poppy*, *Crosswort*, *Radish*, and many others, *Four-Leav'd*; the greater *Number* of *Flowers*, *Five-Leav'd*; of *White Hellebore*, *Tulip*, *Onion*, and most *Plants* with *Bulbous Roots*, *Six-Leav'd*; of *Wild-Crowfoot*, *Seven-Leav'd*; of *French Marigold*, commonly *Eight-Leav'd*; of *Flower-deluce*, *Nine-Leav'd*; of *Chickweed*, *Ladies-Mantle*, *Ten-Leav'd*; of *St. James's Wort*, *Thirteen-Leav'd*; and I think of *Febrifuga*, *Cotula*, *Ageratum*, *Corn-Marigold*, with others; and of *Chamemile*, *Buphthalmum*, and some few more, the *Leaves* are commonly *One* and *Twenty*. In that of *St. James's Wort*, the *Number* is so constant that there is scarce *One Flower* in *Forty*, wherein the *Leaves* are more or fewer than *Thirteen*. Divers of which *Numbers*, seem also to have some relation to the *Number* 5. For 9, is Twice; 13, Thrice; and 25, Five times 5 running into it self.

Tab. 55.

15. §. THE *Constituent Parts* of the *Flower* are the same as those of the *Leaf*, sc. the *Parenchyma* or *Pulp*, and the *Vessels*. But in the *Basis* or bottom of the *Flower*, the *Parenchyma* is commonly much more spongy and dry, than in the *Leaves*; conteining, after the *Flower* is open'd, little or no *Sap*, but only a dry and warm *Aer*. Which standing continually under the *Seed*, hastens the *Maturation* or due *Exiccation* thereof: as we use to dry *Maulted Barly* over a warm *Killn*.

16. §. The *Vessels* of the *Flower*, are both for *Sap* and for *Aer*, as well as in other *Parts*. And both of them sometimes, even in the *Skin* of the *Flower*; as may be argued from its being stained with divers *Colours*; produced as hath formerly been shewed, by the mixed *Tinctures* of the said *Vessels*. These *Colours*, in many *Flowers*, as *Tulips*, as they are in the *Skin* it self, so therein only; the *Pulp* of the *Leaf* being white.

B.2.P.2.
§. 65, 66, 67.

17. §. The *Lignous* or *Sap-Vessels* are fewer, and the *Aer-Vessels* smaller in the *Flower*, than in the *Leaf*. And therefore it is very difficult to observe the latter by *Glasses*; especially the *Proportion* which they hold to the other *Parts*. But if you break the *Leaves* of some *Flowers*, with very great gentleness; they may hereby be *Unroaved* or *drawn out*, as in the *Green Leaves*, to some visible length; and their different *Number* in divers *Flowers* may be discerned.

18. §. THE *Use* of the *Flower* or of the *Foliature* whereof we are speaking, is various; as hath formerly been shewed. I now only add, That one *Use* hereof seemeth to be, for the *Separation* of the more *Volatile* and stronger *Sulphur* of the *Plant*. That so the *Seed*, which lyeth within or next it, may be so much the milder, and the *Principles* thereof more fixed and concentred. And this, both for its better *Duration* till the time of *Sowing*; and also, that its *Fermentation*, when it is sow'n, may not be too hot and precipitate; but suitable to so slow and equal a motion, as is the *Vegetation* of a *Seed*.

B.1.Ch.5.

19. §. And that this *Sulphur* is separated and discharged by the *Flower*, seems evident, not only from the *Strength* of its *Odour*, above that of the other *Parts*; but likewise, in that many times where there is no *Flower*, or that very small, the *Seed*, that is its *Cover*, as in the *Umbelliferous*

belliferous Kind, is the more odorous. And therefore also, the *Vine* hath no *Flower*, partly, that the most *Volatile Spirit* and *Sulphur* might all run into the *Fruit*.

20. §. THE *Figure* of the *Flower*, although it is often much more complex, than that of the *Leaf*: yet there is no doubt, but that the *Measure* hereof may be defined in some way, answerable to that exemplified in the foregoing Part. The difference is only this, That whereas the *Green Leaves*, and the Plain *Leaves* also of the *Flower*, are all measured by the *parts* of several *Circles*: those *Flowers* which are *Bellyed*, and those *Leaves* of the *Flower* which are not *Plain*, but *Convex*, are all measured by the *parts* of several *Spheres*. And as the *Diametres* of those *Circles*, bear a certain proportion to the midle *Stemm* of the *Leaf*; so the *Axes* of these *Spheres*, to an imaginary one in the *Centre* of the *Flower*.

21. §. NOW the reason why the *Figure* of the *Flower* is more *multiplex*, than that of the *Leaf*; may be, *partly*, because it is under the Command and Government of those *Salts*, which are here more *refined* and *depurate*, than in the *Leaf*; and so more free to lay the Foundation of any kind of *Figure*, for which, of their own *Nature*, they are adapted. *Partly*, for that as the *Nitrous* and *Alkaline Salts* are chiefly regnant in the *Leaf*; so in the *Flower*, in which the *Parenchymous Part* hath a greater (*a*) proportion than in the *Leaf*; it is most reasonable, to assign the *Predominion* to the *Acid* (*b*): the *Particles* whereof, both as they are less, and also poynted at both ends, (*c*) seem to be more easily applicable one to another for the making of any Sort of *Line* or *Figure*.

(*a*) §. 17.
(*b*) *Idea*, §. 52.
(*c*) P. 1. Ch. 6. §. 13.

CHAP. III.

Of the Attire, *and first of that sort which may be called* Seminiform.

WITHIN the *Foliature* stands the *Attire*; which is of Two general *Kinds*, every where Various and Elegant; according to the *Description* I have given of them in the *First* Book. I shall here add some further *Remarques*. Ch. 5.

2. §. And first, of that Sort of *Attire*, which may be called *Seminiform*; being usually, as it were, a little *Sheaf* of *seed-like Particles*; standing on so many *Pedicills*, as the *Ear* doth upon the *End* of the *Straw*.

3. §. Of their *Colour* it is observable, That for the most part, they are *White* or *Yellow*; sometimes *Blew*; but never *Red*, let the *Flower* or *Foliature* be of what *Colour* it will. Neither doth their *Colour* allways follow that of the *Foliature*, although that be not *Red*. Whereby it appears, how very Curious and Critical *Nature* is, in the *Separation* of the *Juyces* in *Plants*: that such small *Parts* as these of the *Attire*, and so near the *Leaves* of the *Flower*, should yet receive a different *Tincture*.

4. §.

Tab. 55.

4. §. These *Parts* differ also in their *Position*; standing sometimes double upon each *Pedicil*, as in Toad-flax, Snapdragon, and some others; but usually single, as in *Blattaria, Clematis Austriaca,* &c. Sometimes fastned to their *Pedicils* at their middle, stooping down after the manner of *Poppy* and other hanging *Flowers*; as in *Spanish-Broom, Hysop, Scabeous, Behen,* &c. Sometimes they stand erected, as in *Clematis Austriaca, Ladyes-Looking-Glass, Rape-Crowfoot,* &c. Those of *Coded Arsmart* have no *Pedicils,* but stand upon a large *Base*.

5. §. Of the *Pedicils* themselves, it is to be noted, That they are rarely fastned to the *Top* of the *Repository* or *Case* of the *Seed,* but round about the *Bottom*. Partly, That hereby they may the better intercept and separate the *Incongruous Parts* of the *Sap* from the *Seed*. Yet in the *Coded Arsmart* they stand at the *Top*. Which is not the only thing peculiar in that *Plant*; it being the property thereof, to ejaculate its *Seed,* upon the least touch. Which property seemeth to depend, partly, upon the *Position* of the said *Pedicils,* as shall be shewed in speaking of the *Seed*.

6. §. These *Seed-like Parts* are also of different *Number*. In *Great Celandine, Rose, Rape-Crowfoot,* numerous; in *Great Plantaine,* and some other *Herbs,* much more conspicous than the *Foliature* it self. In *Germander-Chickweed,* they are always *Two,* and no more. Sometimes they follow the number of the *Leaves,* especially in the number 5; as in *Blattaria, Black Henbean,* &c. In *Stichwort* and *Lychnis Sylvestris,* they are 10, just double to the number of the *Leaves*.

7. §. They differ also in their *Bigness,* being in some smaller *Flowers,* large; as in *Borage, Ladys-Looking-Glass,* and others: and in some larger *Flowers,* less; as in the *Rose*.

8. §. But especially in their *Shape,* which is always very Elegant, and with much Variety. In *Borage,* like the point of a *Spear*. In *Blattaria,* like a *Horse-shooe*. In *Clematis Austriaca,* like the *Spatula,* wherewith *Apothecaries* make their *Mixtures*. In *Mallow,* like a *Head-Roll*. In *Hysop,* they have one *Cleft* before; in *Blattaria,* one round about; in *Water Bettony,* one at the *Top*; in *Scabious,* they have a double *Cleft,* one on each side; and so in St. *Johns Wort, Hyoscyamus,* and others; before they open, in the *Shape* of a double *Purse*.

Tab. 56.

9. §. These *Parts*, are all hollow; each being the *Theca* or *Case* of a great many extream small *Particles,* either *Globular,* or otherwise *Convex*; but always regularly *figur'd*. They are all crowded together, and fastned in close *Ranks,* without any *Pedicils,* to the Insides of the *Theca,* like other lesser *Seeds* within a greater; or after the same manner as in *Hyoscyamus* and some other *Plants,* the true *Seeds* themselves grow all round about close to the *Bed* of the *Case*; as in *Clary,* and the *Figures* now referred to, may be seen. And when they are ripe, the *Case* also opens and admits them to the *Aer,* as the *Seed-Case* doth the *Seed*. The whole *Attire,* together with the *Foliature* and *Seed-Case,* See in one Example, amongst the *Figures*.

Tab. 55, 56.

Tab. 57.

10. §. The *Colour* of these small *Particles* conteined in the *Theca,* is also different. But as That is usually *White* or *Yellow,* so are These: sometimes *Blewish*; but never *Red*. And sometimes not of the same *Colour* with that of the *Theca*. Which further shews how scrupulous *Nature* is, in differencing the *Tinctures* of the several *Parts*.

11. §.

11. §. They are also of different *Bigness* and *Figure*. Those in *Snap-dragon*, are of the smallest *Size* I have seen; being no bigger through a good *Microscope*, than the least *Cheese-Mite* to the naked Eye. In *Plantain*, also through a *Glass*, like a *Scurvy-grass-seed*. In *Bears-foot*, like a *Mustard-seed*. In *Carnation*, like a *Turnep-Seed*. In *Bindweed*, like a *Peper-Corn*. In all these of a *Globular Figure*. *Tab.* 58.

12. In *Devils-bit*, they are also *Round*, but depressed, like the *Seed* of *Goos-grass*, or a *Holland Cheese*. In the *Bean* and all sorts of *Puls*, and *Trefoyls*, as also in *Blew-bottle*, &c. they are *Cylindrick*. In *Orange Lilly*, *Oval*, one 5th of an Inch long, like an *Ants-Egg*. In *Deadly-Night-shade*, also *Oval*, but smaller at both Ends. And those of *Pancy*, *Cubick*. In all these and the former, they are *Smooth*. *Tab.* 58.

13. §. But in *Mallow*, *Holyoak*, and all of that kind, they are beset round about with little *Thornes*; whereby each looks like the *Seed-Ball* of *Roman Nettle*, or like the *Fruit* of *Thorn-Apple*, or the *Fish* called *Piscis orbis minor*, or the *Murices*, used antiently in *Wars*. They are also very great, shewing, through a *Glass*, of the bigness of a large *White Pease*; being 200 or 300 times biger than those in *Snapdragon*; of which there are about a *Thousand* in each *Theca*, that is, in the space of about 1000th *Cubical Part* of an Inch. *Tab.* 58.

15. §. In some *Plants*, as in *Deadly Night-shade*, where these *Particles* are *White*, they seem, by a very good *Glass* and advantagious *Position*, to be composed of *Parenchymous* and *Lignous Fibres*, stitched up together, as in the other *Parts*.

15. §. In *Colocynthis*, (and with some *Analogy* in *Wild Cucumer*, and I suppose all of that kind) the *Attire* is very peculiar, not consisting of several little *Thecæ*, upon so many *Pedicils*, as is described; but is all one entire *Part*, like a thick *Columna* in the midst of the *Flower*; having several little *Ridges*, and *Furrows* winding from the *Top* to the *Bottom* round about. In the midle of each *Ridge* runs a *Line*, where the *Skin*, after sometime, openeth into two *Lips*, presenting the *Globular Particles* conteined in the hollow of every *Ridge*.

16. §. Where the *Attire* consists of several *Seed-like Parts*, as is described; there, another *Part* distinct, like a little *Columna* or *Pinacle*, stands on the *Top* of the *Uterus* or true *Seed-Case*. Which is also regularly and variously *Figured*. In *Bindweed*, it hath a round *Head*, like that of a great *Pin*. In the *Common Bell*, St. *Johns wort*, it is divided into *Three Parts*. In *Gerarium*, into *Five*; In *Asarum*, into *Six*. Sometimes, the *Head is Smooth*, and sometimes beset with little *Thorns*, as in *Hyoscyamus*. Of the *Use* of these *Parts*, anon. *Tab.* 56, 57.

E e CHAP.

CHAP. IV.

Of the FLORID ATTIRE.

IN THIS *Attire* there is also much Elegant Variety, according to the *Description* we have given of it in the *First* Book. It always consists of several *Suits*; Ten, Twenty, Fourty, a Hundred, or more, according to the *Bigness* of the *Flower*. And every *Suit* most commonly, of three distinct *Parts*, all of a Regular, but Different *Figure*. The utmost Part, is always like a little *Flower* with Five *Leaves* and a *Tubular Base*, like that of *Cowslip*. So that every *Flower* with the *Florid Attire*, Embosomes, or is, a *Posy* of perfect *Flowers*.

Tab. 59.

2. §. In some *Flowers*, every one of these *Florets*, is encompassed with an *Hedg* of *Hairs*; and every *Hair* branched on both sides almost like a *Sprig* of *Fir*; as in *Aster Atticus*, *Golden-Rod*, and others.

Tab. 59.

3. §. The *Base* of the *Floret* is usually *Cylindrick*, but sometimes Square, as in *French Marigold*. And the *Leaves* hereof which, for the most part, are Smooth on the Inside, in the same *Flower* are all over *Hairy*. And the Edges of these little *Flowers*, are frequently Ridged, or as it were, He m'd, like the Edge of a *Band*.

Tab. 60.

4. §. The midlemost of the Three *Parts*, which I call the *Sheath*, is usually fastened towards the *Top*, or else at the *Bottom* of the *Floret*. This is rather indented, than parted into *Leaves*. The *Surface* seldom Plain or Even, but wrought with Five *Ridges*, and as many *Gutters* running almost Parallel from the *Top* to the *Bottom*.

B.1. Ch. 5. *Tab.* 60, 61, 62.

5. §. The *Inmost Part*, which I call the *Blade*, runs through the hollow of the Two Former, and so is fastned, with the *Floret*, to the convex of the *Seed-Case*. The Head and *Sides* of this *Part*, is always beset round about with *Globulets*, commonly through a *Glass*, as big as a *Turnep-seed*, or a great *Pins-Head*. In some *Plants* growing close to the *Blade*, as in the common *Marigold*; in the *French*, and others, upon *Pedicils* or little slender *Stalks*. These, as the *Blade* springeth up from within the *Sheath*, are still rubed off, and so stand like a *Powder* on them both. And sometimes, as in *Cichory*, they seem to grow on the Inside the *Sheath*, if it be split with a small *Pin*: as also in *Knapweed*, in which they are numerous. Yet in the *Seed-like Attire*, always more numerous, than in the *Florid*.

B.1. Ch. 5.

Tab. 60, 61, 62.

Tab. 58, &c.

6. §. The *Head* of the *Blade* is always divided into *Two*, and sometimes into *Three Parts*, as in *Cichory*; which, by degrees, curl outward, after the manner of *Scorpion-Grass*.

7. §. The *Description* now given, agrees principally to the *Corymbiferous Kind*, as *Tansy*, *Chamemile*, and the like. But in *Scorzonera*, as also *Cichory*, *Hawk-Weed*, *Mousear* and all the *Intybous* Kind, with many

more

more, the *Attire* is not separate from the *Foliature*, so as to stand within that in one entire *Posy*; but every *Leaf* of the *Flower* hath its own *Attire* apart. For the sake of which, the *Basis* of every *Leaf* is form'd into a little *Tube* or *Pipe*, whereby it embosomes its own *Attire* within it self. Consisting commonly of *Two Parts*, a *Sheath* and a *Blade*: the *Leaf* it self answering to the *Floret* in other *Flowers*. *Tab.* 62.

8. §. In some *Plants*, besides the *Attire* or *Posy* in the midle of the *Flower*; the *Leaves* also have each their own to themselves, as in *Marigold*: yet this, as I take it, consisting only of one single *Part*, which answers to the *Blade*; the *Leaf* it self being as the *Sheath*. *Tab.* 61.

9. §. In many *Plants*, this *Florid Attire* is very large; so that not only the *Suits*, but also the several *Parts* whereof every *Suit* consists, being throughly ripe and well blown open, are all visible to the bare Eye, as in *Knapweed*, and all the *Thistle Kind*. This *Attire* is all the *Flower*, that this sort of *Plants* have; being, though *Empal'd*, yet without any *Foliature*. *Tab.* 61.

10. §. And sometimes, there is little or no *Flower* besides this *Attire*, although extream small, as in *Golden Rod*, *Wormwood* and others. Where it may be noted, That the *Medicine* called *Wormseed* or *Semen Santonici*, is no Sort of *Seed*, but the *Buds* of small *Flowers*, or of the *Florid Attire* of that *Plant*.

CHAP. V.

Of the Use of the Attire.

OF the *Secundary Use* hereof, I have spoken in the *First Book*; and particularly, of the *Globulets* or small *Particles* within the *Thecæ* of the *Seed-like Attire*, and upon the *Blades* of the *Florid*, I have conjectur'd, That they are that *Body* which *Bees* gather and carry upon their *Thighs*, and is commonly called their *Bread*. For the *Wax* they carry in little *Flakes* in their *Chaps*: but the *Bread* is a Kind of *Powder*; yet somewhat moist, as are the said little *Particles* of the *Attire*. *Ch.* 3.

2. §. But the Primary and chief *Use* of the *Attire* is such, as hath respect to the *Plant* it self; and so appears to be very great and necessary. Because, even those *Plants* which have no *Flower* or *Foliature*, are yet some way or other *Attir'd*; either with the *Seminiform*, or the *Florid Attire*. So that it seems to perform its service to the *Seed*, as the *Foliature*, to the *Fruit*.

3. §. In discourse hereof with our Learned *Savilian* Professor Sir *Thomas Millington*, he told me, he conceived, That the *Attire* doth serve, as the *Male*, for the *Generation* of the *Seed*.

4. §. I immediately reply'd, That I was of the same Opinion; and gave him some reasons for it, and answered some *Objections*, which might oppose them. But withall, in regard every *Plant* is ἀρρενόθηλυς or *Male* and *Female*, that I was also of Opinion, That it serveth for

the

the *Separation* of some *Parts*, as well as the *Affusion* of others. The sum therefore of my Thoughts concerning this *Matter*, is as follows.

5. §. And First, it seems, That the *Attire* serves to discharge some redundant *Part* of the *Sap*, as a *Work* preparatory to the *Generation* of the *Seed*. In particular, that as the *Foliature* serveth to carry off the *Volatile Saline Sulphur*: So the *Attire*, to minorate and adjust the *Aereal*; to the end, the *Seed* may become the more *Oyly*, and its *Principles*, the better fixed. And therefore the *Foliature* generally hath a much stronger *Odour*, than the *Attire*: because the *Saline Sulphur* is stronger, than an *Aerial*, which is too subtile to affect the Sense. Hence also it is, that the *Colour* of the *Parts* of the *Attire*, is usually *White*, or *Yellow*, never *Red*: the former, depending upon a greater participation of *Aer*; the latter, of *Sulphur*. I add further, That the most *Volatile* and *Aerial Sulphur*; being by means of these *Parts* much discharged; it may hereby come to pass, not only that the *Seed* is more *Oylie*, and its *Principles* more fixed; but also, that the *Body* or *Parenchyma* thereof, is so compact and close: For although it consists of *Bladders*, yet such, as are Twenty times smaller than in any other *Part* of a *Plant* of the like bigness. Whereas, were the *Aer* copiously mixed with the *Sap* here, as in the *Pith*, *Fruit*, and other *Parenchymous Parts*; it would give so quick a *Ferment* to the *Sap*, as to dilate and amplify the *Bladders* of the *Seed*, beyond its present compact and durable *Texture*; and so expose it, either to a precipitant *Growth*, or sudden *Rot*. Wherefore, as the *Seed-Case* is the *Womb*; so the *Attire* (which always stands upon or round about it) and those *Parts* of the *Sap* herinto discharged; are, as it were, the *Menses* or *Flowers*, by which the *Sap* in the *Womb*, is duly qualified, for the approaching *Generation* of the *Seed*.

6. §. And as the young and early *Attire* before it opens, answers to the *Menses* in the *Femal*: so is it probable, that afterward when it opens or cracks, it performs the *Office* of the *Male*. This is hinted from the *Shape* of the *Parts*. For in the *Florid Attire*, the *Blade* doth not unaptly resemble a small *Penis*, with the *Sheath* upon it, as its *Præputium*. And in the *Seed-like Attire*, the several *Thecæ*, are like so many little *Testicles*. And the *Globulets* and other small *Particles* upon the *Blade* or *Penis*, and in the *Thecæ*, are as the *Vegetable Sperme*. Which, so soon as the *Penis* is exerted, or the *Testicles* come to break, falls down upon the *Seed-Case* or *Womb*, and so Touches it with a *Prolifick* Virtue.

7. §. *Consentaneous* hereto it is also observable, That those *Herbs* generally have the *Seed-like Attire*, which either produce a greater Quantity of *Seed*, or a *Perennial Root*: and that there is no *Tree*, with the *Florid Attire*. As if the other, because it contains a far greater Proportion of the abovesaid *Particles*, that is, of *Sperm*; 'tis able to beget a more *Numerous*, *Vivaceous*, or *Gigantick Birth*.

8. §. That the same *Plant* is both *Male* and *Female*, may the rather be believed, in that *Snails*, and some other *Animals*, are such. And the *Parts* which imitate the *Menses*, and the *Sperm*, are not precisely the same: the former, being the External *Parts* of the *Attire*, and the *Sap*, which feeds them; the latter, the small *Particles* or moyst *Powder* which the External inclose.

9. §. And that these *Particles*, only by falling on the *Uterus*, should communicate to it or to the *Sap* therein, a *Prolifick Virtue*; it may seem the more credible, from the manner wherein *Coition* is made by some *Animals*; as by many *Birds*, where there is no *Intromission*, but only an *Adosculation* of *Parts*: And so in many *Fishes*. Neither in others, doth the *Penis* ever enter any further than the *Neck* of the *Womb*. Nor doth perhaps the *Semen* it self: or if it doth, it can by no means be thought, bodily or as to its gross *Substance*, to enter the *Membranes*, in which every *Conception*, or the *Liquor* intended for it, before any *Coition*, is involved; but only some subtle and *vivifick Effluvia*, to which the visible *Body* of the *Semen*, is but a *Vehicle*. And the like *Effluvia* may be very easily transfused from the above said *Particles* into the *Seed-Case* or *Womb* of a *Plant*.

10. §. If any one shall require the Similitude to hold in every Thing; he would not have a *Plant* to resemble, but to be, an *Animal*.

CHAP. VI.

Of the Time of the Generation of the Flower.

HE *Time* in which the *Flower* is Generated or Formed is a Providence in Nature, whereof, I do a little wonder, that no one, amongst so many observers of *Plants*, hath ever yet taken any notice. It is therefore to be remarked, That all *Flowers* are formed or perfectly finished, in all their *Parts*, long before they appear in sight; usually Three or Four Months, and sometimes half a year, or more. And that in all *Perennial Plants*, those *Flowers* which appear and are called the *Flowers* of any one year; are not formed in that year; but were actually in *Being*, and entirely formed in all *Parts*, the year before; as in many *Herbs*, and in all *Shrubs* and *Trees*.

2. §. This will best be seen by some Instances. So the *Flower* of *Mezereon*, which opens in *January*, is entirely formed about the midle of *August* in the year foregoing. At which time, the *Green Leaves* of the *Bud* being cautiously removed, the *Leaves* of the *Flower*, and the *Thecæ Seminiformes* or *seed-like Attire*, encompassing the *Seed-Case*, through an indifferent *Glass*, are all distinctly visible. *Tab.* 63.

3. §. The like may be seen in *Syringa*, and other *Shrubs*, and in *Trees*. In as many of which, as are *Frugiferous*, the *Fruit* also, which answers to the *Seed-Case* in other *Plants*, is about the same time entirely formed.

4. §. And so in *Herbs*; as the *Flower* of *Asarum*, which appeareth in *April* or *May*, is entirely formed in *August* or *July* of the foregoing year. For there are here, as well as in *Trees*, Two Sorts of *Buds*; some which are composed only of *Green Leaves*; and some which also contein a *Flower* and the *Seed-Case*. So in *Bears-foot*, by some called the *January Rose*, the *Flower-Buds*, which open in *January* are all formed in or before the Month of *August* in the year preceding. *Tab.* 64.

5. §.

Tab. 63.

5. §. The same may also be seen about the end of *August* or the beginning of *September* in a *Tulip-Root*. In which, the Two Inmost *Shells* dryer than the rest, stand hollow, with the little young *Flower* (which appears in *March* or *April* following) inclosed now in their *Centre*. Being thus kept *warm* and *dry*, lest it should either perish, or be precipitated upon the *Winter*, by sprouting too soon.

6. §. From hence it is plain, That although the *Flower* appears before the *Seed*; yet if the comparison be made betwixt the *Flower* and *Seed* of the same year; the *Seed* is first formed, and afterward the *Flower*. That is, the *Seed*, for which Nature chooses the Firstborn *Sap*, is formed in the fore part of the year: which work being finished, out of the less *fœcund part* of the *Sap*, the *Flowers* intended for the *Sire* and *Matrix* of the next years *Seed*; is afterwards produced.

7. §. THE true *Time* of the *Generation* of the *Flower* being know'n, it may also be an Inducement to make Tryal, for the bringing of many *Flowers* to grow fairly in *Winter*, which are used to grow, that is, to appear, only in the *Spring* and *Summer*: *sc.* by keeping the *Plants* warm, and thereby enticing the young lurking *Flowers* to come abroad.

The Appendix.

Being a Method *proposed, for the ready finding, by the* Leaf *and* Flower, *to what Sort any* Plant *belongeth.*

ALTHOUGH many have bestowed extraordinary Care and Industry upon the searching out, and Description of *Plants*; and for the reducing of them to their several *Tribes*: yet I will take leave, here to propose a short *Method* whereby Learners, seeing a *Plant* they know not, may be informed to what Sort it belongs, and so be directed where to find it described and discoursed of. For, except they have a Master to conduct them, which few have; they must needs, by seeking at random, lose a great deal of time, which by a regular Enquiry might be saved. Besides, that what is learned by their own Observation, will abide much longer on their mind, than what they are only Poynted to, by another.

2. §. Now the most *Philosophick* way of distinguishing or sorting of *Plants*, were by the *Characteristick* Properties in all *Parts*, both *Compounded, Constituents,* and *Contents*. But of the *Compounded,* the *Seeds,* and some other *Parts,* are oftentimes very *minute*: and the *Roots* always lie hid. As also the *Constituent Parts,* every where, without cuting and the use of *Glasses*. Nor can the *Contents* be accurately observed otherwise. So that for the *Use* here intended, those *Properties* are the fittest to be insisted upon, which are the most *Conspicuous*, and in those *Parts,* where the Learner may the most readily and without any difficulty take notice of them; as in the *Flower* and *Leaf*. The *Flower* hath Varieties enough of it self. But in regard it is often wanting, when the Green *Leaf* is not; it is therefore convenient, that he be assisted

by both, and that the Varieties of both be diſtinctly reduced unto *Tables*. Which may be done, after the following, or ſome other like manner.

3. §. And Firſt for the *Leaves*. The moſt obvious *Varieties* of which, are in their *Poſition, Size* or *Shape*.

4. §. *Leaves* are faſtned with, or without a *Stalk*. Without, only cloſe to the *Branch*, as in *Southiſtle*; or ſurrounding it, as in *Thorow-Wax*.

5. §. Both theſe ways, they ſtand either ſingly, that is, but one at the ſame height; or more together.

6. §. More together, in Even or Odd Numbers. In Even Numbers, commonly Two and Two, as in *Sage, Polium*; Sometimes Four, as in *Croſs-wort, Madder, Herb True-Love, Pomum Majæ*; or more, as, I think, in *Woodrofe*, &c. In Odd Numbers, Three, as in all *Trefoyls, Strawberries*; Five, in *Pentaphil, Caſtanea Equina*; Seven, in *Tormentil*.

7. §. The *Sizes* of *Leaves* are innumerable. It is therefore neceſſary to reduce them to a Standard. And ſo, they may be reckoned, Three; *Small, Mean* and *Great*: with reſpect to the *Length* of the *Leaf*, the *Breadth*, or both. From one Inch and under, all *Leaves* may be accounted *Small*; from one Inch and over, to five Inches, *Mean*; from five and over, *Great*.

8. §. The *Shapes* of *Leaves* are alſo numberleſs. But the moſt obvious diſtinctions which they admit of, are ſuch as theſe;

9. §. *Leaves* are *Membraneous*, as the greater part; *Squameous*, as *Abies*, or *Filamentous*. Which are *ſolid*, as in *Fenil, Meum, Buphthalmum, Chamemile, Groundpine*; or *hollow*, as in *Onion*.

10. §. *Membraneous*, have all their main *Fibres* produced either from the *Stalk*, as in *Holyoak*; or from the middle *Stem* of the *Leaf*, as in moſt. From the midle *Stem*, reciprocally, as in *Scabious*, or oppoſitely, that is, one over againſt another, as in *Roſe*: and both ways, at *Acute Angles*, as in moſt; or *Right*, as in *Dandelion*.

11. §. Again, they are different with reſpect to the *Top*, the *Bottom*, and the *Sides*. The *Top* is *Thorny*, as in *Furz*; or *Unarmed. Unarmed*, either *Produced*, that is, *Poynted*, or at leaſt, *Roundiſh*, as in *Lamium, Ironwort*; or elſe *Reduced*, as in *Woodſorrel*. And ſo the *Bottom*, is either *Reduced* towards the *Top*, as in *Ground-Ivy*; or *Produced* upon the *Stalk*, as in *Poplar, Bay*, &c.

12. §. The *Sides* or *Edges* of the *Leaf*, are either of one and the ſame *Meaſure*, as commonly; or of divers, as in *Doronicum*. Both ways they are Even, as in *Syringa, Mous-ear*; or Uneven. The Uneven, are *Prickly*, as *Holly, Eryngium, Thiſtle*; or Unarmed. Unarmed, are Infected, or Refected. Infected deeply, that is, Lobed, as *Golden Liverwort, Clematis Peregrina*; or with ſhallow *Infections*, as in moſt. And ſo, Indented, or Scallopped: the former, when the *Angle* is made with Straight *Lines*, as in *Dandelion*; the latter, with Crooked, as in *Thalictrum*. Refected, that is, both Lobed, and Infected; or when upon the greater *Infections*, there are other leſſer ones, as in *Wild-Clary, Lovage, Maſterwort*.

13. §. THE moſt Conſpicuous Varieties of *Flowers*, are in their *Poſition, Size, Shape*, and *Colour*.

14. §. Moſt are faſtned with *Stalks*; but many without. Sometimes, they are placed round about the *Branch*, that is, Coronated, as

in *Pulegium*; and sometimes, all on one side; either in *Ranks* only, as in *Bawm*; or in *Rank* and *File*, as in *Foxglove*. In *Saxifraga Aurea*, they grow on the *Leaf*.

15. §. Again, they either stand Singly, as in *Corn Marigold*; or Clustur'd. And so, either all upon one *Branch*, or on several little *Ramificated Sprigs*. On one *Branch*, prolonged like a Tail, as in *Blattaria*; or Contracted. And so, either without *Stalks*, that is, *Capitated*, as in *Scabious*; or with *Stalks*, that is, *Umbellated*, as *Fenil*, &c. On several *Sprigs*, as in *Tanacetum*, *Yarrow*.

16. §. The *Sizes* of *Flowers*, as of the *Leaves*, may be reduced to Three. From ½ an Inch and under, in *Diameter* or *Length*, may be accounted *small*. From ½ an Inch and over to an Inch and ½, may go for *Mean*. And from an Inch and ½ and over, *Great*.

17. In respect of the *Shape*, *Flowers* are *Open* or *Belly'd*. *Open* have both *Leaves* and *Attire*, as most; or else are all *Attire*, as of *Burdock*, *Beta Cretica*.

18. §. The *Open*, consist of a Certain Number of *Leaves*, *One*, *Two*, *Three*, *Four*, *Five*, *Six*, *Seven*, *Eight*, *Nine*, *Ten*, *Thirteen*, or *One* and *Twenty*. Uncertain, commonly called *Double*. Those of a Certain *Number*, either *Uniform*, that is, all of a certain *Size* and *Shape*, as usually; or *Biform*, or *Triform*, as in *Iris*, *Blattaria*. And these again, Even *Edged* or *Notched*; with *Three Poynts*, as in *Marigold*; or *Five*, in *Cichory*.

19. §. The *Belly'd*, are either so in whole; or in Part, that is, with the *Top* divided into *Leaves*, and the *Bottom*, *Hollow*: The former, are also Even *Edged*, as in *Convolvulus*; or *Notched*, as in *Trachelium*. The latter have their *Leaves* distinguished as before. Their *Bottom* or *Base*, either fastned to the *Seed-Case*, as in *Snap-dragon*; or standing below it. And so, either Straight, as I think in *Toad-flax*; or Crooked, as in *Violet*, *Lark-heel*.

20. §. In all these, the *Attire* is either *Seminiform*, or *Florid*. And both, *Clustur'd*, or *Divided*; as in *Mallow*, *St. Johns wort*; *Starwort*, *Hawkweed*.

21. §. The *Colours* of the *Flower*, are *White*, as in *Water-Crowfoot*; *Red*, as *Lychnis*; *Blew*, as *Borage*; *Purple*, as *Stock-July-Flower*; *Black*, as in some *Anemones*; *Yellow*, in *Wall-Flower*; *Tawny*, in *Colus Jovis*; *Green*, in *Laureola*. Which are either *Single*, or *Mixed*: Two together, as in *Butyr-Bur*, *White* and *Red*; in *White Hellibore*, *White* and *Green*; in *Monks Rubarb*, *Red* and *Green*; &c. Or *Three* together, as in *Pancy*, *Yellow*, *Blew*, and *Black*, i. e. *atro-purpureus*.

22. §. How far these, and some other like *Distinctions*, being reduced to *Tables*, would serve for the finding out of any Sort of *Plant*, may be conceived, if we consider, how great a Variety, a few *Bells*, in the ringing of Changes, will produce. And the search will be easy, and successfull, if in every foregoing *Table*, reference be made to those that follow; and in the *Tables* conteining the last *Divisions*, the Names of the *Plants* therein poynted out, be expressed.

THE

THE
ANATOMY
OF
FRUITS,

PROSECUTED

With the bare EYE,

And with the

MICROSCOPE.

Read before the *Royal Society, in the Year* 1677.

The THIRD PART.

By *NEHEMJAH GREW* M.D. Fellow of the *ROYAL SOCIETY*, and of the *COLLEGE* of *PHYSICIANS*.

LONDON,

Printed by *W. Rawlins*, 1682.

THE CONTENTS OF THE Third Part.

CHAP. I.

Of the APPLE; and of the LIMON, and CUCUMER, the Fruits of Plants *vulgarly called POMIFEROUS.*

CHAP. II.

Of the PEAR and QUINCE.

CHAP. III.

Of the PLUM, and some other Fruits *of the same Kindred*

CHAP. IV.

Of the GRAPE, and HAZEL-NUT; with some other Fruits *analogous to each of them.*

CHAP. V.

Of the SEED-CASE or MEMBRANEOUS UTERUS.

CHAP. VI.

Of the USE of the Parts *to the* Fruit.

CHAP. VII.

Of the USE of the Parts *to the* Seed. *And the TIME, in which, the* Uterus *or* Fruit *and* Seed-Case *are formed.*

THE

THE ANATOMY OF FRUITS.

PART III.

CHAP. I.

Of the APPLE; and of the LIMON, and CUCUMER, the Fruits of Plants vulgarly called POMIFEROUS.

THE *Description* and *Use* of *Leaves* and *Flowers*, together with the *Figures* thereto belonging, were presented to this *Honourable Society*, the last year. I shall conclude this Subject with *Fruits* and *Seeds*; beginning with *Fruits*, which will take up the present Discourse.

2. §. And First, I shall describe the Compounding *Parts* of some, more generally known. Which having done, I shall next observe the *Uses* of the same; either for the *Fruit* it self, or for the *Seed*. Some of the *Descriptions*, the Reader may be pleased to compare with those in the *First Book. Ch. 6.* I begin with the *Apple*; to which I shall subjoyn the *Limon*, and *Cucumer*, commonly reduced to the *Pome Kind*.

3. §. AN APPLE, besides the *Skin*, consisteth of a *Parenchyma*, *Vessels*, and *Coar*. The *Parenchyma* or *Pulp*, is the same with that of the *Barque* of the *Tree*. As is apparent, not only from the visible continuation thereof from the one, through the *Stalk*, into the other: but also from the Structure common to them both; being both composed of *Bladders*. In which, notwithstanding, there is this difference,

180   *The Anatomy*   Book IV.

Tab. 65.
That whereas in the *Barque*, they are *spherical*, and very small, most of them, through a good *Glass*, not exceeding $\frac{1}{16}$th of an Inch in *Diametre*, and some of them, less: here, they are oblong and very large, most of them about $\frac{1}{3}$d of an Inch in Length, or more, according to the largeness and tenderness of the *Fruit*; being all uniformly tenter'd or stretched out, by the *arching* of the *Vessels*, from the *Coar* towards the Circumference of the *Apple*.

Tab. 65.
4. §. The *Vessels*, as in the other *Parts* of a *Plant*, are *Succiferous*, and for *Aer*. Both the *Branches* of the former, and the single *Vessels* of the latter, are extream small. They run every where together, not collateral, as *Veins* and *Arteries* do in *Animals*; but the latter, sheathed in the former.

Tab. 65.
5. §. They are distributed into Twenty principal *Branches*. The Ten outmost, a little within the *Apple*, are diverted from a straight *Line*, into so many great *Arches*; from which a few small *Fibres* are without any order dispearsed through the *Apple*. The Five middlemost, and the Five inmost, run in a straight *Line* as far as the *Coar*, and are there diverted into as many lesser *Arches*; the former, at the outer, and the latter at the iner *Angles* of the *Coar*. Upon these Five inmost hang all the *Seeds*.

Tab. 65.
6. §. These Ten, and the other Ten abovesaid, do all meet together at the top of the *Apple*, where originally, they all ran into the *Flower*. But betwixt them, there are scarce any intercurrent *Fibres*; so that they appear every where disjunct from the bottom to the top of the *Apple*.

Tab. 66.
7. §. A LIMON hath a Threefold *Parenchyma*; which seem to be derived one from another: the *Texture*, upon every derivation, being somewhat altered, and so made more close and elaborate. The utmost, called the *Rind*, hath the most open, and the coursest *Texture*; being composed of the largest *Threds*, and those *Threds* woven up into larger *Bladders*. Those little *Cells*, which contein the *Essential Oyl* of the *Fruit*, and stand near the Surface of the *Rind*, are some of the said *Bladders* much more dilated.

Tab. 66.
8. §. From this utmost *Parenchyma*. Nine or Ten *Insertions* or *Lamells* are produced, betwixt as many *Portions* of the *Pulpy Part*, towards the Centre, where they all unite into one *Body*, answerable to the *Pith* in the *Trunk* or *Root* of a *Tree*; and is a conspicuous demonstration, of the communion betwixt the *Barque* and the *Pith*; which there, is much more obscure and difficult to observe. At the bottom, but especially the top of the *Fruit*, the *Pith* is so far expanded, as without the mediation of any *Lamels*, to be joyned to the *Rind*.

Tab. 66.
9. §. Throughout this *Parenchyma*, the *Vessels* are dispearsed. But the chief *Branches* stand on the iner Edge of the *Rind*, and the outer Edge of the *Pith*, just at the two extremities of every *Lamel*. From those *Branches* on the Edge of the *Pith*, other little and very short ones shoot into the *Pulp* of the *Fruit*, upon which the *Seeds* are appendant. In the Centre of the *Pith*, are Eight or Nine, in a *Ring*, which run through the *Fruit* up to the *Flower*.

10. §. Between the *Rind* and the *Pith* and those several *Lamels*, which joyn them together, stands the second Sort of *Parenchyma*, different from the former, in being somewhat closer, and finer wrought Divided, by the *Lamels*, into several distinct Bodies; every one of them a great and entire *Bag*.

11. §.

Book IV.   *of Fruits.*   181

11. §. Within every great *Bag*, is conteined a Third *Parenchyma*, which is alſo a Cluſter of other little *Bags*, about the bigneſs of an *Oate*, all disjoyned one from another, and having their diſtinct *Stalks*, of ſeveral *Lengths*, by which they are all faſtned to the utmoſt Side of the great *Bag*, wherein they are conteined. Within each of theſe leſſer *Bags* are conteined many hundreds of *Bladders*, conſiſting of moſt extream fine *Threds* woven up together into that *Figure*. Within theſe *Bladders* lies the *Acid Juyce* of the *Limon*.   *Tab.* 66.

12. §. A CUCUMER, hath alſo a Threefold *Parenchyma*. The Utmoſt, is derived, from the *Barque*. In this, being expoſed for ſome time to dry, and then cut tranſverſly with a *Raſor*; not only the *Bladders*, but alſo the *Threds* whereof the *Bladders* conſiſt, through a good *Microſcope*, are apparent.

13. §. Throughout this *Parenchyma* the *Sap-Veſſels* are diſperſed; near the Circumference, in Ten or Twelve very large *Branches*. Each of theſe larger *Branches*, emboſoms another of *Aer-Veſſels* in its Centre. Adjacent to the Midle *Parenchyma*, they ſtand in *Cluſtres* of much ſmaller *Branches*, but more numerous.   *Tab.* 66.

14. §. Out of all theſe *Sap-Veſſels*, iſſues a tranſparent and viſcous *Mucilage*; which being dryed, becomes as hard and tough as *Gum Tragacanth*. Analogous to which, I ſuppoſe, is the truly purgative part of *Elaterium*.

15. §. The Midle *Parenchyma* is derived from the *Pith*; and divided into Three *Colums*, ſtanding triangularly, and having each of them a *Triangular Figure*. Within theſe *Colums* ſtand a diſtinct Sort of *Sap-Veſſels*: from whence, ſeveral ſmall and ſhort *Fibres* ſhoot into the Inmoſt *Parenchyma*, whereupon the *Seeds* do hang. So that theſe *Columns* are as it were the *Beds* on which the *Seeds* grow. With each of the *Seed-Branchs* or *Fibres*, goes ſome part of the ſaid *Parenchyma* or *Colum*, out of which, the *Covers* of the *Seed* are formed.   *Tab.* 66.

16. §. The Inmoſt *Parenchyma* wherein the *Seeds* lie, and which anſwers to the *Pulp* of a *Limon*, ſeems likewiſe to be derived from the *Colums*, that is, to be originally thence produced upon the *Seed-Fibres*, and afterwards ſpread and augmented into a *Pulp*. By Three *Inſertions* from the *Colums*, and as many from the Utmoſt *Parenchyma*, and theſe re-inſerted; it is divided into Six *Triangular Bodies*; and every *Triangle*, into Three *Ovals*.   *Tab.* 66.

17. §. A near reſemblance betwixt the *Garden* and *Wild Cucumer*, with reſpect to the Inward *Structure*, as well as the Outward *Figure*, may be obſerved: Both of them having a Threefold *Parenchyma*. Yet with this difference, That the Three *White Triangular Bodies* or *Colums* in the one, is anſwered by a *White Ring* or *Tube* in the other.   *Tab.* 66.

CHAP.

CHAP. II.

Of the PEAR and QUINCE.

Tab. 67.

A PEAR, besides the *Skin*, consisteth of a Twofold *Parenchyma*, of *Vessels*, *Tartareous Knots* or *Grains*, and a *Coar*. The *Skin* is lined with a great number of the said *Tartareous Grains*, through a *Glass*, about the bigness of small *Shot*: whereby it looks withinside, like the *Skin* of the *Scate* and some other *Fishes*. Besides those which grow to the *Skin*, there are also many more standing near adjacent to it all round about the *Fruit*: altogether about $\frac{1}{3}$d of an Inch in thickness, through a *Microscope*; as in a Slice of a *Pear* cut transversly is apparent. Somewhat more or less, as I take it, according to the Delicacy or Harshness of the *Fruit*; as more in a *Burgamy*, or other soft and sweet *Pear*, than in those which are called *Strangulatoria*. As all *Vinous Liquors*, and those especially which are the most *Tartareous*, become more soft and sweet, according as they cast off their *Tartar*, in a greater quantity, upon the Sides of the *Vessel*.

2. §. The Outer *Parenchyma*, is of the same Original, and general Structure, as in an *Apple*. But the *Bladders*, answerable to the Shape of this *Fruit*, not altogether so long, with respect to their Bredth. Throughout this *Parenchyma*, are also dispersed many small *Tartareous Grains*; most of them somewhat round, as those next the *Skin*, and of a like *Size*; but nothing near so numerous.

Tab. 67.

3. §. The *Bladders* here, have also a different *Position* from that they have in an *Apple*: there, they are all so stretched out, as to have respect to one common Centre, which is that of the *Apple* it self. But here, they every where bear a respect to the said *Tartareous Grains*, every *Grain* being the Centre of a certain Number of *Bladders*; like a *Star*, in the midle of its *Vortex*. Whereby, so many of the *Tartareous parts* of the *Sap*, as cannot well be thrown off upon the *Skin*, are more commodiously discharged, upon every little *Knot* or *Grain*, nearer hand.

Tab. 67.

4. §. Throughout this *Parenchyma*, the *Vessels* likewise are dispersed. Of the Two general Kinds, for *Sap*, and for *Aer*. The *Aer-Vessels*, are here extream small, as well as in an *Apple*; yet one degree, larger. They are both together distributed into Fifteen principal *Branches*. The Five Utmost make as many *Arches*, but commonly not near so deep as in an *Apple*. From these, some small *Fibres*, yet a little more numerously than in an *Apple*, are dispersed throughout the *Parenchyma*. The Ten Inmost run along to the *Seed*, and from thence, with the other Five, to the *Flower*.

Tab. 67.

5. §. Next the *Coar*, stands the Inner *Parenchyma*, in divers respects different from the Outer. The *Bladders* of the latter, as hath been said, large and long; of the former, small and round, answerable to those of the *Pith*, of which it seems to be derived. Throughout that,

Book IV.   *of Fruits.*   183

the *Veſſels* and *Tartareous Grains* are diſperſed; in this, there are nei- Tab. 67.
ther. The *Effect* whereof is, that is ſweet, this ſower; for which
reaſon, I have taken leave to name it, the *Acetary*.

6. §. Betwixt this and the outer *Parenchyma*, the ſaid *Tartareous Grains* begin, firſt to ſtand nearer together, to grow biger, and of a more unequal Surface; and by degrees, to unite into a *Body*, in ſome Tab. 67. *Pears*, and eſpecially towards the *Cork*, almoſt as hard as a *Plum-Stone*; which I have thereupon, named the *Calculary*. So that a *Pear*, is Na- B. 1. Ch.6. tures *Preface* or *Introduction* to a *Plum*.

7. §. This *Tartareous Body*, and thoſe ſmall *Grains* above ſaid, I B. 1. Ch. 6. have formerly ſuppoſed, to be precipitated out of the *Sap*, by virtue of the *Veſſels*. Which is not only argued from their growing, where the *Veſſels*, only in the outer *Parenchyma*: but in that the very *Bounds* or *Figure* of the *Calculary*, is determined by the *Situation* of the chief of Tab. 67. thoſe *Veſſels*; as in cuting a *Pear* ſmoothly through the *Centre* and by the *Length*, is apparent.

8. §. The *Coar* as well as the *Acetary*, ſeems to be derived from the *Pith*. And is therefore leſſer here, than in an *Apple*, where the whole *Pith* of the *Stalk*, goes to the making of the *Coar* only.

9. §. In moſt *Pears*, at the bottom of the *Coar*, and a little below the Centre of the *Fruit*, there is a kind of ſmall *Umbelical Knot*; from Tab. 67. whence is extended a ſtraight *Chanel* or *Ductus*, which opens at the midle of the *Cork* or *Stool* of the *Flower*, ſcarce wide enough to admit the ſmalleſt *Pin*. Made for the *Uſe* hereafter mentioned.

10. §. A QUINCE, is nearly allyed to a *Pear*. The diffe- rences betwixt them are theſe; In the *Quince*, the outer *Parenchyma* is more cloſe, that is, the *Bladders* are ſmaller. The *Veſſels* more nu- Tab. 67. merous, and more deeply enarched; the *Calculary* greater, and more ſpread; according to the *Shape* of the *Fruit*: but the *Acetary*, leſs: The *Coar* ſtands higher or nearer to the *Cork*; divided, not into Five, but Four *Cells*. And the *Ductus* from the bottom of the *Coar* to the top of the *Fruit*, much more open and obſervable.

CHAP. III.

Of the PLUM, *and ſome other* Fruits *of the ſame Kindred.*

PLUM conſiſteth of a *Parenchyma*, the Two general Kinds of *Veſſels*, and a *Stone*. All which I have already deſcribed in the *Firſt* Book. I Ch. 6. ſhall here add, and further clear ſome things. And Firſt, it is to be noted, That, in Pro- portion to the *Bulk* of the *Fruit*, there are more *Veſſels* in a *Plum*, than in an *Apple*, *Pear*, or *Quince*. As alſo, That in *Plums*, all the *Veſſels* are braced together into one Uniform Piece of *Net-Work*, every where terminating at an equal diſtance from the *Circumference*, ſc. ⅛th of an Tab. 68. Inch or thereabout. And as for the *Bore* of the *Aer-Veſſels*, although the *Glaſs* I uſed, when I examined this *Fruit*, would not reach it; yet

is it to be presumed, that they bear a just Proportion to those in the *Trunk* of the same *Tree*; and that therefore they are here larger, than in an *Apple* or *Pear*. The *Skin* likewise of a *Plum*, is more *fibrous, thick,* and *tough,* than in those *Fruits*. The Ends of these Diversities, we shall presently speak of.

B.1. Ch.6. 2. §. Of the *Stone,* amongst other particulars wherein the contrivance of Nature is very admirable, I have formerly shewed, That it is composed of Two or rather Three distinct *Bodies*. One of them, the *Lining*; which answers to the *Coar* in a *Pear*. And is originated from the *Parenchyma,* which the *Seed-Branch* brings along with it, through the *Chanel* in the *Side,* and at last into the *Hollow,* of the *Stone*; and is there spread all over it: as when a small *Glass-Pipe,* is blown and expanded into a *Bubble*. Or as if a *Bladder,* being stretch-

Tab. 68. out, and put through the *Neck* of a *Bottle*; were then blown up, so as to be every where contiguous to the *Sides,* and become, as it were, the *Lining* of the *Bottle*.

3. §. The *Foundation* or *Ground* of the Outer and more Bulky Part of the *Stone,* is the Iner Part of the *Parenchyma*; and answers
Tab. 68. to the *Acetary* in a *Pear*. As the *Fruit* grows, the *Tartareous Parts* of the *Sap,* being continually precipitated upon this *Parenchyma,* it is hereby petrify'd. As will best be seen, by comparing the several Ages of the same *Fruit* together. And in some *Stones*; on the Surface
Tab. 68. whereof, some of the said *Tartareous Parts* appear in distinct *Grains*. So that whereas in a *Pear,* the *Calculary* and the *Acetary* are distinct, here in a *Plum,* they are thrown one into the other. Or, as some *Mineral Waters* only make a *Crust* about a Stick or other Bodies immersed in them; but others, by sinking into these Bodies, do hereby petrify them: So in a *Pear,* the *Tartareous Parts* of the *Sap,* only make a *Crust* about the *Acetary*; but in a *Plum,* they sink into the Body thereof, or that Part of the *Parenchyma,* which stands in the place of it, whereby it is converted into a *Stone*. The *Figures* of *Stones* shall hereafter be spoken of, when I come in the next Part, to the Covers of the *Seed*.

4. §. AN APRECOCK is of the *Plum-Kind*. But some things are herein better observed. As first, the *Position* of the *Bladders* of the *Parenchyma*. For the *Tartareous Parts* of the *Sap* not being here dispersed, in little *Grains,* throughout the *Fruit,* as in a *Pear*; but all thrown off into the *Stone*: the *Bladders* therefore are so dispo-
Tab. 68. sed, as not to have respect to several *Centres,* as in a *Pear*; but only the *Stone,* to which they all do most exactly radiate; thereto conveying the *feculent Sap,* in so many little *Streams*. This is best seen, when the *Fruit* is full ripe.

5. §. In this *Fruit,* while it is young, the gradual transmutation of the Inner Part of the *Parenchyma* into a *Stone,* is also more apparent. And so are the Three *Coats,* which serve for the *Generation* of the *Seed*; being now all very distinct; and remarkable, not only for their *Bulk*; but also, the Analogy which they bear to the Three *Membranes* in many *Viviparous Animals*. Whereof I shall give a more particular *Description,* when I come, in the following Part, to the Covers of the *Seed*.

6. §. A PEACH hath a much bigger *Stone,* than either a *Plum,* or an *Aprecock:* and hath therefore, when full ripe, and especially in hot

hot Countries, a more defecated or better fined *Juyce*. For the reason why the *Stone* is so great, is because the *Vessels* run so very numerously through the Body of it; and so cause a more copious precipitation of the *Lees* of the *Sap* thereinto.

7. §. A CHEERY is likewise near related to a *Plum*. But the *Bracement* or *Reticulation* of the *Vessels*, is here carried out further, so as to be all round about contiguous to the *Skin*. And as the *Aer-Vessels* in the *Branch* of a *Cherry-Tree*, are larger than those of an *Apple-Branch*, but less than those of a *Plum-Branch*; so may they be presumed, to bear the same Proportion here in the *Fruit*, Tab. 69.

8. §. A WALNUT, is a *Nuciprune*; or betwixt a *Plum* and a *Nut*, as a *Bat* is betwixt a *Beast* and a *Bird*. For the *Rind*, answers to the *Pulp*; and the *Shell*, as the *Stone*, is also lined. But the *Seed-Vessels*, which in a *Plum* run through a *Chanel* made on purpose in the *Stone*; do here enter, as in a *Nut*, at the *Centre* of the *Shell*. By which means, they are invested with a more fair *Parenchyma*; which Nature hath provided, as her *Cloth*, for the making of the *Coats* wide enough for so vast a *Kernel*.

CHAP. IV.

Of the GRAPE, *and* HAZEL-NUT; *with some other* Fruits, *analogous to each of them.*

A GRAPE, is a *Plum* with two *Stones*; for their thickness, as hard as any other. The *Distribution* of the *Vessels* is also somewhat different. For the principal *Fibres* running up directly betwixt the *Stones*; and the smaller, making only one single *Net*, near the Circumference; they all meet together at the *Top* of the *Grape*. It is also to be noted, That many *Lignous Fibres* are visibly mixed with the *Skin* it self: whereby it becomes very thick and tough. And as the *Aer-Vessels* in the *Trunk* of a *Vine*, are greater than in that of an *Apple, Pear,* or *Plum*: So is it to be presumed, that in a *Grape*, they are greater than in the *Fruits* of those *Trees*. Tab. 69.

2. §. The *Parenchyma* or *Pulp* of a *Grape*, seems to be derived, not from the *Barque*, as in an *Apple*; nor partly from the *Barque*, and partly from the *Pith*, as in a *Goosberry*: but wholly from the *Pith*; at least, as far as the *Reticulation* of the *Fibres*; and the *Skin* only from the *Barque*; whereby the *Pulp* becomes so tender and delicate a *Meat*.

3. §. A GOOSBERRY, hath a Threefold *Parenchyma*. The Utmost is derived from the *Barque*; of a *Greener Colour*, and very *sappy*. The midlemost, from the *Pith*; somewhat *white*, and more *dry*, as the *Diametral Insertions* in some *Roots*. In both of them, the *Bladders* are very conspicuous, above what they are in any *Fruit*, I at present think of; so as to be visible to a good *Eye* without a *Glass*. Tab. 69.

Tab. 69.   4. §. Betwixt these Two *Parenchymas*, do run most of the principal *Fibres*, or *Vascular Threds*. From which several smaller ones are branched into the Inmost *Parenchyma*; upon which, the *Seeds* do hang.

5. §. Each of these smaller *Branches* is invested with some part of the midle or white *Parenchyma*. Serving partly to make the *Covers* of the *Seed*; and partly, the *Pulp*, that is, the Inmost and finest *Parenchyma* of the *Berry*, in which the *Seed* lies.

A *White* CORIN, without taking off the *Skin*, sheweth not unpleasantly how the *Seeds* are fastned. For as the *Trunk* of the *Tree* continues not to any considerable Length, entire, as in a *Plum*, but is presently divided into several *Boughs*; nor are the Edges of the *Leaf* entire, as also in a *Plum*, but slit into several *Lobes*; and the *Fruit*, into a great many *Corins* in a *Bunch*: So again, the *Seeds* do hang upon the *Fibres*, like Two other *Bunches*, in every *Corin*. As by *Refraction*, Objects of all *Sizes* are represented on the *Walls* of the *Eye*. The *Operations* of *Nature* being every where Uniform: and sometimes the same in small, transcribed from a greater *Copy*.

7. §. A NUT, is a *Plum* inverted, or turned inside outward. For the *Shell*, standing naked, includes the *Parenchyma*: the bearded *Cap*, not precisely answering to *that*, but to the *Empalement* of the *Flower*; which likewise in many other *Plants*, out-lives the *Foliature* and Embosomes the *Uterus* of the *Seed*. And whereas the *Stone* of a *Plum* is not Faced, but Lined with a *Parenchyma* derived at second hand from the *Pith*: The *Shell* of a *Nut* is not Lined, but Faced with the iner *Skin* of the *Cap*.

Tab. 69.

8. §. AN AKERN, is the *Nut* of an *Oak*. Yet with this difference; That besides the *Cup*, it stands in, it hath only a *Leathern* or *Parchment Cover* instead of a *Shell*. From whence it come to pass, that whereas the *Kernel* of a *Nut* is sweet; that of an *Akern*, is of a very rough *Tast*: the *Austere Parts* of the *Sap*, which in a *Nut* are drained off into the *Shell*, being here imbibed by the *Kernel* it self.

CHAP. V.

Of the SEED-CASE or MEMBRANEOUS UTERUS.

O the forementioned *Fruits*, I shall subjoyne, in some Examples, the *Description* of the *Seed-Case*, which is analogous to the *Fruit*. For the *Fruit*, strictly so called, is, *A Fleshy Uterus, which grows more moist and Pulpy, as the Seed ripens.* But the *Seed-Case*, whether it be called a *Cod*, *Pod*, or by any other name, is, *A Membraneous Uterus, which as the Seed ripens, still grows the more dry and hard:* as in most *Plants*.

2. §. THE SEED-CASE, is either originally open; Or only when the *Seed* is ripe; Or never opens at all, till the *Seed* be sown: Of the first Sort, is that of *Luteola*; as also of *Clary*, *Sage*, *Hysop*, and the

the like: wherein one and the same *Part*, is both the *Empalement* of the *Flower*, and when that is gone, survives as the *Case* of the *Seed*.

3. §. Of the Last, is that of *Myagrum Monspermon, Lithosperme*, all the *Stones* of *Fruits*, with divers others. And some *Cases*, which are soft, as, I think, that of *Garden Radish*. The former, by cleaving in some *part* or other; these only by roting under Ground.

4. §. THAT of *Garden Radish*, is a Light and Spongy or Pithy *Body*; originally, every where entire. But, as it ripens, breaks within, into several White and Dry *Membranes* round about the Seed. By the Length and about ⅓th of an Inch distant from the *Sides* of the *Case*, do run a pair of little *Vascular Ropes*. Some smaller *Fibres* are from these transmitted to the *sides* of the *Case*; by which they are kept tite and steady. Upon divers others produced towards the *Centre*, hang the *Seeds*, like Two *Ropes* of *Onions*. *Tab.* 70.

5. §. Of those which open so soon as the *Seed* is ripe; some are made to open at the *Top*, as *Popy Heads*; Some on the *Side*, as most *Cods*; and some at the *Bottom*, as that of *Coded Arsmart*.

6. §. THE *Popy-Head*, is a little *Dove Coat*; divided by Eight or Ten *Partitions*, into so many *Stalls*. On both *Sides* the *Partitions*, hangs a most numerous *Brood* of Seeds. The *Partitions* and *Sides* of the *Head*, are made of the *Barque*, and Lined with the *Pith*. While young, they are very thick and spongy; and together with the *Seeds*, do then fill all up. The *Head* is then also every where entire; but as it dries, it gradually opens at the *Top*, into several *Windows*, one for every *Stall*: which are all covered with a very fair *Canopy*. A *Fabrick* designed for several purposes, as shall hearafter be said. *Tab.* 70.

7. §. Of those which open on the *side*; some are made to open, only on One *Side*; some, on both *Sides*; some, with Three *Sides*; some, with more; and some horizontally or round about.

8. §. THE COD of *Garden Bean* (and so the rest of the *Leguminous* kind) opens on one *side*. It hath a Twofold *Parenchyma*. The Utmost derived from the *Barque*: in which stand all the *Vessels*, in several *Parcels*; one whereof, at the *Back* of the *Cod*, is much larger than the rest, shaped like a *Copula* used in *Schemes*; from whence shoot those lesser *Fibres* upon which the *Beans* do grow. *Tab.* 70.

9. §. The Inner *Parenchyma* is derived from the *Pith*. Upon its Nativity, and for some time afterwards, entire and wholly composed of *Bladders*, as the Outer. From the *Base* of the *Cod* they are gradually enlarged, so as to compose this *Parenchyma* into a very soft and delicate *Sponge*. In which (the *Cod* being well grown) the very *Threds* whereof the *Bladders* were woven, are many of them so loose and ample, as easily to be drawn out (as in the uroaving of *Knit-work*) to a considerable Length, fairly visible through an ordinary *Glass*. *Tab.* 70.

10. §. This may further confirm all that I have formerly said of the *Fibrous Texture* of the *Pith*, and of all the other *Parenchymous Parts* of Plants. B. 2. P. 1. Ch. 5. & B. 3. P. 1. Ch. 4.

11. §, THE *Seed-Case* of *Medica*, is a *Cod* wound up: in the *Echinata*, *Spirally*; in the *Tornata*, by an *Helix*. Not finished all together; but, upon the fall of the *Flower*, beginning to wind, continues its *Circles*, till it be come to its full *Growth*.

12. §. THE *Seed-Case* of *Yellow Henbean* opens on both *Sides*. On the *Top*, is erected a *Colum*, about ¼ an Inch long; which, as the *Case* swells, gwrosless, and at last falls off. On the *Sides* of the *Uterus* or *Case*, Two *Vascular Fibres* run oppositely from the bottom to the top, and so into the *Colum*. Along the *Tract* of these *Fibres*, the *Case*, as it ages, gradually cleaves on both *Sides* asunder.

Tab. 70.

13. §. The *Case* is lined with a dry and thin *Parchment*, as smooth as *Glass*. In the *Centre* of the *Case*, stands a great *Parenchymous Boss*, which is, as it were, the *Bed* or *Placentula* of the *Seeds*; which lie all over it, as in a *strawberry*. And so in many other *Plants*. Throughout this *Bed*, the *Vessels* for the *Generation* and *Nourishment* of the *Seeds*, are distributed; one very small *Fibre*, shooting, from the direct ones, obliquely into each *Seed*.

Tab. 70.

14. §. THE *Seed-Case* of *Tulip*, opens with Three *Sides*; being, when young, a *Prism* or long *Triangle*. From the midle of each *Side*, a *Partition* or *Boord* is produced; all three meeting in the *Centre* of the *Case*; and so parting it into Six *Stalls* for the *Seed*. The insides hereof, are, lined with a thin smooth and glossy *Parchment*, like that in *Hen-bean*; derived from the *Pith*; as the outside, from the *Barque*: and so in many other *Seed-Cases*.

Tab. 71.

15. §. The *Vessels*, after they rise above the *Stalk*, are disposed with great *artifice*. For first, they are divided into Three principal *Branches*, which run along the Three *Angles* of the *Case*; where the Three *Sides*, as it ages, gradually cleave asunder. From these chief *Branches*, at the Three *Angles*, divers lesser ones run horizontally, and meet at the midle of each *Side*. From whence again, many yet smaller ones are produced through the bredth of each *Partition* to their Edges in the *Centre* of the *Case*. Where, once more, they are distributed into very fine and short *Threds*, whereupon hang the *Seeds*.

Tab. 71.

16. §. THE *Seed-Case* of *Stramonium* or *Thorn Apple*, is divided into Four *Closets*: Not open one into another, as in *Poppy*, *Tulip*, &c. but so many distinct *Inclosures*. In the midst of each *Closet* stands a *Colum*, joyned to the *Side* of the *Closet* by a *Wall* or *Lamine*. Through the Length of the *Colums* run several greater and lesser *Branches* of *Vessels*, from whence others are obliquely produced, upon which the *Seeds* grow.

Tab. 71.

17. §. THE *Seed-Case* of *Anagallis* or *Pimpernel*, is a little *Globe*; which opens not by its *Meridian* or *Vertically*, as do the former; but by its *Horizon*. For divers very small *Fibres*, being produced from the *Stalk* to the midle of the *Case*; do there fetch a *Circle*, and so divide it exactly into Two *Hemispheres*: the Uppermost of which, when the *Seeds* are ripe, falleth off; and so the wind sowes them.

Tab. 71.

18. §. THE *Seed-Case* of *Coded Arsmart*, neither opens at the *Top*, nor on the *Sides*, as do all the former; but at the *Bottom*. It is composed of Four *Sides*: the Outer Part of which, is softer and more succulent; the Inner a tite and strong *Membrane*. In the *Centre* of the *Case*, is erected a *Pole* or *Colum* upon which the *Seeds* do all hang very loosely.

Tab. 71.

19. §. From this *Mechanism*, the manner of that violent and surprising *Ejaculation* of the *Seeds*, is intelligible. Which is not a motion originally in the *Seeds* themselves; but contrived by the *Structure* of the *Case*. For the *Seeds* hanging very loose, and not on the *Sides* of

the

the *Case*, as sometimes, but on the *Pole*, in the *Centre*, with their thicker end downward, they stand ready for a discharge: and the *Sides* of the *Case* being lined with a strong and Tensed *Membrane*, they hereby perform the office of so many little *Bows:* which, remaining fast at the *Top*, and (contrary to what we see in other *Plants*) opening or being *lett off* at the *Bottom*, forceably curle upward, and so drive all the *Seeds* before them.

CHAP. VI.

Of the USE *of the* Parts *to the* Fruit.

IN the forgoing *Descriptions*, I have already mention'd the *Use* of the *Parts* in some particulars. I shall now a little further explain the manner of their service, both to the *Fruit*, and to the *Seed*.

2. §. And first, the *Vessels* serve for the *Figuration* of the *Fruit*. So in an *Apple*, the Ten great and utmost *Branches* serve not only to nourish and feed it; but also, by the *Arched Lines* they draw, to direct and govern the *Growth* thereof into an orbicular *Figure*. The *Dilatation* of these *Vessels*, not being hindred by any *Braces* or *Conjunction* with the *Interior* ones. By the *Slenderness* of the *Aer-Vessels*, as in the *Root*, so here in the *Fruit*, much promoted. And by their *Saline Principle*, first begun.

3. §. The Five midlemost and the Five Inmost serve together, to figure the *Coar*; the former bounding the Outer, the Latter, the Iner *Angles*. For were they only Five, or were all Ten in the same *Circle*, they would only make a round *Cavity* like that of a hollow *Pith*. Hence it is that *Apples*, in which some small *Threds* of the *Vessels* strike out into the *Circumference*, are very Uneven with divers *Knobs* and *Ridges*. But *Plums, Cherries*, &c. where the *Vessels* all terminate at an Equal distance from the *Skin*, are Even all round about.

4. §. The *Bulk* of the *Fruit* dependeth also on the *Braces* of the *Vessels*. For in *Plums* and *Cherries*, they are more numerous; but in *Apples* and *Pears* they are very loose one from another, and so have liberty left them to spread abroad.

5. §. As also on their *Size*; that is, on the *Size* of the *Aer-Vessels*. Which, the less they are themselves, they serve to make a bigger *Fruit*. As the less they are in any *Root*, they serve to make it the more ample. For the less they are, the more pliable to the *Attraction* of the *Aer*: and in their *Growth* must make so many more spiral *Rings*: by both which means, they make the greater *Arches*. And therefore a *Pear* is commonly a smaller *Fruit* than an *Apple*; a *Plum* than a *Pear*; and a *Grape*, than a *Plum*; in all which the *Aer-Vessels* are still greater and greater.

6. §.

6. §. From the same Cause, it is also most agreeable, That the *Fruit* should not come before the *Leaves* or *Flower*, but last of all. For the *Aer-Vessels*, as hath been often noted, are not exactly *Cylindrick*, but tapered; that is, not only the *Fibres* consisting of divers of these *Vessels*, but the *Vessels* themselves, as they ascend into the *Trunck*, *Branches*, *Leaves*, *Flower*, and *Fruit*, grow still more and more slender. So that the smallest coming last, and being the most pliable; they are also best accommodated for the *Expansion* of the *Parenchyma* into that we call the *Fruit*.

7. §. It is likewise a proper Question to be asked, How it comes to pass, That some *Plants* bear a *Fruit*, and not all? I answer, That as the *Size* of the *Aer-Vessels* conduceth to the *Bulk* of the *Fruit*, and the Order of its Growth: So the *Number* of them, to their being, or not being, any *Fruit* at all. For the *Fruit*, as we have already defined it, is an *Uterus*, which grows moyster and softer, as the *Seed* ripens. The reason therefore, why the *Uterus* in some *Plants*, continues moist and soft after the *Seed* is ripe; and in some, dries up; is, Because in the former, there is a smaller, in the latter, a greater Quanty of the *Aer-Vessels* in proportion to the other *Parts* of the *Uterus*, and so a greater quantity of *Aer*. Which as in the *Pith* of most *Plants*, so here, by degrees excludes the *Sap*, or rendring it more evaporable, comes in the room of it; and so the *Uterus* is dryed up: that is, there is no *Fruit* produced, but only a *Seed-Case*.

8. §. From the *Size*, *Number*, and *Position* of all the *Vessels* in *Fruits* a reason also may be given, for the diversity of their *Tasts*. Some Instances have before been given; to which I shall add one or two more. So the *Rind* of an *Orange*, is bitter; the *Pulp*, sower. Because the former is furnished with many *Lignous Vessels*, the *Sulphureous* or *Oyly Tincture* whereof, being copiously mixed with the *Acid* of the *Parenchyma*, produce that *Tast*. Whereas the *Pulp*, which is very sower, is void of all manner of *Vessels*. But if the *Sap-Vessels* are either less numerous or less *Sulphureous*; they give so mild a *Tincture* to the *Parenchyma*, as not to produce a bitter, but a sweet or soft *Tast*; as in *Apples*, *Grapes*, *Goosberries*, &c. And of a *Goosberry*, it is particularly to be noted, that whereas, in a *Limon*, the *Pulp* only is sower, as being void of *Vessels*: here, on the contrary, the *Pulp* only is sweet, whereinto all the *Vessels* strike, and the *Rind* sower.

9. §. The diversities of the *Skin* it self, have their *Use*. And therefore, the more tender and delicate the *Fruit* is; the *Skin*, on the contrary, is thicker and more tough. So *Apples* have a thicker *Skin*, than *Pears*; *Plums*, than *Apples*; and *Grapes* than *Plums*; those having as it were, only a *Coat* of *Kid*, but this of good thick *Buff*. And therefore some *Fruits*, although tender, yet either not having so rich a *Juyce*, or coming early, and so not being exposed to excessive heats, have a very thin *Skin*, as *Mulberries*, *Strawberries*, &c.

CHAP.

CHAP. VII.

Of the USE *of the* Parts *to the* Seed. *And the* TIME, *in which the* Uterus *or* Fruit *and* Seed-Case *are formed.*

 AND first, for example, in an *Apple*, the Five Inmost *Branches*, do best serve for the *Generation* of the *Seed*; these running into the *Attire* of the *Flower*, and so carrying off the most *Aerial Spirit* from the *Seed*; by which means, it becomes a more compact and denser *Body*, than the *Fruit*, and so more accommodate to the process of *Vegetation*; as *P. 2. Ch. 5. §. 5.* hath formerly been shewed.

2. §. The *Elongation* likewise of the *Seed-Vessels*, in the *Fruit* and *Case*, sometimes directly, as in *Plums* and *Nuts*, and sometimes by several *Ambages* before they shoot into the *Seeds*, as in *Tulip*; shewes a design for the highest refining and maturation of the *Seminal Sap*.

3. §. Chiefly by means of the Inmost *Vessels*, is made that *Chanel* in some *Pears*, and especially in *Quinces*. For these perishing with the *Flower*, the circumjacent *Parenchyma* shrinks up, leaving the said *Chanel* in the midst. Designed for an inlet to the *Aer*, for the better drying of the *Seeds*; which here stand the more in need of it, because encompassed with a *Mucilage*.

4. §. For the better drying of the *seed*, and the disbursing or sowing of it in due time, the opening of the *Case* is, in the same manner, also contrived: either at the *Top*, as in *Popy*; or on the *Sides*, as in *Tulip*, *Pimpernel*; or at the *Bottom*, as in *Codded Arsmart*. All which openings are effected by the running of the *Aer-Vessels* along those places: for by drying the *Parenchyma* next adjacent, they cause it to chop and cleave asunder.

5. §. Of the *Seed-Case* of *Poppy*, it is particularly to be noted, That as the several *Windows*, serve to let in *Aer*, for the drying of the *Seeds*, after their full Growth: So the *Canopy* over them, serves to keep out *Rain*. For here, the *Case* not cleaving down the *Side*, as it usually doth; should the *Rain* get in, it would stand in it, as in a *Pot*, and so rot the *Seeds*. And as the *Canopy* serves to preserve the *Seeds*; so the several *Partitions* or *Walls*, for their better *Stowage*. For by an easie survey of this little piece of Ground, it is plain, that as they stand on both *Sides* every *Wall*, there is as much more Ground for them to stand upon, as if there were no parting *Walls*, but the *Seeds* stuck all round about upon the *Ambit* or *Sides* of the *Case*; or upon a great *Bed* or *Placenta* within it, as in *Hyoscyamus*, *Anagallis*, &c. where there is a less numerous *Brood*.

6. §.

6. §. The *Coar* likewise, by standing betwixt the moyst *Parenchyma* and the *Seed*, and being hollow and so filled with *Aer*; doth much conduce to the ripening and drying of the *Seed*, and its greater fitness both for keeping, and sowing. So the *Parchment Lining* of the *Seed-Case*, as in *Hyoscyamus*, &c. is answerable to a *Coar*.

7. §. The *Parenchyma* serveth, amongst other purposes, for the *Generation* of the *Covers* of the *seed*; as in some instances hath been shewed. For which intent, sometimes the *Exterior Parenchyma*, as in a *Limon*; sometimes the midlemost, as in a *Goosberry* or *Cucumer*, is subservient; both of them, in those *Fruits*, being more white and dry, than the rest, and so fiter to make the *Covers* of the *seed*.

8. §. The *Parenchyma* is also of use for the warmth of the *Seed*; as in the *Seed-Case* of *Garden Radish*. Wherein, as it ripens the *Parenchyma* gradually drys, breaks, and shrinks up into several soft *Membranes*, in which the *Seeds*, in the Centre of the *Case*, lie swadled, as in so many fine *Calico Clouts*.

9. I SHALL conclude with observing the *Time* of the *Generation* of the *Fruit* and *Seed-Case*. This hath hitherto been thought to be initiated upon the opening, I say not, the forming, but the opening of the *Flower*, or not long before. Notwithstanding which, what I have formerly said of the *Flower*; I now do the like, of the *Uterus* it self; *sc*. That in very many *Plants*, 'tis formed, with the *Flower*, the year before it appears and comes to its full *Growth*. As for instance, in *Azarum*, not only all the *Parts* of the *Flower*, but the *Uterus* it self, and there in also the outer *Cover* of the *seed* of any one year, are perfectly formed in *August* or *September* of the year foregoing. The like may be seen in *Tulip*, *Mezereon*, *Corin*, and many other *Perennial Plants*.

Tab. 71.

THE

THE
ANATOMY
OF
SEEDS,

PROSECUTED

With the bare EYE,

And with the

MICROSCOPE.

The Figures presented to the *Royal Society, in the Year* 1677.

The FOURTH PART.

By *NEHEMJAH GREW* M.D. Fellow of the *ROYAL SOCIETY*, and of the *COLLEGE* of *PHYSICIANS*.

LONDON,

Printed by *W. Rawlins,* 1682.

THE CONTENTS OF THE Fourth Part.

CHAP. I.
Of the FIGURES of Seeds.

CHAP. II.
Of the NUMBER and MOTIONS of Seeds.

CHAP. III.
Of the several COVERS of Seeds, *and of the VITELLUM.*

CHAP. IV.
Of the FOETUS or true SEED: and first of the RADICLE and LOBES.

CHAP. V.
Of the BUDS of Seeds. *And of the PARTS of which these, the* Radicle *and* Lobes *are compounded.*

CHAP. VI.
Of the GENERATION of the Seed.

THE ANATOMY OF SEEDS.

PART IV.

CHAP. I.

Of the FIGURES of Seeds.

THE *Figures* of *Seeds*, or rather of their outward *Covers*, are made suitable, Partly to their *Collocation* in the *Uterus*, as the *End*. So those of *Mallow*, standing like a Coronet round the *Stalk*, are of a wedged *Figure*; whereby their sharp Edges do all meet together in one *Centre*. Partly, to the various distribution of the *Vessels* or *Fibers*, as one *Cause*: by which the *Measures* and *Surface* of *Seeds*, as well as of the *Leaves* of *Plants*, are diversified. And partly, to the Nature of the *Saline* and other *Principles* regent in a *Plant*, as another principal *Cause*. And therefore the more *stony*, *brittle*, or full of *salt* the *Covers* of any *Seeds* are, they are generally more *angular*, and their *Figure*, whether *angular* or not, more constantly observed. So the *Tartareous Stone* of a *Plum*, is not only more *angular*, but also more regular than the Husk of the Kernel of a *Pear* or *Apple*.

2. §. For all Stones are measured by several *Circles*, whose *Diametres* hold a certain proportion to the Length of the *Stone*; in the same manner as hath been shewed in the description of the *Leaf.* So *P. 1. Ch. 2.* the *Stone* of the *Pease-Cod-Plum*, is measured by two *Circles*. That of the *Turkey-Plum* with Four. That of the *Aprecok-Plum*, with Two *Tab. 72.* repeated oppositely; being perfectly *Rhomboid*. To which, those also of the *Wheat-Plum*, *Damasceen*, and some others, allude. And some are measured be four *Circles*, and one repeated.

§. 3.

3. §. The *Figures*, not only of the larger fort of *Seeds*, but even of the fmalleft, have much and elegant variety. We will take the pleafure of comparing thefe which follow.

Tab. 73. 4. §. And firft of all, fome are perfectly *Spherick*, and with an even *Surface*; as that of little *Century*. That of *spergula* is alfo *Spherick*; but hath a knobed *Surface*, and is encompaffed with a *Membraneous Rimm*, like the *Horizon* of a Globe. That of little *Celandine* is Circular, but compreffed like a *Cheefe*.

5. §. Others are *Nephroideous*, or as it were *Hemifpherick*. Of which *Figure*, and hereunto approaching, there are a greater number than of any other; as that which agrees with the more frequent *shape* and *Fold* of the *Lobes* and *Radicle* of the *Seed*, as fhall be feen. Yet with *Tab.* 73. fome difference, as to their *shape* and *Surface*. So, that of *Lychnis sylveftris* is figur'd juft like the kidney of a *Cat*; and hath a knobed *Surface*. That of *Poppey* comes near it in *Shape*; but hath a *Surface* exactly like that part of the *Paunch* of a *sheep*, called the *Hony-Coome*. That of great *Celandine*, is a little more oblong; and fo, like the Kidney, not of a *Cat*, but of a *Sheep*: chequered with parallel *Rings* and other fhort *Lines* placed alternately betwixt them.

6. §. Where, by the way, we may fee, as well by the *Seed*, as by the other *Parts*, of how different kinds, the Great and Little *Celandine*, notwithftandig their Names, are to be efteemed.

Tab. 73. 7. §. The *Seed* alfo of *Ben* or *fpatling Poppey* is fomewhat like a *Kidney*: but hath its Circumference raifed up into a double *Ridg*: to which feveral fmall *Ridges* do in fome fort alfo radiate frome one *Centre fc.* the *Bafe* of the *Seed*.

Tab. 73. 8. §. The *Seed* of *Chickweed*, is partly like a *Kidney*, and partly like a little *Retort*. As alfo that of *Pentaphil. fragiferum*. But the former is rough caft with fmall pieces having as it were feet on each fide, like little *Infects*. With which, the *Seed* of *Leuchanthemum* (which may be called, the *Giant-Chickweed*) doth much agree. The latter, hath feveral *Fibrous Ridges*, refembling the *Fibres* in the *Auricles* of the *Heart*; or runing from the nofe to the Circumference, fomewhat like the *Azymuth Lines* on a *Quadrant*.

Tab. 73. 9. §. Some are *Oval*; as that of the little *Bell*, and rough caft with *Fibres* almoft parallel and produced by the Length of the *Seed*. In which latter refpect, the *Seeds* alfo of *Trachelium* and fome other like *Plants*, are agreeable. That of *Brooklime*, is alfo *Oval*, but encompaffed with a thick *Rimm*, narrowing all the way to the *Bafe* of the *Seed*.

10. §. The *Seed* of *Dovesfoot* hath an *oval Cone*, and a flat *Bafe*. Its *Surface* favous, like that of *Poppy*, *Toad-flax*, and fome other *Seeds*. *Tab.* 73. That of *Sedum minus æftivum luteum*, is in a manner the *Figure* of the former inverted, being flat, not at the *Bafe*, but on the *Top*. And whereas that rifes with a blunt *Angle*, this hath only a *Ridg*, raifed above the *Surface* of the *Seed*.

11. §. The *Seeds* of divers forts of *Grafs*, are more *Conick*, as particularly of that, which for the likenefs its *Seed* hath to a *Barly Corn*, may be called *Barley-Grafs*. And I little doubt, but that among the *Tab.* 73. feveral forts of *Grafs*, there are fome which anfwer to all the kinds of *Efculent Grains*, as *Oat-Grafs*, *Rice-Grafs*, *Wheat-Grafs*, *Rye-Grafs*. And accordingly, that they may be more profitably fown in one Ground, than in another; and ufed with diftinction, for the higher, or more

wholfome

wholſome feeding of Cattle. A *Ruſh*, though it ſeems an imperfect *Plant*, yet beſides its *Flower*, hath alſo a plentiful brood of *Seeds* of a *Conick Figure*.

12. §. Some *Seeds* are *Cylindrick*, as that of St. *Johns-wort*, as alſo of *Tutſan*, and ſome other like *Plants*, with ſome little diverſity in the *Shape* or *Surface* of the *Seed*. That of *Vervain*, is in a manner, half a *Cylinder*: the true *Seed* lying in the *Covers*, like a Child, in a Cradle without an head. *Tab.* 73.

13. §. Others are rather *Conico-Cylindrick*, as that of *Jacobæa*; having a Coronet on the top, and ſeveral furrows by the Length round about. Anſwerable to which, is that of *Erygerum*; in *Shape* not unlike to a *Rowling-pin*. *Tab.* 73.

14. §. Some are *Plani-Conick*, as that of *Nettle*, which is ſhaped ſomewhat like the end of a *Speer*. That of *Eye-bright* is more *Eliptick*; with ſeveral *Ridges* running by the Length; and joyned together with ſhort pieces tranſverſly, as in the looping of *Lace*. That of *Worm-wood* not very unlike a little flat *Eſſence-Glaſs*: in which, the *Fibres* are produced by the Length, as the *Ridges* are in *Eye-bright*. And ſo in *Yarrow*, which is alſo encompaſſed with a *Membraneous Rimm*. That of *Dandelyon*, is *Plani-Conick* towards the *Baſe*. And ſo thoſe of *Lettice*, *Sonchus*, and ſome others. To which, thoſe alſo of *Hieraceum*, *Tragopogon*, *Scorzonera*, &c. with reſpect to their *Surface*, do all allude. *Tab.* 74.

15. §. And ſome are *Conico-Triangular*. Of which, that of *Sorrel* is *Conick* at both ends; the ſides equal; and upon every *Angle*, hath a narrow and ſharp *Rimm*. As alſo that of *Anagallis*; but the *Sides* are *Spheri-conick*, and ſo the ends are blunt. They are alſo pounced with many little round *Cavities*. But have no *Rimm* upon the *Angles*. *Tab.* 74.

16. §. The *Seed* of *Nigella* is *Triangular*, and *Conick* only at the *Top*. On every *Angle*, hath a narrow *Rimm*; the three *Sides* equal, and *ſpheri-conick*; ſurrounded with ſeven or eight *Ridges* by the girth, joyned together in ſome places with others tranſverſly. That of *Arſmart*, is alſo *Triangular* and *Conick* at the *Top*. But one of the *ſides* is almoſt equal to the other two; which ſtand low. That of *Knot-Graſs* hath three *ſides*, one leſs than another; being as 5, 3, and 2, or thereabout. *Tab.* 74.

17. §. The next (which I take to be the *Seed* of a ſort of *Bugloſs*) is very oddly *figured*. The *Baſe*, oval; the *Top*, conick; the *Back*, ſwelling and round as an *Egg*; the *Belly* alſo ſwelling, but riſing up into an obtuſe *Angle* higheſt in the midle, ſomewhat like a *Breaſt-piece* of *Armour*: and is encompaſſed with a *Rimm* ſloaped upward. *Tab.* 74.

18. §. That of *Moldavian Bawm*, is *Triangular*, and *Conick* only at the *Baſe*. The place where it is faſtned, ſhaped like the Beard of a *Dart*. Two of the *Sides* are *Plani-conick*, the Third *Sphericonick*, and near as big as both the other two. The *Head* flat, with a *Rimm* erected upon each *Side*, ſo as to make a *Spherical Triangle*. Approahing to this, are thoſe of *Sage*, *Horehound*, *Clary*, &c. *Tab.* 74.

19. §. That alſo of *Bellis Tanaceti folio*, hath two *ſides Planiconick*, and a third *Sphericonick*. The two firſt have ſeveral *Ridges* running to the *Baſe*. Which is not perfectly *conick*, but a little dilated into two obtuſe *Angles*. The *Head Triangular*, with one *Side* convex, *Tab.* 74.

the

the other Two ſtraight, a little hollowd, and having a ſmall pinacle in the *Center*.

Tab. 74.

20. §. That of *Stæchas Arabica*, as the former, ſaving, that the *Head* is oval, and the *Baſe* ſloaped into a little *Triangle*. That of *Wartwort* or *Sun-Spurge*, hath a very complex *Figure*. The *Belly* conſiſteth of two *Planiconick* Sides, as the former; the Back, *Sphericonick*. The whole *Seed*, in a manner, *Conick-oval*. Yet the *Baſe* and *Head* both flat. In the midle of the former, a Peg by which the *Seed* is faſtned; and of the latter, a poynted *Knob*. The midle of the *Belly-Sides*, hollowed, ſo as to make a flat *Rimm* of equal *Bredth*; and the hollows filled up with *Bladders* like thoſe in all the *Parenchymus* Parts of a *Plant*.

Tab. 74.

21. §. Laſtly, there are ſome *Seeds* which are ſquare. Whereof ſome are ſtraight, as that of *Fox-glove*; which hath alſo an even *Surface*: And that of *Blattaria*, in which there are ſeveral little hollows in even Rows. And ſo in *Brounwort*.

Tab. 74.

22. §. And ſome *Convex*, as that of *Chryſanthemum Americ*. 'Tis *Quadrati-conick*, or ſquare and ſharp at the *Baſe*, and big at the *Head*. The *Sides* all plain; and a thin *Rimm* erected upon every *Angle*. As alſo on the four *Sides* of the *Head*, which is flat, with a little *Pinacle* in the midle.

Tab. 74.

23. §. The *Seed* alſo of *Tanſey*, is a *Conick* and bended ſquare not with the *Angle* forward, as the former, but the *Side*. And in the place of every *Rimm*, hath a round *Ridge*. Somewhat like to this, are thoſe of *Febrifuga*, *Mayweed*, and ſome others. Thus far of the *Figures* of *Seeds*.

CHAP. II.

Of the NUMBER and MOTIONS of Seeds.

NATURE hath ſecured the *Propagation* of *Plants* ſeveral ways, but chiefly by the *Seed*: for the *Production* of which, the *Root*, *Leaves*, *Flower*, and *Fruit*, do all officiate, as hath been ſhewed. And according as the *Plant*, or the *Seed* it bears, is more liable to be deſtroyed, Proviſion is made for *Propogation*, either by a greater number of *Seeds*, or other ways. So the *Seeds* of *Strawberry*, being gathered, or eaten by *Vermin*, with the *Fruit*; the *Plant* is therefore eaſily propagated by *Trunk-Roots*. So *Poppy*, being an annual *Plant*, is highly prolifick: for inſtance, the *White Poppy*; which commonly bears about four mature *Heads*, in each of which, there are at leaſt ten *Partitions*, on both ſides whereof, the *Seeds* grow; and upon $\frac{1}{8}$th part of one ſide, about 100 *Seeds*; that is, 800 on one *Partition*: which being multiplied by 10 (the number of *Partitions*) makes 8000; and 8000 again by 4 (the number of *Heads*) makes 32000 *Seeds*, the yearly product of that *Plant*.

2. §.

2. §. So in *Typha major*, the *Seeds* being blow'n off and fow'n (as the Eggs of many *Fishes* fpawn'd) with great hazard, they are ftrangely numerous. For as they ftand altogether upon the *Spike*, they make a *Cylinder* at leaft fix Inches long, and near $\frac{5}{8}$ths of an Inch in Diametre, or an Inch and $\frac{1}{4}$ about. Now 9 of thefe *seeds*, fet fide to fide, as they ftand on the *Spike*, make but $\frac{1}{8}$th of an Inch; fo that 72 make a line of an Inch in Length. But becaufe upon the *spike*, the *Hairs* belonging to the *seeds* come between them; we will abate 10, and count but 62. To which $\frac{3}{4}$ths of 62, that is (without the Fraction) 46. being added, makes 108 for the Circuit of the *Cylinder*. And the *Cylinder* being fix Inches long, there are fix times 62, that is, 372, for a Line the length of the *Cylinder*. Which number being multiplied by 108, produceth 40176 the number of *Seeds* which ftand upon one *Stalk*; and fo, upon three *stalks*, which one *Plant* commonly bears, there are in one year, above a hundred and twenty Thoufand *Seeds*.

3. §. SO SOON as the *Seed* is ripe, Nature taketh feveral Methods for its being duly fow'n: not only in the opening of the *Uterus*, as in fome Inftances (*a*) hath already been feen; but alfo in the make of the *Seed* it felf. For *Firft*, the *Seeds* of many *Plants*, which affect a peculiar *Soil* or *Seat*, as of *Arum*, *Poppy*, &c. are heavy and fmall enough, without further care, to fall directly down into the Ground: and fo to grow in the fame place where themfelves had their *Birth*. (*a*) P. 3. Ch. 5. Tab. 70, & 71.

4. §. But if they are fo large and light, as to be expofed to the wind, they are often furnifhed with one or more *Hooks*; To ftay them from ftraying over far from their proper place, till by the fall of *Leavs* or otherwife, they are fafely lodged. So the *Seeds* of *Avens* have one fingle *Hook*, thofe of *Agrimony* and *Goofe-grafs*, many; both the former, loving a *Bank* for warmth, the latter, a *Hedge* for its fupport. Tab. 72.

5. §. On the contrary, many *Seeds* are furnifhed with *Wings* or *Feathers*. Partly, with the help of the Wind to carry them, when they are ripe, from off the *Plant*, as thofe of *Afh*, *Maple*, *Orach*, &c. leaft ftaying thereon too long, they fhould either be corrupted, or mifs their feafon. And partly, to enable them to make their flight, more or lefs, abroad: that fo they may not, by falling together, come up too thick; and that if one fhould mifs a good *Soyl* or *Bed*, another may hit. So the *Kernels* of *Pine* have wings not unlike to thofe of fome *Infects*; yet very fhort, in refpect of the weight of the *Seed*; whereby they flye not in the *Aer*, but like domeftick *Fowls*, only flutter upon the Ground. But thofe of *Typha*, *Dandelion* and moft of the *Pappous* kind, with many more, have very long and numerous *Feathers*, by which they are wafted every way, and to any diftance neceffary for the aforefaid purpofes. Tab. 72.

6. §. Again, there are fome *Seeds*, which are fcattered not by flying abroad, but by being either *Spurted*, or *Slung* away. The firft are thofe of *Woodforrel*; which having a running *Root*, Nature fees it fit to fow the *Seeds* at fome diftance. The doing of which is effected by a white thick and fturdy *Cover* of a *Tendinous* or *Springy Nature*, in which the *Seed* lies within the *Cafe*. This *Cover*, fo foon as it begins to drye, burfts open on one fide, in an inftant, and is violently turned infide outward, as you would turn the *Gizard* of a *Fowl*; and fo fmartly throws off the *Seed*. Tab. 72.

7. §.

200　　　　　　　　　*The Anatomy*　　　　Book IV.

Tab. 72.

7. §. The *Seeds* of *Harts-tongue*, and of all that *Tribe*, are *Slung* or *Shot* away. The doing of which is performed by the curious contrivance of the *Seed-Case*; as in *Codded Arsmart*, and some other like *Plants*. Only there, the *spring* moves and curles up inward; but here it moves outward. I shall describe it, as well as the *Weather* (which when I observed it was cloudy) would permit. Every *Seed-Case*, as it appears through a good *Glass*, stands upon a *Pedicle* from ½ an Inch to an Inch or more in Length; at the bottom about as thick again as a *Horse-hair*, and a little thicker at the *Top*, on which stands the *Case*, of a *Silver Colour*; about the bigness of a *Cherry-stone*, of a spherick *Figure*, and girded about with a sturdy *Tendon* or *Spring*, of the *Colour* of Gold: the whole *Machine* looking not much unlike a little *Padlock*. The *Surface* of the *Spring* resembles a fine *Screw*, or some of the *Aer-Vessels* in the *Wood* of a *Plant*. So soon as by the *Innate Aer* of the *Plant*, or otherwise, this *Spring* is become stark enough, it suddenly breaks the *Case* into two halfs, like two little *Cups*, and so flings the *Seed*.

Tab. 72.

8. §. These *Cases* grow in oblique *Furrows* or *Trenches* on the back side the *Leaf*, from ¼ of an Inch to an Inch in Length, and about 1/8th of an Inch broad. In one of these *Trenches* an Inch long are more than 300 of the *Cases* above described; and allowing but 10 *Seeds* to every *Case*, above 3000 *Seeds*. Which being multiplied by the number of *Furrows* in one *Leaf*, with allowance for the lesser *Furrows*; and that summ by the number of *Leaves* commonly growing upon one *Root*, comes to above Ten Hundred Thousand *Seeds*, the annual product of this *Plant*. The *Seed* is of a *Tawny Colour*, through a good *Glass* about 1/12th of an Inch long, flat, and somewhat oval. Of these, ten Thousand are not so big as a white *Pepper Corn*.

CHAP. III.

Of the several COVERS of Seeds, and of the VITELLUM.

HE next step of *Natures Managery*, relates chiefly to the Growth of the *Seed* when it is sow'n. For which purpose, the outer *Covers* are somewhere furnished with *Apertures* sufficient for the reception of *Alimental Moyster* from the Ground; and *Divisions*, for the *shooting forth* of the young *Root* into it. As in the *Seed* of a *Gourd*, at the Bottom; in a *Bean*, on the *Side*; and in a *Chesnut*, at the *Top*: in which places the *Radicle* or young *Root* always lies and puts forth, in the said several *Seeds*. And the *Seed* of *Palma Christi*; which falls to the Ground not only in the usual *Covers*, but also in the *Seed-Case*, for the more plentiful admission of *Aliment*, hath a double *Aperture*. Not much unlike to this, is that found sometimes

Tab. 72.

Tab. 72.

in

in larger parcels of *Euphorbium*; for which *Cause*, I suspect it to be the *Gumm* of a *Plant* of the *Tithymal* kind.

2. §. If the *Cover* of the *Seed* be stony and very hard, it is also distinguished into several *Pieces*; whereby they easily cleave asunder without much resistance to the eruption of the *Root*. So the *Shell* of a *Hazel-nut* easily cleavs on the edg; and the cleft begins best at the poynt, where the *Root* stands and shoots forth. The *Shell* of some *Walnuts* cleavs into three *Parts*; and the *Stone* of the *Bellerick Myrobalan* into five: that so, being very thick and hard, if one piece should not yield, another may not fail to do it. And the *Covers* or *Husks* of some forts of *Grain*, as of *Millet*, are only folded or laped one over another, the better to give way to their tender *Sprouts*. *Tab.* 75.

3. §. Besides the *Kernels* of *Plums* and some other *Fruits*, there are very many *Seeds*, even of the smaller sort, which have also stony *Covers*; as of *Carthamum, Myagrum monospermon*, *Beet*, *Borage*, *Lithosperme, Amaranthus, Violet*, &c. Sometimes, for the reception of the harsher and less matured *Principles* from the *Seed*, in its *Generation*, as in *Borage*. Commonly, to keep it warmer before and after its sow'n. For which purpose, the outer *Covers* of some *Seeds*, are as it were Lined with *Fur*: in that of *Great Maple*, Short; of *Gossipium*, Long. And if the *Seed* requires a longer stay under ground, the hardness of the *Cover* serves to stint the *Aliment*; left too much, should either rot it, or cause it to germinate, before its proper season, or full time for a more *Masculine* Growth.

4. §. On the contrary, many *Seeds*, as those of *Clary, Garden-Cress*, and others of that *Tribe*, have their upper *Covers* faced with a *Mucilage*: which being easily receptive of any *Moysture* in the Ground, gradually swells, till it lies like a *Gelly* round about the *Seed*. Either for a more plentiful supply of *Aliment*; or at least, to soften the *Covers*, the better to accelerate the Growth of the *Seed*.

5. §. The process of Nature in the several steps of the *Vegetation* of the *Seed*, hath formerly been explained. (*a*)

6. §. THE COVERS of all, or at least the far greater number of *Seeds*, are Three; some way or other derived from the *Pith*: as shall hereafter be seen. And sometimes, Four: even those of ston'd *Fruits*, have Three, besides the *Stone*. In that of *Gossipium*, there are Two *Coats* under that lined with the *Cotton*. The *Seeds* of *Cucumer, Goats-beard, Broom, Scabious, Lettice*, &c. although so small, have plainly Three *Coats*. But in some of these, and many more, there are only Two distinctly visible, except in the State of *Generation*. (*a*) B. 1. Ch. 1.

Tab. 75.

7. §. In the Upper *Coat*, the *Seed-Vessels* are disseminated. The Second, is first a meer *Pulp*; but afterwards shrinks up and sticks close to the upper. The Third or Inmost is more dense; and if it be thin, for the most part, transparent; whereby the *Seed* seems sometimes to be naked while it lies therein; as in *Almonds, Cucumers*, and the like. *Tab.* 76. For this sticks not to the midle *Coat*, as that doth to the outer; but commonly, remains entire, after those are stripp'd off, being as it were, the *Smock* of the *Seed*.

8. §. In *Melissa* and some other small *Seeds*, it comes finely off upon soaking in warm Water or on the *Tongue*. In *Fenugreek*, 'tis soft, and of an *Amber-Colour*; and being moystened, looks almost like fine *Glew*. But commonly, tis a prety tough *Membrane*, and often with

some thickness, as in *Plums, Borage, Scabious*. Yet always extream thin at the *Tip* of the *Radicle*; the more easily to break and yield to it, as the *Secundine* to the *Fœtus*, when it first shoots into the Ground. And sometimes, as in the *Seeds* of an *Orange*, it hath at one end, the resemblance of a *Placenta*. But of this, and the two upper *Coats*, I shall give a further Description in the last *Chapter*.

Tab. 75.

9. §. AS ALL *Seeds* are *ex Ovo*; so there are many with thin *Covers*, as of *Orach, Spinage, Beet*, and the rest of that *Tribe*, &c. which besides the *Albumen* or clear *Liquor* out of which they are bred; have also, a *Vitellum*, or a *Body* thereunto *Analogus*: being neither part of the *Seed*, nor part of the *Covers*, but distinct from them both. With respect to the *Bulk* of the *Seed*, very large, as white as *Starch*, and pretty friable, like good *Rice* or *Barley*: of a roundish *Figure*, and grooved on the *Girth*, so as to have a double *Edge*; Whereby the *Seed*, which is long and slender, lies round it, as a *Sack* of *Corn* upon a *Pack-Saddle* or a *Rope* upon a *Pully-wheel*. Upon my first notice hereof, it seemed to answer to a *Placenta*. But upon further consideration, the *Analogy* doth not hold betwixt them. For the *Placenta* lies without the *Membranes* in which the *Fœtus* is conteined: whereas this body lies within the *Covers* contiguous to the *Seed*, and so becomes its first and finest *Aliment*, as the *Yelk* doth to the *Chick*. For which purpose, as in the *Generation* of the *Seed*, it is a pure *Milky Chyle*; So in its *Vegetation*, it is converted into the like again.

Tab. 75.

10. §. The same Body for Sustance, is observable in the *seeds* of *Rhapontick, Dock, Sorrel*, and the rest of that kindred, with this difference; That whereas in *Orach*, &c. the *Seed* only lies upon it; here, the main *Body* or *Lobes* of the *Seed* are immersed therein, the *Radicle* standing naked or above it. So that the said *Lobes*, and therein the *Seminal-Root* are beded herein, as in a *Tub* of *Meal* or a little pot of pure refin'd *Mould*, necessary for the first *Vegetation* of the *Radicle*.

Tab. 75.

11. §. BY THESE midle *Steps*, Nature proceeds from the *Thiner Covers of Seeds*; or those, which after the *Generation* of the *Seed* is finished, shrink up; to the *Bulky Kind*, or those which keep their *Bulk* after they are dry. Wherein, not only the *Lobes*, as in *Dock*, but the whole *Seed* is immediately lodged. Different in *Substance, Shape* and *Bulk*; but always many times biger than the true *Seed* within it: for which it is commonly mistaken; but is no more the *Seed*, than is the *Stone* of a *Plum*, the *Kernel*.

12. §. In the *Barbado Nut*, 'tis *White, Soft, Conick-oval*, and taking all its *Dimentions*, 8 or 10 times bigger than the *Seed* within it. In *Ashen Keys*, 'tis of a sad *Colour*, hard, yet somewhat *Oyly, Oval* and flat, and of the same *Bigness* as in the *Barbado Nut*, with respect to the *Seed*. In the *Fruit* commonly called *Nux Vomica Officinarum*, 'tis of the *Colour* and *Hardness* of a *Cows-Horne*; and makes almost the whole *Body* of the *Fruit*, being about 14 or 15 times the *Bulk* of the *Seed*. In *Goosgrass* or *Cliver* 'tis of the like *Horny Substance*, but shaped somewhat like a *Bonet* with the *Rimm* tuckt in. And so in a *Coffee-Berry*; but rowled or foulded up into a kind of *Oval Figure*, with a *Notch* or *Rima* through the Length, where the two *Ends* meet. With other diversities which will best be understood, when I come presently to the *Description* of the *Seed* herein contained.

Tab. 76.

Tab. 77.

13. §.

13. §. With respect to the use of this *Cover*, it is observable, that where there is a *Stone* or *shell* over it, as in the *Barbado Nut*, it is *soft*; but where there is none, as in *Nux Vomica*, *Ash*, &c. 'tis hard; and so it self instead of a *Stone*. As also, That it becomes *hard*, only by the proper Nature of its *Parenchyma*, and the exquisite *smallness* of the *Bladders* of which it consists. Whereas a *Stone*, is also hardened by the *Lees* or *Tartar* of the *Sap* which sinks into it, and thereby *petrifies* it (*a*) as hath been said. So that whereas a *Stone*, as it lies in the *Ground*, only cleavs in certain Places, but continues hard: This *Cover*, like some *Horns*, upon the due accession of *Moisture*, doth gradually become *soft*. Whereby, as while it is *hard*, it performs the Office of a *Stone*, in guarding the *Seed* til the proper *Season* for its *Growth*: So afterwards when it is *soft*, it answers, as in *Orach* or *Dock*, (*b*) to a *Vitellum*, from whence the *Seed* receiveth its first and purest *Aliment*.

(*a*) P. 3. Ch. 3.

(*b*) Ch. 3. §. 8, 9.

CHAP. IV.

Of the FOETUS or true SEED: and first of the RADICLE and LOBES.

HAVING discoursed of the *Covers*, I come next to the *Seed* or *Fœtus* it self. Of the *Shape* and *Posture* whereof, I shall give some *Examples*, first, among those with the thinner sort of *Covers*; and then, of those with the *Bulky* one: where I shall speak only of the *Lobes*, or *Main Body*, and the *Radicle*. Next, I shall describe the several sorts of *Nodes* or *Buds* of *Seeds*. And lastly, the several Parts, of which the *Lobes*, *Radicle*, and *Buds* are compounded.

2. §. Among *seeds* with the *Thinner Covers*, are those of all sorts of *Corn* and *Grass* Of a different make, from that of most other *Seeds*: The *Main Body* being not divided into *Lobes*, but one entire Piece, doubled in the form of a Pair of *Lipps*. And whereas commonly, the whole *Seed* is very *Soft* and *Oyly*; here, only those two minute *Parts*, which become the *Root* and *Stalk*, are so: The *Main Body* being of a different *Substance*; when the *Corn* is ripe, *hard* and *friable*; but when it is sown, easily *colliquable* into a kind of *Milk* or *Chyle*, so that, in some respects, it hath a near *Analogy* to a *Vitellum*. For as that is gradually melted into a sort of *Chyle*, and by the *Branches* of the *Ductus Intestinalis* carryed into the *Bowels* of the *Chick*: So is this, into a like *Substance*, and by the *Branches* of the *Seminal Root* (formerly describ'd) conveyed to those *Parts*, which become the future Plant. B. 1. ch. 1.

3. §. Of Relation to this Kind, the *Seeds* of *Dates*, and of some other like *Plants*, may be esteem'd. For that which is commonly called the *Stone*, seems indeed to be the *Main Body* of the *seed*, doubled or folded up in the same manner as in *Corn*. To which that *Part* which becomes the *Plant*, is annexed. But whereas in *Corn*, 'tis placed at Tab. 75.
the

the Bottom of the *Main Body*; here it lies in a small round *Cavity* in the middle of the *Back*. The *Stone*, or *Main Body*, where this *Part* grows to it, is not so hard, as more remote from it: and is therefore probably in some part dissolved, by lying in the *Ground*, as in *Corn*.

Tab. 75.

4. §. But for the most part, the *Main Body* is divided, as hath been said, into two *Lobes*; and those in Substance *Homogeneous* to the other *Part* or *Parts*, plainly distinguished in most *Kernels* and other large *Seeds*; and not difficultly in many lesser ones, as in that of *Viola Lunaris*, *Scabious*, *Doves-Foot*, &c. if slipped out of their *Covers* before they are full ripe.

Tab. 75.

5. §. In *Hounds-Tongue*, they are of a circular figure, and very large in Proportion to the *Radicle*. In *Cucumer*, oblong, with some visible *Branches* of the *Seminal Root*; and the *Radicle* somewhat bigger. But in *Scorzonera*, very long, like the Leggs of a Pair of *Compasses*: and the two first, or *dissimular Leavs* of the *Plant* into which they are converted, are of the same *Shape*. Of these and many more, the *Radicle* is short and pointed; and lies in one straight Line with the *Lobes*.

Tab. 75.

6. §. In *Viola Lunaria*, they are very large; and the *Branches* of the *Seminal Root*, fairly apparent, so as to resemble a Pair of *Leavs*. The *Radicle* pretty long, equally thick from end to end, and couched down upon the two *Lobes*, each of them having a little *Shoulder* for it to lie upon. In *Woad*, where it hath the like *Posture* and *Shape*, as also in *Chamælina*, *Eruca*, and many others, it is very *Bulky*, being bigger than both the *Lobes* put together.

7. §. Of this Part, I think it may be observed, That commonly those *Seeds*, wherein it is very small with respect to the *Lobes*, produce a *Perennial* Plant: And so, *vice versa*, where it is very large, an *Annual* one. In the latter, the *Seminal Virtue* being more vigorous, and so tending more hastily to the Business of *Generation*, followed with the *Death* of the Plant.

Tab. 75.

8. §. IN THE former *Seeds*, the *Lobes* lie flat one against another. But in *Garden-Radish*, they are folded up, so as to receive the *Radicle* into their *Bosome*: as when a *Chicken* tucks his Head under his Wing.

Tab. 75.

9. §. In *Holyoak*, the *Lobes* are plated upwards, and re-plated down again. Being most agreeably composed to the *Shape* of the *Covers*, as those are to their *Posture* on the Plant. In *Maple*, they are plated one over another, and so rouled up.

Tab. 75.

10. §. In the *Cotton-Seed*, which consisteth almost wholly of two very broad and thin *Lobes* or *Leaves*, the *Folds* are yet more numerous; all curiously reduced to an exact and solid *Oval*.

11. §. It happens now and then, that instead of two, there are three *Lobes*, as in the *Kernels* of *Plums*, *Apples*, and other *Fruits*, and the smaller sorts of *Seeds*, will spring up sometimes with more than two *dissimilar Leaves*, originally the *Lobes* of the *Seed*. These are observed by some, more frequently to produce a double *Flower*, which may be, because the *seminal Virtue* in such Seeds, is increased by a third Part.

12. §. IN

12. §. IN many *Seeds*, the *Radicle* is of one and the same *Colour* from end to end. But in others, as in the *Lupine*, it is observable, That the upper and greater half, is *White*; the Lower to the *Point*, hath a kind of *Horny Gloss*, and seems to be of a somewhat different make. *Tab. 75.* Whereby it comes to pass, that after the *Radicle* is shot forth a little way, only this lower half descends and becomes the *Root*: The upper half is produced or raised above ground, as a *Pillar* upon which the *Lobes*, or *dissimilar Leaves* are erected.

13. §. This *Seed*, on the out side of each *Lobe*, and near the *Radicle*, hath a very small and round *Node*, like a *Navel*; whereof, in the first Book: the whole *Seed* looking not much unlike a *Pidgeons Head*; *Ch. 7.* the *Radicle* resembling the *Bill*, and the *Navel* the *Eye*.

14. §. IN the *Seed* of *Garden-Orach*, both the *Radicle* and *Lobes* *Tab. 75.* are very long and slender, and lie almost in a compleat Circle round about the *Vitellum* before describ'd. The *Lobes* of *Rhapontick* are shaped like the *Bitt* of a *Spade*; and the *Radicle* stands erected above *Ch. 3.* them like the *Handle*.

15. §. OF SEEDS also with the *Bulky Cover*, there are many not divided into *Lobes*; being in a manner, all one *Piece*; as all of the *Bulbous-Kind*. In some of which, though the inmost *Cover* be thin; yet compared either with the other *Covers*, or with the *Seed* it self, it may very well be accounted of the *Bulky- Kind*.

16. §. In *Flag*, it is above twenty times bigger than the *Seed* within it. Consisting of *Bladders* all *Radiated* towards the *Seat* of the *Seed*. The *Seed* it self is shaped somewhat like a *Penknife*. The lower Part *Tab. 76.* which becoms the *Bulb*, as the *Haft*, is thick, and cometh near to a *Cylindrick* Figure, and the end, round. The upper Part which becomes the first years *Leaf*, as the *Blade*, is rather flat, double edged, and pointed, and the Point a little bent. The *Fibers* and *Bladders* of which it consists, are all disposed into Parallel Lines running by the length. In *Lily*, where this *Cover* is thinner and more *Transparent*, without being cut, but only held up against the Light, the *Seed* may be seen within it.

17. §. BUT THE greater number of *Seeds* also with the *Bulky Cover*, are divided into two *Lobes*; which, for the most part, resemble a pair of little *Leavs*. In the *Purging Nut* of *Angola*, the *Shell* being taken off, the upper *Covers* (dry'd and shrunk up) seem to be but one. *Tab. 76.* In these, the *Spermatick Vessels* are Branched. Under these, lies the Thick and Inmost *Cover*; which being cut down the middle, exhibits the true *Seed*: Consisting of a couple of fair *Leavs*, Veined, and as white as *Milk*, joyned together with the *Radicle* at their *Base*; and let into a Hollow, made in the *Cover*, of an answerable shape. The like is observable in the *Barbado-Nut*, *Ricinus Americanus*, and some other *Indian Fruits*; with some little difference in the Shape of the *Root* and *Leavs*.

18. §. IN the foregoing *Fruits*, the *Bulky Cover* is very soft. But in the *Nux Vomica Officinarum*, 'tis near as hard as a *Date-stone*. *Tab. 76.* In this, besides the hollow made for the reception of the *seed*, or the two *Leavs* and *Root*; the *Sides* are separated or distinct almost to the *Edge* of the *Cover* round about, especially towards the *Root*: So that it may not unaptly be compared to a little *Pouch* with the *Sides* clapt together.

Tab. 77.

19. §. IN this and the *Nuts* above mentioned, the *Seeds* are all very large. But in some other *Plants*, they are extream small, so as to be hardly visible without a *Glass*; as in *Staphisagria*, *Peony*, &c. In *Staphisagria*, the *Thick* or *Inmost Cover*, is commonly a *Spherical Triangle*, *conick* towards the *Base*. At the poynt of which, there is a little *Cavity*, in which the *Seed*, about as big as a small pins head, is lodged. The *Root* whereof is a little poynted, and the two *Lobes* rounded at the *Top*.

Tab. 77.

20. §. In *Peony*, the same *Cover* is *Soft*, *White*, and of an *Oval Figure*; the part used in *Medicine*. Usually thought to be the *Seed* it self. But is near two hundred times biger than the true *Seed*, which is almost invisible. It lies in a little *Cavity* near the bottom of the *Cover*; with a thick and blunt *Root*, and two poynted *Lobes* or *Leavs*.

Ch. 3.
Tab. 77.

23. §. IN the *Coffee-Berry*, the *Seed* lies in the *Inner* or *Cartilaginous Cover* (formerly described) where one would not expect to find it, *sc*. near the *Top* or *Surface* of the *Back*. The *Lobes* of the *Seed* are veined like two very minute *Leaves*, and joyned to a long *Root* like a *Stalk*. The end of which comes just to the bottom of the *Cover*, ready for its *exit* into the *Ground*.

Tab. 77.

22. §. In *Goosgrass*, where the *Inner Cover* is also *Cartilaginous* or *Horney*, the *Seed* is postured in much a like manner, and looks just like a couple of poynted *Leavs* with a very long *Stalk*.

Tab. 77.

23. §. THE *Seed* of *Stramonium*, is also inclosed in a *Bulky Cover*. Which being soaked in warm water, and very warily cut about the edges, with a *Rasor*, the *Seed* may be taken out of it entire. Shaped like that of *Orach*, but much longer. For the *Reception* whereof, the *Cover* is formed with a hollow, which runs round about it near the *Edge*; where in the *Seed* lies like a little winding *Snake*.

CHAP. V.

Of the BUDS of Seeds. And of the PARTS, of which these, the Radicle, and Lobes are compounded.

ROM between the two *Lobes*, rises up the *Stalk* of the *Plant*. The original whereof, either to the naked *Eye*, or by a good *Glass*, is always visible in the *Seed*.

2. §. In many *Plants*, *Nature* sees fit only to lay the foundation hereof in a small round *Node*; whereupon the *Leavs*, in the *Vegetation* of the *Seed*, are superstructed: as in *Viola Lunaria*, and others.

3. §. But in the greater number of *Seeds*, is formed a true *Bud*, consisting of perfect *Leavs*; different from those, which grow upon the *Stalk*, only in *Bigness*; and so far in *Shape*, as the same *Parts* of an Animal *Fœtus*, in its several ages in the *Womb*. In many seeds, as well

small

small as great, and as well of *Herbs* as *Trees*, it is very apparent. But oftentimes lyeth so deep between the *Lobes* as to be almost undiscernable, as in *Maple*.

4. §. The *Leaves* of the *Bud*, in different *Plants*, are of a different *Number*; in some, Two; in others, Four, Six, and sometimes more. In the *Bay-Berry*, they are only two; very small, but thick or fat, and finely veined. In the *Seed* of *Carduus Benedictus*, they are also Two; almost invisible; broad at the *Bottom*, poynted at the *Top*, thick or fat, yet plated inward, and postured a little distant one from the other; for the two next to rise up between them. The like may be seen in *Carthamum*; and so, I suppose, in all the *Carduus Kind*. *Tab.* 78.

5. §. In some *Herbs*, although the *Bud* consisteth but of two perfect *Leaves*, yet they are very conspicuous. Not only in larger *Seeds*, as in the *Phaseolus* or *French Bean*; but in those which are small, as in the *Seed* of *Hemp*. In this, the two *Leaves* are both plated, and so set *Edge* to *Edge*, with mutual *Undulations*. Of that Length, as to be extended beyond a third part of the *Lobes*. *Tab.* 78.

6. §. In the *Seed* of *Sena*, the *Bud* consisteth of Four *Leaves*; of which, the greater pair is the outer, and guards the less. Shaped not much unlike those in the *Seed* of *Carduus*; but are a little more visible. *Tab.* 78.

7. §. In the *Bud* of an *Almond*, we may easily count six, or eight *Leaves*, and sometimes more; the Inermost being laid bare by a dexterous *Separation* of the Outer. These are by much the greatest, doubled Inward, and so laped one over another; whereby they embosome all the rest, as a *Hen* spreads her *Wings* over her *Chickens*. The like is observable in many other large *Kernels*, as also in the *Garden Bean*, and some other *Plants*. With respect to which, I have taken leave (*a*) to call this *Part* the *Plume*. *Tab.* 78. (*a*) B. 1. Ch. 1.

8. §. THE LOBES of the *Seed*, and so likewise the *Radicle* and *Bud* consist of a *Skin*, *Parenchyma*, and *Branched Vessels*: all which I have formerly described. (*b*) I shall now add the following Remarques. (*b*) B. 1. Ch. 1.

9. §. And first of the *Skin*, which in some *Seeds*, as the *French-Bean* may easily be separated from the *Parenchyma*: especially if the *Bean* be soaked in water for some days; for then it will slip off, like the *Skin* in any part of ones *Body* where it is blistered. 'Tis woven into *Bladders*, as the *Parenchyma*; but into smaller ones, and upon the *Lobes* of a *Garden Bean*, all radiated towards the *Center*. With these *Bladders*, there are also mixed a sort of *Lignous Fibres*, incomparably small, which give a *Toughness* to the *Skin*, and by which the *Bladders* are directed into *Rays*. *Tab.* 79.

10. §. The *Bladders* of the *Parenchyma*, as is said, are much larger than those of the *Skin*, especially in the *Lobes*. In those of a *Garden Bean*, somewhat oval, about $\frac{1}{4}$ of an Inch Diametre by their *Bredth*, and directed towards the *Branches* of the *Seminal Root*. In the *Radicle*, they are twenty times smaller, than in the *Lobes*: and so in the *Plume*. *Tab.* 79.

11. §. Throughout the *Parenchyma* run the *Branched Vessels*, which in the *Lobes* make the *Seminal Root*; in the *Radicle* and *Plume*, the *Wood* of the *Root* and *Stalk*. In all of them, distributed as hath been (*c*) formerly shewed. *Tab.* 79. (*c*) B. 1. Ch. 1.

Tab. 79.

12. §. I shall here further note, That the utmost divisions are no where extended to the Circumference of the *Lobes*, but are all inosculated together at a considerable distance from it, as in the *Leaves* of some *Plants*.

Tab. 79.

13. §. In the *Lobes* they all meet in one solid *Nerve*. But in the *Radicle*, are dilated into a hollow *Trunk*, filled up with a *Pith*; composed of *Bladders* somewhat bigger than those which make, as it were, the *Barque* of the *Radicle*. In the *Radicle* of a *French Bean*, the *Pith* is very conspicuous.

Tab. 79.

14. §. The *Vessels* are of two kinds, as in the other *Parts* of a *Plant*; for *Sap*, and for *Aer*. Not running collateral, as *Arteries* and *Veins*; but the latter every where sheathed in the former. From the *Aer-Vessels* it is, that if a *Bean* be steeped in water, and then the *Radicle* cut transversly and pressed, it will yield *Bubles* as well as *Liquor*. These *Vessels* are admirably small, yet through a very good *Glass* become visible.

15. §. The *Liquor* conteined in the *Seed*, when full ripe is chiefly *Oyl*; generally, found in a greater proportion here, than in any other part of a *Plant*. Being as the *Pickle*, in which the *Seminal Virtues*, *i. e.* the more *volatile* and *active Principles* of the *Seed*, are immersed for their *Preservation*: and to curb them from too great a *Luxuriance* in the *Vegetation* of the *Seed*.

CHAP.

Book IV.   *of Seeds.*   209

CHAP. VI.

Of the GENERATION *of the* SEED.

AS I made choice of a *Garden-Bean,* to shew the manner of the *Vegetation* of the *Seed:* so I shall take an *Aprecock,* as very apt and convenient, to observe and represent the *Method* which *Nature* taketh in its *Generation.*

2. §. In order to this, the first thing that is to be done, is to make a fit *Uterus.* Both to keep the *Membranes* of the *Fœtus* warm, and succulent, till it be formed: and to preserve and secure the *Fœtus* it self afterwards, till it comes to be born into the *Ground.*

3. §. For this purpose, the *Pulp* and *Stone* of the *Fruit* are both necessary; but primarily the *Stone:* the *Meat* or *Pulp* being no otherwise necessary, but because the *Stone* cannot be made without it; the petrifying of that *Parenchyma* which is the *Ground* of the *Stone,* being effected, by the sinking of the *Tartar* from the *Pulp* thereinto.

4. §. And that, at the first, the *Ground* of the *Stone,* is a distinct, but soft *Parenchyma*; is evident in the cuting of a young *Aprecock.* Of which, also a slice cut off, with a *Rasor,* and viewed through a good *Glass,* sheweth it to be composed of *Bladders,* as the *Pulp* it self. Only, whereas many of those of the *Pulp* are large, now about as big as a white *Pepper-Corn:* these are no bigger than a *Mustard-Seed.* But as the *Parenchyma* hardens into a *Stone,* these *Bladders* are all gradually filled up, and disappear. *Tab.* 82.

5. §. This *Parenchyma* is derived immediately from the *Pith,* as the *Pulp* is from the *Barque*; and makes the far greater part of the *Stone.* 'Tis covered all over within, with a very thin *Lining*; derived, not from the *Pith* but the *Parenchyma* which covers the *Seed-Branch,* upon its first entrance within the hollow of the *Stone.* This *Lining* is of a close substance; yet composed of *Bladders,* exquisitely small and hardly visible. By which means, it soon becomes a very hard and dry *Body*; and is hereby fitted, both to promote the induration of the rest of the *Stone*; and the seasonable drying, and so, the shrinking up, of the *Covers* of the *Seed,* to make room for its *Growth.* *Tab.* 80.

6. §. The *Stone* being made hard and dry; it could never be so sufficiently softned by lying under ground, but that, it would keep the *Seed* a perpetual prisoner, unless it were also made pretty easily to cleave in two. For which purpose, the *Skin* of the *Fruit* doth observably conduce. For in a *Slice* of a young *Aprecock* cut transversly with a very sharp knife, it may be seen, especially with the help of a *Glass,* to be doubled inward from the two *Lips* of the *Fruit,* and so to be continued, *Tab.* 80.

K k

tinued, not only through the *Pulp*, but also through the *Stone* it self, into the hollow of the same, where it meets, and is united with the *Lining* thereof. Whereby, as it further helps to the drying and hardning of the *Stone*; so also renders it cleavable in that part, where it runs through it. And therefore, whereas towards the *Stalk*, it goes no farther than to the *Seed-Branch*, and so but half way through the *Stone*: towards the *Top* of the *Fruit*, where the *Radicle* stands, and where the *Stone* begins to cleave, it runs quite through it.

7. §. *Nature* having thus provided a convenient *Uterus*, She next taketh care about the *Membranes* of the *Fœtus*. These are *Three* apparently distinct, and in many respects different one from another.

8. §. The outer *Membrane* is derived from the *Parenchyma* which surrounds the *Seed-Branch*; which, upon its entry into the hollow of the *Stone*, is expanded, as it were, into two *Bladders*, one within another; whereof, one becomes the *Lining* of the *Stone*; the other, this outer *Membrane*: as is best seen by cuting a young *Aprecock*, when it is about half an Inch long, down through the midle, or from the *Seat* of the *Flower* to the *Stalk*, between the two *Lips*.

Tab. 80.

9. §. This outer *Membrane*, at this age, hath a good full and frim *Body*, about $\frac{1}{7}$.th of an Inch thick, or through an ordinary *Glass*, half an Inch, where it is thickest, as at the *Sides* and the greater end: the *Poynt* being thinner, for the more easy eruption of the *Radicle* into the *Earth*. Composed of *Bladders*, through an ordinary *Glass*, about as big, as a *Colewort-Seed*.

Tab. 80.

10. §. Throughout this *Membrane*, the *Vessels* conteined in the *Seed-Branch* are distributed. Beginning a little below the smaller end of the *Coat* or *Membrane*, they thence fetch their circuit both ways round about, just beneath the *Surface* of the *Membrane*, and at last, meet in the midle of the greater end, where they are all inosculated, so as to make a kind of *umbilical Node*. From whence they strike deeper into it, and at last, into the midle *Membrane*, in which they presently become invisible. By these *Vessels*, the *Sap* is brought and spewed into the midle *Membrane*. So that the outer *Membrane* seemeth, in some respects, to be answerable to the *Placenta* in *Animals*.

Tab. 80.

11. §. The midle *Membrane*, is derived from the bottome of the Outer. From whence especially, but also round about, the *Bladders* hereof (all angular) are more and more amplified towards the *Centre*; most of them being at least two hundred times biger, than those of the Outer *Membrane*: whereby it looks, through a *Glass*, not unlike a *Coome* full of *Hony*; or in regard of their great transparency, like a company of little *Crystal Pans* full of a pure *Lympha*.

Tab. 80.

12. §. This Midle *Membrane*, is properly so called, from the state and condition it hath, upon the *Augmentation* of the *Seed*, at which time, it obteins the nature of an *Involucrum*. But originally, it is every where entire, without any *Hollow*, filling up the *Cavity* of the Outer *Membrane*, like a soft and delicate *Pulp*. After a short time, there

there appears in it a small *Ductus* or *Chanel*; which runs from the bottom to the top, like an *Axis*, through the midle of it. At first, Tab. 81. no wider than to receive the *Hair* of a *Mans Head*; not visible, except in a slice hereof cut transversly, and viewed in a *Glass*. Being grown a little wider, it may be seen, if the *Membrane* be dexterously cut by the length. At which time, it is also dilated into two *Oval Cavities*, one at each end: which are as two little *Cisterns*, whereinto a most pure *Lympha* continually owzeth, and is therein reserved for the nourishment of the *Seed*; and through the *Chanel* which runs between the *Cisterns* is emptied out of one *Cistern* into another, according as the *Seed* or the Inmost *Membrane* hath need of it; *i. e.* as the *Weather* and other Circumstances do more or less accelerate their *Growth*, and so render the *Lympha* useful to them.

13. §. A few days after this, the Innermost *Membrane* begins to appear; growing, like a soft *Node* or *Bud*, out of the upper *Cistern*; to the lower end of which it is joyned by a short and tender *Stalk*, from Tab. 81. whence it is produced into a *Conick-oval Figure*, answerable to that of the *Cistern*.

14. §. This *Membrane*, though soft and full of *Sap*, yet being compared with the midlemost, is a close and compact *Body*, composed of *Bladders* above 300 times smaller than they are in that. Whereby, as the *Seed* is so well guarded, as not to be supplyed with any part of the *Lympha*, but the purest: so neither with any more of this, than will suffice, without the danger of making an *Inundation* out of so great a *Lake*.

15. §. This *Membrane*, if it be pulled with a most steady hand, and very gently, upwards, it will draw a small transparent *string* after it to the bottom of the Midle *Membrane*: The said *String* though for the greater part, *Parenchymous*, yet being strengthened with the admixture of some *Lignous Fibres*; no otherwise visible in either of these two *Membranes*. So that they seem, to be a small portion of those which are inosculated at the bottom of the *Outer Membrane*, and thence produced through the midlemost, underneath the *Chanel*, till at last they break forth into the upper *Cistern*, where they form this Inner *Membrane*: a piece of close-wrought *Work*, suitable to the incomparable fineness of all the *Stuff* out of which it is made.

16. §. The same *Membrane* is originally entire, as the Midlemost: but being grown to about the bigness of a *Carvi-Seed*, becomes a little Tab. 81. hollow near the *Cone*. And the *Lignous Fibers* abovesaid, fetching their compass from the *Base*, shoot forth into the *Cone*; and so make a very small *Node* therein, for the first *Essay* towards the *Generation* of the *Seed*. The said *Fibers* being thus spun out, to the utmost degree of fineness for this purpose.

17. §. This *Node*, being grown about ⅓th part as big as a *Cheese-Mite*; it begins next to be divided by a little indenture at the *Top*. Tab. 81. Which growing by degrees still deeper, the *Node* is hereby at length distinguished into two *Lobes* or thick *Leavs*.

18. §.

Tab. 81.

18. §. So soon as these are finished, their *Basis* begins afterwards to be contracted, and so to be formed into a *Radicle* or that part of the *Seed* which becomes the *Root*. As the *Stalks* of *Fruits* do grow lesser, while the *Fruits* themselves are expanded. So that in this estate, the *Radicle* is, as it were, the *Stalk* of the *Seed*.

19. §. At this time, the *Seed* being extream small, the *Lobes* are not so manageable as to be separated one from the other. But it is most reasonable to suppose that so soon as the *Radicle* is finished, the next step, is the pushing forth of another *Node*, between the *Lobes*, in order to the making of a *Bud*, and so the perfection of the *Seed*.

Tab. 81.

20. §. This being done or in doing, the *Radicle* or *Stalk* of the *Seed*, contracting still more and more at the bottome, hangs at the Inner *Membrane*, only by an extream small and short *Ligament* or *Navel-String*. Which at last, also breaks; and so the *Seed*, as *Fruits* when they are ripe, falls off and lies loose in the Iner *Membrane*; this gradually shrinking up and so becoming more hollow, to make room for the further *Growth* of the *Seed*.

Several

Several
LECTURES
Read before the
ROYAL SOCIETY.

By *NEHEMJAH GREW* M.D. Fellow of the *ROYAL SOCIETY*, and of the *COLLEGE* of *PHYSICIANS*.

LONDON,

Printed by *W. Rawlins*, 1682.

THE TITLES
Of the following
LECTURES.

I. Of the Nature, Causes, and Power of MIXTURE. The second Edition. 221.

II. Of the LUCTATION arising upon the Mixture of several Menstruum's with all sorts of Bodies. The second Edition.

III. An Essay, Of the various Proportions, wherein LIXIVIAL SALTS are found in Plants.

IV. Of the ESSENTIAL and MARINE SALTS of Plants.

V. Of the COLOURS of Plants.

VI. Of the Diversities and Causes of TASTS; chiefly in Plants. With an Appendix, Of the ODOURS of Plants.

VII. Experiments in Consort, upon the SOLUTION of SALTS in Water.

TO THE
Right Honourable
WILLIAM
Lord Vi-Count *BROUNCKER*,
PRESIDENT
OF THE
Royal Society.

MY LORD,

NE *Reason* why I Dedicate the following Discourses *to Your* Lordship, *is, For that by Your great und undeserved Respects, You have obliged me to do no less.*

Another, my Lord, *is, Because I could not but Publickly return Your* Lordship *Thanks, for minding the* Royal Society *of so good a Way, they are lately resolved upon, for the Management of a great part of their Business. Wherein,* my Lord, *I do more than presume, that I also speak the Sense of the whole Society; I think, not any one excepted.*

I may with the same Confidence intimate, my Lord, *how happy they account themselves, in having a* Person *so fit to preside their Affairs, as Your* Lordship. *The Largeness of your Knowledge, the Exactness of Your Judgment, the Evenness of Your Comport; being some of those necessary Qualifications, which His* Majesty *had in His Eye (as right well understanding what He did) when He fixed His Choice upon Your* Lordship.

I know, my Lord, *that there are some men, who have*

just

just so much Understanding, as only to teach them how to be Ambitious: The Flattering of whom, is somewhat like the Tickling of Children, till they fall a Dancing. But I also know, that Your Lordship unconcerneth Your self as much, in what I even now spake; as Cæsar did himself, when his Souldiers began to style him King. For as he said, Non Rex, sed Cæsar: So let Your Lordship be but once nam'd, and all that follows, is but a Tautology to what You are already known to be. Your being President of the Royal Society, Your being the First that was Chosen, and Chosen by so Knowing a Prince; becomes so real a Panegyrick to Your Lordship, as leaveth Verbal ones without any sound.

Whence, my Lord, I have a third Reason most naturally emergent, which is, That I dare to submit my self, as to what I have hereafter said, to Your Lordships Censure. You being so able and just an Arbiter betwixt the same and all those Persons therein concern'd; that You can neither be deceived, nor corrupted, to make a Judgment in any Point, to the Injury of either.

And truly, my Lord, were it only from a Principle of self-Interest, yet I could not desire it should be otherwise. For the World, if it lives, will certainly grow as much more knowing than it is; as it is now more, than it was heretofore. So that we have as little Reason to despise Antiquity; as we can have willingness, that we our selves should be despised by Posterity.

Yet some difference there is to be made; viz. betwixt those of all Ages, who have been modestly ignorant; and those who have thought, or pretended, that they were Omniscient. Or if knowing and acknowledging that they were Ignorant; have yet not been contented to be so; unless, with as good manners, as sense, they did conjure all Mankind not to offer at the knowing any more than themselves.

Upon the whole, my Lord, I desire not You should be a Patron, any further than You are a Judge. For if this small Essay hath deserved the least acceptance, I am sure, that in being one, You will be both.   I am,

My Lord,
Your *Lordships* most Faithful
and Obedient Servant,

NEHEMJAH GREW.

A DISCOURSE

Read before the

ROYAL SOCIETY

Decemb. 10. 1674.

Concerning the

NATURE, CAUSES, and *POWER*

O F

MIXTURE.

AVING the honour to perform the Task of this day; I shall endeavour to conform to the *Phylosophy*, which this *Society* doth profess; which is, *Reasoning grounded upon Experiment, and the Common Notions of Sense*. The former being, without the latter, too subtle and intangible; the latter without the former, too gross and unmanageable: but both together, bearing a true analogy to our selves; who are neither Angels, nor meer Animals, but Men.

The Subject I have chosen to speak of, is *Mixture*. Whereof, that our *Discourse* may be the more *consistent*, and the better *intelligible*; all I have to say, shall be ranged into this Method; *viz.*

1. First, I shall give a brief account of the received *Doctrine* of *Mixture*.

2. Next, lay down some *Propositions* of the *Principles* whereof all *Mixed* Bodies consist.

3. Then, open the true *Nature* of *Mixture*; or say, *What* it is.

4. And then enumerate the *Causes* of *Mixture*; or say, *How* it is made.

5. Lastly, I shall shew the *Power* of *Mixture*; or, *What* it can do.

CHAP.

CHAP. I.

Of the received Doctrine of Mixture.

IRST, As to the received *Doctrine* of *Mixture*; not to trouble you with tedious quotations of what *Aristotle, Galen, Fernelius, Scaliger, Sennertus, Riverius,* and other Learned men say hereof; we may suppose the whole summed up in that *Definition* which *Aristotle* himself hath given of it, and which the greater number of his Followers, have almost religiously adhered to; *viz.* that 'tis, τῶν μικτῶν ἀλλοιωθέντων ἕνωσις. that 'tis, *Miscibilium alteratorum unio.* Which *Definition,* as it is usually explicated, is both *Unintelligible,* and *Unuseful.*

Lib. 1. *de Generat. & Corrupt. Cap.* ult.

2. §. Two things are *unintelligible;* what they mean by *Alteration;* and what by *Union.* In this *Alteration,* they say, That the very *Forms* of the *Elements* are *altered.* And therefore lay it down for an *Axiom, Quod in Mixto, Formæ Elementares tantum sint in potentia,* But let us see the consequence. For if in a *mixed* body, the *Forms* of the *Elements* are but *in potentia;* then the *Elements* themselves are but *in potentia:* for we all say, *Forma dat esse.* And if the *Compounding Elements,* are only *in potentia;* then the *Compounded Body* it self can be only *in potentia:* yet to say it is no more, is most absurd.

3. §. As for the *Union of Elements* in a *mixed Body;* they make it such, as brings them at last to assert, the *Penetration of Bodies,* and that the *Union* of *mixed Bodies* is nothing else. For they say it is made in such sort, that every particle of the *mixed Body,* partaketh of the *Nature* of the whole. Which *Nature,* ariseth from the contemperated *Qualities* of the four *Elements.* Whence they conclude, That every particle of the *mixed Body,* containeth in it self all the four *Elements.* Which is plainly to assert a *penetration* of *Bodies.* For every *Element* is, at least, one particle; if therefore every particle of the *mixed* Body, containeth four *Elements*; then four particles are but one. I conclude then, That the received *Doctrine of Mixture* is *Unintelligible.*

4. §. Whence it follows, That it is also *Barren* and *Unuseful.* For who can make any use of that which he understandeth not? And the experience of so many years, wherein it hath been ventilated by the disputes of men, proveth as much: Scarce any of them, except the Learned *Sennertus,* daring to venture upon Experiment, for fear they should come to understand themselves.

5. §. It is confessed, that many gallant things have been found out by artificial *Mixture.* But no thanks to this *Definition* of it. For as an *Ignorant* Man may make bad *Work,* and a good *Rule* be never the worse; so one that is *Ingenious* may make good *Work,* and a bad *Rule* be never the better. The question is not, what have men done? but what have they done upon this foundation, *Quod Mixitio sit miscibilium alteratorum unio.* Had this ever taught them to do any thing, even so much as to make the *Inke* wherewith they have wrote, all their *Disputes*; I confess, they would have had something to shew for it. But the truth is, their *notions* of *Mixture,* have been so far from doing us any good, that they have done us much harm: being, through their seeming subtlety, but real absurdity, as so many phantastick *Spectrums,* serving only to affright men from coming near them, or the Subject whereof they treat.

6. §.

§. 6. I shall therefore endeavour to open the true *Nature* of *Mixture*. And I shall build my *Doctrine* upon the *Common Notions* of *Sense*: which none can *deny*; and every one may *conceive* of. In order to which, I shall take leave to lay down some *Propositions*, of the *Principles* of all *mixed Bodies*.

CHAP. II.
Of the Principles of Bodies.

AND first, by *Principles*, I mean *Atomes*, or certain *Sorts* of *Atomes*, or of the *simplest* of *Bodies*. For otherwise they would not be *Principles*; for a compounded *Principle*, in strict speaking, is a *Contradiction*. Even as *Fives*, *Threes*, or *Two's* are not the *Principles* of *Number*, but *Unites*.

2. §. Whence, secondly, it follows, that they are also *Indivisible*. Not *Mathematically*; for the *Atomes* of every *Principle* have their *Dimensions*. But *Physically*; and so, what is but *one*, cannot be made *two*. If it be asked, Whether a Stick cut with a Knife, be not of one, made two? I say, that a Stick, is not *one* Body, but *many millions* of Bodies; that is, of *Atomes*; not any one whereof is *divided* within it self, but only they are *separated* one from another, where the Knife forceth its way. As in the drawing of a mans Finger through a Heap of Corn; there is no *Division* made in any one *Grain*, but only a *separation* of them one from another, all remaining still in themselves entire. I say, therefore, that what is *Physically* one, is also most firm, and *Indivisible*, that is, *Impenetrable*: for *Penetration* is but the *Separation*, not the *Division* of *Atomes*.

3. §. Hence, thirdly, they are also *Immutable*. For that which cannot be *divided*, cannot be *chang'd*. So that of the whole World of *Atomes*, not any one hath ever suffer'd, or can suffer the least *mutation*. Hereupon is grounded the *Constancy* of *Causes* and *Effects*. So that, in all *Generations*, it is not less certain, that the self same *Principle* is still propagated from the same; than, that *Man* is from *Man*. Wherefore, compounded *Bodies* are generated; but *Principles* are not, but only *propagated*; that is, in every *Generation*, they pass, in themselves unaltered, from one Body, into another.

4. §. If *Principles*, or *Atomes* are all *Immutable*; it again follows, That they are of *Divers Kinds*. For one and the same *Principle*, or *Kind* of *Atomes*, will still make the *Same* Thing, and have the *same Effect*: so that all *Generations* would then be the *Same*. Wherefore, since they are *Immutable*, they must be *Divers*.

5. §. This *Diversity*, for the same reason, is not small, but very *Numerous*. For as the *World*, taken together, is *Natures Shop*; so the *Principles* of Things are her *Tools*, and her *Materials*. Wherefore, as it speaks the *goodness* of a *Shop*; so the *Perfection* of the *Universe*, That it is furnished with many *Tools* wherewith, and many *Materials* whereupon to *work*. And consequently, that *Philosophy* beareth best its own name; which doth not strain all to two or three *Principles*, like two or three

Bells

Bells in a Steeple, making a pitiful *Chime*: but tryeth to rise up to *Natures* own *Number*, and so to *ring* all the *Changes* in the World.

6. §. Yet doth not this vast *Diversity* take away the *Regiment* and *Subordination* of *Principles*. There being a certain lesser *number* of them, which either by their greater *quantity*, or other ways, have *Rule* and *Dominion*, in their several *Orders*, over all the rest. For where-ever the *Subject* is *Multitude*, *Order* is part of its *Perfection*. For *Order* is *Proportion*. And how can *Nature* be imagin'd to hold *Proportion* in all things else, and not here? Wherefore, as certainly, as *Order* and *Government* are in all the Parts of the *Rational*; so certainly, of the *Material World*. Whence it is, That although the *species* of *Principles* be very *numerous*; yet the *Principles* called *Galenical*, *Chymical*, or any others, which do any way fall under the notice of Sense, are notwithstanding *reduceable* to a *smaller number*: *viz.* according to the *number* of *Predominant Principles* in *Nature*; or, rather in this part of the *Universe* which is *near and round about us*. To the *Power* and *Empire* whereof, all other *Principles* do submit. Which *Submission*, is not the *quitting* of their own *Nature*; but only their appearance under the external Face or *Habit* of the said *Predominant Principles*.

7. §. As there can be no *Order* of *Principles*, without *Diversity*; so no *Diversity*, but what is *originally* made by these two ways; *sc.* by *Size* and *Figure*. By *these* they may be exceeding different: and all other *Properties* besides, whereby they differ, must be *dependent* upon these *Two*.

8. §. Nor therefore, can they be of any other *Figures*, than what are *Regular*. For *Regularity*, *is a Similitude continu'd*. Since therefore all kinds of *Atomes* are *divers* only by their *Size* and *Figure*; if the selfe same *Size* and *Figure* were not *common* to a certain number of *Atomes*, they could not be said to be of any *one kind*: and consequently, if there were no *similitude* of *Atomes*, there could be no *Distinction* of *Principles*.

9. §. Hence also, these two *Modes* of *Atomes*, *viz.* their *Size* and *Figure*, are the true, and only *original Qualities* of *Atomes*. That is, an *Atome* is *such* or *such*, because it is of such a certain *Size* and *Figure*.

10. §. Lastly, As these two *Modes*, taken severally, are the *Qualities* of an *Atome*: so consider'd together, they are its *Form*. A *substantial Form* of a *Body*, being an unintelligible thing. I say of a *Body*; for although the *Rational Soul* be a *substantial Form*, yet is it the *Form* of a *Man*, and not of a *Body*. For the *Form* of a *Body*, we can conceive of no otherwise, than as of the *Modification* of a *Body*, or a *Complexion* of all the *Modes* of a *Body*. Which also agrees with that *Definition* of a *Form*, which amongst the *Peripatetick Philosophers* is well enough accepted, *viz. Quod sit, Ratio ejus Essentiæ, quæ cuique Rei competit.* Which *Ratio*, if it be referred to a *Body*, what is it, but the *Modification* of that *Body*? Having thus proposed a Summary of my *Thoughts* about *Principles*; I shall next proceed to shew what their *Mixture* is.

CHAP. III.

Of the NATURE of Mixture.

AND first of all, from the *Premisses*, we arrive at this *Conclusion*; *sc.* That the *Formation* and *Transformation* of all Bodies, can be nothing else, but the *Mixture* of Bodies. For all *Principles* are *immutable*; as we have above proved: and therefore not *generable, formable,* or *transformable.* And the *Forms* of *Principles*, being but their *Modes*, are also *immutable.* So that the whole *Business* of the *Material World*, is nothing else, but *Mixture.* *Ch. 2. §. 3. Ch. 2. §. 10.*

2. §. Again, as *Nature* worketh every where only by *Mixture*; so is this *Mixture* every where but *one thing*, and can be but *one*. For whether it be the *Mixture* of *great* Bodies, or of *small*; of *Compounds*, or of *Atomes*; it is every where *Mixture*, and the *Mixture* of *Bodies.* Wherefore, *Mixture* is either an *intelligible Affection* of *all* Bodies, or of *none*; which later, no man will say. As many ways therefore, as we can *see*, or *conceive* the *Mixture* of any *gross* Bodies, which we hold in our hand; so many ways, we may, of the *subtilest Mixtures* which *Nature* maketh, or of *Atomes* themselves; and no other ways. *Ch. 2. §. 2.*

3. §. Now all the ways we can distinguish *Mixture* by, are, in general, these *Two*; either in respect of the *Bodies Mixed*, or else of the *Modes* of the *Mixture* it self.

4. §. In respect of the *Bodies Mixed*, Mixture is distinguished also two ways; viz. by *Conjugation*, and by *Proportion*.

5. §. By *Conjugation*, I mean, a *Mixture* of some certain *Principles, and not of others.* Which is *threefold*. *First*, As to *Number:* as when one Body may be compounded of *two* Principles, another of *three*, a third of *four*, a fourth of *five*, and so on. *Secondly*, As to *Kind*: where, though there be a conjunction of the same *Number*, yet not of the same *Kind.* *Thirdly*, When they differ from one another both in *Number* and *Kind.* So many ways the *Principles* of Bodies may be conceived to be *Conjugated*; and therefore are: for here, that which *may be, is.* The Consequence is clear. For *first*, Nature hath various *Materials* wherewith to make these *Mixtures*; as we have shewed. *Secondly*, By these *Mixtures* she *may*, and without the concurrence of any imaginary *Forms, must* produce all the varieties in the *material world*; as likewise hath been said. Wherefore, since all imaginable *Mixtures may be made*, and that to *some purpose*; if they should not be *so*, Nature would be *Imperfect*: because we our selves can think, how she might put her *Materials* to further use, then *so* she would do. To think therefore, that all *Kinds* of *Principles*, or all *Elements* go to make up every *Compounded Body*, as by the *Peripatetick Philosophy* we are taught; is a conceit, no more to be credited, than one that should tell us, all *Kind* of *Wheels* and other *Ch. 2. §. 5. Ch. 3. §. 1.*

parts

parts of a *Watch*, were put into a *Clock*; or that there were no other *Materials* wherewith to build an *House*, then for a *Tent* or a *Ship*. For why should *Nature*, the great *Artificer* by which all *perfect Works* are made, be feigned to cram and ram *all things into one*, which we our selves look upon as absurd?

6. §. *Secondly*, The *Mixture* of *Principles* is diversifi'd, as by *Conjugation*, so also by *Proportion*. That is, by the divers *Quantities*, of the several *Principles* or *Parts mixed* together. As if the *Quantity* of one, were as *five* to *ten*; of a second, as *five* to *fifteen*; of a third, as *five* to *twenty*, &c. Or if that of one, be as *five* to *six*; of a second, as *six* to *seven*; of a third, as *seven* to *eight*. By which, and by other *Proportions*, *Mixture* may be varied innumerable ways.

7. §. *Again*, As *Mixture* is varied with respect to the *Bodies Mixed*; so likewise in respect of the *Mixture* it self, which I call the *Location* of *Principles*, or the *Modes* of their *Conjunction*. Which may be various, as well as their *Conjugation* and *Proportion*. Yet are they all reduceable unto *two* general *Modes*: all *Bodies*, and therefore all *Principles*, being *mixed* either by *Mediation*, or by *Contact*.

Ch. 2. §. 2.

8. §. Now all *Contact*, whether of *Compounds*, or of *Atomes*, can be no other way, than such as is answerable to their *Figures*. Whereof, therefore, we can conceive but *three* general ways, viz.

First, By *Contract* in a *Point*, or some *smaller part*: as when *two Atomes* meet, which are *globular* or otherwise *gibbose*. *secondly*, By *Contact* in a *Plain*: as in the conjunction of the *sides* of *Triangular* or *Quadrangular Atomes*, or otherwise *flat*. *Thirdly*, By *Contact* in a *Concave*: as when one *Atome* is admitted into the *Concave* or *hole* of another; as a Spigot is into a Fosset. The *first* may be called, *Apposition*; the *second*, *Application*; the *third*, *Reception* or *Intrusion*.

9. §. In the *two last* ways, *Atomes* may be joyned by *Mediation*; but best of all the *last*. As when the *two extreams* of one *Atome* are received into the *Concaves* or the *holes* of two others.

10. §. And these are all the *general* ways, whereby we can conceive *Bodies* to be *Mixed* together; *sc.* by their various *Conjugation*, *Proportion* and *Location*. So that the *Composition* of *Atomes*, in *Bodies*; is like that of *Letters*, in *Words*. What a Thunderclap would such a *Word* be, wherein all the four and twenty *Letters* were pack'd up? One therefore is compounded of more, another of fewer: this of some, and that of others: and both the *Conjugation*, *Proportion*, and *Location* of *Letters* is varied in every *Word*: whereby, we have many thousands of *differing Words*, without any *alteration* at all, in the *Letters* themselves; and might have ten times as many more. In like manner, therefore, or in the self same analogous way, as the *Letters* of the *Alphabet*, are the *Principles* of *Words*; so *Principles*, are the *Alphabet* of *Things*.

11. §. What we have said of *Principles*; and of *Mixture* as consequent thereupon; may be a *foundation* for an *intelligible* account, of the *Nature* and *Cause* of most of the Intrinsick *Properties*, and *Qualities* of *Bodies*: as of *Gravity*, *Levity*, *Fixity*, *Fluidity*, *Angularity*, *Roundness*, *Heat*, *Cold*, *Blackness*, *Whiteness*, *Sowerness*, *Sweetness*, *Fragrancy*, *Fetidness*, and very many more. I say an *intelligible* account; *sc.* such as is grounded upon the *Notions* of *Sense*, and made out *Mechanically*. But the exemplification hereof, being too large a field

for

for this, or any one *Lecture*, I shall, before I come to the *Causes* of *Mixture*, only deduce from the *Premises*, these following *Corollaries*.

12. §. *First*, That there is no *alteration* of *Principles* or of *Elements*, in the most *perfect Mixture* of Bodies. It *cannot* be; for *Principles* are *Immutable*, as we have said. And if it could be, yet it *needeth not* to be: for they are also *many*, and *compoundable infinite* ways; as hath been shewed. So that we have no need to perplex our selves with any of those difficulties, that arise from the *Doctrine* of the *Alteration* of *Elements*. The ground of which conceit, is that, of three being but four *Elements*, and all in every particle of the *mixed Body*. And so men being puzeled, how from thence to make out the infinite *variety* of Bodies, they feigned them to be alterable, and *altered*, upon every *perfect Mixture*. Not considering, that if their four *Elements* be *alterable*; as few as they are, no fewer then *three* of them may be spared: for *one Element*, if alterable, may be made any. *Ch. 2. §. 3. Ch. 2. §. 5. Ch. 3. §. 10.*

13. §. Hence, *Secondly*, may be solved that great *Dispute*, Whether such as we call *Lixivial Salts*, are *made* by the *fire*? For *first*, No *Principle* is made by the *fire*: all *Principles* being *unalterable*; and therefore *unmakable*. *Secondly*, We must therefore distinguish betwixt the *Principle*, and its various *Mixture* with other *Principles*; from whence it may receive different *Shapes* and *Names*. Wherefore, a *Lixivial Salt*, qua *Lixivial*, is certainly *made* by the *fire*. But *quatenus Salt*, it is not: that *Principle* being *extractable* out of most Bodies; and by *divers* other *ways*, then by the *fire*. For whether you *Calcine* a body, or else *Ferment* it, (after the manner shewed by the curious *Improver* of *Chimical* Knowledg, Dr. *Daniel Cox*) or *putrifie* it under ground, or *drown* it in the Sea; it still yieldeth *some kind* of *Salt*. All which *Salts* are *made*, not by *making* the *Saline Principle*; but only by its being differently *Mixed*, by those several ways of the *Solution* of Bodies) with other *Principles*: from which its different *Mixture*, it receives the *various Denominations*, of *Marine*, *Nitrous*, *Volatile*, or *Lixivial*. *Ch. 2. §. 3.*

14. §. Hence, *Thirdly*, the most *perfect Mixture* of Bodies, can go no higher than *Contact*. For all *Principles* are *unalterable*; and all *Matter* is *impenetrable*; as hath been said. In the most *visible* and *laxe Mixture*, there is *Contact*; and in the most *subtile* and *perfect*, as in *Generation* it self, there is *nothing more*. *Ch. 2. §. 3. Ch. 3. §. 2.*

15. §. Hence, *Fourthly*, we easily understand, how divers of the same *Principles*, belonging both to *Vegetables* and many *other* Bodies, are also *actually* existent in the Body of *Man*. Because even in *Generation* or *Transmutation*, the *Principles* which are translated from one Body to another, as from a *Vegetable* to an *Animal*, are not in the least alter'd in themselves; but only their *Mixture*, that is, their *Conjugation*, *Proportion* and *Location*, is varied.

16. §. Hence also the difference of *Mixture*, arising from the difference of *Contact*, is intelligible; *sc.* as to those *three degrees*, *Congregation*, *Union*, and *Concentration*.

Congregation, and *Inconsistent Mixture*, is when the several *Atomes* touch but in a *Point*, or *smaller part*. In *which* manner, I have divers arguments, inducing me to believe the *Atomes* of all *Fluid Bodies*, qua *Fluid*, do touch; and in *no other*. *Ch. 3. §. 8.*

Union

Ch. 3. §. 8. *Union*, is when they touch in a *Plain*. As in the *Cryſtals* and *Shootings* of all *Salts*, and *other* like Bodies. For if we purſue their divided and ſubdivided parts, with our eye, as far as we can; they ſtill *terminate*, on every ſide, in *Plains*. Wherefore, 'tis intelligible, That their very *Atomes* do alſo *terminate*, and therefore *touch*, in *Plain*.

Ch. 3. §. 8. *Concentration*, is when two, or more *Atomes* touch by *Reception* and *Intruſion* of one into another: which is the *cloſeſt*, and *firmeſt Mixture* of all; as in any *fixed unodorable*, or *untaſtable* Body: the *Atomes* of ſuch Bodies, being not able to make any *Smell* or *Taſte*, unleſs they were firſt *diſſolved*; that is to ſay, *unpin'd* one from another.

17. §. Hence, *Sixthly*, we underſtand, how in ſome caſes, there ſeemeth to be a *Penetration* of *Bodies*; and in what *ſenſe* it may be admitted: *viz.* if we will mean no more by *Penetration*, but *Intruſion*. For the *Intruſion* of one *Atome* into the *Concave* or *hole* of another, is a *kind* of *Penetration*; whereby they take up leſs room in the *mixed* Body, then they would do by any other way of *Contact*. As a naked knife and its ſheath, take up almoſt double room, to what they do, when the knife is ſheathed. Whence we may aſſign the *reaſon*, Why many *Liquors* being *mixed*; take up leſs room or ſpace, then they did *apart*; as the *Ingenious* Mr. *Hook* hath made it to appear by *Experiment*, that they do. I ſay the plain *reaſon* hereof, or at leaſt one reaſon, is the *Intruſion* of many of their *Atomes* into one another. Which yet is not a *Penetration* of Bodies ſtrictly ſo called.

Ch. 3. §. 1.
Ch. 3. §. 14. 18. §. *Seventhly*, If all that *Nature maketh*, be but *Mixture*; and all this *Mixture* be but *Contact* 'tis then evident, That *Natural* and *Artificial* Mixture, are the *ſame*. And all thoſe *ſeeming ſubtilties* whereby *Philoſophers* have gone about to *diſtinguiſh* them; have been but ſo many *Scarcrows* to affright *Men* from the *Imitation* of *Nature*.

19. §. *Eighthly*, Hence it follows, That *Art* it ſelf may go far in doing what *Nature* doth. And who can ſay, how far? For we have nothing to *Make*; but only to *mix* thoſe *Materials*, which are already *made* to our hands. Even *Nature* her ſelf, as hath been ſaid,
Ch. 3. §. 1. *Maketh* nothing *new*; but only *mixeth* all things. So far, therefore, as we can govern *Mixture*, we may do what *Nature* doth.

20. §. Which that we may ſtill the better underſtand; let us before, and in the next place, ſee the *Cauſes* of *Mixture*. For ſince
Ch. 3. §. 18. *Natural* and *Artificial Mixture* are the *ſame*; the *immediate Cauſes* of both, are and muſt be the *ſame*.

CHAP.

CHAP. IV.

Of the CAUSES of Mixture.

NOW all the *Causes* of *Mixture* we can conceive of, must, I think, be reduced to these *six* in general; viz. *Congruity, Weight, Compression, Solution, Digestion,* and *Agitation*.

1. §. *Congruity,* or aptitude and *respondence* betwixt the *Sizes* and *Figures* of *Parts* to be *mixed*: whereby Bodies may be truly called the *Instrumental Causes* of their own *Mixture*. As when a *Plain* answers to a *Plain*, a *Square* to a *Square*, a *Convex* to a *Concave*, or a *Less* to a *Greater* or an *Equal*, &c. according to which *Respondencies* in the *parts* of Bodies, they are more or less easily *mingleable*.

2. §. *Weight,* by means whereof all *Fluid* Bodies, upon supposition of the *Congruity* of their parts, must unavoidably *mingle*.

3. §. *Compression;* which either by the *Air*, or any other Body, added to *Weight*, must, in some degree, further *Mixture*. Because, that *Weight* it self, is but *Pression*. For further Proof of all the said *Causes*, I made this *Experiment*; Let *Oyle* of *Aniseeds*, and *Oyl* of *Vitriol* be put apart into the *Receiver* of an *Air-Pump*. And, having exhausted it of the *Air*, let the *two* said *Oyls* be then affused one upon the other. Whereupon, *First*, It is visible, that they here *mix* and *coagulate* together; that is, their parts are *wedged* and *intruded* one into another, without the *usual compression* of the *Air*; for that is *exhausted*, and therefore only by the *Congruity* of their receiving and intruding parts; and by their *Weight*; by *which* alone they are so *compressed*, as to make that *Intrusion*. *Secondly*, It is also evident, That although they do *Coagulate*; yet not altogether so much, as when poured together in the same manner, and quantity, in the *open Air*. Wherefore, *Compression*, whether made by the *Air*, or any thing else, as it doth further the *Dissolution* of some Bodies, so the *Mixture* of others, and the greater the *Compression*, the more.

4. §. *Solution;* For all Bodies *mix* best, in *Forma fluida*. And that for two reasons. *First*, Because the *parts* of a Body are not then in a *state* of *Union*, but of *Separation*; and therefore, in a more capable *state*, for their *Mixture* and *Union* with the *parts* of *another* Body. *Secondly*, because then they are also in a *state* of *Motion*, more or less; and therefore, in a continual tendency towards *Mixture*; all *Mixture* being made by *Motion*. Wherefore all *Generations*, and most *perfect Mixtures* in *Nature*, are made by *Fluids*; whether *Animal, Vegetable,* or *Mineral*. Which is also agreeable to the *Doctrine* of the *Honourable* Mr. *Boyle,* in his *Excellent Treatise* of the *Nature* and *Vertues* of *Gems*. And it is well known, That Bodies are ordinarily *petrified,* or *Stones* made, out of *Water*. That is, out of *petrifying parts* dissolved *per minima* in *Water*, as both their *Menstruum* and their *Vehicle*. Wherefore, if we will talk of *making Gold*; it must not be by the Philosophers *Stone*, but by the Philosophers *Liquor*.

5. §. *Digestion.* For which there is the same reason, as for *Mixture,* by *Solution.* For, *First,* All heat doth *attenuate,* that is, still further *separate* the *parts* of a Body; and so render them more *mingleable* with the *parts* of *another.* And therefore, *Secondly*, Doth also add more *Motion* to them, in order to their *Mixture.*

6. §. *Agitation.* Which I am induced to believe a great and effectual means of *Mixture,* upon divers Considerations. As, *First*, That the *making* of Blood in the *Bodies* of *Animals,* and the *mixing* of the *Chyle* therewith, is very much promoted by the same means; *sc.* by the *Agitation* of the parts of the *Blood* and *Chyle,* in their continual *Circulation.* Again, from the *making* of Butter out of *Milk*, by the same means: whereby alone is made a *separation* of the oleous parts from the *Whey*, and Conjunction of the *Oleous* together. *Moreover,* From the great *Effects* of *Digestion*; well known to all that are conversant in *Chymical Preparations.* Which *Digestion* it self, is but a *kind* of *insensible agitation* of the *parts* of *digested* Bodies. 'Tis *also* a known *Experiment*, That the readiest way to dissolve *Sugar* in *Wine* or other *Liquor*; is to give the *Vessel* a *hasty turn,* together with a *smart knock*, against any *hard* and *steady* Body: whereby all the parts of the *Sugar* and *Liquor,* are put into a vehement *Agitation,* and so the *Sugar* immediately dissolved, and *mixed* with the *Liquor.* And I remember, that having (with intent, to make Mr. *Matthews's Pill*) put some *Oyl* of *Turpentine* and *Salt* of *Tartar* together in a Bottle, and sent it up hither out of the *Country*; I found, that the continual *Agitation* upon the *Road*, for three or four days, had done more towards their *Mixture*; than a far greater time of *Digestion* alone had done before. And it is certain, That a vehement *Agitation,* especially, if continu'd, or joyned with *Digestion*; will accelerate the *Mixture* of some Bodies, ten times more, than any bare *Digestion* alone; as may be proved by many *Experiments.* I will instance in this one. Let some *Oyl* of *Turpentine* and good *Spirit* of *Nitre* be stop'd up together in a Bottle, and the Bottle held to the Fire, till the *Liquors* be a little heated, and begin to bubble. Then having removed it, and the Bubbles by degrees increasing more and more; the two *Liquors* will of themselves, at last fall into so impetuous an *Ebullition,* as to make a kind of *Explosion*; sending forth a *smoak* for the space of almost *two yards* high. Whereupon, the *parts* of both the *Liquors,* being violently *agitated,* they are, in a *great portion, incorporated* into a *thick Balsam* in a *moment*: and that without any *intense* heat, as may be felt by the Bottle. And thus much for the *Causes* of *Mixture.*

CHAP.

CHAP. V.

Of the *POWER* and *USE* of *Mixture*.

HAVING enumerated the *general Causes*, we shall, lastly, enquire into the *Power* and *Use* of *Mixture*; or, into what it can *Do* and *Teach*. And I shall Instance in *six* particulars. *First*, to Render all Bodies *Sociable*, whatsoever they be. *Secondly*, To *Make Artificial* Bodies in Imitation of those of *Natures* own production. *Thirdly*, to *make* or *imitate* the *sensible Qualities* of Bodies; as *Smells*, and *Tasts*. *Fourthly*, To *make*, or *imitate* their *Faculties*. *Fifthly*, It is a *Key*, to discover the *Nature* of *Bodies*. *Sixthly*, To discover their *Use*, and the *Manner* of their *Medicinal Operation*.

INSTANCE I.

FIRST, To render all Bodies *Sociable* or *Mingleable*: as *Water* with *Oyl*, *Salt* with *Spirit*, and the like. For *Natural* and *Artificial Mixture*, are the same; as we have before proved. If therefore *Nature* can do it, as we see in the *Generation* of Bodies she doth, 'tis likewise in the *Power* of *Art* to do it.   *Ch.3.§.18.*

2. §. And for the doing of it, two *general Rules* result from the *Premisses*, *sc.* The *Application* of *Causes*, and the *Choice* of *Materials*. As for the *Causes*, they are such as I have now instanc'd in. And for   *Ch. 4.* the *Application* of them, I shall give these *two Rules*.

3. §. *First*, That we tread in *Natures* steps as near as we can; not only in the *Application* of such a *Cause*, as may be most proper for such a *Mixture*; but also in allowing it *sufficient time* for its *effect*. For so we see *Nature* her self, for her more *perfect Mixtures*, usually doth. She maketh not a *Flower*, or an *Apple*, a *Horse*, or a *Man*, in a *moment*; but all things *by degrees*; and for her more *perfect and elaborate Mixtures*, for the most part, she requireth *more time*. Because all such *Mixtures* are made and carri'd on *per minima*; and therefore require a greater time for the compleating of them.

4. §. A *second Rule* is, Not only to make a due *Application* of the *Causes*; but sometimes to *Accumulate* them. By which means, we may not only, *imitate Nature*, but in some cases go beyond her. For as by adding a *Graft* or *Bud* to the *Stock*, we may produce *Fruit sooner*, and sometimes *better*, than *Nature* by the *Stock* alone would do: So here, by *accumulating* the *Causes* of *Mixture*, that is, by joyning *two*, *three*, or *more* together; or by *applying more* in some Cases, where *Nature* applyeth *fewer*; we may be able to make, if not a more *perfect*, yet a far more *speedy Mixture*, than *Nature* doth. As by joyning *Compression*, *Heat*, and violent *Agitation*, and so continuing them all together, by some means contriv'd for the purpose, for the space of a

Week, or *Month*, or *longer*, without ceſſation. Which may probably produce, not only *ſtrange*, but *uſeful Effects*, in the *Solution* of ſome, and the *Mixture* of other Bodies. And may ſerve to *mix* ſuch Bodies, as through the *ſmall number* of their *congruous* parts, are hardly *mingleable* any other way. *Agitation* being, as carrying the *Key* to and fro, till it hit the *Lock*; or within the *Lock*, till it hit the *Wards*.

5. §. Secondly, For the *Choice* of *materials*, if they are not *immediately*, that is, of themſelves, *mingleable*; we are then to turn one *Ch. 3. §. 9.* Species of *Mixture* into a *Rule*; which is, To *mix* them by *mediation* of ſome *third*, whether more *ſimple* or *compounded* Body, which may be congruous *in part* to them *both*: as *Sulphureous Salts* are to *Water* and *Oyl*; and are for that reaſon *mingleable* with *either* of them. Or, By any *two* congruous Bodies, which are alſo, *in part*, congruous to *two others*: and other like ways. Whereby the *parts* of Bodies, though never ſo *heterogeneous*, may yet be all *bound* and *lock'd* up together. Even as *twenty Keys* may be *united*, only by *uniting* the *two Rings* whereon they hang.

6. §. The Conſideration of theſe things, have put me upon making ſeveral *Experiments*, for the *mingling* of *heterogeneous* Bodies. I ſhall give two Examples of Tryal; the one upon *Fluid*, the other upon *conſiſtent* Bodies.

7. §. For the *firſt*, I took *Oyl* of *Aniſeeds*, and pouring it upon another Body; I ſo order'd it, that it was thereby turned into a perfect *milk-white Balſam*, or *Butyr*. By which means the ſaid *Oyl* became *mingleable* with any *Winy*, or *Watery Liquor*; *eaſily*, and *inſtantaneouſly diſſolving* therein, in the form of a *Milk*. And *note*, That this is done, without the *leaſt alteration* of the *Smell*, *Taſt*, *Nature*, or *Operation* of the ſaid *Oyl*. By ſomewhat the like means, not only *Oyl* of *Aniſeeds*, but any other *ſtillatitious Oyl*, may be transformed into a *milk-white Butyr*; and in like manner be *mingled* with *Water* or any other *Liquor*. Which is of *various uſe* in *Medicine*; and what I find oftentimes very convenient and advantageous to be done.

8. §. Again, not only *Fluid* but *conſiſtent* Bodies, which of themſelves will *mix* only with *Oyl*; by due *mixture* with other Bodies, may be render'd *eaſily* diſſoluble in *Water*; as may *Roſin*, and all *reſinous* and *friable Gums*. As alſo *Wax*: and this without changing much of their *Color*, *Taſt*, or *Smell*. Whereof likewiſe, whatſoever others may do, the *Phyſician* may make a manifold *Uſe*.

INSTANCE II.

BY *Mixture* alſo, we may be *taught* to *Imitate* the *Productions* of *Nature*. As to which, from what we have before ſaid of *Mixture*, we may conclude; That there is no *Generation* of Bodies *unorganical*, but what is in the *Power of Mixture* to imitate. As of *Animals*, to Imitate *Blood*, *Fat*, *Chyle*, *Spittle*, *Flegm*, *Bile*, &c. Of *Vegetables*, to Imitate a *Milk*, *Mucilage*, *Roſin*, *Gum*, or *Salt*. Of *Minerals*, to Imitate *Vitriol*, *Allom*, and other *Salts*; as alſo *Metals*, and the like.

2. §. I do not ſay, I can do all this: yet if, upon good *Premiſſes*, we can conclude this poſſible to be done; it is one ſtep to the doing of it. But I will alſo give an *Inſtance* of ſomewhat that may be *done* in every *kind*. And,

3. §.

3. §. *First*, For the *Imitation* of an *Animal Body*, I will instance in *Fat*. Which may be *made* thus; Take *Oyl Olive*, and pour it upon high *Spirit* of *Nitre*. Then *digest* them for some days. By degrees, the *Oyl* becomes of the *colour* of *Marrow*; and at last, is *congealed*, or hardned into a *white Fat* or *Butter*, which *dissolveth* only by the *fire*, as that of *Animals*. In converting *Oyl* thus into *Fat*, it is to be *noted*, That it *hardens* most upon the *exhalation* of some of the more *Sulphureous* parts of the *Spirit* of *Nitre*. Which I effected, well enough for my purpose, by unstopping the glass after some time of *digestion*; and so suffering the *Oyl* to *dissolve* and *thicken* divers times by successive *heat* and *cold*. Hence, The true *Congealing Principle*, is a *Spirit of Nitre separated from its Sulphur*. For the better doing whereof, the *Aer* is a most commodious *Menstruum* to the said *Spirit* of *Nitre*. Whence also, if we could procure such a *Spirit* of *Nitre*, we might *congeal Water* in the midst of *Summer*. We might also *refrigerate Rooms* herewith *Artificially*. And might *Imitate* all *frosty Meteors*. For the *making* of *Fat*, is but the *Durable Congelation* of *Oyl*: which may be done without *frost*, as I have shewed how.

Hence also it appears, That *Animal Fat* it self, is but the *Curdling* of the *Oyly* parts of the *Blood*; either by some of its own *Saline* parts; or by the *Nitrous* parts of the *Aer* mingled therewith.

Hence likewise it is, That some *Animals*, as *Conies*, and *Fieldfares*, grow *fatter* in *frosty* weather: the *oily* parts of the *blood*, being then more than ordinarily *coagulated* with a greater abundance of *nitrous* parts received from the *Aer* into their *bodies*.

For the same reason it is, That the *Fat* of *Land-Animals* is hard; whereas that of *Fishes* is *very soft*, and runs all to *Oyl, &c*. Because the *Water*, wherein they live, and which they have instead of *breath*, hath but very few *nitrous* parts in it, in comparison of what the *Aer* hath.

4. §. *Secondly*, For the *Imitation* of a *Vegetable Body*, I will give three *Instances*; In *Rosin, Gum*, and a *Lixivial Salt*. The *first* may be made thus; Take good *Oyl* of *Vitriol*, and drop it upon *Oyl* of *Anise-seeds*; and they will forthwith *incorporate* together; and by degrees, will harden into a *perfect Rosin*; with the *general* and *defining Properties* of a truly *Natural Resinous Gum*. Being not at all *dissoluble* in *Water*; or at least, not any more, then any natural *Rosin* or *Gum*: yet very *easily* by *fire*: as also higly *inflamable*: and exceeding *friable*. Although this *Artificial Rosin*, be the result of *two Liquors*, both which very strongly affect the *Sense*: yet being well *washed* from the *unincorporated parts*, (which is to be done with some care) it hath scarce any *Tast* or *Smell*.

The *Concentration* of these *two Liquors*, is likewise so *universal*; that the *Rosin* is not made by *Precipitation*, but almost a *total Combination* of the said *Liquors*; and that with scarce so much, as any *visible* fumes.

5. §. *Again*, Having taken a certain *Powder* and a *Saline Liquor*, and mixed them together in a bottle, and so *digested* them for some time; the *Powder* was at last transmuted to a perfect *Oily Gum*; which will also *dissolve* either in *Oyl* or in *Water*; in the self same manner, as *Galbanum, Ammoniac*, and the like will do.

6. §. And *Lastly*, A *Lixivial Salt* may be *imitated* thus; Take *Nitre, Oyl* of *Vitriol*, and high *spirit* of *Wine*, of each a like quantity. Of these *three* Bodies, not any *two* being *put together*, that is to say

neither the *Nitre* with the *Oyl*, nor the *Oyl* with the *Spirit*, nor the *Nitre* with the *Spirit*, will make the least *Ebullition*: yet all *three mingled together*, make a very *conspicuous* one. The *Spirit* of *Wine* being as the *Sulphur*; and so that, and the *Nitre* together, standing, as it were, in the stead of an *Alkalizate*, that is, a *Sulphurious Salt*, against the *Oyl* of *Vitriol*. Divers other *Experiments* may be shew'n of the like Nature.

7. §. In the last place, for the *Imitation* of a *Mineral Body*, I will instance in *two, sc. Nitre* and *Marine Salt*; if I may have leave to reckon them amongst *Mineral Bodies*. As for *Nitre*, by mixing of *four Liquors* together, and then setting them to *shoot*; I have obtained *Chrystals* of *true* and *perfect Salt*; which have had much of a *nitrous tast*; and would be *melted* with a *gentle Heat*, as *Nitre* is; and even as easily as *Butyr* it self: I mean not, by the addition of any sort of *Liquor*, or any other Body, to *dissolve* it; but only by the *fire*.

8. §. And as for a *Sea-Salt*, that I might *Imitate Nature* for the *making* thereof, I consider'd, That the said *Salt* is nothing else but that of *Animals* and *Vegetables*, freed from its true *spirit* and *sulphur*, and some *Saline particles, specifically Animal* or *Vegetable*, together with them. For both *Animal* and *Vegetable* Bodies being continually carried by all *Rivers* into the Sea; and many likewise by *shipwrack*, and divers *other ways immersed* therein: they are at last *corrupted*, that is, their *Compounding* parts are *opened* and *resolved*. Yet the *Resolution* being in the *Water*, is not made *precipitately*, as it is in the *Air*; but by degrees, and very *gently*; whence the *Sulphurious* and other *Volatile* parts, in their *Avolation*, make not so much *haste*, as to carry the more *fixed Saline* parts along with them; but leaveth them behind in the *Water*, which *imbibeth* them as their proper *Menstruum*.

And the *Imitation* of *Nature* herein, may be performed thus; Put as much of a *Lixivial Salt* as you please, into a wide-mouth'd Bottle, and with fair *Water* make a strong *Solution* of it; so as some part thereof may remain *unresolved* at the bottom of the Bottle. Let the Bottle stand thus for the space of about half or three quarters of a year, all the time unstopped. In which time, many of the *Sulphurious* and *other Volatile* parts gradually flying away; the top of the *unresolved Salt* will be *incrustate*, or as it were *frosted* over, with many small and hard *Concretions*, which, in their nature, are become a true *Sea-Salt*. Whereof there is a double Proof; *First*, In that most of the said *Concretions* are of a *Cubical*, or very like *Figure*. Especially on their *upper* parts; because having a *fixed Body* for their *Basis*, their *under* parts, therefore, contiguous thereto, are less *regular*. Whereas the parts of the *salt* in the Sea, being environed on all sides with a *Fluid*; their *Figure* is on all sides *regular*. *Secondly*, In that a strong *Acid Spirit* or *Oyl* being poured upon a *full body'd Solution* hereof; yet it maketh herewith no *Ebullition*, which is also the *property* of *Sea-Salt*. And thus much for the more *General Imitation* of *Bodies*.

INSTANCE. III, & IV.

FROM the aforesaid *Premisses*, and by the aforesaid *Means*, there is no doubt to be made, but that also the other *sensible Qualities* of Bodies may be *Imitated*, as their *Odors*, and *Tasts*. And that not only the *general* ones, as *Fragrant*, or *Astringent*: but also those which are *specifical* and *proper* to such a *species* of Bodies.

2. §. Thus for Example, by *mixing* Spirit of Nitre or Vitriol with *rectified Oyl* of *Turpentine*, and some other *Vegetable Oyls*, severally, and in a due *Proportion* and *Time*, I have *Imitated* the *Smells* of *divers Vegetables*; as of *Tansy*, of *Lignum Rhodium*, and *others*. And I conclude it feasable, To *Imitate* the *Tast* or *Smell* of *Musk*, or *Ambergreece*, or *any other* body in the world.

3. §. Hence also we may be *Taught*, How to *Imitate* the *Faculties*, as well as other *Qualities* of Bodies. The reason is, because even *these* have no dependance upon any *substantial Form*: but are the meer result of *Mixture*; effected by the same *Causes*, whether in *Nature* or *Art*; as I think I have made to appear in the foregoing *Idea*. And as in the *Premisses* of this *Discourse* hath been shew'd.

Id §. 55.
Ch. 2. §. 10.
Ch. 3. §. 10.

INSTANCE V.

FROM whence, *again*, it is likewise a *Key to Discover* the *Nature* of Bodies. For how far soever we can attain to *Mingle*, or to *Make* them, we may also know *what they are*.

2. §. For Bodies are *mingleable*, either *of themselves*, or by some *Third*. As to those which *mingle of themselves*, we may certainly conclude, That there is a *congruity* betwixt them, in some respect or other. So upon various Tryals I find, That *Essential Oyls* do more easily *imbibe* an *Acid*, then an *Alkaly*. Whence it is evident, That there is some *Congruity* and *Similitude* betwixt *Essential Oyls*, and an *Acid*, which there is not betwixt the said *Oyls* and an *Alkaly*.

3. §. As to those that mingle only by some *third*; we may also certainly conclude, That though the *two extreams* are *unlike*; yet that they have both of them some *congruity* with that *third*, by which they are *united*.

4. §. Moreover, We may make a *Judgment* from the *manner* or *Degree* of *Mixture*. Thus the *Acid Spirit* of *Nitre*, as is said, will *coagulate Oyl-Olive*, and render it *consistent*. Whence it might be thought, That any other strong *Acid* will do the like; and that therefore, there is no great difference in the *Nature* of the said *Acid Liquors*. But the contrary hereunto, is proved by *Experiment*. For having *digested* the same *Oyl* in the same *manner*, and for a much *longer* time, with strong *Oyl* of *Sulphur*; although it thence acquired some change of *Colour*, yet not any *Consistence*.

5. §. Again, Because the said *Spirit* of *Nitre coagulates Oyl-Olive*; it might be expected, it should have the same effect upon *Oyl* of *Aniseeds*; or, at least, that if other *Acids* will *Coagulate Oyl* of *Aniseeds*, that this should do it *best*. But *Experiment* proveth the contrary. For of all I have tryed, *Oyl* of *Vitriol* is the only *Acid* that doth it *instantaneously*. *Oyl* of *Sulphur*, if very strong, will do it; but not so soon,

nor so *much Aqua fortis*, and *Spirit of Salt*, for the present, *do not* at all *touch* it. And *Spirit of Nitre* it self will not *coagulate* it, under *eight* or *ten* hours at least.

INSTANCE VI.

Lastly, and *consequently*, It is a *Key* To *Discover* the *Medicinal Use* and *Operation* of Bodies. Thus, for *Example*, by the *Imitation* of *Rosins* and *Resinous Gums*, we certainly know what all of them are, and *when*, and *wherefore* to be *used*. For what are *Mastick*, *Frankincense*, *Olibanum*, *Benzoin*, and *other like Rosins*, or *Resinous Gums*, for their *principle* and *predominant* parts, that is, *quà Rosins*; but Bodies resulting from *Natural*, in like manner, as I have shewed, they may be made to result, from *Artificial Mixture?* That is to say, the *Oleous*, and *Acid* parts of *Vegetables*, being both *affused* and *mingled* together, *per minima*, in some one sort of *Vessels* in a *Plant*, they thus *incorporate* into one *consistent* and *friable* Body, which we call *Rosin*.

2. §. Now from hence it is, That the said *Rosins*, and *Resinous Gums*; as also *Amber* and *Sulphur* for the same Reasons; are of so great and effectual *Use* against most *thin* and *salt Rheums*; *sc.* as they are *Acidoleous* Bodies. For by their *Acid* parts, which in all these Bodies are exceeding *copious*, they *mortifie* and *refract* those *Salt* ones, which feed the *Rheum*. And by their *oleous* parts, the same *Salt* ones are also *Imbibed*. Whence, they are all, in some degree, *incorporated* together; that is, The *Rheum* is *thickned*: which is the desired *effect*.

3. §. Whereas, on the contrary, if the *Cough* proceed not from a *thin*, and specially a *Salt Rheum*, but from a *Viscous Flegm*; the use of *many other* Bodies which are also more *oleous*, and abound not so much with an *Acid* as these do, especially *some* of them, is more proper: such as *these*, in this Case, proving sometimes not only *ineffectual*, but *prejudicial*. Since the very *Cause* of the said *Viscousness* of *Phlegm*, is chiefly some great *Acidity* in the *Blood*, or in some other *part*, as may be proved by divers Arguments.

4. §. Many more *Instances* might be hereunto subjoyned: and may hereafter be offered to the acceptance of such, who are inquisitive into *matters* of this *Nature*. If I shall not herein anticipate, or reiterate the *Thoughts* and *Observations*, of those two *Accurate* and *Learned Persons* Dr. *Willis*, and Dr. *Walter Needham*, as to what the one hath already *published*, and both have put us in *Expectation* of. But the *Instances* already given, are sufficient to evidence what I have said. And, I hope, this present *Discourse* to prove, in some measure, thus much; That *Experiment*, and the *Common Notions of Sense* are *prolifick*; and that nothing is *Barren*, but Phansie and *Imagination*.

An *Appendix* to the precedent *discourse* of *Mixture*.

HAVING, in the first Edition of the foregoing Discourse, made mention of the preparation of *Essen-* *tial Oyls*, so as to become easily mingleable with any *unoyly Liquor*. I shall here acquaint the Reader, That this may be done, by digesting any of the said *Oyls* with about an equal quantity of the *Yelk* of an *Egg*, with a very soft heat, like that of the *Meridian Sun* in *Summer*, continued for the space of three Weeks or a Month; and in the mean time, to be now and then stirred a little together. The *Yelk* will by degrees, imbibe the *Oyl*, and at length be incorporated with it, and become a *Balsam*, as white as *Milk*, easily dissoluble in any *watery* or *winy Liquor*.   *Ch.* 5. *Inst.* 1. §. 8.

2. §. I confess, that it will be very difficult to prepare any good quantity for use, this way. But this being a sufficient proof of the possibility of such a *Mixture*; I considered, whether the application of some other forementioned *Cause of Mixture*, might not supply the defect of this: and hereupon, have made several successful tryals; not only for the mixing of the said *Oyls*, but likewise of all sorts of *Rosins* and *Gums* with any *winy* or *watery Liquor*, in great quantities, in a short time, and without much trouble. But for the mixing of some of them, the *Yelk* of an *Egg* alone will not serve, without the intervening of some other sociable Body, according to one of the *Rules* given in the foregoing *Discourse*.   *Ch.* 5. *Inst.* 1. §. 6.

3. §. In the same *Discourse*, upon certain premises, I have laid down this following conclusion.   *Ch.* 5. §. 5.

—— By accumulating the *Causes of Mixture*, that is, by joyning "two or three or more together; or by applying more in some cases, "where *Nature* applyeth fewer; we may be able to make, if not a more "perfect, yet a far more speedy *Mixture*, than *Nature* doth. As by joyn-"ing COMPRESSION, *Heat*, and *violent Agitation*, and so con-"tinuing them altogether, by some means contrived for the purpose, for "the space of a Week or Month, or longer without *Cessation*. Which "may probably produce, not only strange, but useful effects, in the "SOLUTION of some, and the *Mixture* of other *Bodies*.

4. §. For the proof whereof, and that I had throughly weighed what I have said, Mr. *Pappin* hath since given us an ingenious Instance, in his new *Digester*. Which is, a *Balneum Mariæ clausum*: all *Infusions* and *Digestions* made with *Double Vessels*, having hitherto been made with the outer *Vessel*, open. So that whereas by the old way of *Digestion*, their is no other *Power* made use of but that of *Heat*: in this way, that also of *Compression* is joyned therewith.

EXPERIMENTS IN CONSORT OF THE LUCTATION

Arising from the Affusion of several

MENSTRUUMS

Upon all sorts of

BODIES,

Exhibited to the *Royal Society*, *April* 13. and *June* 1. 1676.

Ch.5.Inst.5

THE intent of the following Experiments *is twofold.* The one, to be as a Demonstration of the Truth of one, amongst other Propositions, laid down in the precedent Discourse of Mixture, &c. That it would be a Key to let us easily into the knowledge of the Nature of Bodies.

The other, and that consequently, To be as a Specimen of a Natural History of the Materia Medica : *that is to say, a multifarious Scrutiny into the intrinsick Properties of all those Materials, which have been, or may be used in* Medicine : *for the performance whereof, the following Method is exibited as one, amongst others, necessary to be insisted upon. For what Dominion a* Prince *hath over the* Moral, *that a* Physician *hath, as one of God Almighty's* Vice-Roys, *over the Corporeal World. Whom therefore nothing can more import, than a particular knowledge of the Genius of all his Subjects, those several Tribes of Matter, supposed to be under his Command.*

There

Lect. II.

There are some known Observations *of this nature : but there is no Author, I think, who hath given us a* Systeme *of Experiments upon the Subject : The performance whereof is here intended.*

The Experiments *may seem too numerous to be of one make. But no less a number would have answered the design of an* Universal Survey; *which, though less pleasing, proves the more instructive in the end : not being like angling with a single Hook; but like casting a Net against a shole : with assurance of drawing up something. Besides the advantage of comparing many together ; which being thus joyned, do oftentimes, like* Figures, *signifie ten times more, then standing alone, they would have done.*

How far the Corollaries *all along subjoyned have made this good, is left to the* Reader *to judge. And also, to add to them, so many more, as he pleases : for I make my own Thoughts no mans* Measure.

CHAP. I.

What is generally to be observed upon the Affusion *of the* Menstruum; *and what, particularly of* Vegetable Bodies.

THE Bodies whereupon I made tryal, were of all kinds, *Animal, Vegetable,* and *Mineral.* Amongst *Vegetables,* such as these, *scil.* Date-stones, Ginger, *Colocynthis, Pyrethrum,* Hawthorn-stones, *Staphisagria, Euphorbium,* the *Arenulæ* in Pears, *Semen Milii Solis,* Tartar, Spirit of *Scurvygrass,* Spirit of *Wine,* &c.

2. §. Amongst *Minerals*, several sorts of *Earths, Stones, Ores, Metals, Sulphurs,* and *Salts.*

3. §. Amongst *Animals*; such as these, *scil.* Hairs, Hoofs, Horns, *shells,* and *shelly Insects,* Bones, Flesh, *and the several* Viscera, Silk, Blood, Whites *and Yelks of Eggs,* sperma Ceti, Civet, Musk, Castor, Gall, Urine, Dungs, *animal Salts and Stones.*

4. §. The *Liquors* which I poured hereupon severally, were these, *sc.* Spirit *of salt Armoniac,* Spirit *of Harts-Horn,* Spirit *of Nitre, Aqua fortis,* Oyl *of Salt,* Oyl *of Sulphur, and* Oyl *of Vitriol*; commonly so called.

5. §. In the *Mixture* of these Bodies, two things, in general, are all along to be observed, *viz. First,* which they are, that make *any,* or *no Luctation.* For, as some which seem to promise it, make none: So, many, contrary to expectation, make a considerable one.

6. §. Next, the *manner* wherein the *Luctation* is made ; being with much variety in these *five* sensible *Effects.* 1. *Bullition*; when the Bodies mixed produce only a certain quantity of froth or bubbles. 2. *Elevation*; when, like Paste in baking, or Barm in the working of Beer, they swell and huff up. 3. *Crepitation*; when, they make a

kind

kind of hiſſing and ſometimes a crackling noiſe. 4. *Effervescence*; then only and properly ſo called, when they produce ſome degree of heat. 5. *Exhalation*; when not only fumes, but viſible ſteams are produced.

7. §. Of all theſe, ſometime one only happens, ſometimes two or more are concomitant. Sometimes the *Luctation* begins preſently upon mixture, and ſometimes not till after ſome intermiſſion. In ſome bodies, it continues a great while; in others, is almoſt inſtantaneous: Examples of all which I ſhall now produce; beginning with *Vegetables*, as affording the leaſt variety.

8. §. And *first*, if we take Spirit or *Oyl of Salt*, *Oyl of Vitriol*, Spirit of Nitre, or *Aqua fortis*, and pour them ſeverally upon the ſeveral parts of Vegetables, as Roots, Woods, Stones, &c. we ſhall find, that they are, generally far leſs apt to make a *Luctation*, than either *Animal*, or ſubterraneal Bodies. Whence, as from one argument, it ſeemeth evident, That in moſt *Vegetables*, and in moſt of their parts, the predominant *Salt* is an *Acid*. But that, on the contrary, the predominant *Salt* in moſt *Minerals*, and parts of *Animals*, is an *Alkaly*: in the former, uſually a *fixed*; in the latter, a *volatile Alkaly*.

9. §. Again, although the *Luctation* which moſt *Vegetables*, and moſt of their parts make with *Acids*, be but ſmall, yet ſome they make; eſpecially with ſome *Acids*, as with Spirit of *Nitre* and *Aqua fortis*. Whence it ſeemeth plain, That there is an *Alkaline Salt* exiſtent in many *Vegetables*, even in their *natural eſtate*; and that it is not made *Alkaline*, but only *Lixivial*, by the *fire*. Or, there is ſome quantity of a *Salt*, call it what we will, in the ſaid Bodies, which is ſo far different from an *Acid*, as to make a *Luctation* therewith. But to give particular inſtances of the ſeveral *proportions*, or *manner* of *Mixture*, wherein it appears to be in ſeveral *Plants*.

10. §. And *firſt*, of all vegetable Bodies, *Date-ſtones* are amongſt the leaſt apt to make a *Luctation* with *Acids*, if they may be ſaid to make any at all. Hence they are not ſo potent *Nephriticks*, as many other Stones, which make a more ſenſible *Luctation*.

11. §. *Ginger makes a ſmall Bullition* with *Aqua fortis*, only obſervable by a *Glaſs*. Hence the *pungency* of Ginger lyeth in a *ſulphureous* and *volatile Salt*, which yet is very little *Alkalizate*.

12. §. *Scurvygraſs-ſeeds make a very ſmall Bullition with Aqua fortis*, like that of *Ginger*. So doth alſo the *Seed of Purſlane*. Hence, although there is much more of a certain kind of *volatile Salt* in Ginger or Scurvygraſs, than in *Purſlane*; yet there is little more of an *Alkaly* in any one, than in an other.

13. §. *The Pulp of Colocynthis, Fruit-Stones, the ſtony Covers of the Seeds of Elder, of white Bryony, of Violets,* and others, with *Aqua fortis* make a *Bullition* juſt perceivable without a *Glaſs*. Hence it appears, That the great *Cathartick* power of *Colocynthis* lieth not ſo much in an *Alkaly*, as an *Acid*; as making a much leſs *Bullition*, than ſome other *vegetable Bodies*, which are leſs *Cathartick*. For which reaſon likewiſe it is, That the beſt Correctors, or Refractors of the force of *Colocynthis*, are ſome kinds of *Alkalies*, as particularly that of *Urine*, as *Riverius* hath ſomewhere obſerved.

14. §. *The Root of Pyrethrum, with Aqua fortis, makes a Bullition and huff, in a ſhort time*. Hence, the Cauſe of a *durable Heat*, upon the Tongue, is an *Alkalizate Sulphur*. For the Heat of Ginger, though

greater;

greater; yet abideth *nothing near so long* as that of *Pyrethrum*; which, as is said, maketh also a more sensible *Bullition* with *Acids*.

15. §. *Kermes-berries, commonly, but ignorantly, so called, with the said Liquor, huff up to an equal height, but in a somewhat longer time.* Hence they are gently *astringent*; *scil.* as their *Alkaly* binds in with some preternatural *Acid* in the stomach.

16. §. *Hawthorn-stones, with Aqua fortis, huff up equally with the former Body; but the Bullition is not so visible. The like is also observable of Medlar-stones.* Hence, as they contain a middle quantity of an *Alkaly*, they are not insignificantly used against the *Stone*.

17. §. *Seeds of Staphisagria, with Aqua fortis, make a Bullition still more visible. But it quickly ends.* This confirms what was said before, *sc.* That the cause of a *durable* Heat is an *Alkaline Sulphur*; these Seeds producing a *durable* Heat, as doth the Root of *Pyrethrum*.

18. §. *The Seeds also of red Roses, Borage, and Comfrey do all with Aqua fortis make a considerable Bullition and huff; and that very quickly.* So that amongst all Shells and Stones, those generally make the greatest *Bullition*, which are the hardest and the brittlest, and so the fullest of *Salt*.

19. §. *Euphorbium makes a Bullition yet more considerable, with much froth, and very quickly.* From which Experiment, compared with two of the former, it appears, That *Euphorbium* is not an *Acid*, but an *Alkaline Gum*. As also, that the cause of its so very *durable* Heat, is an alkaline *Sulphur*, as of *Pyrethrum* and *Staphisagria* hath been said. It seems also hence evident, that the power of all great *Sternutatories* lyeth not in their *Acid*, but their *Alkalies*.

20. §. *The Arenulæ or little stones in Pears, cluster'd round about the Coar, with Aqua fortis, presently huff up, and make a great Bullition and Effervescence, much greater than do any of the Bodies above-named.* Whence, although, so far as I know, they have never yet been used in *Medicine*; yet it is probable, that they are a more potent and effectual *Nephritick*, than any of the Bodies aforsaid, some of which are usually prescribed. It is hence also manifest, That, according to what I have elsewere said, for the sweetning of the Fruit and Seed, the *Tartareous* and *Alkaline* parts of the *Sap*, are *precipitated* into their Stones, stony parts, and Shells. *Anat. of Plants, B. 1. Ch. 6.*

21. §. The last *Instance* shall be in the shells *of the Seeds of Milium Solis; which not only with Aqua fortis, but some other Acids, make a greater and quicker Bullition and Effervescence, than any other vegetable Body, upon which I have yet made tryal, in its natural estate.* Hence, as well as from divers of the last fore-going Instances, we have a clear confirmation of what I have, towards the beginning of this Discourse, asserted; *sc.* That there is some kind of *Alkaline Salt* in *Plants*, even in their *natural estate*. As also, that they are as significantly used against the Stone, *quatenus alkalizate*, as *Millipedes, Egg-shells*, or any other *testaceous* Bodies of the same strength. To these I shall subjoyn one or two Examples of Vegetable Bodies which are more or less altered from their *natural estate*.

22. §. *Neither Crystals of Tartar, nor Tartar it self (although they have some store of alkaline mixed with their acid parts) make any Effervescence with Acids, but only with Alkalites, as spirit of Harts-Horn, &c.* Hence the *calculous sediment* or *Arenula* in *Urine*, may not so properly be called

the *Tartareous* part of the *Urine*; the events following the mixture hereof with the aforesaid *Salts*, being quite contrary; as will be seen in the *Last Chapter.*

23. §. *Spirit of Scurvy-grass maketh no Luctation with any Acid.* Hence (as from a former Experiment was above-noted) it seems, That there may be a kind of *volatile Salt*, which is neither *acid*, nor *alkaline*; such as this of *Scurvygrass* and other like *Plants* seems to be: yet contrary to an *acid*; as experience shews in their efficacy against the acid *Scurvy*.

24. §. *Rectified Spirit of Wine, both with Spirit of Nitre, and with Oil of Vitriol, severally, maketh a little Luctation.* Which argues, that there is contained, even in this Spirit, some portion of a *volatile Alkaly.*

25. §. *Spirit of Wine, and double Aqua fortis, as the strongest is called, make an effervescence so vehement, as plainly to boil.*

26. §. Besides the vehemency hereof, there is another surprizing circumstance. For whereas all other *Liquors* which make an *Effervescence* together, will do it in any proportion assigned, although but one drop to a thousand: these two, *sc.* rectified *Spirit of Wine* and *Aqua fortis*, require a certain proportion the one to the other. For if, suppose, into six drops of *Spirit of Wine* you put but two or three of *Aqua fortis*, they stir no more than if you put in so much *Water*: but drop in about seven or eight drops of *Aqua fortis*, and they presently boil up with very great vehemency. Hence we may conceive the reason of the sudden access of an acute *Disease*, and of its *Crisis*. These not beginning gradually with the *Cause*; but then, when the *Cause* is arrived unto such an ἀκμή, or such a certain *Proportion*, as is necessary to bring Nature to the contest. And these may serve for Examples upon *Vegetables.*

CHAP. II.

What may be observed of MINERALS.

HAVING given several Instances of tryal upon *Vegetables*; I next proceed to *Minerals*, which, for some orders sake, I shall distribute into five or six sorts, *sc. Earths, Stones Ores* and *Metals, Sulphurs,* and *Salts.*

2. §. First for *Earths. Oyl of Vitriol* upon *Fullers Earth*, doth not stir it, or cause the least *Bullition.* Nor upon yellow *Oker.* Nor upon the *Oker* which falls from green *Vitriol.* The same *Oyl of Vitriol* and *Spirit of Harts-Horn* poured severally upon *Bolus Armena* of two kinds, and upon one kind of *Terra sigillata*, stir none of them. Hence *Bolus*'s are the *Beds*, or as it were, the *Materia prima*, both of opacous *Stones*, and *Metals*; into which the said *Bolus*'s are transmuted, by being *concentred* with divers kinds of *Salts* and *Sulphurs*, which successively flow in upon them.

3. §.

3. §. *Aqua fortis*, and *Oyl of Vitriol* being poured severally upon another sealed Earth, which was vended by the name of *Terra Lemnia*; they both made a very considerable *Effervescence* herewith. Whence it appears, That there is no small difference in the nature, and therefore the operation of *Bolus Armena* and *Terra Lemnia*. As also, betwixt the sealed Earths themselves, one making a great *Effervescence*, another none at all. Whereto those that use them, are to have regard.

4. §. Next *for Stones*. And first, *Irish Slat*, with Spirit of *Hartshorn*, maketh a small, yet visible *Bullition*: and it presently ceaseth. So that it seems to be nothing else but a *Vitriolick Bole*. As is also argued from its taste, which is plainly acid, and somewhat rough. Whence also it is with good reason given upon any inward Bruises. Because by coagulating the Blood, it prohibits its too copious afflux into the affected part. Yet being but gently astringent, and so the Coagulations it makes, not great; they are likewise well enough carried off from the same part in the Circulation; by both which means an Inflammation may be either prevented, or the better over-ruled.

5. §. *Lapis Hæmatites* maketh no *Effervescence* at all either with *Alkalies* or *Acids*.

6. §. Powder of the green part of a *Magnet* with *Oyl of Vitriol* maketh some few bubbles, yet not visible without a Glass. But the powder of the black part of a *Magnet*, which is the said stone fully perfect, stirreth not with any acid. Neither doth the calcined *Magnet*. Hence there is some considerable difference betwixt *Iron* and the *Magnet*.

7. §. *Lapis Lazuli*, with *Oil of Vitriol*, and especially with *Spirit of Nitre*, maketh a conspicuous *Bullition*. Hence its *Cathartick* virtue lyeth in an *Alkaly*. For which reason it is also appropriate, in like manner as *Steel*, to the cure of *Hypochondriacal Affections*; originated from some kind of fermenting *Acid*.

8. *Osteocolla*, with *Spirit of Nitre* maketh yet a greater *Effervescence*. How it comes to be so great a knitter of broken Bones, as it is reputed, is obscure. It seemeth, that upon its solution by a *Nitrous Acid* in the body; it is *precipitated* upon the broken part, and so becomes a kind of *Cement* thereto.

9. §. *Lapis Tuthiæ*, with *Spirit of Nitre*, maketh an *Effervescence* much alike. And with *Oyl of Vitriol* very considerably. But *Lapis Calaminaris* with *Oyl of Vitriol* grows stark; as the powder of *Alabaster* doth with water. With *Spirit of Nitre* it maketh a little *Bullition*, and quickly. But with *Aqua fortis*, a great one; beyond any of the Stones above named. Hence both *Tutty* and *Calamy* are *Ophthalmicks* from their *Alkaly*. Which is also confirmed, from the efficacy of some *Alkalies* of the like use. Hence also *Calamy* seemeth to partake somewhat of the nature of *Silver*: as by tryal made upon that also, will hereafter better appear.

10. §. *Chalk* and *Oil of Sulphur* or *Vitriol* make as strong an *Effervescence* as any of the rest. Whence it is sometimes well used against a *Cardialgia*.

11. §. *Whiting* makes as great an *Effervescence* as *Chalk*. So that it seems the saline parts are not washed away with the water, wherein the *Chalk*, for the making of *Whiting*, is dissolved.

12. §. *Talk* will not stir in the least either with *Spirit of Nitre*, or *Oyl of Vitriol*. But the *Lead-Spar* maketh a considerable *Effervescence* with both of them severally. Hence, however this be also called *English Talk*, yet there is no small difference betwixt this, and true *Talk*.

13. §. *To these Stones may be added petrified bodies*. As petrified wood; which (that upon which I made tryal) no acid stirreth in the least. Petrified shells; upon four or five several sorts whereof, *Oyl of Vitriol* being poured, produceth a great *Effervescence*. The Root or rougher part of the Stone called Glossopetra, with *Spirit of Nitre*, makes a conspicuous *Bullition*. Asteria, the Stone so called, and found in some places in England, with *Oyl of Vitriol*, maketh an *Effervescence* at the same degree. So doth the *Belemnites*, or *Thunder-Stone*, both the larger and the lesser kinds. So that none of these are *acid*, or *vitriolick*, but alkalizate Stones.

14. §. *Coraline*, with *Oyl of Vitriol*, makes a conspicuous *Bullition*, yet mild and gentle; that is, with very little, if any heat, and without any visible *Fumes*. And red and white *Coral* do the like. Hence they are all of a very gentle operation, and fit for Children, as the case requires.

15. §. *Magistery of Coral* (prepared the ordinary way) stirreth not in the least, either with *Alkalies* or *Acids*. Whence it is evident, That its active Principles are in its preparation destroyed and washed away: that is to say, It is an elaborate Medicine good for nothing. And thus far of Stones.

16. §. *I next come to Metals and Ores. And first for Lead;* upon which *spirit of Salt Spirit of Nitre*, or *Aqua fortis* being dropped, it stirreth not in the least with any of them: but with *Oyl of Sulphur*, and especially with *Oyl of Vitriol* it maketh a slow *Bullition* and froth. Hence it seemeth to be *the most alkalizate Metal*. Which is also confirmed by a foregoing Experiment upon the *Lead-Spar*, which maketh a considerable Effervescence with any sort of *acid*. And which likewise, being calcined, yieldeth a good quantity of *Lixivial Salt*.

17. §. *Lead-Ore stirreth not at all with Aqua fortis or Oil of Vitriol*. But *Spirit of salt* makes it bubble, and *Spirit of Nitre* makes it boil. Hence there is a considerable difference betwixt the perfect Metal and the Ore.

18. §. *Burnt Lead and red Lead, make a very small Bullition with Oyl of Vitriol*, with *Spirit of Nitre* a far greater.

19. §. *Mercury, with Oyl of Vitriol, will not stir, nor with Oyl of Sulphur*. But with *Spirit of Nitre* presently boyls up. Hence *Mercury* is a *subacid Metal*; *Spirit of Nitre* being a *subalkaline Acid*.

20. §. *The filings of Iron or Steel, with Oyl of Vitriol, make a fair Bullition*, like that of *Minium*. But *Spirit of Nitre* makes them boil with much celerity. Hence Iron is likewise a *subacid Metal*.

21. §. *Steel prepared with Sulphur maketh a far less Effervescence with the same Spirit of Nitre*, than do the *filings*. Hence there is a great difference in their strength. So that ten grains of the filings unprepared, will go as far as fifteen grains or more of those which are prepared, as above-said. Yet in some cases the weaker and milder may be the better.

22. §. *There is one Circumstance in the mixture of Steel and Aqua fortis, which is surprizing; and that is this, That strong Aqua fortis, dropped upon Steel, will not, of it self, make the least Bullition: but if*

hereto

hereto you only add a drop or two of *Water,* they presently boil up with very great vehemency. The Cause is obscure; yet it is well known, that *Water* it self will dissolve *Iron*: so that it appeas, as well by this, as by some other Experiments, that even in common Water, as mild as it is, there is some kind of corrosive Principle.

23. §. *Antimony* with Spirit of *Nitre,* and *Aqua fortis* severally, maketh an *Effervescence*; somewhat lower than Iron. With Oil of *Vitriol* the *Bullition* is so small, as difficulty to be perceived with a *Glass.* Hence it seemeth to be of a very compounded nature; if I may so call it, a subacid-alkaline Metal.

24. §. *Antimonium Diaphoreticum,* with Spirit of *Nitre* and *Oil of Vitriol* severally, makes a considerable *Effervescence.* Wherefore it is not an useless *Preparation*; as from the *Calcination* and *Ablution* used therein, some have thought.

25. §. *Bezoardicum Minerale,* (that upon which I made tryal) stirreth not at all either with *Alkalies* or *Acids.* To which, let those who make use of it, have regard.

26. §. *Tin,* with Spirit of *Nitre,* makes so hot and vehement an *Effervescence,* that it turns presently, as it were, into a *Coal.* It makes also a fair *Bullition* with *Oyl of Vitriol.* And a gentle one with Spirit of *Salt.* Wherefore, it hath something of the nature both of *Iron, Lead,* and *Copper.*

27. §. The like remarkable circumstance is seen in the mixture of *Aqua fortis* with *Tin,* as with *Iron.* For *Tin* and strong *Aqua fortis* of themselves will not stir; but add a few drops of water to them, and they boyl up with the greatest vehemency.

28. §. *Copper,* with Spirit of *Salt,* and *Oyl of Vitriol* severally, stirs not at all. Spirit of *Nitre,* and *Aqua fortis,* both boil it up vehemently. Neither Spirit of *Harts-horn,* nor Spirit of *Salt Armoniac* maketh any *Bullition* therewith. But both of them, by a gentle solution, that is, gently separating its *Sulphur* from its *Salts,* turn it blue. Hence *Copper* hath a greater proportion of *acid* than *any* of the forementioned *Metals.*

29. §. *Silver,* neither with Spirit of *Salt,* nor *Oyl of Vitriol* makes any *Bullition.* With Spirit of *Nitre* it makes one, but tis soon over: and then continues to dissolve slowly into white *Coagulations.* It also maketh with Spirit of *Harts-horn,* or of *Salt Armoniac,* a full and deep blue. Hence there is a greater proportion of *acid* in *Silver,* than in *Lead, Mercury, Iron, Antimony, Tin,* or *Copper.*

30. §. *Litharge* of *Silver* maketh the greatest *Effervescence* with *Oyl of Vitriol.* Yet some with Spirit of *Nitre.* And with Spirit of *Salt Armoniac* maketh some little huff or elevation. And being mixed with Spirit of *Nitre* and Spirit of *Salt Armoniac* both together, produceth a faint blue. Hence, although the far greater part of this *Litharge* be but *Lead*; yet, it seems, it hath some small mixture of *Silver.* But that of *Gold* seemeth, for contrary reasons, not to have any *Gold.*

31. §. *Gold* maketh no *Effervescence* with any single Salt I know of. But it is commonly dissolved with *Aqua Regis,* which is known to be an alkaline Liquor. Whence it seemeth, That as *Lead* is the most alkalizate, so *Gold* the most acid of Metals.

32. §. These things considered, and other observations added hereunto, may possibly give some directions, not only for the ordering and using, but even for the making, imitating and transmuting of *Metals*. Thus far of *Metals*.

33. §. I will next give one or two *Instances* of tryal upon *Sulphurs*. And *first Sulphur vive*, with *Aqua fortis*, maketh an apparent *Bullition*, but it is some time, before it begins. But the *factitious* or common *Brimstone*, maketh scarce any, if any at all. So that there is no small difference betwixt them.

34. §. *White* and *yellow Arsenick* make no *Bullition* either *with Alkalies* or *Acids*. Wherefore the strength of its operation on the Body, lies more in a *Sulphur* than a *Salt*; or in a *Salt* drowned in its *Sulphure*.

35. §. The *ashes* either of *Pit-Coal*, or *Sea-Coal*, make no *Effervescence* with *Alkalies* or *Acids*. Whence the saline Principle is altogether volatile, and sublimed away by the fire.

36. Lastly for *Salts*. And *first* of *all*, *Borax* maketh no *Effervescence* nor any *Fumes* with *Oyl* of *Vitriol* or *Spirit* of *Nitre*.

37. §. *Oyl of Vitriol* and *Nitre* make fumes or steams, though no *Effervescence*.

38. §. *Green Vitriol*, with *Spirit of Harts-Horn*, is scarcely moved. *White Vitriol*, with the same *Spirit*, maketh a conspicuous huff. And *Roman Vitriol* a vehement *Effervescence*. Whence the *former* is the *least acid*, and the *latter the most of all*. Which also confirms what I said before of the like natures of the several *Metals* to which they belong.

39. §. *Salt of Vitriol*, though a *fixed Salt*, and made by *Calcination*, yet maketh *no Effervescence with the strongest acid*; but only with *Alkalies*; as may be seen upon their mixture, but much better heard by holding the mixture to ones ear. Hence, there are *fixed Acids*. Which further confirms what I have above asserted concerning the nature of *Gold, &c.* That the *predominant Salt* thereof is a *fixed Acid*.

40. §. *Sal Martis*, with *Spirit of Harts-horn*, maketh a considerable huff. Hence it is much more *acid* than *green Vitriol*; and is therefore a cooler body.

41. §. *Alum* and *Spirit of Harts-horn* make a plain *Effervescence*.

42. §. *Saccharum Saturni*, with *Oyl of Vitriol*, stirs not at all. With *Spirit of Salt*, huffs a little. With *Spirit of Nitre* much more. Hence the *acid* of the *Vinegar*, and not the *Alkaly* of the *Lead*, is the predominant Principle.

43. §. *Common Salt* stirs neither with *Spirit of Salt*, nor with *Spirit of Nitre*; nor with *Aqua fortis*. But with *Oyl of Vitriol* it maketh a great *Effervescence* with noise and steams. Hence, even common *Salt*, though it be not reckoned amongst *alkaline Salts*, yet is far nearer in nature to *that*, than to an *acid*. Hence also the *Spirit of Salt* is a *subalkaline Acid*, and of a very different nature from *Oyl of Sulphur* or *Vitriol*.

44. §. *Salt Armoniac*, with *Spirit of Nitre*, stirreth not. But with *Oyl of Vitriol* it maketh a great *Effervescence*. Hence *Spirit of Nitre* is a *subalkalizate Spirit*.

45. §. *Oyl of Vitriol* and *Spirit of Nitre*, though both acids, yet make a great *smoak*; greater than that which the *Spirit* maketh of it self. Which confirms the last precedent *Corollary*.

46. §. *Oyl of Vitriol* and *Spirit of Salt*, though both *acids*, yet make a strong *Effervescence*, with *noise* and *fumes*. Which further confirms, what was noted before, *sc.* that Spirit of Salt is a *subalkaline Acid.*

47. §. *Spirit of Salt Armoniac*, with *Oyl of Vitriol*, makes an *Effervescence* so extraordinary quick, and as it were *instantaneous*, that nothing seemeth quicker. Whence it is probable, That if *Gun-powder* were made of *Salt Armoniac*, instead of *Nitre*, or with *both* mixed together; it would be *far stronger*, than any kind now in use. And thus far for *Minerals.*

48. §. I have only one *Corollary* to add, from the whole; which is, That whoever doth undertake the *Natural History* of a *Country*, (such as that the Learned Dr. *Plot* hath exceedingly well done of *Oxfordshire*) the foregoing *Method*, seemeth so easie, cheap, and indeceitful, for the finding out and well distinguishing the natures of all kinds of *Metalls*, *Ores*, *Salts*, *Earths*, *Stones*, or other *subterraneal Bodies*; as cannot, I think, be supply'd, but by others of greater difficulty and expence.

CHAP. III.

What may be observed of the PARTS *of* Animals.

NOW proceed to the several *Parts of Animals*; as *Hairs, Hoofs, Horns, Shells* and *shelly Insects, Bones, Flesh* and the several *Viscera, Silk, Blood, Eggs, Musk, Castor, Gall, Urine, Dungs, Salts* and *Stones.*

2. §. And *first* of all, the *Hair* of a *mans head*, with *Oyl of Vitriol*, maketh no *Bullition* at all. Nor yet with *Spirit of Nitre*. So that although it contains a good deal of *volatile Salt*; yet it seemeth either not to be *alkaline*, or else is *centred* in so great a quantity of *Oyl*, that the *acid menstruum* cannot reach it.

3. §. *Hares Fur*, with *spirit of Nitre*, maketh, although *a short*, yet very plain *Bullition and huff*. Hence the *Hair*, and therefore the *Blood*, of some *Animals*, is fuller of *Salt*, at least of an *Alkaline Salt*, than that of some others. And perhaps the *Hair* of some men, as of *Black's*, may be so full of *Salt*, as to make a Bullition like *Hares Fur*.

4. §. *The shavings of Nails* stir not at all, either with *Oyl of Vitriol*, or *Spirit of Nitre*: only with the latter they turn yellow. But *Elks Claws*, with *spirit of Nitre*, make a *small and slow Bullition.*

5. §. *Horses Hoof*, with *Oyl of Vitriol*, stirs not of many hours. But with *spirit of Nitre*, allowing it some time, makes a very plain *Bullition*, and *huffs* up very *high*.

6. §. *Cows Horn*, neither with *Oyl of Vitriol*, nor with *Spirit of Nitre*, maketh any *Bullition*, only turneth to a *yellow colour.*

7. §. *Rams Horn* stirs not with *Oyl of Vitriol*; but with *Spirit of Nitre*, makes a *small* and *slow Bullition*.

8. §. *Harts-Horn* makes a confiderable *Bullition* and *huff*, even with *Oyl of Vitriol*, which the reſt of the *Bodies* abovefaid, will not do. But with *Spirit of Nitre*, it makes yet a greater. From the foregoing Experiments, and almoſt all that follow, what is before aſſerted of the *Salts* of *Vegetables* and *Minerals*, is here alſo evident concerning that of *Animals*, ſcil. That it is not *made*, but only *ſeparated* by the fire. It likewiſe hence appears, That the *proportion* of *Salt* in the forementioned *parts* is very different; and that therefore ſome of them are never, and none of them but with good diſcretion, to be ſubſtituted one for another in Medicine. As alſo, that there is a different proportion of *Salt* in the ſeveral *Animals* themſelves, to which the ſaid *Parts* belong.

9. §. Next for *ſhells*; as thoſe of *Lobſters*, *Eggs*, *Snails* and *Oiſters*: all which make an *Effervefcence*, both with *Oyl of Vitriol*, and *Spirit of* tre. But with *Spirit of Nitre* the greateſt. *Lobſter-ſhells* make a confiderable *Bullition* and *huff*, but no *noiſe* nor *ſteams*. *Egg-ſhells* make a *Bullition* and *huff*, with ſome *noiſe*, but no *ſteams*. *Snail-ſhells* make an *Effervefcence* with *noiſe* and *ſteams*. *Oyſter-ſhells make* one with the grateſt *noiſe* and thickeſt *ſteams*. Hence we may judge, in what caſe to adminiſter *one* more appoſitely than *another*. As alſo *in what proportion*, according to their different ſtrength. Some may be better for *Children*, as being milder. Or for a Body whoſe very ſharp *Blood* or other *Humors*, are more eaſily kindled into *Ferments*. Or elſe may be ſafeſt, to avoid a ſudden *precipitation* of the *Humors*; or for ſome other cauſe.

10. §. *Oyſter-ſhells*, and the *reſt above-ſaid*, make a quicker *Effervefcence*, not only *with Spirit of Nitre*, but even with *Spirit of Salt*, than they do with *Oyl of Sulphur*, or *Oyl of Vitriol*. So that *theſe bodies*, as well as *Metals*, have their proper *Menſtruums* whereby they are be diſſolved.

11. §. *Egg-ſhells calcined*, make with *Oyl of Sulphur*, or *Oyl of Vitriol*, or *Spirit of Nitre*, a greater *Effervefcence*, than when *uncalcined*. As alſo with *ſteams*, which *uncalcined*, they produce not. The like is ſeen in calcined *Oyſter-ſhells*. And the longer the *Calcination* is continued, the quicker and ſtronger will be the *Effervefcence*. This I tryed at ſeveral terms, from a quarter of an hour, to five hours. So that after ſo long a *Calcination*, they make an *Effervefcence* almoſt *inſtantaneous*. The reaſon hereof is, Becauſe the ſeveral *Principles* whereof the *Shells* conſiſt, being relaxed, and the *Sulphur* for the greateſt part, driven away by the fire; the remaining *Salt* lies now more open and naked to the attaque of the *Menſtruum*, ſo ſoon as ever they are mixed together. From hence it is plain, That *Egg-ſhells*, and the others above-ſaid, being *burnt*, are far *ſtronger* Medicines, than when *unburnt*. It is hereby likewiſe evident, That a great portion of their *Salt*, is not a *volatile*, but a *fixed Alkaly*. To theſe may be ſubjoyned all kinds of ſhelly *Inſects*. I will inſtance in three or four.

12. §. And firſt *Bees*, with *Oyl of Vitriol*, ſtir not in the leaſt. With *Spirit of Nitre* they make an exceeding ſmall *Bullition*, without any elevation.

13. §. *Cochinele* (the *Neſt* of an *Inſect*) makes ſome *Bullition* with *Oyl of Vitriol*, but very ſmall: for the bubbles are not to be ſeen without a *Glaſs*. But with *Spirit of Nitre* the *Bullition* is more viſible, and joyned with ſome elevation.

14. §. *Cantharides* make no visible *Bullition* with *Oyl of Vitriol*. But with *Spirit of Nitre* they do, and huff up rather more than *Cochinele*. Yet is this done very slowly, and comparatively with many other bodies, is not much. Hence it is not the *quantity*, but the *quality* of their *volatile Salt*, which makes them so strong an *Epispastick*. For most of those Bodies above, and hereafter named, make a greater *Bullition*, and yet are neither *Caustick* nor *Epispastick* in the least. It is hence also evident, as hath been before suggested, That there are divers kinds of *Volatile Salts*, eminently different; some being highly *alkaline*, others very *little*, and some scarce any thing so: such as those of *Scurvy-grass*, *Anemone*, *Crowfoot*, and many the like *Plants*; to whose *Salts*, this of *Cantharides* seemeth to be very near of kin.

15. §. *Millepedes* make a *Bullition* and *huff, much greater and quicker*, than any of the *Insects* above-named: and that both with *Spirit of Nitre*, and *Oyl of Vitriol* it self. Yet is this *Insect* of *a very temperate nature*. Whereby is further demonstrated, That the being *simply alkaline*, is not enough to make a body to be *Caustick*.

16. §. Again, although *Millepedes* make a *Bullition*, greater than any of the *Insects* above named: yet is it much less, than that of *Oyster, Snail*, or even *Egg-shells*; and of divers other bodies above, and hereafter mentioned. Hence, being given to the same intent, as any of those bodies; it is the mildest and gentlest in its operation of them all.

17. §. *Millepedes* likewise calcined, makes a *stronger Effervescence*, than when uncalcined, as do the *Oyster-shells, &c.* So that it appears, That *all Testaceous salts*, are at least in part, *fixed Salts*.

18. §. I next proceed to Bones. And first *Whale-bone* maketh no *Bullition* at all with any acid. A *Cartilage*, with *Spirit of Nitre*, makes some very small bubbles, not to be seen without a *Glass*.

19. §. The Bone in the Throat of a *Carp*, makes a *little and slow Bullition* with *Spirit of Nitre*. The *Spina* of a *Fish* (that which I used was of a *Cod-fish*) maketh a *Bullition* one degree higher.

20. §. All sorts of *Teeth*, as *Dogs*, *Boars*, the *Sea-horse*, *Elephant*, make the like. As also the *Bone of an Oxes heart*. So that all these are very gentle in their operation, and fit for *Children*.

21. §. *Sheeps* and *Calves* Bones both of them make a *Bullition* yet a little higher, especially with *Spirit of Nitre*. *Cocks Bones* somewhat higher than the former. *Cranium humanum* a little higher than all the rest.

22. §. *Bones* likewise, being calcined, make a *Bullition* with *Acids*. And so doth also calcined *Harts-Horn*. But in neither of them, is the *Bullition* advanced by Calcination, any thing comparable to what it is in *shells*. Whence it appears, That the *Salt of Horns* and *Bones*, is much more *volatile*, than that of *shells*.

23. §. Next for *Flesh* and the several *Viscera*. And first, dryed and powdered *Mutton*, with *Oil of Vitriol*, stirs not at all. But with *Spirit of Nitre* makes a small *Bullition* and *huff*. *Sheeps Heart* doth the like somewhat more apparently. *Vipers flesh* produceth a froth, but huffs not, Powdered *Earthsworms* make a great froth, and huff a little. Powdered *Tripe* makes only a little *Bullition*. *Lamb-stones* do the like. *Kidney*, *Spleen*, and *Liver*, with some elevation. *Lungs*, with bubbles very large; because extraordinary slowly. Dryed *Brain* makes also a little

and flow Bullition. Hence there is a greater proportion of *Sulphur* or *Oyl*, and less of *an Alkaly in all these parts*, than there is in *Bones, Shells*, and divers other *parts* hereafter mentioned. And in some of them, as in the *Brain*, that *Alkaline Salt* which there is, may rather be lodged in some *sanguineous* parts mixed with them, than in their own proper substance.

24. §. I proceed to instance in all sorts of Animal *Contents*. And *first, raw Silk*, with *Spirit of Nitre*, *makes a very small Bullition*, but *the elevation is considerable*.

25. §. *The grumous part of the Blood dryed, with Oyl of Vitriol, stirs but little.* But with *Spirit of Nitre* it huffs up considerably.

26. §. *Serum of Blood dryed, with the same Spirit makes a plain elevation, with a little Bullition.* Herewith may be reckoned the *White of an Egg*, which is nothing but a pure *Crystalline Serum separated from the common stock*. This being dryed, with *Spirit of Nitre*, huffs up rather more than even the grumous part of the *Blood*, the *Bubbles* are much larger, break oftner, and the elevation sooner made. Whence it seemeth, that there is a greater quantity of a *volatile Alkaly* in proportion to the *Sulphur*, requisite to the *Generation*, than to the *Nutrition* of an Animal.

27. §. *The Yelk of an Egg is scarce moved with Spirit of Nitre, producing only a very few Bubbles.* The *Salt* being either little *alkalizate*, or else *immersed* in so great a quantity of *Oyl*, that the *Menstruum* cannot reach it. For the same reason *Sperma Ceti stirs not with any Acid*. Neither doth *Civet*.

28. §. *Russian Castor, with Oyl of Vitriol, stirs not.* But with *Spirit of Nitre* makes *a considerable huff and froth. Yet it requires time.* Wherefore it seemeth, That *Castor* by virtue of its *alkaline Sulphur*, becomes so good a *Corrector* of the *acid-alkaline Sulphur* of *Opium*: so I take leave to call it, having some reasons to believe it such.

29. §. *Musk, with Oyl of Vitriol, stirs not.* But *with Spirit of Nitre it makes a considerable and quick Bullition, with large bubbles, which often break and rise again.* Whence there is a very eminent difference betwixt *Musk* and *Civet*. Hence also, *Musk* is Cordial, not only from its *Sulphur*, but its *Alkaly*; by both directly opposite to preternatural *Acidities*.

30. §. *Dryed Gall with Spirit of Nitre, for some time, is still: but at length it makes a considerable Bullition and froth.* The reason why it is so long before it begins, is because the *Salt*, (as was observed of some other *Parts*) is locked up in so great a quantity of *Oyl*. The abundance whereof is manifest, not only from *Destillation*, but also from hence, In that the dryed *Powder*, in lying by, incorporateth all together into one body, as *Mirrh*, and some other softer and oily *Gums* are used to do.

31. §. *Extract of Urine, with spirit of Nitre, makes a Bullition with some Effervescence, which continues for a considerable time; and at last it huffs up with great bubbles. The Bullition begins presently: the Salt being copious, and the Oyl but little.*

32. §. *The same Extract of Urine makes a considerable Bullition and froth, not only with Spirit of Nitre, but even with Oyl of Vitriol.* Hence the *Salt of Urine* is more *alkaline* than that in most of the aforesaid *Contents*.

Contents. From this and some of the following *Experiments*, it also appears, That the *Salt* which concurs to the generation of *Gravel* or of a *Stone* in the *Kidneys* or *Bladder*, is of a very different nature from the *Salt* of *Urine*.

33. §. Next for *Dungs*. And first, dryed *Goats-dung* makes with *Spirit of Nitre*, a small *Bullition*, but no elevation. That of *Mice* the like. And that of *Cows*. So that of all I have tryed, these three stir the least.

34. §. *Goose-dung*, with *Spirit of Nitre*, makes a very small *Bullition* and some elevation. But it requires time. *Oyl of Vitriol* stirs it not.

35. §. *Album Græcum*, with *Spirit of Nitre*, besides innumerable small bubbles, rises up with some great ones, exactly resembling the huffing up of *Yest* or *Barm*. Also with *Oyl of Vitriol* it maketh some little froth, but slowly. So that it should seem, that the *Bones* are a little opened by some *acid Menstruum* in the Dogs stomach (as the body of *Steel* is in its preparation with *Sulphur*) whereby it becomes a good mild *Topick* in *Quinzies*.

36. §. Hens dung, with *Spirit of Nitre*, makes a very great *bullition* and huff: greater and quicker, than any of the rest above-named.

37. §. But of all I have tryed, *Pigeons dung*, with the same *Spirit*, maketh the greatest and the quickest *Effervescence* and huff; and that not without steams. Yet neither the same *Dung*, nor that of *Hens*, is moved in the least with *Oyl of Vitriol*. The Cause of so great an *Effervescence* in these, more than in the rest, is that *white part* which is here mixed in a great quantity with the *Dung*. Which *white part*, descendeth not from the *Stomach*, but is an *Excrement* separated from the *Blood* (as the *Urine* in other *Animals*) by a peculiar *Organ*, which evacuates it into the *Intestinum rectum*; whence, together with the *Stercus* it is excluded. Hence it is evident, That in the said *white part* of *Hens*, and especially *Pigeons dung*, is contained a great quantity of a *volatile Alkaly*.

38. §. I proceed to *Salts*. And first *Salt of Blood* and *Urine* both make a more durable *Effervescence* with *Acids*, than doth *Salt of Wormwood*, or *Salt of Fern*. Hence the former are more *alkaline*, than the latter.

39. §. Again, though divers other *Animal Salts* will not stir with *Spirit of Salt*, or with *Oyl of Sulphur* or *Vitriol*; yet the *Salt of Blood* will make an *Effervescence* with all kinds of *Acids*. Whence it is further argued to be highly *alkaline*, and very proper for the *correction*, of all sorts of *preternatural Acids* in the body. There is little doubt, but that *Spirit of Harts-horn* will do the like.

40. §. The *Gravel* which is precipitated out of *Urine*; with *Oyl of Vitriol* makes no bullition in the least. Nor with strong *Spirit of Salt*. But with *Spirit of Nitre*, it makes a very great one, with *Effervescence* and steams. From hence it appears, That there is much difference to be made in the use of *acid Diureticks*, *Nephriticks*, &c.

41. §. And that I may not altogether omit to mention, what may be so much for the good of mankind, I do here declare, That for preventing (I say not, the breaking, but preventing) the generation of the *Stone*, either in the *Kidneys*, or in the *Bladder*, there are not bet-

ter *Medicines* in the world, than some certain *Preparations* of *Nitre*, duly administred. Whoever shall think that any kind of *acid*, as *Oyl of Sulphur*, *Oyl of Vitriol*, *Spirit of Salt*, or the like, will have the same effects, will find themselves much deceived in their practice.

42. §. *I conclude with Stones.* And first, *Spirit of Nitre* droped upon a *Stone* of the *Kidneys* or *Bladder*, produceth the very same effect, as upon the *Gravel* in *Urine*. That is to say, it makes it boil and huff up, until at length it is perfectly dissolved into a soft *Pulp*; which neither *Oyl of Sulphur*, nor *Oyl of Vitriol*, nor *Spirit of Salt* will do; nor give the least touch towards its dissolution. This confirms what I said before of the use of *Nitre* and *Nitrous Spirits*, if duly prepared and administred, above any other *Acids*, against the breeding of the *Stone*.

43. §. *Pearls*, with any *Acid*, make the like *Effervescence*, as do *Oyster-shells*. But *Magistery of Pearls*, as usually prepared, stirs not at all, with any *Alkaly* or *Acid*. So that as to the effect frequently intended by it, it is very insignificant; as of that of *Corals* hath been said.

44. §. *Crabs Eyes*, with any *Acid*, make an *Effervescence*, almost as quick as that of *Oyster-shells*.

45. *Crabs Eyes* likewise calcined, make a stronger *Effervesnence*, than when uncalcined. So that these, as well as *Shells*, contain a *fixed Alkaly*.

46. §. The *Stones* in *Whitings heads* make a strong *Effervescence* like that of *Oyster-shells*.

47. §. *Stone of humane Gall*, stirs not with *Oyl of Vitriol*. But with *Spirit of Nitre* maketh a little bullition just upon mixing, and after a considerable time, a little froth. Much less than what was observed before of the *Gall* it self. So that it seemeth to be generated of the *Gall* coagulated by some *Acid*, which hath already *refracted* the *Alkaly* wherewith the *Gall* abounds. This confirms the use of those *Medicines* in the *Jaundies*, or any other bordering *Disease*, which destroys those *Acidities* by which the *Gall* is *curdled* or *coagulated*, and so rendred more difficulty separable into the *Guts*.

48. §. Since the first publishing of these Observations, Mr. *William Matthews* an *Apothecary* in *Ledbury*, sent me part, as I take it, of a *Stomach-stone*, as big as a *Wallnut* of the largest Size, voided by a woman about 82 years of age, sometime after an *Autumn Fever*. It consisteth of the same *Striæ*, as the *Bezoar Stone*; and maketh some *Bullition* with *Spirit of Nitre*.

49. §. *Bezoar*, neither the *Western* nor the *Eastern*, doth stir at all with *Oyl of Vitriol*.

50. §. *Western Bezoar*, with *Spirit of Nitre*, makes a very little thin froth, and that's all; and that it doth very slowly. But *Oriental Bezoar*, with *Spirit of Nitre*, after some time, maketh a very great *Effervescence*, froth, elevation, noise, and steams (as if you poured *Oyl of Vitriol* upon *Salt of Tartar*) till it be wholly dissolved by the affused *Spirit*, and turned into almost a blood-red. Hence it may seem to be no mean *Remedy* against such *fretting* and *venenate acids*, as oftentimes in *Fevers*, and other *Distempers*, lye about the *stomach*, and are thence frequently translated to the *Heart, Brain, Nerves*, and other *parts*. The difference likewise betwixt the *Western* and the *Eastern Bezoar*, is so great, that in any case of danger, and where the *Bezoar* is relyed upon, it is an unpardonable

able fault, for the *Apothecary*, or any *Perſon*, to *ſubſtitute* the one for the other: unleſs he will take *ten times* as much, or *ten times* as little of the one, as he would have done of the other: if that will ſerve turn.

51. §. The *Stones* already mentioned, (except the great *Stomach-ſtone*) are ordinarily generated in the bodies of *Animals*. I have one Inſtance more of ſome other *Stones* which are extraordinary. In the City of *Hereford* lives a Maid, who often voids theſe *Stones*, and in the ſpace of ſome years laſt paſt, hath voided ſeveral pounds, of ſeveral *Colours* and *Sizes*, not only *per vias urinarias*, but alſo by vomit, and by ſtool. The firſt mention made to me of them, was by Mr. *Diggs*, a worthy Gentleman of that City, as a thing that was there much wondred at. And ſome of them, upon my deſire, were ſent me by Mr. *Wellington*, an Apothecary in the ſame place. *I have tryed what ſeveral acid Menſtruums will work upon them; and find, That with Oyl of Vitriol, and eſpeially with Spirit of Nitre the great ones make a very quick and conſpicuous Effervescence. But the ſmall ones, neither the white, nor the grey, make any Bullition in the leaſt:* for in truth, they are no other but little *Pebbles* and *Grit-ſtones*.

52. §. This being conſiderd, and the various *colours* and *mixture* of any one of the great *Stones*, being well obſerved; it ſeemeth plain, That although ſhe be ſomewhat old (above thirty years) yet may ſhe have a kind of μαλακία, or *diſeaſed Appetite* to *Stones*, *Bones*, *Wood-aſhes*, *Tobacco-Pipes*, *Chalk*, and ſuch like things; which ſometimes ſwallowing in little *lumps*, ſometimes groſly, or finely ground betwixt her teeth; they are in her *Stomach* and *Bowels*, more or fewer of them, *cemented* together, either with a *pituitous*, *bilious*, or ſome other more or leſs *glutinous ſubſtance*. And that by virtue alſo of the ſaid *Cement*, or any of the ſaid, or other like *alkalizate Bodies*, the greater *Stones*, which conſiſt of thoſe partly, do make an *Effervescence* with *acid Liquors*. Thus far of *Inſtances* upon the *parts* of *Animals*. I ſhall cloſe with ſome *Corollaries* deduced from the whole.

53. §. And firſt, ſince we find, that amongſt all the *Menſtuums* we have made uſe of, *Spirit* of *Nitre*, or any very *Nitrous Spirit*, is the moſt *univerſal diſſolver* of all kinds of *Animal Bodies*; the *beſt diſſolver* of many others both *Vegetable* and *Mineral*, and the *only diſſolver* of ſome: Hence it is probable, That the great *ſtomachick Menſtruum*, which either *diſſolves*, or *opens* almoſt all *Bodies* which come into the *Stomach*, is a kind of *Nitrous Spirit*.

54. §. Again, *Spirit* of *Nitre* being a *ſubalkaline Acid*, and working more evidently upon *Animal bodies*, than other *ſimpler Acids* do, which yet are as *ſtrong*; It hence follows, That moſt of the *Salts* of *Animals* are *ſubacid Alkalies*. How far this *concluſion* may further inſtruct us, I ſhall have occaſion to ſhew in another *Diſcourſe*.

55. §. Laſtly, there being ſo many, ſay twenty or thirty *degrees*, from the *ſloweſt* to the moſt *vehement*, in the *Bullition* of *mixed Bodies*; it ſeemeth, That *Fermentation* it ſelf, as to the *formal notion* of it, is

nothing

nothing elfe: or that from the common *Luctation* of *mixed Bodies* whereof we have now been fpeaking, it differs not *in fpecie*, but only in the manner of its *caufation*, and in *degree:* the *Aer*, or fome certain *Menftruum* lodged therein, being of no greater *ftrength*, than to produce a *Bullition* or *Luctation* of that *low* and *foft degree*, which we call *Fermentation*.

56. §. I have thus endeavoured to prove, by various *Inftances*, how inftructive this moft eafie, plain and fimple *Method* in the *Mixture* of *Bodies*, may become to us: and that meerly by obferving the *Luctations* which thence arife betwixt them. How much more then, if a diligent remarque be made of all thofe various *Colours*, *Smells*, *Taftes*, *Confiftencies*, and other *Mutations* thereupon emergent?

AN ESSAY

OF THE

Various Proportions

Wherein the

LIXIVIAL SALT

Is found in

PLANTS.

Read before the *Royal Society*, March, 1676.

CHAP. I.

Of the QUANTITIES afforded by several Plants calcined in gross.

IT is the part of a *Physician*, knowingly and artificially to use and govern *Nature*. And therefore by every likely *Method*, to inspect the *State* and *Properties* of all sorts of *Bodies*. One *Method*, is that I have taken in the foregoing *Experiments*; *sc.* by mixing them with several *Menstruums* or *Liquors*: whereby we may be assisted to judge, both of the *Kinds* and the *Proportions* of *Principles* in any *Body*; and of the manner of their *Mixture* in the same.

Another is by *Calcining* them; or, as it were, by *mixing* them with the *Fire*, a potent and almost universal *Menstruum*. I shall here only set down some *Tryals* for an *Essay*, upon *Plants*; chiefly noting, The different *Proportions* of their *Lixivial Salts.* Of these *Tryals*, some

were

were made upon the whole *Plant*, or some *Portion* of it wherein several *Parts* are mixed together: And others, upon some one *Part* of a *Plant* distinct from the rest. All of them answering to such *Queries*, as may seem proper to be proposed.

Query 1. As first, *Whether Trees or Herbs and Bushes, quantity for quantity & cæteris paribus, yeild the most Lixivial Salt?*

For this I took *Ash-Barque* and *Rosemary* of each ℔j. The latter yielded 5 *Scruples*; the former but 32 *Grains*; which is three times less. I took also the same quantity of the *Barque* of *Black-Thorn*, and of *Agrimony*. The latter yielded 5 *Scruples* and 6 *Grains*; the former, not above 1 *Scruple* and 5 *Grains*; which is four times less.

Although the *Barque* of a *Tree* be compounded of *Pithy* and *Lignous Parts*; yet to answer the *Query* exactly, the *Wood* of these *Trees* should be taken with the *Barque*, that there may be some portion of every *Part* of the *Tree*, as well as of the *Herb*.

But thus far the *Experiment* is *conclusive*, That the same quantity of *Lixivial Salt*, doth not always follow the same *Generical Tast*. For the *Barque* of *Ash* and *Rosemary*, are both equally *Bitter*; and the *Barque* of *Black-Thorn* and *Agrimony* are both *Astringent* and *Bitter*.

Quer. 2. *Whether any Plant growing in a Garden or the Field, doth not yield a lesser quantity of Lixivial Salt, than another of the same kindred growing on the Sea-Coast; and with what difference?*

For this, I took *Garden* and *Sea-Scurvygrass*, of each ℔j. The former yields 2 *Drachms* and 1 *Scruple*; the latter, being well washed, 9 *Drachms*, which is more than 4 times as much. The like may be tryed upon others.

Quer. 3. *Whether the same specifick Plant affords more Lixivial Salt, being only dryed, and then calcin'd, or after it hath first been distilled, it is then dryed and calcin'd?*

For this, was taken ℔j of *Mint* only dryed and then calcin'd; and another first distilled. The former yielded $\frac{1}{2}$ an Ounce and $\frac{1}{2}$ a Drachm of *Salt*; the latter, 5 Drachms and a Scruple; which is almost $\frac{1}{5}$th more. This also should be tryed on other *Plants*.

Quer. 4. *How far the proportion follows the different Tasts of Plants?* The first Experiment, relates to the same *Tast* in several *Plants*; this, to several *Tasts*. And so,

Of *Majorane*, which is *Aromatick*, ℔j affords but one Scruple of *Lixivial Salt*; which is but the 384th part of the whole pound.

Of *Oak-Barque* which is *Astringent*, ℔j yields $\frac{1}{2}$ a Drachm of *Salt*; or the 256th part of the whole.

Of *Liquirish*, which is sweet, ℔j yields about the same quantity. But *Anise Seeds* ℔j yields 2 Scruples or a 192d part.

Of *Sorrel*, which is sower, ℔j yields one Drachm, or the 128th part.

Of *Garden Scurvygrass*, which is *Hot*, ℔j yields 2 Drachms and $\frac{1}{2}$ a Scruple; or the 59th part.

Of *Mint*, which is *Hot* and *Bitter*, ℔j yields 5 Drachms and a Scruple, or the 24th part.

Of *Sea Scurvygrass*, which is *Salt*, ℔j yields 9 Drachms and a Scruple or 28 Scruples; which is near $\frac{1}{13}$th part of the whole. A greater proportion of *Salt*, than in any other *Plant* upon which I have hitherto made Tryal: Or even in *Tartar* it self. Yet is it not a *Marine*, but true *Lixivial Salt*: as is evident, both from its *Taste*; and in that it

maketh

maketh an *Effervescence* with *Spirit* of *Salt*; which *Sea-Salt* will not do.

For the Experiment to be fully adequate to the *Query*; the *Tryals* shouldbe made, either all on *Trees*, or all on *Herbs*; all on *Roots*, or all on *Stalks*, &c. Yet thus much is evident, That *sorrel* yields Thrice as much as *Majorane*; *Sea-Scurvygrass*, Eight and Twenty times as much: *Mint*, Five times as much as *Sorrel*; and Sixteentimes as much *Majorane*, &c.

Quer. 5. *How far the Proportion follows the Faculties of Plants?* And so, it appears, that

Majorane, a *Cephalick*, hath a greater *Proportion* of *Volatile Parts*, than any of the *Plants* above mentioned, and so far, is more agreeable to the *Animal Spirits*, and *Genus Nervosum*.

Agrimony, (a) an *Aperient*, yields above Five times as much *Lixivial Salt*, as *Majorane*. Yet much less than many other opening *Plants* which are stronger. (a) *Quer.* 1.

Mugwort (℔j) yields two Drachms and two Scruples; or above half as much more as *Agrimony*. So that this *Plant*, though it hath no considerable *Taste*, and in that respect promiseth but little; yet yielding a good quantity of *Lixivial Salt*, seems no contemptible *Medicine* to subdue those *Acidities* which either by causing *Obstructions*, or immoderate *Fermentations*, frequently disorder the *Female Sex*.

Mint, yieldeth still a greater quantity; and is therefore, partly for the same cause so excellent a *Stomachick*: And *Rosemary*,(b) which is appropriated both to the *Head* and *Stomach*, yieldeth a midle quantity of *Salt*; more than the chief *Cephalicks*, and less than the chief *Stomachicks*. (b) *Quer.* 1.

Common Mallow (℔j) yields 5 Drachms and 2 Scruples. *i. e.* the 23ᵈ part of the whole. So that this *Plant*, though of a very mild *Taste*, yet yields more *Salt* than *Mint* it self a *Bitter Plant*. Whereby it no longer seems strange, that a *Plant* of so soft a *Taste*, should be very *Diuretick*, and so evidently affect the *Reins*.

Rhubarb (2 Ounces) yieldeth scarce any fixed *Salt*, so far as can be judged by the *Taste* of the *Ashes*, not more than a Grain or two. So that its *Salt* is, in a manner, wholly volatile; and thereby apter to operate upon the *Bilious* parts of the *Blood*; which contein a far greater proportion of *Volatile Salt*, than do the *Serous*.

Of the *Caput Mortuum* or meer *Earth*, it is observable, that it was near $\frac{1}{2}$ an Ounce or $\frac{1}{4}$th part of the whole; Which is almost Six times as much as the *Caput Mortuum* of *Common Dock*: and much more than that of any other *Root* I have yet calcin'd. Whereby it seemeth probable that *Rhubarb* looseth much of its *Volatile Part*, and therefore of its *Virtue*, before it comes to our *Shops*.

Sena (℔j) yields 4 Scuples and $\frac{1}{2}$ of *Salt*; or the 85th part.

Jalap (℔j) yields but one Drachm and 15 Grains, or 102ᵈ part.

Colocynthis (℔j of the *Pulp*) yields an Ounce and half of *Caput Mortuum*, which is almost all *Salt*. Yet allow half an Ounce of the *Salt*, and *Earth* to be wasted in filtring &c. the remaining Ounce is no less than $\frac{1}{8}$th part of the whole. Which is more than in any of the above named *Plants*, except the *Sea-Scurvygrass*.

CHAP. II.

Of the QUANTITIES afforded by the Parts of several Plants distinctly calcin'd.

I SHALL next set down some *Tryals*, upon one *Part* of a *Plant*, as well *Organick*, as *Content*, separated from the rest; in answer to these supposed *Queries*.

Quer. 1. *What Proportion doth the Lixivial Salt of the Pith or Pithy Part of a Plant, bear to that of the Fibrous, or of the Woody Part? Or whether is there a Fixed Salt always found in either of them?* A sufficient *Answer* to which, must be built upon many *Tryals*. At present I shall mention only Two; one upon *Starch*, answerable to the *Pithy Parts*; the other upon *Flax*, consisting almost wholly of the *Nervous* or *Towy Fibres*: of the *Volatile parts* whereof, chiefly, I have given some account in the foregoing *Idea*.

§. 50, 51, 52.

Of *Starch*, ℔j yieldeth about ℔¼ not of *Ashes*, but of *Black Coal*. For though it be exposed in a *Calcining Furnace* to a vehement *fire*, for 5 or 6 hours, which is longer then will serve to calcine most *Bodies*: yet would it not in the least part, be reduced to *Ashes*; but to the last continued (though the fierceness of the *Fire* consumed part of it) as *black*, as when it was first burnt. So strangely was the remaining part of the *Sulphur* fixed to the *Earth*; that in flying away, it did volatilize and carry that away with it. In this *Coal* or *Cinder*, there is not the least of a *Lixivial* or other *Taste*. And although, upon *Tryal* I find, That the *Pith* of many *Plants*, as of a *Cabbage Stalk*, will yield some quantity of *Lixivial Salt*; yet it is probably, that generally, it yields less than the *Wood*.

Of *Flax*, ℔j yields not above 50 Grains of *Caput Mortuum* or *white Ashes*, which are *Salt*. According to vulgar conceit, it would seem to be a very dry Body: yet of 153 parts, 152 are *volatile*, and being distilled would have been collected into *Liquor*. Hence also appears the great and unexpected *Variety* in the *Proportion* of the *Earthy Parts*, as well as the other *Principles* of *Bodies*. Or else, that there are divers kinds of *Earths*, even in *Plants*, of which, as well as of *Salts* &c. some are *volatile*. For of ℔j of this *Plant*, there remaineth fixed but 50 Grains: whereas of ℔j of *Rhubarb*, there will remain near 1920 Grains, *i. e.* 88 times as much as the former.

Quer. 2. *In what proportion is the Lixivial Salt found in the Gumms of Plants? and whether is it yielded, more or less, by all?* For answer to which, I caused the Eleven following, of each two Ounces, to be calcin'd, and so observed,

That *Common Rosin*, yields but one Grain and ½ of *Caput Mortuum*. So that ℔j will yield but 12 Grains. In this *Caput Mort.* there is not the least particle of *Salt*, it being altogether insipid.

Mastick yields gr. 12 of *Cap. Mort.* But not the least part of *Salt*. Of this *Rosin*, it is observable; That being set, in a *Crucible*, within the

the fire, before it comes to have thick fumes, it boyls up with a very great foame or froath; and is the only *Gum* or *Rosin* (of the Eleven) that hath this property. So that I suspect, there is a great quantity of some kind of *volatile Spirit*, which then flies away; and so, in breaking through the *Oyly parts*, huffs them up to so great a froath.

Olibanum yields half a Drachm of *Caput Mortuum*. But it is to be noted, That the weight is encreased by certain little *Spar-Stones*, which in the burning of several parcels, I always found mixed with this *Gumm*. These being picked clean out, the *Cap. Mort.* weigheth not much more than that of *Mastick*. And is in like manner insipid, when the said *Stones* are picked out.

From hence it appears, how proper these *Gums* are for the *Concoction* of *Salt Rheums*; according to what I have formerly suggested from another Experiment. *Discourse of Mixture Cap. Ult.*

It may also be noted, that *Rosin* and *Mastick*, seem to be more purely *Acidoleous Gums*; not only from their consistence which is uniform; and their *Smell*, which is less strong and more pleasant: but also from the *Acid Liquor* they yield by *Distillation*; and in that the young *Leavs* of *Fir*, and especially of *Pine*, are sower; and tis probable that those of *Mastick* are so likewise. Whereby these, and other like *Gums* are more especially fitted for the abovesaid purpose. But *Olibanum* seems, besides its *Acidity*, to contein some *Volatile Alkaly*, and so to be an *Acid-Alkaline Gum*. For as it hath a stronger *Smell* than the former, so a hotter *Taste*; both the ordinary effects of an *Alkaline Sulphur*. And being infused in several *Menstruums*, appears to consist of two Bodies, one of them more *Resinous* than the other. Of which, it is probable, that the one is made by the *Acid* parts as the other by the *Alkaline*. Whereby it is very well adapted in some *Cases*, as in a *Pleuresie*, for removing the *Coagulations* of the *Blood*, or its disposition thereunto.

Asa fœtida yileds no less than half its weight or an Ounce of *Caput Mort.* that is 8 times as much as that of the other *Gumms*, and 48 times as much as that of some of them. Yet doth it not contein one grain of *Salt*, so far as can be judged by its *Tast*. Yet the *Strength* and *Loathsomness* of the *smell* and *Tast* of the *Gumm* do argue it to be highly impregnated with some kind of *Volatile Alkaly* proper to arrest those offensive *Vapours* (to use the vulgar word) which flying, either by the *Blood* or *Nerves*, from part to part, do often prove so troublesome.

Gum Arabick yields one Scruple of *Cap. Mort.* whereof, by the *Taste*, about $\frac{1}{3}$d part is *Salt*.

Euphorbium yields one Drachm of *Caput Mort.* of which, by the strength of the *Taste*, two Scruples seem to be *Salt*. Which confirms a former conjecture (*a*) of its being an *Alkaline Gumm*. *(a) Of the Luctation of Bodies, Ch. 1.*

Myrrh also yields a Drachm of *Cap. Mort.* and at least two Scruples of *Salt*. Of the Eleven, these two *Gums* have the greatest quantity of a fixed *Alkaly*.

Opium yields half a Drachm of *Cap. Mort.* whereof the one half is *Salt*.

Aloe yields a Drachm of *Cap. Mort.* conteining about one Scruple of *Salt*.

Scammony

Scammony yields Two Scruples of *Cap. Mort.* of which, about half a Scruple is *Salt*.

Gutta Gamba yields but half a Scruple of *Cap. Mort.* of which four or five Grains are *Salt*.

So that confidering the *Dofe* of any *Cathartick Gumm*, the quantity of the Fixed *Alkaly*, is extream fmall with refpect to the *Volatile parts*: In which, therefore, its *Crthartick Power* doth chiefly refide.

Yet none of the *Cathartick Gumms* are without fome portion, more or lefs, of a Fixed *Alkaly*; though fome of the reft are. Which feemeth to prove, That the Fixed it felf, hath fome Intereft in the bufinefs of *Purgation*: as by being a *Clog* to the *Volatile*, and fo preventing its being deleterious; or fome other way. But the manner of their *Operation* will better be underftood, when the *Volatile Parts* have likewife been examined.

It may alfo be of good import, to know, what different quantities of *Salt*, are afforded by the *Tartars* of all forts of *Wines* Whereby, partly, as well as by the quantity of the *Tartar*, we may be enabled the better to judge of the *Nature of Wines*.

A DISCOURSE

Concerning the

ESSENTIAL and MARINE

Salts of Plants.

Read before the *Royal Society*, *December* 21. 1676.

CHAP. I.

In which is shewed the way of making both an ESSENTIAL and a MARINE Salt, out of the LIXIVIAL Salt of a Plant.

OMETIME since, I took the boldness to present my thoughts to this Honourable and Learned Body in a *Discourse* concerning *Mixture*. Wherein I have endeavoured to lay such a *Foundation*, as might hereafter reduce the *Doctrine* hereof to *Experience* and *Practise*; and to demonstrate, the *Power* and *Use* of *Artificial Mixture*. And in further proof of what is therein asserted, I have since made a continuation of *Experiments* upon the same Subject, in Two *Methods*. One in the *Mixture* of several *Menstruums*, both *Acid* and *Alkaline*, with all Sorts of *Bodies*. The Other, by calcining them, or, as it were, *mixing* them with the *Fire*.

2. §. I shall now proceed to a Third, which is, the *mixing* them with the *Aer* or exposing them to it; another of Natures grand *Menstruums*; which goes sometimes further than the *Fire* it self, in the dissolution

Discourse of Mixture Ch. 5. Inst. 2.

solution of Bodies. This I have formerly mentioned for the Imitation of *Nature*, in producing a *Marine* or *Muriatick Salt* out of the *Lixivial Salt* of a *Plant*. But some Learned Persons then present, seeming to doubt of the Experiment; I thought it requisite to prosecute the same a little further; that so, if possible, it might become clear and unquestionable. And because that *Method* was imperfect, and required half a year, or a longer time: I bethought my self of an other way; which proved far better, and more expedite. And which, withall, afforded me, not only a true *Marine Salt*, out of the *Lixivial Salt* of a *Plant*; but also another kind of *Salt*, different from them both: which may not be improperly called, an *Essential Salt* or *Nitre of Plants*. The History or manner of the production of them both, is as follows.

3. §. *December* 15. 1675, I took about half a pound of a strong *solution* of the *Lixivial Salt* of *Firne*: and pouring it into an *Earthen Pan*, well glazed, broad and shallow, exposed it therein to the open *Aer*, in a Chamber Window, to evaporate of it self.

Tab. 83.

4. §. This *Solution* or *Lee*, although it was very clear before, and having stood corked up in a bottle many days, had no sediment: yet standing now in the open *Aer*, within the space of 4 or 5 days, it began to let fall a very *white sediment*, like fine *Chalk*; which encreased daily for 8 or 10 days; amounting at last to about half a Drachm of white, light and meer *Earth*, altogether insipid, and when it was well washed, stirring not upon the *Affusion* of *Acids*.

5. §. Within the space of a day or two after this *white Sediment* began to fall to the bottom; there was also gatherd on the top, a kind of soft *Scum* or *Cremor*, wherewith the *Solution* was covered all over.

6. §. Within 8 or 9 days after the first exposing of the *Liquor*, or 2 or 3 days after the gathering of the *Cremor*; that *Salt*, which I take leave to call, an *Essential Salt* of *Plants*, began to appear; shooting into several little *Crystals*. These *Crystals*, as they grew bigger, began to sink, and at last fell down to the bottom of the *Pan*.

Tab. 83.

7. §. Upon their first generation or shooting, the said *Cremor* presently breaks, leaving a bare space round about each *Crystal*; and upon the bounds of every space is indented; the space growing bigger and bigger together with the *Crystal* in the Centre. And so, by that time the *Crystals* are grown to a considerable number and bigness, the *Cremor* vanishes away, the several *Circles* or bare places breaking at last one into another all over the *Surface* of the *Lee*. After which, it never comes again.

8. §. From whence it seemeth, That the several *Circles* or bare *Spaces* about the *Crystals*, are made for the more free admission of the *Aer*, requisite to their *Generation*. For as there is no *Crystal* begins to be formed before there is a breach made in the *Cremor*: so that breach is enlarged together with the *Crystal*. So that as the falling of the *Sediment* and the gathering of the *Cremor*, sheweth that the *Aer*, as a *Menstruum* separates some part from the *Lee*: so the breaking of the *Cremor* afterwards, that as a *Vehicle*, it brings something to it: both in order to the *Generation* of the *Crystals*. Nature taking a *Method* for the *Generation* of simpler Bodies, as well as of those which are *Compounded* and *Organical*.

9. §.

9. §. The *Figure* of these *Crystals* is angular and oblong, most of them about the fifth, sixth or seventh of an Inch; but none of them very regular. Yet we are not hence to conclude, but that with the help of some Circumstances which might be wanting in the shooting of these; some portion of regular ones may be obteined from this, as well as other *Lixivial Salts* hereafter mentioned.

10. §. They are somewhat transparent, and of a dark *Ambar Color*, or like that of brown *Sugar-Candy*. Of a quite different *Taste* from that of the *Solution* or *Lee* out of which they are bred; being not at all *Lixivial*, but very weak and mild; not *Salt*, but *Bitter* in a good degree.

11. §. It is also observable, that *Alkaline* and *Acid Salts* being both poured severally upon these *Crystals*, they stir not, nor are any way affected with either of them. So that these *Crystals* are no sort of *Tartar*, or *Tartareous Salt*. As is plain, from the manner of their *Generation*; *Tartar* being still bred in close *Vessels*; these never, but by exposing the *Liquor* to the *Aer*. As also from their *Taste*, being not sower, in the least, but bitter. And in that *Tartar* will make a *Bullition* with *Alkaline Salts*, which these will not do. Upon which accounts it appears, that they are a *Salt* different in Nature from all other *Salts* hitherto known, or a new *Species* added to the *Inventory* of *Nature*.

12. §. These *Crystals* within the space of about a fortnight after their first *Generation*, did also cease to shoot any more, but only increased a little in their *Bulk*. After which time, I dayly expected to see the production also of a true *Marine Salt*. And about two months after the said *Essential Crystals* had done shooting, and not before, this also began to shoot, in many small *Crystals*, and at the top of the *Solution*, as the other did, still falling to the bottom as they grew biger.

13. §. The *Size* of most of them was near that of the *Flakes* or *Grains* of *Bay-Salt*. The *Colour* of some of them *white*, of others *transparent*; and of others *white* in the *Centre*, with transparent *Edges*; as is also usual in the *Crystals* of *Common Salt*.

14. §. The *Figure* of most is a perfect *square*, and of very many coming near to a *Cube*; which is also the *Figure* of *Common Salt*, and seldome an exact *Cube*. An exact *Cube*, being the constant property of no *Marine Salt*, that I know of, except that of the *Dead Sea*. Divers *Tab. 83*. of them were also raised as it were by several steps from a deep Centre to the Top: as is often seen in the common shooting of *Common Salt*; and not in any other. Their *Taste* is neither *Lixivial*, as that of the *Solution* out of which they shoot; nor bitterish, as that of the *Essential Crystals*; nor sowerish, as that of *Tartar*; but the perfect *Taste* of *Common Salt*.

15. §. It is also to be noted, That if *Oyl* of *Vitriol*, and some other strong *Acids*, be poured upon this artificial *Sea-Salt*, they make an *Effervescence* together: but if *Spirit* of *Salt* or *Spirit* of *Nitre* either be poured on it, though it be never so strong it stirreth it not. In both which, and all the formentioned respects, it answers to the *Properties* of a *Marine* or *Common Salt*, which no other *Salt* doth. I conclude it therefore to be a true *Marine Salt* produced by *Art* in the imitation of *Nature*.

CHAP.

CHAP. II.

Wherein is shewed, That the said ESSENTIAL *and* MARINE *Salts of* Plants *are both of different Sorts.*

AVING made the Experiment, that both an *Essential* and *Marine Salt* may be produced out of the *Lixival Salt* of a *Plant*. I thought it probable, that neither the one nor the other, was always the same, but that as they had their general properties which made them to be of two general kinds; so they might have some special property, for the distinguishing of each kind into several Sorts. And withall, that in a warmer season, than before taken, the *Tryal* hereof might be finished in a shorter time.

2. §. For the making of which, I conceived it requisite to remove an Opinion which seemed to lye in my way; *sc.* That there is little or no difference between the several *Lixivial Salts* of *Plants*, as some Learned men have thought. But either there is a difference, or not: if not, it should be proved: and if there be, it should then be justly stated, what that difference is. For the doing of which, I chose this *Method*. I took an equal quantity of the whitest and purest *Salts* of divers *Plants*, all made by an equal degree of *Calcination*; and dissolved them all severally in an equal quantity of water. And pouring likewise an equal quantity, as about 10 or 12 drops of each into a spoon, I tasted them severally. Whereby it was very evident, that they were not all of one *Tast*, but of very different ones, both as to strength and kind: and therefore different in Nature also. The *Salts* I made tryal of were those of *Sorrel, Anise, Wormwood, Mallow, Ash, Tartar* and others: and upon half a Drachm of each I poured ℥ijß of water. The *Solutions* are here present to be tasted. By which the differences will easily be observed, and particularly that the *Salt* of *Wormwood* or *Scurvygrass*, is almost as strong again as the *salt* of *Anise*, or *Sorrel*: and that the *Salt* of *Ash* is above twice as strong, and that of *Tartar* above thrice as strong, as that of *Sorrel*, and almost thrice as strong as that of *Wormwood* or *Scurvygrass*. So that he who shall give half a Scruple, suppose of *Salt of Tartar*; instead of half a Scruple of *Salt of Wormwood*, or other like *Salt*; he may as well give a Scruple of *Rosin* of *Jalap*, for a Scruple of the powder, or almost three Drachms of *Rhubarb*, or other like *Purge*, instead of one. And the like is to be said of other *Lixivial Salts* in their degrees.

3. §. Having observed thus much, I proceeded to repeat the former Experiment, with some of the aforesaid, and some other *Vegetable Salts*, the best calcin'd, and the purest, that could be made for this purpose, being these Six *Salts, sc.* of *Rosemary, Garden Scurvygrass,*
Black

Black Thorn, Common *Wormwood*, *Aſh*, and *Tartar*. All which diſſolved ſeverally in fair water, I expoſed in a Chamber window, and not in Winter, as before, but in the heat of Summer, ſc. on the 19 of *July*, to evaporate of themſelves.

4. §. The Effect was, That the third day after their being expoſed, the *Eſſential Cryſtals* began to ſhoot in three of the *Solutions*, ſc. in that of *Roſemary*, of *Garden Scurvygraſs*, and of *Black Thorn*. On the fouth day, in that of *Wormwood*. On the fifth day, in that of *Aſh*. In that of *Tartar*, not at all.

5. §. Theſe *Eſſential Cryſtals* began, in all, to ſhoot at the top, and then to fall to the bottome; as in the Experiment before. But as there was very little of the *white Sedement* before mentioned, that preceded; So no *Scum* or *Cremor* at all. Which although a more perfect *Calcination*, it ſeems, did here almoſt prevent; yet did not in the leaſt deſtroy the aforeſaid *Eſſential Salt*, but rather make way for its more ſpeedy and copious *Production*: exhibiting likewiſe a diſtinct *Species* in ſeveral of the *Solutions*.

6. §. For firſt, the *Cryſtals* of *Roſemary* (the largeſt of them) were about the bigneſs of a *Rice-Corn*. In *Figure* almoſt like a *Tip-Cat*, which Boys play with, ſplit down the midle. Each *Tip* being cut into 5 ſides all ending in a poynt: the midle part divided into 7, all drawn by parallel *Lines*; the topmoſt with the lowermoſt but one, on each ſide, beeing three exact *Squares*. Tab. 83.

7. §. The *Cryſtals* of *Black Thorn* are moſt of them poynted with juſt ſix ſides of Equal *Meaſure*: very like to the ſhooting of true *Cryſtal* it ſelf. From the topmoſt of which ſix Sides, a *Line* being drawn out, runs parallel to a broad *Baſe*, whereon each *Cryſtal* ſtands. So that they are in ſome ſort of a *Rhomboid Figure*. Tab. 83.

8. §. The *Cryſtals* of *Scurvygraſs* have alſo a very elegant and regular *Figure*, which is in a manner compounded of the two former now deſcribed. But they are nothing near ſo bigg, the largeſt of them, being no biger than a *Grain* of that which we call *Pearl Barley*.

9. §. The *Cryſtals* of *Wormwood* have alſo very many of them a regular *Figure*; but quite different from that of the *Cryſtals* before mentioned; each *Cryſtal* being a little *Cylinder*, ſaving that it is conſtantly ſomewhat ſmaller at one end, than the other: as it were one half of a *Rowling-pin*. And not evenly *Circular*, but cut out by Six *Sides* of equal *Meaſure*: almoſt as in the *Cryſtal* of *Nitre*. So that contrary to what is ſeen in the forementioned *Cryſtals*, the ends of theſe of *Wormwood* are not poynted, but *flat*; and cut at *Right Angles* with the *Sides*. Tab. 83.

10. §. The *Cryſtals* of *Aſh*, though by their properties they appear likewiſe to be *Eſſential*; yet are nothing near ſo regularly figur'd, as all the forementioned.

11. §. The *Colour* alſo of the ſaid *Cryſtals* is ſomewhat different: Thoſe of *Aſh* being of a *brown* tranſparency, almoſt like thoſe of *Firne*. Thoſe of *Wormwood* being alſo *browniſh*, but *paler*. Thoſe of *Roſemary* and *Scurvygraſs* having ſome little *Tincture*, yet very *clear*. But thoſe of *Black Thorn* without the leaſt *Tincture*, and as *clear* as *Cryſtal* it ſelf.

12. §. None of these *Essential Crystals* have any hot fiery *Taste*, but are very *mild*, and sensibly *Bitter*; especially, about the *Root* of the *Tongue*: as is also observable of some *Plants* hereafter mentioned, in speaking of the different *Tastes* of *Plants*.

13. §. *Oyl* of *Vitriol* droped upon these *Crystals* doth not affect them in the least: yet droped into the several *Solutions* out of which the *Crystals* are produced, immediately causeth a great *Effervescence*.

14. §. Of the *Solutions* above named, that, of *Salt of Tartar* was the 6th. Whereof it is remarquable, That having waited several Months together, I could not observe the least *Essential Salt* to be therein produced in all that time. Whether there be any other *Vegetable Salts*, besides this of *Tartar*, which will not yield the *Essential* above described, I have not yet experimented.

15. §. In the mean time, from the *Premises* it is very probable, that most of them afford more or fewer of the said *Crystals*. In regard they are *Plants* of a very different kind, which I made tryal upon: as *Garden Scurvygrass*, very *Hot*; *Rosemary*, very *Aromatick*. *Wormwood* very *Bitter*; *Black Thorne*, *Astringent* and *Sower*. And it is also plain, That the said *Essential Salts* contained in the *Lixivial*, are not altogether one and the same, but of divers Sorts.

16. §. ABOUT 7 or 8 days after the *Essential Crystals* were produced; the *Marine Salt* did also begin to shoot; first in *Rosemary*; quickly after, in *Scurvygrass*; Next, in *Black Thorn* and *Wormwood*, &c. after the space of a week or 10 days more. And in all of them with some difference of *Size* and *Figure*.

Tab. 83.
17. §. The plainest of all, was that produced out of the *Salt* of *Black Thorn*, consisting for the most part of very small *Crystals*, not above the 15th of an Inch square, as also *thin*, shaped like a *Duch Tile* used for *Chimnies*. Many others were very *thick*, and near to a *Cube*. Most of which were a little hollowed in the midle, like a grinding *Marble* or *Salt-Celler*; and the hollow bounded by 4 plain and equal *Sides*, all descending a little towards the *Centre*; and measured by two cross *Lines*, which staid upon the four *Angles* of the *Square*, and so cut one the other at *Right Angles*. Both which *Properties* are likewise usually seen in the *Crystals* of *Common Salt*.

Tab. 83.
18. §. In *Wormwood*, many of these *Crystals*, besides the plain ones, were figur'd crossways like a *Dagger-Hilt*. Which was sometimes naked, and sometimes inclosed in a square and almost *Cubical Box*. Many others were figur'd into *Sprigs* made up of four chief *Branches* standing crosswise, and those *subbranched*; and all the *Branches* made up of little square *Crystals*, clustered together in that *Figure*. The *Sprigy Figure* of these *Crystals* is not accidental, but hath constantly come after they had been three times dissolved, and the *Solution* exposed to evaporate.

Tab. 83.
19. §. The *Marine Salt* of *Rosemary* hath also some variety. For besides the plain ones above described, there are some thick *Squares*, which have also a square hollow descending by five, six, or seven narrow steps, towards the *Centre*; being in *Figure*, saving these *Steps*, somewhat like the *Hoper* in a *Mill*.

Tab. 83.
20. §. Upon a second *Solution* of the same *Salt*, there shoots another sort of square; which is not plain on the edges, as the above-named, but scalloped or florid all round about, not unlike the *Leaves* of some *Plants*.

21. §.

21. §. The *Cryſtals* of *Marine Salt* of *Scruvygraſs* are ſomewhat like to thoſe of *Roſemary* now deſcribed.

22. §. As for the *Lixivial Salts* of *Aſh* and *Tartar*, though in a Month or Five Weeks Space, they yield ſome *Cryſtals* of very clear *Salt*: yet of *Marine Salt* neither of them yieldeth the leaſt particle. So that of theſe Six *Lixivial Salts*, ſc. of *Roſemary*, *Scurvygraſs*, *Black Thorn*, *Wormwood*, *Aſh* and *Tartar*, all, but that of *Tartar*, yielded an *Eſſential Salt*. And all, but thoſe of *Aſh* and *Tartar*, yielded a *Marine*, ſuch as is above deſcribed. All which *Salts* both *Eſſential* and *Marine*, together with their *Models*, made of *white Alabaſtre*, I have here ready to be ſeen.

23. §. Of thoſe that yield theſe *ſalts*, or either of them, it is further to be noted, That there is a conſiderable difference in the *Proportion* or *Quantities* which they yield. The *Roſemary* yields ſtore both of *Eſſential* and *Marine*, but more *Eſſential*. *Wormwood* and *Scurvygraſs* more *Marine*. *Black Thorn* leſs of Either. The *Aſh* no *Marine*, and the *Tartar* neither the *Eſſential* nor *Marine*, as hath been ſaid.

24. §. From what hath been ſaid, I deduce only at preſent theſe Three *Corallaie*. *Firſt*, That a *Lixivial Salt*, is not only a compounded Body ſc. of *Salt*, *Sulphur*, *Aer* and *Earth*; but even a *Compounded Salt*, containing both a *Vegetable Nitre*, and a true *Sea Salt*.

25. §. *Secondly*, That the *Expoſing* of *Bodies*, in the manner above ſhewed, may juſtly be accounted one *Part of Chymiſtry* hitherto *Deficient*, and much farther to be improved for the *Diſcovery* of the *Nature* of *Bodies*. For as *Nature* chiefly compoundeth *Bodies* by *Digeſting* them, and ſo either ſhutting out or keeping in the *Aer*: So ſhe *Diſſolveth* them by *Expoſing*, and ſo neither ſhutting in the *Aer*, nor keeping it out, but leaving it free to come and go; and thereby to bring, and carry off whatſoever is neceſſary for the *Separation* or *Solution* of *Bodies*. For the *Sea* it ſelf (to confine the fimilitude to our preſent caſe) is but as a *Great Pan*, wherein all kinds of bodies being long expoſed, are throughly reſolved, ultimately yielding from the reſt of their *viſible Principles*, that which we call *Sea Salt*.

26. §. *Laſtly*, if by *Expoſing* and *Diſſolving* we can make one *Salt*; then by *Compounding* and *Digeſting* we may make another, yea any other *Salt*; either a *Fixed* of a *Volatile*, or a *Volatile* of a *Fixed*. That is to ſay, a *Volatile Salt* may be ſo ſeparated from other *Bodies*, as to become *Fixed*; or a *Fixed Salt* may be ſo *mixed* with other *Bodies* as to become *Volatile*. For that any *Salt* ſhould of it ſelf become *Fixed* or *Volatile*, is a *Fixion* not grounded upon *Experiment*.

27. §. As for the *Virtue* of the *Eſſential Salts* above deſcribed, I believe they will be found upon tryal, not contemptible in ſome Caſes. For which amongſt other reaſons, I have been the more punctual

in relating the manner of their *Generation*; that others also may have the opportunity of making proof hereof.

28. §. When I made the *Experiments* for this and the foregoing *Discourse*, not having so good conveniency at home for making the *Salts* I used: I procured them all (except that of *Firne*, which I made my self) to be purposely prepared by Mr. *John Blackstone* a *London Apothecary*, who assured me of his great care herein; and particularly, that he added no *Nitre* to whiten any of the *Salts* with, as is commonly done for that of *Tartar*.

I do declare, That all the *Lixivial Salts* mentioned in this and the foregoing *Discourse* except that of *Firne*, were faithfully prepared by me

John Blackstone.

Lect. V.

A DISCOURSE OF THE COLOURS OF PLANTS.

Read before the *Royal Society*, *May* 3. 1677.

CHAP. I.

Of the COLOURS of Plants in their Natural Estate.

HAVING formerly made some *Observations* of the *Colours* of *Plants*; I shall now crave leave to add some more to them of the like *Nature*. None of which, nor any of the *Conclusions* thence deduced, will, if duly considered, appear contrary to the *Hypothesis* and *Experiments* of Mr. *Boyle*, Mr. *Des Cartes*, Mr. *Hook*, Mr. *Newton*, or any other, concerning *Colours*. As not having respect to the *Colours* of all *Bodies* in general. Nor to the *Body* of *Colour*, which is *Light*; Nor to the formal notion of *Colours* (*ad extra*) as the *Rays* of *Light* are moved or mixed: But to those *Materials*, which are principally necessary to their *Production* in *Plants*. Concerning which, the present *Discourse* shall be reduced to these Three general *Heads*, scil.

Idea, §. 27. and *Anat.* of *R.P.* 2. §. 65, &c.

2. §.

2. §. *First*, Of those several *Colours*, which appear in *Plants* in their *Natural Estate*.

3. §. *Secondly*, As they appear upon the *Infusion* of *Plants* into several Sorts of *Liquors*.

4. §. *Thirdly*, As upon the *Mixture* of those *Infusions*, or of any one of them with some other *Liquor*, or other *Body*.

5. §. As they appear in the *Plants* themselves, it may be observed in the first place, That there is a far less variety in the *Colours* of *Roots*, than of the other *Parts*: the *Parenchyma* being, within the *Skin*, usually *White*, sometimes *Yellow*, rarely *Red*. The *Cause* hereof being, for that they are kept, by the *Earth*, from a free and open *Aer*; which concurreth with the *Juyces* of the several *Parts*, to the *Production* of their several *Colours*. And therefore the upper parts of *Roots*, when they happen to stand naked above the Ground, are often deyed with several *Colours*: so the tops of *Sorrel Roots* will turn *Red*, those of *Mullen*, *Turneps* and *Radishes*, will turn *purple*, and many others *green*. Whereas those parts of the same *Roots* which lie more under Ground, are commonly *White*.

6. §. As *Roots* are most commonly *White*; so the *Leaves*, *Green*. Which *Colour* is so proper to them, that many *Leaves*, as those of *Sage*, the young *Sprouts* of St. *Johns-wort*, and others, which are *Redish* when in the *Bud*; upon their full *Growth*, acquire a perfect *Green*.

7. §. The *Cause* of this *Colour*, is the *action* of the *Aer*, both from within, and from without the *Plant*, upon the *Juyces* thereof, whereby it strikes them into that *Colour*.

8. §. By the *Aer* from without, I mean that which surrounds the *Body* of the *Plant*: which is the Cause of its *Greeness*, not meerly as it is contiguous to it, but as it penetrates through the *Pores* of the *Skin*, thereinto; and so mixing with the *Juyces* thereof, plainly deys or strikes them into a *Green*.

9. §. By the *Aer* from within, I mean, that which entring, together with the *Aliment*, at the *Root*, thence ascends by the *Aer-Vessels*, into the *Trunk* and *Leaves*, and is there transfused into all the several *Juyces*, thereby likewise concurring to their *Verdure*. Whence it is, that the *Parts* of *Plants* which lie under *Water*, are *Green*, as well as those which stand above it; because, though the ambient *Aer*, conteined in the *Water* be but little, yet the want of it is compensated, by that which ascends from the *Root*.

10. §. And therefore it is observable, that the *Stalks* of *Marsh-Mallow*, and some other *Plants*, being cut transversly, though the *Parenchyma* in the *Barque* be *white*, yet the *Sap-Vessels* which lie within that *Parenchyma*, are as *Green* as the *Skin* it self; *scil.* because they stand close to the *Aer-Vessels*. The *Parenchyma*, I say, which is intercepted from the *Aer*, without, by the *Skin*; and from the *Aer* within, by the *Sap Vessels*, is *white*: but the *Skin*, which is exposed to the *Aer* without, and the *Sap-Vessels* which are next neighbours to that within, are both equally *Green*. So likewise if a *Carrot* be plucked up, and suffered to lie sometime in the open *Aer*; that part which standeth in and near the *Centre*, amongst the *Aer-Vessels*, will become *Green* as well as the *Skin*, all the other *Parts* continuing of a *Redish Yellow*, as before. The *Aer* therefore, both from without, and from within the *Plant*, together with the *Juyces* of the *Plant*, are all the concurrent *Causes* of its *Verdure*.

11. §.

11. §. BUT how doth the *Aer* concur to the *Greeness* of *Plants*? I answer; Not as it is meerly either *cold* or *dry*, or *moist*, nor yet *quatenus Aer*; but as it is a mixed, and particularly, a *Saline Body*: that is, as there is a considerable quantity of *Saline Parts* mixed with those which are properly *Aereal*. It being plain from manifold Experience; That the several kinds of *Salts*, are the grand *Agents* in the *Variation* of *Colours*. So that, to speak strictly, although *Sulphur* be indeed the *Female*, or *Materia substrata*, of all *Colours*; yet *Salt* is the *Male* or *Prime Agent*, by which the *Sulphur* is determined to the *Production* of one *Colour*, and not of another.

12. §. If then it be the *Aer* mixed with the *Juyces* of a *Plant*, and the *Salt* of the *Aer*, that makes it *Green*; It may further be asked, what kind of *Salt*? But this is more hard to judge of. Yet it seemeth, that it is not an *Acid*, but a *Subalkaline Salt*; or at least some *Salt* which is different from a simple *Acid*, and hath an *Affinity* with *Alkalies*.

13. §. One reason why I so judge, is, Because that although all *Plants* yield an *Alkaly*, or other *Salt* different from an *Acid*, and some in good quantity; yet in most *Plants*, the *Prædominant Principle* is an *Acid*. So that the *Supply* of an *Acid Principle* from the *Aer*, for the *Production* of a *Green Colour*, as it would be superfluous; So also ineffectual: a different *Principle* being requisite to the striking of this, together with the *Sulphur*, into a *Green Colour*.

14. §. I suppose therefore, That not only *Green*, but all the *Colours* of *Plants*, are a kind of *Precipitate*, resulting from the concurrence of the *Saline Parts* of the *Aer*, with the *Saline* and *Sulphurious Parts* of the *Plant*; and that the *Subalkaline*, or other like *Saline Part* of the *Aer*, is concurrent with the *Acid* and *Sulphurious Parts* of *Plants*, for the *Production* of their *Verdure*; that is, as they strike altogether into a *Green Precipitate*. Which also seemeth to be confirmed by divers *Experiments* hereafter mentioned.

15. §. THE *Colours* of *Flowers* are various; differing therein not only from the *Leaf*, but one from another. Yet all seem to depend upon the general *Causes* aforesaid. And therefore the *Colours* of *Flowers*, as well as of *Leaves*, to result not solely from the *Contents* of the *Plant*, but from the concurrence likewise of the *ambient Aer*. Hence it is, that as they gradually open, and are exposed to the *Aer*, they still either acquire, or change their *Colour*: no *Flower* having its proper *Colour* in the *Bud*, (though it be then perfectly formed) but only when it is expanded. So the *Purple Flower* of *Stock-July Flowers*, while they are in the *Bud*, are *white*, or *pale*. So *Butchelors Buttons*, *Blew Bottle*, *Poppy*, *Red Daisies*, and many others, though of divers *Colours* when blown, yet are all *white* in the *Bud*. And many *Flowers* do thus change their *Colours* thrice successively; as the youngest *Buds* of *Ladys-Lookinglass*, *Buglofs* and the like, are all white, the larger *Buds* are *purple* or *murrey*, and the open *Flowers*, *blew*: according as they come still neerer, and are longer exposed, to the *Aer*.

16. §. But if the *Colour* of the *Flower* dependeth on the *ambient Aer*; it may be asked: How it comes to pass then, that this *Colour* is various, and not one, and that one, a *Green*? that is to say that all *Flowers* are not *Green*, as well as the *Leaves*? In answer to this Three things are to be premised.

17. §.

17. §. *First*, What was said before, is to be remembred, that here the *Aer* is not a *solitary*, but concurrent *Cause*. So that besides the *Efficacy* of this, we are to consider that of the several parts of the *Plant*, by which the *Contents* both *Aereal* and *Liquid* are supplied to the *Flower*.

18. §. *Secondly*, That in the *Lymphæducts* of a *Plant*, *Sulphur* is the *predominant Principle*, and much more abounding than in any other part of a *Plant*, as also hath been formerly shewed.

19. §. *Thirdly*, That it appears, according to what we have observed in the *Anatomy* of the *Flower*, That the quantity of *Lymphæducts* with respect to the *Aer-Vessels* is greater in the *Flower* than in the *Leaf*.

20. §. It semeth therefore, that the *Aer-Vessels*, and therefore the *Aer*, being *predominant* in the *Leaf*; *Green*, is therein also the *predominant Colour*. I say *predominant*, because there are other *Colours* lye vailed under the *Green*, even in the *Leafe*, as will hereafter appear more manifest.

21. §. On the contrary, the *Lymphæducts*, and therefore the *Sulphur*, being more, and the *Aer-Vessels* and therefore the *Aer*, less, in the *Flower* than in the *Leaf*; the *ambient Aer* alone is not able to controle the *Sulphur* so far, but that it generally carrys the greatest port in the *Production* of the *Colour*. Yet in different degrees; For if the proportion betwixt the *Lymphæducts* and the *Aer-Vessels* be more equal, the *Flower* is either *White* or else *Yellow*, which latter *Colour* is the next of kin to a *Green*. If the *Sulphur* be somewhat *predominant*, the *Flower* will shew it self *Red* at first; but the *ambient Aer* hath so much power upon it, as gradually to turn the *Red* into a *Blew*. But if the *Sulphur* be much *predominant*, then the *Acid* of the *ambient Aer* will heighten it to a fixed *Red*.

22. §. Hence it is, that *Yellows* and *Greens* are less alterable, upon the drying of *Plants* than other *Colours*; *sc*. Because the *Aer* being *predominant* in their *Production*, they are the less lyable to suffer from it afterwards. Whereas *Reds* and *Purples*, in the *Production* whereof *Sulphur* is predominant, are very changeable. So the *Red* Flowers of *Lysimachia*, upon drying, turn *Purple*, and the young *purple* Flowers of *gloss* turn *Blew*. So likewise the *Purple* of *Bilberries*, and the *Crimson* of baked *Damascens*, both turn *Blew*. For being gathered, and so wanting a continued supply of fresh *Sulphur*, to bear up the *Colour* against the force of the *Aer*; it strikes it down at last from *Red* to *Purple* or *Blew*. I conclude therefore, that one *Principal Cause* of the *Variety* of *Colours* in the *Flower*, is the over proportion of the *Lymphæducts* to the *Aer-Vessels*, and therefore the dominion of the *Sulphur* over the *Aer*, therein.

23. §. If it be objected, that the *Aer* doth not deepen, but highten the *Colour* of the *Blood*: I answer, *First*, That I am not now speaking of *Animal*, but of *Vegetable Bodies*; the same *Aer* which hightens the *Colour* of *Blood* one way, may deepen that of a *Flower*, another: nay and may highten that of some *Flowers* too, some other way.

24. §. And therefore, *Secondly*, it is to be considered, That as there is not one only, but divers *Saline Principles* in the *Aer*; so are there also in the several *Parts* of one *Plant*; as in the *Root*, of one sort; in the *Leavs*, of another; in the *Flower*, of another; and so in the other *Parts*. For since the *Figuration* of the *Parts* of a *Plant* dependeth

chiefly

chiefly upon the *Saline Principles*: and that the *Flower* hath a different *Figure* from that of the *Leaf*: it follows, that there is some *Saline Principle* in the one, which is not in the other, especially, all in such *Flowers*, whose *Figures* are cut out by a greater *Variety* and *Complication* of *Lines*. The *Leavs* therefore, though variously shaped, yet agreeing so far in one common *Figure*, as usualy to be *flat*; it therefore seemeth plain, that there is a *Saline Principle* in them all, so far *one*, as to be the chief *Cause* of that common *Figure*: and in concurrence with the *ambient Aer*, to be likewise the chief *Cause* of one common *Colour, sc.* a *Green*.

25. §. Whereas the *Figure* of the *Flowers*, and therefore their *Saline Principle*, being more various, and commonly distinct from that of the *Leaf*; it will easily concur with as a great *Variety of Salts* in the *Aer*, whether *Acid, Alkaline, Nitrous, Urinous, Armoniacal*, or any other therein existent, to the *Precipitation* of the *Sulphur* into the like *Variety* of *Colours*. Thus far of the *Colours* of *Plants* as they appear in their *Natural Estate*.

CHAP. II.

Of the *COLOURS* of Plants *by* Infusion.

THE next general *Inquiry*, proposed to be made, was this, After what manner the *Colours* of *Plants* shew themselves, upon their *infusion* into *Liquors*. The *Liquors* I made use of for this purpose, were three, *sc.* *Oyl* of *Olives*, *Water*, and *spirit* of *Wine*. The *Water* I used was from the *Thames*, because I could not procure any clear *Rain Water*, and had not leasure at present to distill any. But next to this, that yields as little *Salt*, as any.

2. §. As for *Oyl*, it is known, that most *Plants* either by *Coction* or long *Infusion*, will give it their *Green Colour*. I have likewise tryed some *Yellows*, and find they will do indifferently well; as *Saffron*, which, by *Infusion* in *Oyl*, gives it a light golden *Tincture*.

3. §. Divers *Aromatick Plants*, as *Mint*, *Majorane*, &c. being dryed and infused in *Oyl* give it a double *Tincture*, both *green* and *yellow*; one drop of the *Oyl* shewing *green*; but a good quantity of it held up against a candle looketh *redish* or of a deep *yellow*.

4. §. But there is no *Vegetable* yet known which gives a true *Red* to *Oyl*, except *Alkanet Root*: with which, some colouring either common or other *Oyl*, vend it under the name of the *Red Oyl* of *Scorpions*.

5. §. These things confirm what we have said concerning the *Causes* of *Colours* in the *Leavs* and *Flowers* of *Plants*, upon this twofold Consideration. *First*, that *Oyl* is the most proper *Menstruum* of *Sulphur*. Secondly, that *Oyls* have a greater congruity with *Acids* than with *Alkalies*; as I have formerly shewed.

Discourse of Mixture Ch. 5. Inst. 2. §. 3. & Inst. 5.

6. §.

6. §. I say therefore, that in *Blews*, *Purples* and especially *Reds*, the *predominant Principles* being *Sulphur* and *Acid*, the *Oyl* either abstracts the *Sulphur* of it self, or at least, unlocks it from the *Acid Parts*; whereby both of them are bestowed seperately to their like parts in the *Oyl*; upon which their disunion the *Colour* vanishes: that depending, not upon either of them alone, which of themselves are *Colourless*, but upon both united together.

7. §. On the contrary, a *Green Colour* not depending on a *predominant Acid*, but an *Alkaly*, or some *Saline Principle* different from an *Acid*; this will not so easily be imbibed separately, into the *Pores* of the *Oyl*, but only by mediation of their *Sulphur*. So that being both imbibed without any disunion, they still retein the same *green Colour* they had before in the *Plant*.

8. §. Hence also it is, that *red Roses* being dryed and infused some time in *Oyl of Anise Seeds*, a more potent *Menstruum* than *Common Oyl*; they wholly lose their own *Colour*, and turn *white*; the *Oyl* remaining *limpid*, as at the first. That is the *Sulphur* or that part of it on which cheifly the *Red* depended, is absorbed separately by the *Oyl*, and so the *Colour* vanishes.

9. §. A SECOND *Menstruum* I made use of, was *Water*. And *First*, *Alkanet Root*, which immediately tinctures *Oyl* with a deeper *Red*, will not colour *Water* in the least.

10. §. Next it is observable, That *Water* will take all the *Colours* of *Plants* in *Infusion* except a *Green*. So that as no *Plant* will by *Infusion* give a perfect *Blew* to *Oyl*; so their is none, that I know of, which, by *Infusion* will give a perfect *Green* to *Water*.

11. §. But although the *Green Leavs* will not give their *visible Colour*, by *Infusion* in *Water*; yet they will give most other *Colours*, as well as the *Flowers* themselves. So the *Green Leavs* of *Cinquefoyl*, give a *Tincture* no higher than to resemble *Rhenish Wine*; those of *Hyssop*, *Canary*; of *Strawberrey*, *Malaga*; of *Mint*, *Muscadine*; of *Wood-Sorrel*, *Water* and some drops of *Claret*; of *Blood-wort*, *Water* and a dash of *Claret*; and those of *Bawm* make a *Tincture* near as red as ordinary *Claret* alone. All *Aromatick* hot *Plants*, give a *yellow-red Tincture*, or *colorem ex luteo rubrum*. All *Plants* with a *yellow Flower* give either a pale *citrine* or *yellowish Tincture*; and the like. Yet all give not their *Tincture* in the same space of time; some requiring a fortnight, others a week, others five, three or two days, and some but one, or half a day. From hence it appears, that the *Colours* of most *Flowers* are begun in the *Leavs*; only *Green* being therein the *predominant Colour*, as a *veil* spred over them, conceils all the rest. But passing on into the *Flower*, where the *Aer-Vessels*, as is aforesaid, are under the dominion of the *Lymphæducts* they shew themselves distinctly.

12. §. A THIRD and the last *Menstruum* I made use of, was *Spirit of Wine*. And here it is to be remarqued; That as *Oyl* rarely takes a *Red*, there being but one known *Instance* of it; nor *Water*, a *Green*: So neither *Spirit of Wine*, a *Blew*. I have tryed with several *blew Flowers*, as of *Lark-heel*, *Violet*, *Mallows*, *Burrage*, and others, whereof it will not take the least *Tincture*.

13. §. Again though no *Blew Flowers*, that I know of, will give a *Blew Tincture* to *Spirit of Wine*: yet having been for some days infused

in the said *Spirit*, and the *Spirit* still remaining in a manner *Limpid*, and void of the least *Ray* of *Blew*; if you drop into it a little *Spirit* of *Sulphur*, it is somewhat surprizing to see, that it immediately strikes it into a full *Red*, as if it had been *Blew* before: and so, if you drop *Spirit* of *Sal Armoniac* or other *Alkaly* upon it, it presently strikes it *Green*. Which further confirms what have been before said of the *Causes* of *Vegetable Colours*.

14. §. It is also observable, That the *Green Leaves* of *Bawm*, which give a *Muscadine Red*, with some *Rays* of *Claret*, to *Water*, gives a pure and perfect *Green* to *Spirit of Wine*: and is the only *Plant* of all that I have yet tryed, which doth the like.

15. §. It is likewise to be noted, That both *Yellow* and *Red Flowers* give a stronger and fuller *Tincture* to *Water*, than to *Spirit of Wine*; as in the *Tinctures* of *Cowslip*, *Poppys*, *Clove-July-Flowers* and *Roses*, made both in *Water* and *Spirit of Wine*, and compared together, is easily seen. So that for *Tinctures* made with *Flowers*, whether for *Medicines*, or other purposes, *Water*, with respect to the *Colour*, is the better *Menstruum*. I say for *Tinctures* made with *Flowers*; for there are some other *Parts*, especially *Gumms*, as *Gamboja*, *Myrrh* and *Aloes*, which give their *Tinctures* full and clear, only to *Spirit of Wine*. Some of which are used by *Leather-Gilders*, and others, for the washing over of *Silver*, so as to give it the *Colour* of *Gold*. Thus far of the *Colours* of *Plants* as they appear upon *Infusion*.

CHAP. III.

Of the COLOURS of Plants *produced by their* Mixture *with other Bodies.*

THE last general Enquiry proposed to be made, was this, After what manner they would exhibite themselves upon the *Mixture* of those *Infusions*, or of any one of them with some other *Liquor*.

2. §. A strong *Infusion*, or the *Juyce* of the *Leavs* of *Rose-Tree*, *Raspis*, *Strawberry*, *Cynquefoyle*, *Goosberry*, *Primrose*, *Jerusalem Cowslip*, *Bearseare*, *Bearsfoot*, *Peony*, *Bistort*, *Lawrel*, *Goats-beard*, droped upon *Steel*, make a *Purple Tincture*. But that of *Vine Leaves* scarce maketh any *Tincture* at all. So that there is something else besides *Sowerness* concurring to the *Purple* upon *Steel*.

3. §. *Saccharum Saturni* droped on a *Tincture* of *Red Roses*, turneth it to a *faint pale Green*.

4. §. *Salt of Tartar* droped upon the same *Tincture*, turneth it to a *deeper Green*.

5. §. *Spirit* of *Harts Horn* droped upon a *Tincture* of the *Flower* of *Lark-heel* and *Borage* turn them to a *verdegreese Green*.

6. §. *Spirit* of *Harts Horn* droped on most *green Leavs* doth not change them at all. The like Effects have *Aq. Calcis*, and *Spirit* of *S. Armoniac*.

7. §. These *Experiments* seem to confirm, That it is some *Alkaline* or other like *Salt* in the *Aer*, which is *predominant* in the production of *Green* in the *Leavs* of *Plants*.

8. §. *Salt* of *Tartar* droped on the *white Flowers* of *Daisy*, changeth them into a *light Green*. Which as it further confirms the aforesaid *Position*; so likewise argues, That *Whiteness* in *Flowers*, is not always from the defect of *Tincture*: but that there may be *White*, as well as *Yellow*, *Green*, *Red* or *Blew Tinctures*.

9. §. *Spirit* of *Sulphur* droped on the *green Leavs* of *Adonis Flower*, *Everlasting Pease*, and *Holy Oak*, turns them all *Yellow*.

10. §. *Spirit* of *Sulphur* on a *Tincture* of *Saffron* changeth it not.

11. §. *Spirit* of *Sulphur* on the *Yellow Flower* of *Crowfoot* alters them not. Neither are they changed by the *Affusion* of *Alkalies*.

12. §. So that it seemeth, that in all *Yellows*, the *Sulphureous Acid* and *Alkaline Parts* are all more equal.

13. §. *Spirit* of *Sulphur* on a *Tincture* of *Violets* turns it from *Blew* to a true *Lacke*, or midle *Crimson*.

14. §. *Spirit* of *Sulphur* upon a *Tincture* of *Clove-July-Flowers* makes a bright blood *Red*. Into the like *Colour*, it hightens a *Tincture* of *Red Roses*.

15. §. So that as *Alkalys*, or other *Analogous Salts*, are predominant in *Greens*, so *Acids* in *Reds*, especially in the brighter *Reds*, in the *Leavs* and *Flowers* of *Plants*. Hence it is, that *Spirit* of *Nitre* droped upon the *Blew Flower* of *Ladies Looking-Glass*, *Larkspur*, *Borage*, turns them all *Red*, *&c.* into the *Red* of *Common Lychnis*. But (which is particularly to be noted) being droped on the said *Red Flowers* of *Lychnis*, alters them little or nothing: because, that very *Colour* is therein produced by a copious admixture of the like *Principle*.

16. §. The Summ therefore of what hath now been said, of the *Causes* of *Vegetable Colours*, is this: That while their *Sulphur* and *Saline Principles*, only swim together, and are not as yet united into one *Precipitate*, no *Colour* results from them, but the *Contents* are rather *Limpid*; as usually in the *Root*, and many other *Parenchymous Parts*.

17. §. When they are united, and the *Alkaline* are predominant, they produce a *Green*.

18. §. When the *Sulphur* and the *Alkaline* are more equal, they produce a *Tauny*.

19. §. When the *Sulphur*, *Acid* and *Alkaline*, there a *Yellow*.

20. §. When the *Sulphur* predominant, and the *Acid* and *Alkaline* equal, there a *Blew*.

21. §. When the *Sulphur* and *Acid* are predominant to the *Alkaline*, then a *Purple*.

22. §. When the *Sulphur* predominant to the *Alkaline* and the *Acid* to them both, a *Scarlet*.

23. §.

23. §. *Lastly*, When the *Acid* predominant to the *Alkaline*, and the *Sulphur* to them both, a *Blood-Red*: which is the highest and most *Sulphurious Colour* in Nature.

24. §. From the *Premises*, divers Rules do also result for the making of *Tinctures*, either for *Medicines*, or for any other purposes.

25. §. I shall only add one or two Notes. As first, that of all *Colours*, *Yellows* are the most fixed and unfading. As for instance, if you drop either a *Solution* of *Tartar*, or of *Spirit* of *Sulphur* upon a *Tincture* of the *Yellow Flowers* of *Crowfoot*, of *Adonis*, or of *Saffron*, neither of them will alter their *Colour*. Which shewes the strength of most *Yellows*, to resist all manner of impressions from the *Aer*.

26. §. Again, that the use of *Salts*, is not only to highten or deepen *Colours*, but also to fix and make them permanent. As for Instance, The *Tincture* of *Clove-July-Flowers*, made either with *Water* or *Spirit* of *Wine* being exposed to the *Aer*, will often turn into a *Blackish Purple*. But the addition of a few drops of *Spirit* of *Sulphur*, doth not only highten the *Colour*, but renders it stable and permanent.

27. §. Likewise, of *Salts* themselves there is choice to be made. For there are some, which although they fix the *Colour*, yet, will a little *give*, as we say, and not hold throughly dry; as most *Lixivial Salts*, and *Stillatious Acids*. But there are some *Salts*, which will not *give* in the least, as *Alum*, that in *Lime-Water* and some others; which latter, is so far from being moystened, that it is rather petrified by the *Aer*. For which reason I take it to be one of the best *Liquors* for a stable and permanent *Green*, and some other *Colours*.

28. §. Amongst all *Water-Colours*, the rarest, and most difficult to make clear bright and permanent, is a *Blew*. There are many *Flowers* of an excellent *Blew*, as those of *Buglofs*, *Lark-heel* and others; but they easily fade. And there are very few *Flowers* that will strike into a *Blew* by any *Liquor*; being almost all changeable into *Green*, *Purple* or *Red*. Yet some few there are, in which this *Colour* may be produced. As for instance, the *Flower* of *Lathyrus* or *Parseverlasting*; which upon the affusion of *Spirit* of *Harts-Horn* is changed from a *Peach*, to as pure a *Blew*, as the best *Ultramarine*: that which hitherto is, I think, wanting in *Water Colours*. *Spirit* of *Harts Horn* was the *Liquor* I used; but I question not, but that other *Alkalies*, and particularly *Lime-Water*, will have the like Effect, and so render it the more stable.

29. §. From what hath been said, we may likewise be confirmed in the use of the already known *Rules*, and directed unto others yet unknown, in order to the variation of the *Colours* of *Flowers* in their Growth. The effecting of this, by putting the *Colour* desired in the *Flower*, into the *Body* or *Root* of the *Plant*, is vainly talked of by some: being such a piece of cunning, as for the obteining a painted face, to eat good store of *white* and *Red Lead*.

30. §. The best known *Rules* are these Two; First, that the *Seed* be used above any other part, if the variation of the *Colour* be intended. One reason whereof is, because that part being but very small, the *Tinctures* of the *Soyl* will have the greater over proportion to those of the *Seed*. Besides, the tender and *Virgin Seed*, being committed to the *Soyl*, will more easily take any peculiar *Tincture* from it, then an

other *Part*, which is not so susceptive, and hath been tinctur'd already. All the strange varieties in *Carnations*, *Tulips*, and other *Flowers* are made this way.

31. §. The other *Rule* is, To change the *Soyl*, or frequently to transplant from one *Bed* to another. By which means, the *Plant*, is as it were, *superimpregnated* with several *Tinctures*, which are prolifick of several *Colours*; which way is taken for *Roots* and *Slips*.

32. §. The consideration whereof, and of the foregoing *Experiments*, may direct us not only in changing the *Bed*, but also in compounding the *Soyl*, as by mixing such and such *Salts*, or *Bodies* impregnated with such *Salts*, I say by mixing these Bodies in such a proportion, with the *Soyl*, as although they have no *Colour* in themselves, yet may be effectual to produce a great variety of *Colours* in the *Plants* they nourish; supplying the *Plants* with such *Tinctures*, as shall concur with the *Aer*, to strike or precipitate their *Sulphur* into so many several *Colours*, after the manner above explicated: and so to bring even Natures Art of *Painting*, in a great part, into our own power.

A DISCOURSE OF THE DIVERSITIES and CAUSES OF TASTS CHIEFLY IN PLANTS.

Read before the *Royal Society*, March 25. 1675.

CHAP. I.

Of the several Sorts of SIMPLE and COMPOUNDED Tasts; and the DEGREES of both.

 HAVE formerly published some Notes, concerning *Tasts*. Since then, I have made other *Observations* upon the same Subject: and these have produced further *Thoughts*. I will summ up all in giving an account, First, of the *Diversities*; and then, of the *Causes of Tasts*, chiefly in *Plants*.

Idea, §.29. & Anat. of R. P. 2. §. 68, &c.

2. §. The *Diversities of Tasts* are so many, and so considerable; that it seemeth strange, to see the matter treated of both by *Philosophers* and *Physicians*, with so much scantness and defect. For the *Subject* is not barren, but yieldeth much and pleasant *Variety*. And doth also appear to be of great import unto *Medicine*. Besides, it is preposterous to discourse of the *Causes of Tasts*, before we have taken an account of their *Diversities*; Whereof therefore I shall in the first place, exhibit the following *Scheme*.

3. §.

3. §. TASTS may be distinguished by these Three general ways. First, with respect to the *Sensation* it self. Secondly, with respect to its *Duration* and *Terms*. Thirdly, with respect to its *Subject*.

4. §. The *Sensation* it self is differenced two ways, by its *Species*, and by its *Degrees*. With respect to the *Species*, *Tasts* are *Simple*, or *Compounded*. By *Simple Tasts*, I mean not such, as are never found in conjunction with other *Tasts*: but the *Simple* or *Single Modes* of *Tast*, although they are mixed with divers others in the same *Body*. As for example, the *Taste* of a *Peppin*, is *Acidulcis*; of *Rhubarb*, *Amarastringens*; and therefore *Compounded* in both. Yet in the *Peppin*, the *Acid* is one *Simple Taste*, and the *Sweet* another; and so in *Rhubarb*, the *Bitter* is one *Simple Taste*, and the *Astringent* is another.

5. §. Two faults have here been committed; the defective *Enumeration* of *Simple Tasts*; and reckoning them indistinctly among some others which are *Compounded*.

6. §. SIMPLE *Tasts*, (of which, properly so called, there are commonly reckoned but Six or Seven Sorts,) are, at least Sixteen. *Fist*, *Bitter*, as in *Wormwood*: to which, the contrary is *Sweet*, as in *Sugar*. *Thirdly*, *Sower*, as in *Vinegar*: to which, the contrary is *Salt*. *Fifthly*, *Hot*, as in *Cloves*: whereto, the contrary is *Cold*. For we may as properly say, a *Cold Taste*, as a *Hot Taste*: there being some *Bodies*, which do manifestly impress the *Sense* of *Cold* upon the *Tongue*, though not by *Touch*. So doth *Sal Prunellæ*, although the *Liquor* wherein it is dissolved, be first warmed.

7. §. Seventhly, *Aromatick*. For it doth not more properly agree to an *Odour*, than a *Taste*, to be *Aromatick*. And that an *Aromatick Taste*, is distinct from an *Hot*, is clear; In that, there are many *Bodies* of a *Hot Taste*, some meanly and others vehemently *Hot*; which yet are not in the least *Aromatick*: as amongst others, is apparent in *Euphorbium*. So that although an *Aromatick Taste* be often conjoyned with *Heat*; yet it is not that *Heat* it self, but another distinct *Sense*.

8. §. Eighthly, *Nauseous* or *Malignant*, contrary to the former. Such as is perceived, together with the *Astringent* and *Bitter*, in *Rhubarb*; or with the *Bitter*, and *Sweet*, in *Aloes*. It may be called *Malignant*, because distastful although mixed in a low degree with other *Tasts*: whereas other *Tasts* will render one another grateful.

9. §. Again, *Tasts* may properly be said, to be *Soft* or *Hard*. A *Soft Taste*, is either *Vapid*, as in *Watery Bodies*, *Whites of Eggs*, *Starch*, *Fine Boles*, &c. Or *Unctuous*, as in *Oyls*, *Fat*, &c.

10. §. A *Hard Taste* is Fourfold, *sc. Penetrant, Stupifacient, Astringent, Pungent*. Contrary to a *Vapid*, are *Penetrant* and *Stupifacient*.

11. § *Penetrant*, is a kind of *Taste*, which worketh it self into the *Tongue* (as some *Insects* into the *Skin*) without any *Pungency*; as in the *Root* and *Leavs* of *Wild Cucumer*.

12. §. *Stupifacient*, as in the *Root* of *Black Hellebore*. Which being chew'd, and for sometime reteined upon the *Tongue*; after a few minutes, it seemeth to be benum'd and affected with a kind of *Paralytick Stupor*; or as when it hath been a little burnt with eating or supping of any thing too hot.

13. §. Contrary to an *Unctuous Taste*, are *Astringent*, and *Pungent*; as in *Galls*, and *Spirit* of *Sal Aromanick*.

14. §.

14. §. Again, *Tasts* are either *Continual*, as most commonly: or *Intermittent*; as that of *Dracontium*, especially in the *Root*. For after it seems to be lost and extinguished; it will then again (chiefly upon the *Collision* of the *Tongue* and *Goomes*) be plainly heightened and reviv'd.

15. §. Lastly, *Tasts* are either *Still*, as usually; or may be called *Tremulous*, as the *Heat* produced by *Pyrethrum*. *Distinct* from that of *Cloves*, *Ginger*, and many other *Hot Bodies*, in that there the *Heat* is *still*; but here in *Pyrethrum*, 'tis joyned with a kind of *Vibration*: as when a *Flame* is brandished with a *Lamp-Furnace*. Thus far of the Sorts of *Simple Tasts*.

16. §. COMPOUNDED *Tasts* are very numerous; being made by the various *Conjunction* of *Simple Tasts*, as *Words* are of *Letters*. Sometimes of two, as in *Saccharum Saturni*, of *Astringent* and *Sweet*. Sometimes three, as in *Aloes*, *Malignant*, *Bitter* and *Sweet*; in *Rhubarb*, *Malignant*, *Astringent* and *Bitter*. Sometimes four, as in *Agarick*, *Malignant*, *Astringent*, *Bitter* and *Sweet*. And in some Bodies, five or six *Species* may be joyned together.

17. §. For the more accurate *Observation* whereof, there are these easie *Rules*. That not too many be tasted at one time: least the *Tongue* being surcharged, become less critical. That the *Mouth* be washed with warm water betwixt every tasting. And that those things be first tasted which produce a less durable *Taste*; that so one may be throughly extinguished, before another be try'd.

18. §. Of the numerous *Conjunctions* of *Tasts*, which may thus be observed, there are only Six to which the penury of *Language* hath allowed (if I may call them) *Proper Names*, sc. *Acerbus*, *Austerus*, *Acris*, *Muriaticus*, *Lixivus* & *Nitrosus*. Most of which are commonly taken in to make up the number of *Simple Tasts*. But very improperly; being all of them *Compounded* and *Decompounded Tasts*: to which Class they ought therefore to be refer'd. For

19. §. *Austere*, is *Astringent* and *Bitter*; as in the green and soft *Stones* of *Grapes*.

20. §. *Acerb*, properly so called, is *Astringent* and *Acid*; as in the *Juyce* of unripe *Grapes*.

21. §. *Acris*, is also *Compounded*. For first, simply *Hot*, it is not: because there are many *Hot Bodies*, which are not *Acria*; as the *Roots* of *Zedoary*, *Yarrow*, *Contrayerva*. Nor Secondly, is it simply *Pungent*, because there are also *Bodies*, which are *Non-acria pungentia*; of which kind is the *Root* of *Arum*. Wherefore *Acritude*, is *Pungency* joyned with *Heat*.

22. §. *Muriatick*, is *Saltness* joyned with some *Pungency*, as in common *Salt*.

23. §. *Lixivial*, is *Saltness* joyned with *Pungency* and *Heat*.

24. §. *Nitrous*, is *Saltness* joyned with *Pungency* and *Cold*.

25. §. Besides these Six, or perhaps one or two more, there are, as is said, a great number of *Conjunctions*, for which we have no *Proper Names*. For admit that there were but *Ten* species of *Simple Tasts*, sc. these *Ten*; *Amarus*, *Dulcis*, *Acidus*, *Salsus*, *Calidus*, *Frigidus*, *Aromaticus*, *Malignus*, *Astringens*, *Pungens*. And of these *Ten*, but *Two*, or at most, but *Three* to be compounded together in any one *Body*. If only *Two*, they produce 45 *Compounded Tasts*. For the *First*, may

be compounded with all the 9 following; the *Second*, with all the 8 following; and so, the rest: which together make 45. But if the same *Ten* be compounded by *Threes* together; they produce no less than 120 *Variations*: as by the *Table* made of them all doth plainly appear.

26. §. Some few of the *Conjunctions* therein set down, may not be found actually existent in *Nature*. The abatement of which, will be much more than compensated two ways. *First*, by the other *Six Species* of *Simple Tasts*, which are also sometimes compounded. And by other more complex *Conjunctions*, as of many *Quadruples*, and perhaps some *Quintuple* or *Sextuple* ones. Thus far of the *Simple Species*, and *Conjunctions* of *Tasts*.

27. §. THE DEGREES of *Tasts* are also numerous; and each *Species*, in every *Conjunction*, capable of *Variation* herein. For the more accurate observing whereof, it will be best, To take those *Bodies*, whose *Tasts* are, as near as may be, the same in *Specie*: and that those be first tasted, which are less strong; whereby the true Degree will be more precisely taken.

28. §. The *Tasts* of *Bodies* will thus appear to be varied, in most *Species* unto *Five Degrees*; and in some of them, unto *Ten*. So the Root of *Turmerick*, is bitter in the *First Degree*; of *Gentian*, in the *Tenth*. The Root of *Carduus Benedictus*, is *Hot* in the *First Degree*; the *Green Pods* or *Seed-Cases* of *Clematis peregrina*, in the *Tenth*. So that, allowing some to vary under *Five*; yet by a moderate estimate, we may reckon every *Species*, one with another, to be varied by at least *Five Degrees*. Which being added to the several *Species of Tasts*, in all the *Treble Conjunctions* of the aforesaid *Table*, come to 1800 sensible and defineable *Variations* of *Taste*. And these are the *Diversities* of *Taste*, with respect to the *Sensation* it self.

CHAP. II.

Of the DURATION and several TERMES of Tasts.

THE next general way of dinguishing *Tasts*, is by their *Duration*, and their *Terms*, or their *Motion* of *Intension* and *Remission* from one *Degree* to another. For there are many *Tasts*, which have their *Motions* analogous to those of *Diseases*; and by those may be distinguished in the same manner. For as of *Diseases*, so of *Tasts*, there are Four Times, as *Physicians* call them, or *Terms* of *Motion*; sc. *Principium, Augmentum, Status, & Declinatio*.

2. §. For the distinct observing of which, those *Bodies* which are hard, and so their tastable parts less easily extractable by the *Tongue*, should be reduced to a fine *Powder*: otherwise, the true measure of the

Principium will be loft. And for the precife meafuring of all the *Four Termes*, it fhould be done by a *Minute-Watch* or a *Minute-Glafs*. For fo it will appear, that the *Variations* of each, are divers and remarquable.

3. §. To inftance firft in thofe of the *Principium*. Which I call, That fpace of time, betwixt the firft *Contact* of the *Body* to be tafted, and the firft manifeft *Perception* of the *Tafte*. For Example, thofe *Bodies* which are *Acid*, or *Bitter*, as *Vinegar* or *Wormwood*, are prefently perceiv'd, *quatenus Acid* or *Bitter*, upon the firft *Contact*; or have *Principium breviffimum*. Thofe *Bodies* which are *Acria*, have their *Principium* fomewhat longer. So the *Seed-Cafes* of *Clematis peregrina*, although they have a vehement *Acritude*, even in the Tenth *Degree*; yet is not that *Acritude* fo foon tafted, as the *Bitternefs* of *Rofes*, which is but in the fecond. But the *Principium* of *Hot Tafts*, is generally longer than that of any other. So the *Bitternefs* of the *Root* of *Black-Helebore*, which exceedeth not the fecond *Degree*, is yet prefently tafted: but the *Heat* proceeding from the fame *Root*, and which afcendeth to the third *Degree*, is not perceived at all, till after two full *Minutes*. And fo the *Bitternefs* of *Enula*, which exceedeth not the 4th *Degree*, yet is fooner tafted than its *Heat*, which afcendeth to the 8th.

4. §. Next, in thofe of the *Augment*. Which I call, That fpace, betwixt the firft *Perception* of the *Tafte*, till it be come to the heighth. So the *Heat* of *Galangale*, is not only prefently perceived, but arifeth to the heighth within half a *Minute*. But the *Heat* of the *Root* of *Enula*, comes not to the heighth till after a whole *Minute*. And the *Heat* of *Black-Hellebore*, not till after four full *Minutes* from the firft *Contact*.

5. §. The *Status*, or fpace wherein the *Tafte* continues in its heighth, is alfo divers. So the *Heat* of the *Seed-Cafe* of *Helleborafter*, comes to its heighth, and begins to decline within half a *Minute*; that of the *Root* of *Garden-Scurvygrafs*, not till after a *Minute*; and that of the *Root* of *Afarum*, not till after two full *Minutes*.

6 §. And *Laftly*, the *Declination*, or the fpace betwixt the firft *Remiffion* of the *Tafte*, and its total *Extinction*. For inftance, The *Leavs* of *Millefolium*, are *Bitter* in the 4th *Degree*, and *Hot* only in the 1^{ft}. yet the *Heat* continues for fometime, and the *Bitter* prefently vanifhes. *Calamus Aromaticus*, is *Bitter* in the 4th *Degree*, *Hot* in the 1^{ft}, and *Aromatick* in the 3^{d}: yet the *Bitter* quickly vanifhes, the *Heat* continues two *Minutes*, and the *Aromatick* feven or eight. The *Heat* of the *Root* of *Contrayerra*, is extended, almoft to two *Minutes*; the *Purgency* of *Jalap*, almoft to fix; the *Heat* of *Garden Scurvygrafs*, to feven or eight. And even the *Bitterefs* of *Wild Cucumer*, to near a quarter of an hour. But the *Heat* of *Euphorbium* dureth much longer, as alfo that of *Black Hellebore. &c.* above half an hour.

7. §. So that the *Augmentum*, is feldom extended beyond Four or Six *Minutes*, from the firft *Contact*: but the *Declination*, fometimes to Thirty, Fourty, or more. Thus far of the *Terms* of *Taft*, or the manner of their *Intenfion* and *Remiffion*.

CHAP. III.

Of the SUBJECT or SEAT of Tasts.

THE *Third* and *Last* way of distinguishing *Tasts*, is by their *Subject*, or the *Part* or *Parts* where they are either wholly or chiefly perceived. And so, *Tasts* are either *Fixed*, or *Movable*.

2. §. A *Fixed Tast*, is that which keepeth within the compass of some one *Part*, all the time of its *Duration*; as upon the *Tip*, or the *Root* of the *Tongue*, or other *Part*.

3. §. A *Movable Taste*, is either *Diffusive* or *Transitive*.

4. §. A *Diffusive Taste*, I call that, which by degrees spreads abroad into divers *Parts*, and yet in the mean time, adheres to that *Part* in which it is first perceived. So the *Bitterness* in the dryed *Roots* of *Black Hellebore*, is first felt on the *Tip* of the *Tongue*; from whence it spreads it self to the midle of the same. And the *Bitterness* of the *Leavs* of *Wild Cucumer*, spreads from the *Tip*, to the *Root* of the *Tongue*.

5. §. A *Transitive Taste*, is that, which after sometime, wholly quitting the *Part* wherein it is first perceived, is thence transfered into some other *Part*: as the *Bitterness* of *Gentian*, imediately from the *Tip*, to the midle of the *Tongue*. And most of the *Diffusive*, are also *Transitive*.

6. §. The several *Parts* which these ways become, and with some latitude may be called, the *Seats* of *Tasts*, are, the *Lips*, *Tongue*, *Palate*, *Throat* and *Gulet*.

7. §. Upon the *Lips*, the *Root* of white *Hellebore*, as also of *Pyrethrum*, being chewed, make a sensible *Impression*; which continues (like the flame of a *Coal* betwixt in and out) for 9 or 10 *Minutes*. But the *Heat* in other *Parts* much longer.

8. §. Upon the *Tongue*, *Tasts* are perceived in Three places, as hath been intimated. On the *Tip* or *Cone* of the *Tongue*; as most commonly. On or near the *Basis* of the *Tongue*; where the *Taste* of the *Leavs* of *Wild Cucumer* chiefly fixeth it self. Or on the *Vertex* or midle of the *Tongue*; in which place it is observable, that the *Tast* of *Gentian*, *Colocynthis*, and divers other *Bodies*, is then considerably strong, when not at all perceived at the *Tip* of the *Tongue* or in any other *Part*.

9. §. Upon the *Palate* or *Roofe* of the *Mouth*, the *Root*, as I take it, of *Deadly Nightshade* maketh its chief *Impression*; and there continues about four *Minutes* in some degree.

10. §. The *Throat*, or the *Uvula*, *Larinx* and other adjacent *Parts* are oftentimes the *Seat* of *Taste*. For there are many *Bodies*, which although they have scarce any *Taste* upon the *Tongue*, or any other of the aforesaid *Parts*, yet make a strong *Impression* on the *Throat*: as the *Leavs* of little *Daisy*, little *Celandine*, and of *Pimpinel*; as also the *Roots* of *Jalap*, *Mercury*, *Asparagus* and others. Which being chewed makelittle or no *Impression* on the *Tongue*, but their *Juyce* being swallowed,

lowed, causeth a kind of pricking in the *Throat*; as when one is provoked by a sharp *Rheum*.

11. §. And that this *Taste* or *Sense*, is truly distinct from either the *Heat, Pungency*, or *Acritude* upon the *Tongue*, it is hence further manifest; In that *Pyrethrum*, which is very *Hot*, and *Cortex Winteranus* which is very *Pungent* upon the *Tongue*; yet their *Juyce* being swallowed, causeth no *Heat, Pungency* or *Exasperation* in the *Throat*.

12. §. *Lastly*, if we will take the word (*Tast*) in a larger sense, the *Oesophagus* it self may be said to be sometimes the *subject* thereof; as of the *Heat* produced by the *Root* of *Common Wormwood*. For of this *Heat* it is remarquable, that being first perceived on the *Tip* of the *Tongue*, it thence maketh its *transit* to the *Root* of the *Tongue*, and so into the *Throat*, and by degrees descends into the very *Gulet*; where it seemeth to warm the *Stomach*; and so continues, in some degree, almost ¼ of an hour. And the *Transition* and *Descent* of this *Heat* is made, although none of the *Juyce* be swallowed. And in this maner *Tasts* are distinguished with respect to their *Subject*.

13. §. So that the general *Diversities* of *Tasts* are these. With respect to their *Species*, they are *Simplices vel Compositi*; To their *Degree, Remissi vel Intensi*; To their *Duration, Breves vel Diuturni*; To the *Terms* of their *Motion, Celeres vel Tardi*; and lastly, To their *Subject, Fixi, Diffusivi & Transitivi*.

14. §. I HAVE thus endeavourd to draw up a *Scheme* or *Inventory* of the several sorts of *Tasts*. In which, some may think, that I have over done: and that as *Galen* hath been censured for being too curious in the *Distinctions* of *Pulses*; so have I been, in these of *Tasts*. Not to enquire now, how far the *Differences* of the *Pulse* may be extended, or be fit to be taken notice of; I shall only say. That we have not so much reason to censure him, if he hath given us some few which are coincident; as we have to thank him, for observing so many which are really distinct.

15. §. By the *Scheme* of *Tasts* here represented, we may be able, so to enumerate the *Modes* of any *Tast*, as to make a *Scientifick Definition* of it. Which is pleasant *Instruction* to any inquisitive mind; these things being all matter of sense and demonstration; wherein lyeth, though not always the most plausible, yet the most satisfying *Philosophy*; and where men, after they are grown weary with turning round, are oftentimes contented to rest.

16. §. But the usefulness of this *Schem* will further appear, in two respects; *sc.* In conducting us to a cleerer and more particular *Explication* of the *Causes* of *Tasts*: and the *Investigation* of the *Virtues* of those *Bodies* in which they reside. Whereof in the following *Chapters*.

CHAP.

CHAP. IV.

Of the CAUSES of Tafts.

O fpeak of the *Caufes* of *Tafts*, before we have well enumerated and diftinguifhed them; is to provide *Furniture* for a *Houfe*, before the *Roomes* have been counted and meafured out. But the *Varieties* of *Tafts* having been firft laid down; it will induce us to believe, and inveftigate as great a variety in their *Caufes*.

2. §. Now the *Caufes* of *Tafts*, particularly of the *Tafts* of *Plants*, whereof we chiefly fpeak, are, in general, thefe Four or Five, *fc.* The *Bed* out of which they grow; The *Aer* in which they ftand; The *Parts* of which they confift; The feveral *Fermentations* under which their *Juyces* pafs; And the *Organs* by which their *Taftable Parts* are perceiv'd: as will appear upon Inftance.

3 §. But the immediate *Caufes*, befides the *Organs* of *Tafte*, are the *Principles* of *Plants*. As many of which, as come under the notice of *Senfe*, we have already fuppofed to be thefe Seven, *Alkaline, Acid, Aer, Water, Oyl, Spirit* and *Earth*. The *Particles* both of *Alkaline* and *Acid Salts*, are all *angular* and *poynted*. Thofe of *Aer*, properly and ftrictly fo called, are *Elaftick* or *Springy*; and therefore alfo *Crooked*; as I have likewife formerly conjectured. And I find the Learned *Borelli*, in a Book of his fince then publifhed, to be of the fame Opinion. Thofe of all *Fluid Bodies*, *quà Fluid*, and therefore of *Water*, *Oyl* and *Spirit*, I conceive to be *Globular*, but *hollow*, and with holes in their Sides. Thofe of *Water*, to be larger *Globes*, with more *holes*; thofe of *Oyl*, to be leffer, with fewer *holes*; and thofe of *Spirit* the leaft. Laftly, that the particles of *Earth* are alfo *Round*; yet angular; and nearer to a *folid*.

4. §. Thefe *Principles* affect the *Organs* of *Senfe*, according to the variety of their *Figures*, and of their *Mixture*. So thofe which are fharp or poynted; and thofe which are *fpringy*; are fitted to produce any ftronger *Tafte*: and thofe which are round, are apt, of their own *Nature*, to produce a *weaker* or *fofter* one. And fo by the diverfities of their *Mixture*; not only with refpect to their *Proportion*, but alfo the very *Mode* of their *Conjunction*. Hence it is, that many *Bodies* which abound with *Salt*, as *Ambar* with an *Acid*, and the *Bones* of *Land-Animals* with an *Alkaline*, have notwithftanding but a weak *Taft*; the *Saline Parts* being in the former drowned in the *Oyl*, and in the latter alfo buried in the *Earth*.

Of Mixt. Ch. 5.

5. §. The fame is further confirmed by an *Experiment* mentioned in a former *Difcourfe*; *fc.* the *Tranfmutation* of *Oyl* of *Anife-Seeds*, with the help of *Oyl* of *Vitriol*, into a *Rofin*. For both thofe *Liquors*, though fo ftrongly tafted, apart; yet the *Rofin* made of them, being well wafhed, hath a very mild *Tafte*, and without any fmatch of that

in

in either of the *Liquors*. Whence it follows, that the very *Mode* of *Mixture* is sufficient, not only for the variation of the *Degrees* in any one *Species* of *Taft*; but also for the destroying of one *Species*, and the introducing of another.

6. §. THESE things being premised, I conceive, That as an *Unctuous Taft* dependeth upon *Oyl*; so a *Vapid* either on *Water*, or *Earth:* or upon such an intimate *Mixture* of other *Principles*, as renders them indissoluble by the *Saliva*, and so, in a manner, untastable.

7. §. That a *Pungent*, is made either by an *Alkaly* or an *Acid* sharpned or whetted; that is, cleared from the soyl of other *Principles*; as in the *spirit* of *Sal Aromoniac* or of *Sulphur*. And so in those *Plants* which have a *Pungent Taft*; whose *Juyces* or *Tinctures*, although they consist of divers *Principles*, yet all so loosely mixed, that being dissolved by the *Saliva*, the *Saline* are hereupon left naked. Wherefore biting *Plants*, quà biting, are *Nitrous Plants*. So that the *Juyce* of such *Plants*, is a kind of *Spirit of Nitre*, made by the several *Parts* of the *Plant*. Hence *Arum* grows best under an *Hedg*; where the Ground, not being exposed to the *Sun*, but the *Aer* only, like those *Rooms* in *Houses*, which are covered, is impregnated with a greater quantity of *Nitrous Salt*. And those *Roots* which are *Biting*, have but few or but small *Aer-Vessels*; whereby fewer parts of the *nitroaereal Sap* are carryed off into the *Trunk*. For the same *Cause*, it is no wonder, that many *Aquaticks* are *Biting*; *Water* being, though it self cold, yet the *Menstruum* by which all *Salts* are imbibed most easily, and in laxer state of *Commixture* with other *Principles*.

8. §. *Penetrant* (something flower than *Pungent*) is made by any *Salt* that is also foiled or guarded with *Earth*. *Sower*, by an *Acid* only foyled with *Earth*. *Salt*, by an *Acid* guarded by an *Alkaly*, and foyled with *Earth*. *Cold*, by an *Acid* drowned in *Water*, and foyled with *Earth*.

9. §. In all these, the *Salts* are predominant; In *Heat* the *Oyl* or *Sulphur*. The particles whereof being *Spherick* and bored with *holes*; those of *Salt* stick in them, as the *Spokes* do in the *Hub* of a *Wheel*, or as the *Quills* in the *Skin* of a *Porcupine*. Whereby, as in *Common Fire* the *Sparks* of *Sulphur* being agitated and whirled about by the *Aer*; with the help of the *Salts*, which stick in them, tear in pieces all kinds of *Bodies*: so here, being agitated by the *Circulation* of the *Blood*, they make a kind of hurry or combustion; and so, according to the degree and strength of their *Motion*, tear in pieces fewer or more of the *Fibers* of the *Tongue*; and in a greater quantity, would raise a *Blister* upon it; the common *Effect* of *Fire*, or any strong *Epispastick*. So that a *Hot Taft*, is produced by *Sulphur* toothed or armed with *Salts*. Wherefore all *Stillatitious Oyls* are *Hot*; being strongly impregnated or armed with the *Essential Salts* of the *Plants* from whence they are distilled. And as those *Plants* which are very *Parenchymous*, from the predominancy of their *Volatile Acid*, are *biting*: So those which are *Lignous*, that is, have a good quantity of *Lympheducts*, from the dominion of their *Sulphur* are commonly *Hot*. For the same reason it is, that many both *Biting* and *Hot Plants*, as the *Roots* of *Dragon*, *Garden-Radish*, *Onion*, *Iris*, *Rape-Crowfoot*, &c. being corked up in a bottle with *Water*, and set in a *Cellar* or other cool place; they do all of them turn *Sower* in a few days: The same *Fermentation*, at once

fullying

fullying the *Salts* of the one, and difarming the *Sulphurs* of the other. But fome, wherein the *Sulphureous* parts are more copious, will hardly ever become *Sower*. Hence alfo, fome *Plants*, whofe *Roots* are neither *Hot*, nor of any ftrong *Tafte*, as thofe of *Wild Anemone* ; yet their *Leaves* and *Flowers* are plainly *Cauftick* : So that it feems, that as their *Juyces* rife up into the *Trunk* or *Stalk*, and are therein further fermented, the *Sulphureus Parts* thereof, are at the fame time relaxed from the other *Principles*, and acuated with an *Aereal Salt*.

10. §. A *Stupifacient Taft* (as the *Impreffion* which fome *Hot Plants* make upon the *Tongue* may be called) is in fome fort, analogous to the mortifying of any part of the Body by the application of a *Cauftick*. For as there the mortification fucceeds the burning pain, fo here, the *Stupifaction*, neither comes before, nor with the *Heat*, but follows it.

11. §. *Sweetnefs* is produced, fometimes by an *Alkaly* ; fmoothed either by a *Sulphur*, as in *Lime-Water* ; or by both a *Spirit* and a *Sulphur*, as in the *Stillatitious Oyls of Animals*. But moft commonly, by a fmoothed *Acid* ; as in *Malt*, *Sugar*, *Hony*. Hence a *Sweet Tafte*, is generally founded in a *Sower*; So Sower *Apples*, by mellowing, and harfh *Pears*, by baking become fweet ; the *Spirit* and *Sulphur* being hereby at once feparated from the other *Principles* and brought to a nearer union with the *Acid*. So the Sower *Leaves* of *Wood-Sorrel*, being dry'd, become fweet: and thofe of a fower *Codlin*, while they hang on the *Tree*, and even of a *Crab-Tree*, are neither *Aftringent*, nor fower, but fenfibly fweet. And fo commonly, wherever the faid *Principles* are a little exalted by a foft *Fermentation* ; as in the *Juyce* of the *Stalk* of *Maze* or *Indian Wheat*, which is a fweet as *Sugar* ; and in the green *Stalks* of all forts of *Corn* and *Grafs*, in feveral degrees. So likewife *Tulips* and fome other *Roots*, being taken up, in open weather, fometime before they *fprout* ; if tafted, are as fweet as *Liquirifh* or *Sugar*; and at no other time : not only *Fruits*, but many *Roots*, *Seeds*, and other *Parts*, upon their firft or early *Germination*, acquiring a curious *Mellownefs*, wherein, all their *Principles* are refolved, and their moft *Spirituous Parts* exalted and fpread over the *Acid*. Wherefore alfo moft *Roots*, which are not meerly long, but grow deep in the ground, have at leaft fome of their *Juyces* of a fweet *Taft* ; as *Liquirifh*, *Eryngo*, *Hounds-Tongue*, *Garden-Parfnep*, *Black Henbeane*, *Deadly Nightfhade*, *&c.* Even the *Juyce* of *Horfe Radifh*, which bleeds at the *Lympheducts*, is of a fweet *Tafte*. And of the fame kindred thofe which grow the deepeft, are the fweeteft ; as a *Parfnep* is fweeter than a *Carroot*, efpecially if you taft the bleeding *Sap* ; and the *Root* of *Common Tall Trefoyl* tafteth fomewhat like *Liquirifh*, but is not near fo fweet. For all deep *Roots*, are fed with a lefs *Nitrous Aliment :* and being remoter from the *Aer*, their *Juyces* pafs under much more foft and moderate *Fermentations*.

12. §. *Bitternefs* is produced by a *Sulphur* well impregnated, either with an *Alkaline*, or an *Acid Salt*, but alfo fhackled with *Earth*. And therefore the *Bittereft Plants*, commonly yield the greateft quantity of *Lixivial Salt*. So alfo many *Stillatitious Oyls* digefted with any ftrong *Acid*, will acquire a *Bitter Tafte*. Wherefore this *Tafte* is often founded either in a *Hot Tafte*, or a *Sweet*. Hence it is, that the *Leaves* of all fweet *Roots* are *Bitter*. And that the *Fig-Tree*, which bears a fweet *Fruit*, bleeds a *Bitter Milk*. So likewife thofe *Plants*, which bear a

Bitter

Bitter Stalk, have not *Bitter*, but *Hot Roots*, as in *Yarrow, Primrose, Wormwood, Rue, Carduus benedictus* &c. is manifest. So the *Coats* of the *Seeds* of *Viola Lunaria* are of a *hot* and *biting Taft*; but the *Seeds* themselves, in which the *Salts*, though copious, yet are also immersed in a greater quantity of *Oyl*, are *Bitter*. And that the *Earthy Parts* do also contribute something more to this, than to most of the forementioned *Tafts*, is argued from its being more *Fixed*; that is, the *Body* in which it resides, is either more *Fixed*, or else flyeth not away in that same state of conjunction, by which it maketh a *Bitter Tafte*. For whereas *Hot, Biting*, and divers other *Plants* lose the strength of their *Tafte*, by drying; most of those which are *Bitter*, do hereby increase it. And although the *Extract* of *Dandelion* and some other *Roots*, which are very *Bitter*, hath scarce any *Taft*; yet generally, they are *Bitter Plants*, which are best for the making of *Extracts*. And the distilled waters of *Plants* which are *Hot* and *Bitter*, notwithstanding that they always taft high of the *Heat*, yet rarely and very faintly of the *Bitter*.

13. §. *Aftringency*, is made, partly, by the further increase and more intimate union of the *Earth*. And therefore this is seated still in a more *Fixed Compofition*, than a *Bitter*. And partly, by the diminution of the *Sulphur*. And therefore the *Acid Parts* ingredient to it, either by *Fermentation* or otherwise, are easily exposed. *Aftringency* being the *Womb* or *Bud* of a *Sower*. For all or most *Aftringent Roots* bear a sower *Leaf*, or a sower *Fruit*; as those of all *Docks* and *Sorrels, Black-Thorn, Dog-Rose*, and others. Wherefore also, *Aftringency* is often found in conjunction with *Bitter, Sweet*, or *Sower*; but scarce ever with *Pungent*, or *Hot*.

14. §. An *Aromatick Taft*, seems to be produced, chiefly, by a *spirituous, acid*, and *volatile Sulphur*; as in *Ambar-griese, Cardamon-Seeds*, many *Stillatitious Oyls* &c. A *Naufeous*, by a *Sulphur* less *Spirituos* and *Volatile*, and more *Alkaline*; as in the *Root* of *Dog-ftones, Sheep-fcabious*, the young and green *Leaves* of *Coriander*, or the *Seeds* of *Cumine*. The *Spirit*, as it enters the *Nerves*, carrying the *Alkaline Sulphur* along with it; as when a City is betrayed by one of its Inhabitants to an Enemy.

25 §. An *Intermittent Tafte*, as in *Arum*, seems to have its dependance upon a simple and very pure *Nitre*, which by its subtilty enters into the very *Concaves* of the *Nervous Fibers* of the *Tongue*: and so being lodged there, is little affected or stirred, by the *Motion* of the *Blood*; but only when the *Tongue* it self is moved, at which time it causeth a kind of pricking *Tafte*.

16. §. A *Tremulous Tafte*, as in *Pyrethrum*, dependeth probably, upon an *Aereal Sulphur*; which being agitated by the *Blood* in its *Circulation*, the *springy Motion* or *Vibration* of the *Aereal Parts* produce that *Tafte*.

17. §. A *Taft* is *Lingual, Guttural*, &c. according to the grofnefs or finenefs or other difference of the *Membranes* into which the *taftable parts* are admitted. For *Tafts* are made not meerly by the outward *Contact*, but the *Ingrefs* of the *taftable parts*. Now the outer *Skin* of the *Tongue*, which is commonly obferved to pill off in boyling, like the *Cuticula* in other *Parts*, hath either no fenfe, or much lefs than that which lies under it; and is therefore, but a *Seive* or *Strainer* to the *taftable parts*. So that being of different finenefs in the feveral parts of the

U u

Tongue;

Tongue; it hereby comes to pass, that according as the *tastable parts* of any *Plant* are more or less penetrant, subtle, or dissoluble, they are admitted into one part of the *Tongue*, and not another. And in the *Throat*, the outer *Skin* it self, seems to be the immediate *Sensory*; and so, to be evidently affected with the *Juyces* of some *Plants*, from which the *Tongue* receiveth little or no sensible *Impression*.

18. §. When the *Tast* is *Permanent* and *Fixed* in some one *Part*; it is a sign, either that the *Gustable Parts* are less dissoluble; or more subtle, so as to enter the *Concaves* of the *Fibers*; and that there is an admixture of an *Aereal Salt*, or a like *Sulphur*; some of the parts whereof, being crooked, hang like *Hooks* on the *Fibers* of the *Tongue*. For the reception of such a *Tast*, is not to be looked upon as a wound made with a *Lancet*, and so the *Lancet* taken away; but with the *Lancet* sticking in the wound; until in time, 'tis carryed off by the *Circulation* of the *Blood*; which like the *Stream* of a *River* in a *Flood*, carries all before it, but those things last, which stick in the *Mud*.

19. §. But when the *Tast*, though *Permanent*, yet is *Diffusive* or *Transitive*; it seems probable, that as there is a less admixture of *Aer*; so a greater subtlety of the *Tastable Parts*, whereby they are conveyed, through the *Nervous Fibers*, from one *Part* to another.

CHAP. V.

Of the Judgment which may be made of the VIRTUES *of* Plants, *from their* Tasts.

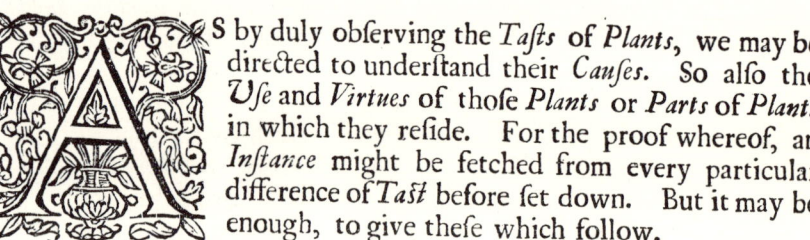

S by duly observing the *Tasts* of *Plants*, we may be directed to understand their *Causes*. So also the *Use* and *Virtues* of those *Plants* or *Parts of Plants* in which they reside. For the proof whereof, an *Instance* might be fetched from every particular difference of *Tast* before set down. But it may be enough, to give these which follow.

2. §. And first, we may make no ill guess *ex Analogia*, or where we find the same *Tast*, that there the same *Virtue* in some kind, and in some degree, may reside. So *Jalap*, *Mercury*, and *Daisy*, have all of them that exasperating *Tast* in the *Throat* before described; and they are all three more or less *Cathartick*. Wherefore, we may believe, that other *Plants* which make the like *Impression* on the *Throat*, and there are many others which do, that they are in some degree alike *Cathartick*. Those *Plants* which are reckoned amongst the chiefest *Cephalicks*, cause rather a durable, than a vehement *Heat* upon the *Tongue*, as *Pyrethrum*, *Euphorbium*, *Black-Hellebore*, &c. It seemeth therefore reasonable to rank with these, any other *Plant*, though not used, which produceth the like durable *Heat*. The young *Roots* of *Yarrow*, or *Millefolium*,

have

have the same *Taste*, as the *Root* of *Contrayerva*: and may therefore be used for the same purpose, with a probability of the like success; if not a better, because they may be gotten fresher. But by drying the *Root*, the *Tast* and *Virtue*, which lie in its exhalible parts, are much lost. The *Seeds* of the lesser *Cardamom*, and of *Zedoary Root*, if sound, have both a smatch of the *Tast* of *Camphire*. They may therefore all, so far, reach the same Case.

3. §. Again, as we may make no ill conjecture from the sameness of *Taste* in *Plants* of several *Tribes*; so from the diversity of *Taste*, in those of the same. So the *Flowers* of all the *Docks* are evidently *Astringent*, and not *Sower*; except those of the *Rha-pontick*, which are extream sower, even in the 5th degree. Which is no mean *Signature* of some more than ordinary *Virtue* in it, besides what it hath in common with the rest of the *Tribe*. The *Flowers* of *Pancy* have a kind of fulsome *Tast*, plainly different from that of *Violets*: and in some *Hypochondriacal Cases* may be more useful.

4. §. It likewise importeth much, to observe the difference of *Taste* in the several *Parts* of the same *Plant*. So the *Barque* of *Sassafras* is three times as strong, as the *Wood*: and the like may be observed in any other commonly known *Tree*. If therefore we could obtein the *Barques* of *Santalum*, *Lignum Rhodium*, *Lignum Aloes*, &c. they would doubtless, most of them, be of much greater use. And as the *Taste* is sometimes stronger; so, much more grateful, in one *Part* than in another: as in the *Flowers* or *Yellow Attire* in the *Heads* of *Carduus Benedictus*; which being infused in *Spirit* of *Wine*, or other convenient *Liquor*, make a pleasant *Cordial*. Nature having laped up the *Virtue* in the *Leavs*, as in a brown *Paper*; but in the *Flowers*, as in *Leaf-Gold*.

5. §. As also, how far the *Taste* of any *Plant* may alter, either in preserving, or preparing it. So the *Root* of *Arum*, when taken fresh out of the ground, is notably *Pungent*: but being throughly dryed, and especially kept for some time, hath no more *Taste*, and therefore in all likelyhood, no more *Virtue*, than a *Lump* of *Starch*. The like we are to judge of all other *Plants*, whose *Virtue* lieth in their exhalible *Parts*. The *stillatitious Oyls* of many *Plants*, are stronger than the *Leavs* or other *Parts* from whence they are drawn: but some there are, which are weaker; as is that of *Euphorbium*, in which the *Heat* is neither pertinaceous, as in the *Gum* it self, nor so great.

6. §. We may make, moreover, a jugdment from the *Nature* of the *Tast*. So those *Roots* which are *Bitter*, and not *Hot*, as of *Cichory*, and the rest of the *Intybous* kind, may be accounted *Nitro-Sulphureous*; and so, to be *Absterfive* without any *Heating Quality*. The *Marum Austriacum*, which is extream *Pungent*, as well as *Aromatick*, may be looked upon as the best *Cephalick* of that *Tribe*. Because we find, that *Jalap* hath a special property of imitating the *Glandulous Parts* of the *Mouth*, and *Throat*; we may gather, That it is a better *Purge* to all the other *Glandulous Parts*, than most other *Catharticks*. Which is also one reason of its operation, for the most part, with at least a tendency to vomit; the *Stomach* it self being *Glandulous* as well as the *Throat*, and thereby answerably affected with it. A strong *Infusion* of *white Sarzaparilla* in *Water*, botled up, and kept in a *Cellar* for the space of two months, becomes extream sower; far beyond any thing observed in the *Tasts* of the *Juyces* and *Infusions* of divers other

Plants kept as long and in the same manner. Which shews, how well Nature hath adapted a *Plant* of so mild a *Taste*, either by similitude of parts, for the carrying off of any *preternatural Acid*; or by contrariety, for the curbing of an exorbitant *Salt*. The *Barque* of the *Root* of common *Wormwood*, which impresseth a pertinaceous and diffusive *Taste*, which descendeth from the *Tongue* into the *Gulet*, as is before described; may be justly ranked with the most excellent *Stomachicks*; and upon tryal, I find it one of the best: besides, that it is neither unpleasant, nor affecteth the *Head*, as the *Leavs*. Yet the *Gardener*, and every Body throws it away, as good for nothing.

7. §. I shall conclude with one note, which is this; That the *Specifick Virtue* of *Medicines*, which some *Physicians* positively deny, and most dispute; from some of the forementioned *Differences* of *Taste*, as well as for other reasons, may seem, at least, to be probable. For why should not a *Medicine* make an *Impression* upon one *Part*, and not upon another, within the *Body*, as well as we find it doth within the *Mouth*? especially, since the *Parts* of the *Mouth*, are of a less different *Nature*, than some of the *Viscera*.

An *Appendix*.

Of the ODOURS of Plants.

THE *Senses* of *Tasting* and *Smelling* being so nearly ally'd; many things already explained concerning the *Diversities* and *Causes* of *Tasts* in *Plants*, may easily be transferr'd to those of their *Odours*. I shall now therefore only remarque some particulars, not commonly taken notice of hitherto, and leave them as a *Specimen* to be Improved by other Hands.

§. 2. The *Root* of *Rape-Crowfoot* being cut, and held to the *Nose*, when it is newly taken out of the *Ground*, smelleth almost like *Spirit* of *Sal Armoniac*, or fresh *Scurvygrass Juyce*. And hath the property of making the *Eyes* to water, as *Onions* do. *Horse-Radish Root* is not so *Pungent* to the *Nose*, but gets pretty much into the *Eyes*. But that of *Dragon*, doth neither affect the *Eyes*, nor the *Nose*.

3. §. The *Succulent Roots* of *Dogstones*, and most of that *Tribe*, have a ranck *Smell*. And that of *Crown Imperial*, being rub'd a little, smells as like a *Fox*, as one *Fox* smelleth like another.

4. §. The *Root* of *Patience* digested with *Water*, in a warm *Room*, for the space of three weeks, smels like *Spirit of Harts Horn*, or other *Urinous Spirit*. Of *Red Dock*, almost like *Aqua fortis* or *Spirit of Nitre*. That of *Dragon* bottled up with *Water*, and set in a *Cellar*, about a *Month*, stinks like the *pus* of the most *Fetid Ulcer*. At the end of five Months, more abominably, than either to be endured or expressed.

5. §. The *Leavs* of *Mountain Calamint*, smell like *Peny Royal*. Those of *Ulmaria*, like *Walnut Pills*. Of *Yellow Lamium*, like a *Balsame*. Of *Sena*, a good quantity being held to the *Nose*, of a rank

Smell betwixt that of *Sweat* and *Urine*. Of *Coriander*, when green and young, ſtink ſo baſely, that they can hardly be endur'd. Sometimes the *Leavs* have a ſtronger *Smell*, than the *Flower*, as in *Borage*, and ſometimes the *Stalk*, a ſtronger than the *Leavs*, as in *Ulmaria*.

6. §. *Rue Leavs* corked up in a bottle and ſet in a *Cellar* for about ten weeks, ſmell like *Spirit* of *Harts Horn*, or of *Urine*. The green *Leaves* of *Roſes* infuſed in water, have a mild, but pleaſant *Smell*. Neither is that of *Savine* unpleaſant, upon the like *Infuſion*.

7. §. *Scurvygraſs Juyce* kept about ¾ of a year in bottles, with the green *Sedement*, in a warm *Room*, ſtinks like Humane *Excrements*. And *Scurvygraſs Wine*, made only of the *Juyce*, ſmells like ſome *Iſſues*.

8. §. The *Flowers* of *Yarrow*, ſmell not much unlike to thoſe of *Southernwood*. And the *Flowers* of *Crowfoot* almoſt like thoſe of *Scurvygraſs*. Some *Flowers* are of a weaker *Smell* in the *Bud*, as thoſe of *Mallow*. But many have a ſtronger, than when they are blown open; as thoſe of *Lavender*, *Roſemary*, &c.

9. §. The *Buds* of *Vervaine Mallow*, while they are young, and the *Flowers* unſeen, have a very pleaſant *Smell*, like that of *Geranium Moſchatum*: but when afterwards they are opened they have an unpleaſant *Smell*. Common *Mallow Flowers* dryed and bottled up for ſome time, acquire, though not a ſtrong, yet very noyſom *Smell*.

10. §. The *Purple Pouch* of *Dragon* which covers the *Seed*, being broken, ſmells juſt like a *Lobſter*. But permitted to lie in a warm *Room* for ſome days, ſmells exactly like *Carrion*; and ſcents the *Room* with the ſame *Smell*.

11. §. Some *Seeds* as thoſe of *Cumine*, *Daucus*, being powdered and laped up only in *Papers*, do notwithſtanding retein their *Smell*. But many others, as of *Sweet Fenil*, in a ſhort time, loſe it. Some *Seeds*, when they firſt begin to ſprout, become *Odorous*, which were not ſo before; as the *Garden Bean*.

Tabula, quâ perspicuè videre est, quot Triplicati Sapores, ex solummodo decem Simplicibus numerantur.

AMARUS	Am.du.ac.						
	am.du.sa.	am.ac.sa.					
	am.du.ca.	am.ac.ca.	am.sa.ca.				
	am.du.fr.	am.ac.fr.	am.sa.fr.	am.ca.fr.			
	am.du.ar.	am.ac.ar.	am.sa.ar.	am.ca.ar.	am.fr.ar.		
	am.du.ma.	am.ac.ma.	am.sa.ma.	am.ca.ma.	am.fr.ma.	am.ar.ma.	
	am.du.as.	am.ac.as.	am.sa.as.	am.ca.as.	am.fr.as.	am.ar.as.	am.ma.as
	am.du.pu.	am.ac pu.	am.sa.pu.	am.ca.pu.	am.fr.pu.	am.ar.pu.	am.ma.pu.
							(am.as.pu.

DULCIS	Du.ac.sa.						
	du.ac.ca.	du.sa.ca.					
	du.ac.fr.	du.sa.fr.	du.ca.fr.				
	du.ac.ar.	du.sa.ar.	du.ca.ar.	du.fr.ar.			
	du.ac.ma.	du.sa.ma.	du.ca.ma.	du.fr.ma.	du.ar.ma.		
	du.ac.as.	du.sa.as.	du.ca.as.	du.fr.as.	du.ar.as.	du.ma.sa.	
	du.ac.pu.	du.sa.pu.	du.ca.pu.	du.fr.pu.	du.ar.pu.	du.ma.pu.	du.as.pu.

ACIDUS	Aci.sal.cal.				
	aci.sal.fri.	aci.cal.fri.			
	aci.sal.aro.	aci.cal. ar.	aci.fri.ar.		
	aci.sal.mal.	aci.cal.ma.	aci.fri.mal.	aci.ar.mal.	
	aci.sal.ast.	aci.cal.ast.	aci.fri.ast.	aci.ar.ast.	ac.ma.ast.
	ac.sal.pu.	aci.ca pun.	aci.fr.pun.	aci.ar.pun.	ac.ma.pu. ac. ast.pu.

SALSUS	Sal.cal.fri.				
	sal.cal.aro.	sal. fri. aro.			
	sal.cal.mal.	sal.fri.mal.	sal.aro.mal.		
	sal.cal. ast.	sal. fri. ast.	sal.aro.ast.	sal.ma.ast	
	sal.cal.pun.	sal.fri.pun.	sal.aro.pun.	sal.ma.pu.	sal.ast.pun.

CALIDUS	Cal.fri.aro.			
	cal.fri.mal.	cal.aro.mal.		
	cal.fri. ast.	cal.aro.ast.	cal.mal.ast.	
	cal.fri. pun.	cal.aro.pun.	cal.mal.pun.	cal.ast.pun.

FRIGIDUS	Fri.aro.mal.		
	frig.aro.ast.	fri.mal. ast.	
	fri.aro.pun.	fri.mal.pun.	fri.ast.pun.

AROMATICUS	Aro.mal.ast.	
	aro.mal.pun.	aro. ast. pun.

MALIGNUS

ASTRINGENS

PUNGENS

Tabula,

Lect. VI.   295

Tabula, quæ Genericas omnes Saporum differentias comprehedit.

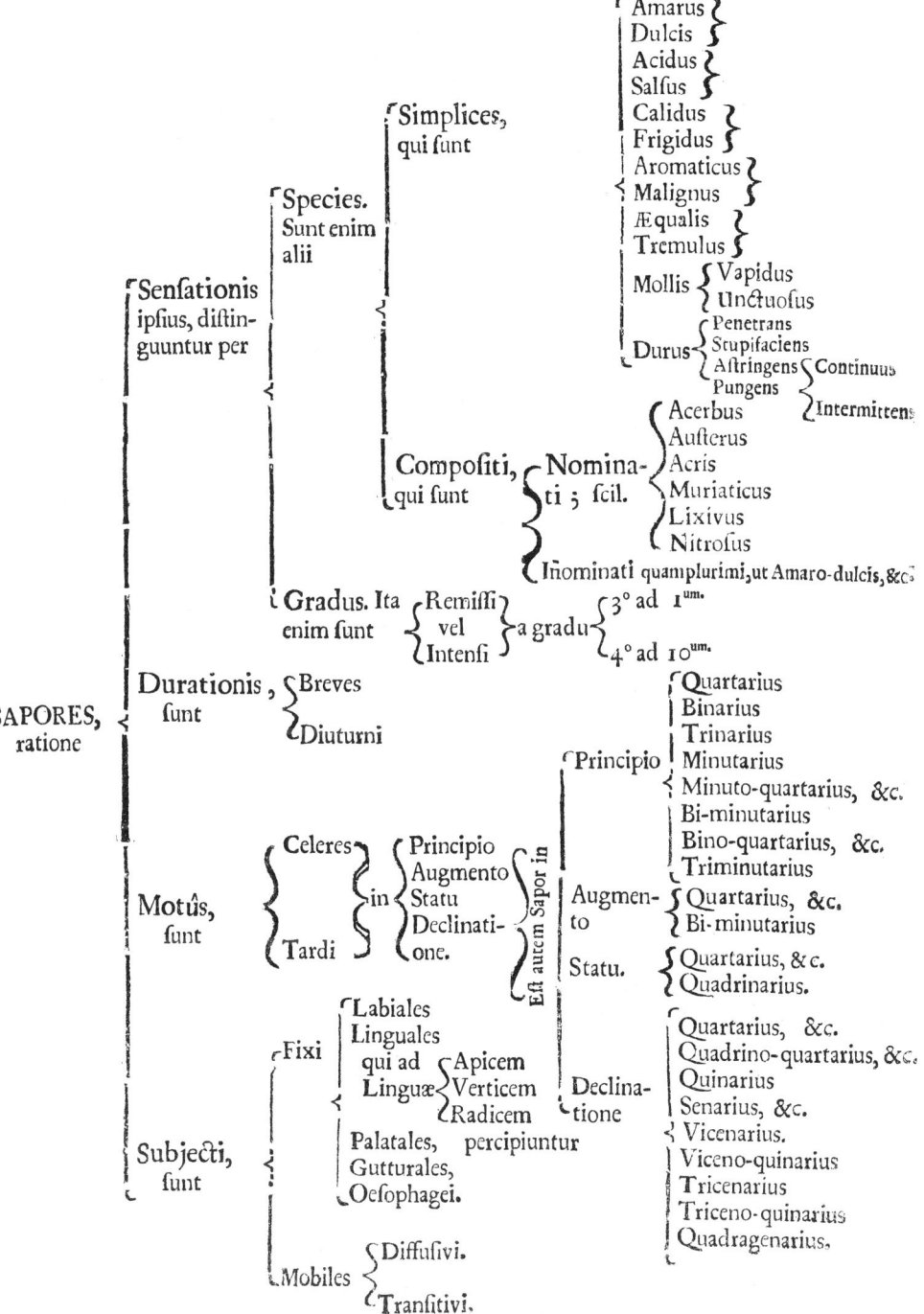

EXPERIMENTS IN CONSORT UPON THE Solution of Salts IN WATER.

Read before the *Royal Society*, *January*, 18. 167$\frac{6}{7}$.

CHAP. I.

In which is shewed, the Compleat or Utmost Impregnation of WATER with several kinds of Salt, both together, and apart.

IN discourse upon a *Lecture* formerly read, concerning the *Lixivial Salts* of *Plants* ; It was mentioned, as a thing asserted by some *Phylosophers*, That *Water* having been fully impregnated with one kind of *Salt*, so as to bear no more of that kind; it would yet bear, or dissolve some portion of another; and so of a third. And it was referred to Me by this Honourable Chair, to examine and produce the *Experiment*. The doing whereof brought into my mind divers other *Experiments* hereunto relating.

2. §. As next, With what difference of quantity this *Superimpregnation* would be made, upon the *Solution* of different *Salts* ?

3. §. Thirdly, Whether the *Solution* of a smaller quantity of several *Salts*, doth consist with the *non-increase* of the bulk of the *Water* ? Because this also is affirmed by some.

4. §.

4. §. *Fourthly*, What quantity of the several kinds of *Salt*, may be diffolved feverally, in the fame quantity of *Water*?

5. §. *Fifthly*, Whether by diffolving a *Salt* in *Water*, there be any *Space* gained, or not? That is, whether the *Bulk* of the *Water* be greater, before the *Salt* lying in it be fully diffolved, than it is afterwards? Or if a *Cubick Inch* of *Salt* be diffolved in nine *Cubick Inches* of *Water*; Whether the *Water* will then fill a *Veffel* of ten *Cubick Inches* content?

6. §. *Sixthly*, Whether the *Space* be equally gained, by an equal encreafe of the fame *Salt*?

7. §. *Seventhly*, Whether upon the *Solution* of feveral kinds of *Salts*, be gained fo many feveral quantities of *Space*? That is, if the *Solution* of common *Salt* gains, fuppofe, an *Inch*, whether the *Solution* of *Salt Armoniack* gains as much, or more, or lefs? and fo for other *Salts*.

8. §. *Eighthly*, What that juft fpace may be, which any *Salt* gaineth with refpect to its own *Bulk*, or that of the *Water*?

9. §. And firft, for the *Superimpregnation* of *Water*; I put into a bottle ℥ij of fair *Water*; adding thereto, firft half an Ounce of *Nitre*; and afterwards more, as the *Water* would diffolve it; and (that I might be fure the *Impregnation* was full) fome portion above what the *Water* would bear. Then having feparated this remaining portion; I put to this *Solution* of *Nitre*, two Drachms of *Sal Armoniac*; which wholly and eafily diffolved in the faid *Solution*; though it would not bear a grain more of *Nitre*. I then added a third Drachm of *Sal Armoniac*, after that a fourth, and a fifth; all which, within the fpace of half an hour, were perfectly diffolved in the faid *Solution*, without any precipitation of the *Nitre*.

10. §. In the making of this Experiment, two things, to render it infallacious, are to be noted. That the faid *Salts* were not diffolved by the help of *Fire*, but only by a ftrong and continued *Agitation*. And that this was done upon a warm day: which I mention, becaufe that even the changes of the weather will fomewhat alter the *Solubility* of the *Salts*.

11. §. Having made the Experiment upon two *Salts*, I proceeded to repeat it upon three. And firft I diffolved as much common *Salt* in ℥ij of *Water*, as that quantity would bear. Then having feparated the fubfiding portion; I put to the *Solution*, no lefs than five Drachms of *Nitre*, which by a continued *Agitation*, was wholly diffolved therein, neither the *Nitre* nor the common *Salt* being in the leaft precipitated. Then adding a Scruple more, it would not diffolve, but fubfided. This fecond fufiding portion, I again feparated; and then put to this *Superimpregnation*, near ℨj of *Sal Armoniac*, which was alfo diffolved as the former. And if as many more *Salts* had been added, tis probable that the fame *Water* would have born fome quantity of them all.

12. §. From this Experiment, it is a Conclufion demonftrated, That not only the vifible *Cryftals*, but the very *Atomes* of every *Salt*, at leaft thofe *Particles* which are ultimately diffolved in *Water*, have a different *Figure* one from another. Becaufe that if they were all of one *Figure*; there would be no *Superimpregnation*, but the *Pores* of the fame *Water*, would imbibe as much of one *Salt*, as anfwers to the total of two more *Salts* imbibed: that is to fay, it would as well imbibe two Ounces of common *Salt*, as one Ounce of common *Salt* and

another of *Nitre:* which yet is contrary to the Experiment. And it is the same thing, whether we suppose the *Pores* of *Water* to be also different, or not. Because, that if the *Figure* of all the said *Atomes* be the same; then their respect to the *Pores* of the *Water* must be the same, how different so ever those *Pores* be: which is also contrary to the Experiment. Besides it is a great presumption, to say, that the *Pores,* and therefore the *Atomes* of *Water* have different *Figures*; and yet not those of *Salts.*

13. §. From the same Experiment we may go upon good ground in *Compounded Infusions*; whether of *Purgative,* or other *Materials.* As not doubting, but that the same *Menstruum* may be highly impregnated with several *Ingredients* at once, whose operative parts may be therein copiously dissolved, without hindring either an *Extraction,* or causing a *Precipitation* one of an other.

14. §. The *Second* Enquiry is, With what difference this *Superimpregnation* of *Water* is made? which I find considerable. For a *Solution* of above five Drachms of *Nitre* may be *superimpregnated* with no less quantity of *Sal Armoniac.* And a *Solution* of five Drachms of common *Salt,* may be *superimpregnated* with as much *Nitre.* Yet neither a strong *Solution* (as of five Drachms) of common *Salt,* will bear above two Scruples of *Sal Armoniac:* nor will a strong *Solution* (as of five Drachms) of *Sal Armoniac,* bear above a Drachm of common *Salt:* for if above the said quantities of either of them be mixed together: they are both copiously and forthwith precipitated to the bottom of the *Glass.*

15. §. Whence, notwithstanding the former Experiment, yet are we admonished, not to infuse all manner of *Ingredients* in any proportion. Because though some do not, yet others will precipitate one another.

16. §. The *Third* Enquiry was this, Whether the *Solution* of a smaller quantity of several *Salts,* doth consist with the *non-increase* of the *Bulk* of the *Water?* For this I took a *Bolthead* with a slender *Neck,* conteining about a pint and a quarter of *Water;* and dissolved therein about ʒjß of *Nitre.* And marking the place to which the *Water* ascended in the *Neck* of the *Bolthead:* I then dissolved in the same *Water* about a Drachm of *Sal Gemmæ:* which little quantity raised the *Water* above half an Inch higher then it was before. The like I observed in the addition of *Nitre* to a *Solution* of *Sal Armoniac.* So that to suppose the variation of the *Salt* doth prevent the increase of the *Bulk* of the *Water,* is a manifest Error.

17. §. From the same Experiment it also appears, That the ascent of the *Water* upon a *superimpregnation,* is the same, by whatsoever *Salt* the first *Impregnation* be made. For instance, Let a *Solution* of *Nitre* ascend in the *Neck* of the *Bolthead,* suppose, to 10 Inches, then add ½ an Ounce more of *Nitre,* so as to raise the *Water,* suppose, 12 Inches or more, or less, according to the *Bore* of the *Neck.* In like manner, let a *Solution* of *Sal Armoniac* reach to ten Inches: then add again half an Ounce of *Nitre*; and it will reach just 12 Inches, or more or less, as before.

18. §. The *Fourth* Enquiry is, What quantity of the several kinds of *Salt,* may be dissolved severally in the same quantity of *Water:* that is to say, by agitation alone, without the help of fire, as I noted before.

before. And upon tryal it appears, Firſt, that two Ounces of *Water* will diſſolve three Ounces of *Loaf-Sugar* and no more, except the *Water* be heated.

19. §. The ſame quantity of *Water* that is, two Ounces will diſſolve above two Ounces of *Salt of Tartar*. I ſay above, for how much more, want of a greater quantity of *Salt* which I could confide in, made me that I could not finiſh the Experiment.

20. §. The ſame quantity, *ſc.* two Ounces of *Water*, diſſolveth an Ounce and a Drachm of *Green Vitriol*.

21. §. The like quantity diſſolveth ſix Drachms and a Scruple or above ¾ of an Ounce of common *Salt*.

22. §. Of *Nitre*, Five Drachms two Scruples and an half.

23. §. Of *Sal Armoniac*, five Drachms and two Scruples.

24. §. Of *Alum*, not above two Drachms and a Scruple.

25. §. And of *Borax*, not above a Drachm and half a Scruple.

26. §. Of theſe note, That although Common *Salt* be very diſſoluble, and will preſently catch the moyſture of the *Aer:* yet a much greater quantity not only of *Salt* of *Tartar*, but even of *Loaf Sugar*, and of *Green Vitriol* it ſelf, may be diſſolved in *Water* than of Common *Salt*.

27. §. Again, as the great *Solubility* of ſome, ſo the leſs *Solubility* of other *Salts* is alſo obſervable, as of *Alum*, and *Borax*. For the ſame quantity of *Water* will diſſolve near four times as much of *Green Vitriol*, as it will of *Alum*. And of *Sugar* more than ten times as much. Of *Green Vitriol* near eight times as much as of *Borax*; and of *Sugar*, twenty times as much.

28. §. From this Experiment we are likewiſe cautioned, not only in the *Infuſion* of ſeveral *Ingredients* together, but of any one ſingly; that ſuch a proportion thereof to the *Menſtruum*, be not exceeded. For all that is over and above what the *Menſtruum* will bear, is either not extracted, or will be precipitated. As is evident not only in the *Diſſolution* of the *Salts* above named, but in the *Infuſion* of *Plants* themſelves: as, for inſtance; of *Senna*; two Drachms whereof will impregnate four Ounces of *Water* as ſtrongly, as if twice the quantity were infuſed; becauſe the *Water* will bear no more of the *Purgative Parts* of that *Body*.

29. §. There is only one *Salt* more remaineth to be ſpoken of under this Experiment; and that is, the *Cryſtals* of *Tartar*. Whereof, it is ſomewhat ſtrange to obſerve, that it will ſcarce at all diſſolve in *Water:* not more, than even divers *Reſinous Gums*, as *Maſtick*, *Tolu*, *Tacchamahacca*, and ſome others will do. For if two Drachms, ſuppoſe of theſe *Cryſtals*, of *Tartar* (commonly ſold for *Cremor Tartari*) be put to one Ounce of *Water*, ſcarce five Grains thereof will, by *Agitation*, be therein diſſolved.

CHAP. II.

In which is shewed, that by the Solution *of Salts in* Water, *some certain space, more or less, is gained. That the space is different according to the Nature of the* Salt. *And what the just space is, which is gained.*

HE *Fifth* Enquiry is, Whether by dissolving of a *Salt* in *Water*, there be any space gained, or not. That is, whether the *Bulk* of the *Water* be greater before the *Salt* lying in it be fully dissolved, than afterwards. For tryal whereof, I took a *Bolt-head* with a slender *Neck*, holding somewhat more than a pint; and filling it up to a certain place in the *Neck*; I then put in an Ounce or two of *Salt*. And observing the hight of the *Water*, both before it was dissolved, and afterwards; It plainly appeared, that there was some, and that a considerable space, gained by the *Dissolution*; the *Water* thereby sinking several Inches below the place, where it stood after the *Salt* was first put into it.

2. §. From this Experiment it is plain, that there are *Vacuities* in *Water*. That is to say, that all the parts of *Water* are not contiguous, but that either betwixt, or in the *Atomes* of the *Water* themselves, there are certain *Pores*, either absolutely void, or at least filled up with another more subtile body which is easily excluded by the particles of *Salt*: by possessing the room of which the above said space is gained.

3. §. The *Sixth* Enquiry is, Whether the space be equally gained, by an equal encrease of the same *Salt*.

4. §. For this I made two tryals; the first was this. Two half Ounces of *Salt Armoniac*, being successively dissolved in the same *Water*; both of them raised up the *Water* in the *Neck* of the *Bolt-head*, equally; the first 3 Inches ⅞, and so the second.

5. §. The other was this. Four half Ounces of *Nitre*, being successively dissolved in the same *Water*, they all of them raised up the *Water* in the *Neck* of the *Bolt-head*, equally; the first a little above two Inches, and the 2ᵈ, 3ᵈ, and 4ᵗʰ, just as much.

6. §. The *Seventh* Enquiry is, Whether upon the *Dissolution* of several kinds of *Salts*, be gained so many several quantities of space. For this I made tryal upon Eleven several *Salts*, sc. *Salt of Tartar*, *Common Salt*, *Sal Gemmeus*, *Roman Vitriol*, *Nitre*, *White Vitriol*, *Green Vitriol*, *Alum*, *Borax*, *Loaf-Sugar*, and *Sal Armoniac*; of all which, I dissolved an equal quantity sc. two Ounces, in an equal quantity of *Water*, severally; that is, taking fresh *Water* for every *Solution*. The success was, That the *Sal Armoniac* raised the *Water* 15 Inches. The *Loaf-Sugar*, 13 Inches and ⅜ths. The *Borax*, a Foot. The *Alum* 11 Inches, and ⅝ths. *Green Vitriol*, 9 Inches and ⅝ths. *White Vitriol*, 9 Inches and ⅛th. *Nitre*, 8 Inches, and ⅜ths. *Roman Vitriol*, 7 Inches and

and ⅝ths. *Sal Gemmæ*, 6 Inches, and ⅝ths. *Common Salt*, 6 Inches and ⅜ths. *Salt of Tartar*, not above 4 Inches and ⅝th. All which differences are plain, and moſt of them very remarquable: Two Ounces of *Sal Armoniac* raiſing the *Water* near four times as high, as the ſame quantity of *Salt of Tartar*.

7. §. From this and the fourth Experiment, compared, it alſo appears, That the ſeveral ſpaces gained by the ſeveral *Salts*, though ſometimes they do, yet do not always anſwer to the *Solubility* of the ſaid *Salts*. As to give ſome Inſtances; *Loaf-Sugar* is the moſt diſſoluble of any other *Salt*; yet it gaineth leſs ſpace than all the reſt, ſave only *Sal Armoniac*. So *Green Vitriol* is more diſſoluble then either *Nitre* or *Common Salt*; yet gaineth leſs ſpace than either, eſpecially than the latter. And *Sal Armoniac*, which is more diſſoluble than *Alum* or *Borax*, yet gaineth leſs ſpace than either of them. The *Cauſe* whereof is not eaſily aſſigned.

8. §. Note alſo, that by the ſame Experiment, as well as by the *Taſte* and other Circumſtances, it is plain, That *Sal Gemmæ* is nothing elſe but *Common Salt*, coagulated or *Cryſtalliz'd* under *Ground*.

9. §. Again, as the Fifth Experiment ſheweth, That there are *Vacuities* in *Water*: ſo doth this Laſt, that thoſe *Vacuities*, are of differing kinds. Becauſe, otherwiſe, it ſhould ſeem, That the *Bulk* of the *Water* would increaſe, more or leſs, according to the *Solubilitie* of every *Salt*, and not be alternately differenced as it is; Some *Salts*, more diſſoluble, increaſing the *Bulk* of the *Water* leſs, and others leſs diſſoluble, increaſing it more. I ſay, that this difference dependeth not only upon the different *Figures* of the *Atomes* of *Salt*; becauſe then every *Salt* which is more diſſoluble, would (quantity for quantity) take up leſs room in the *Water*: which is contrary to the Experiment.

10. §. From the ſame Experiment, howſoever *paradoxical* it may ſeem, yet is it alſo manifeſt, That although *Water* be a *Fluid*, yet the *Particles* thereof are *hard* and *conſiſtent*, and unalterable in their *Figure*. Otherwiſe it is plain, That all manner of *Salts* would be diſſolved in the ſame manner, and take up the ſame room in the *Water*. For let the *Figures* of the *Salts* be never ſo various, yet if the *Particles* of *Water* were themſelves *Fluid* or *Inconſiſtent* and *Alterable*, they would always ſo conforme to thoſe *Figures*, as to fill up all *Vacuities*; and ſo upon the *Solution* of ſeveral *Salts*, if of equal quantity, the *Water* would ſtill retein an equal *Bulk*. As ſuppoſe an Ounce of *Iron* were drawn into *Wyer*, another beaten into *Plates*, a third made into *Hooks*, a fourth into *Needles*, a fifth into *Nails*; every one of theſe five Ounces, being put ſeverally into *Water* will encreaſe its *Bulk* equally. I conclude therefore, That the *Atomes* of *Water* are hard and unalterable.

11. §. The *Eighth* Enquiry was this, What that juſt ſpace might be, which any *Salt* gaineth upon *Diſſolution*, with reſpect to its own *Bulk*, or the *Bulk* of the *Water*? For the making of this Experiment, *Water* will not ſerve, nor yet *Spirit of Wine*; becauſe they both of them diſſolve more or leſs of thoſe *Salts* which are put into them; whereby the obſervation of the true *Bulk* of the *Salt*, and conſequently of the juſt ſpace it gaineth by *Diſſolution* is loſt. I took therefore *Oyl of Turpentine*, and pouring it into a *Bolt-head*, marked the place of its aſcent in the *Neck*. Then pouring likewiſe into it two Ounces of

Common

Common Salt, I marked the second ascent of the *Oyl*; and found it to be 10 Inches and 6 eighths. Repeating the Experiment in like manner with two Ounces of *Nitre*, I found the ascent of the *Oyl* to be 11 Inches and $\frac{1}{8}$th. Repeating it again with two Ounces of *Alum*, the ascent of the *Oyl* was 13 Inches and $\frac{3}{8}$ths. And making it once more with *Sal Armoniac*, the *Oyl* ascended to 15 Inches: the said several ascents of the *Oyl* being the true spaces which the Four abovesaid *Salts* take. From which, the space which the same *Salts* take up upon *Dissolution*, being deducted; the remainder is the space gained by that *Dissolution*. And so it appears, first, that *Sal Armoniac* gaineth nothing; being the only *Salt* of all I have tryed, which causeth the equal ascent both of the *Water* and the *Oyl sc.* just 15 Inches in both. *Alum* causeth the ascent of the *Oyl* to 13 $\frac{3}{8}$ths, of the *Water*, to 11 Inches and $\frac{5}{8}$ths: So that it gains about 1 Inch and $\frac{1}{2}$ out of 13. *Nitre* causeth the ascent of the *Oyl*, to 11 Inches and $\frac{1}{8}$th; of the *Water*, to 8 Inches and $\frac{3}{8}$ths. So that *Nitre* by *Dissolution* gets almost the space of 3 Inches in 11. *Common Salt* causeth the ascent of the *Oyl*, to 10 Inches and $\frac{6}{8}$ths; of the *Water*, 6 Inches and $\frac{3}{8}$ths. So that *Common Salt* gains by *Dissolution* 4 Inches in 10, which is very considerable.

12. §. By this way the *Specifick Gravity* of all kinds of *Salts* may be easily taken, and the difference betwixt them is somewhat surprizing. For it appears by the Ascent of the *Oyl*, that *Nitre*, quantity for quantity, is about a 22th part lighter than *Common Salt*. *Alum* about a 6th part lighter. And *Salt Armoniac*, almost a 4th part lighter than *Common Salt*. The like estimate may be made of the *Gravity* of all other *Salts*.

13. §. By the same Experiment it also appears, That according to the *Specifick Gravity* of *Salts* they are many times at least more or less *Volatile*; as in the four last *Salts* is plain. For *Common Salt* which of all the four is the most fixed, is also the heavyest. *Nitre* which is somewhat less fixed is somewhat lighter. But *Alum* which is still less fixed is much lighter. And *Sal Armoniac* which is wholly *Volatile*, is the lightest of all the *Salts* above mentioned.

CHAP. III.

Wherein, from the Experiments in the foregoing Chapter, is shewed, the Cause of the Motion of the Mercury in the BAROMETER.

OR the doing of this, it will first be acknowledg'd, That not only several sorts of *Sulphur*, but also of *Volatile Salts*, are continually sublimed from most *Bodies* into the *Aer*. So *Lightning*, from the celerity of the accension, appears to be made of a *Meteor*, which is *Nitro-Sulphureous*. *Snow* dependeth upon a *Mixture* of *Nitrous*, and other *Salts*; as is evident, from the regularly and differently *Figur'd Parts*, which compose the whole Body of a *Snowy Cloud*, before it clusters into *Flakes*. And one reason, why *Rain* is the best *Water* for any *Soyl*, is because it is impregnated with divers *Volatile* and *Fruitful Salts*. And so from other *Meteors*.

2. §. And next, that these *Salts*, are not always in the same *Quantity*, *Proportion*, and *State*, in the *Aer*: but that sometimes they are more copious; at others, less: sometimes, one more copious, than another: sometimes, more plentifully dissolved; at others, more sparingly: and that, either as they are more or less pure and dissoluble; or according to the quantity of the *Vaporous Parts* in the *Aer*, in which they are incorporated or dissolved.

3. §. Thus much being granted, from the *Experiments* in the foregoing *Chapter* compared together, we may resolve our selves about some *Phænomena* in the *Barometre*. Which seems to vary, not so much with the meer *Weight* of the *Aer*, which hitherto hath been supposed: as by the different pressure it makes, in being *crowded* more at one time, than at another. That is, according as certain *Nitrous*, or other *Saline Bodies*, take up less *space* in the *Aer*, when dissolved in the *Watery Parts* therein, than while they are undissolved.

4. §. And therefore it is especially to be observed, That as the *Mercury* commonly riseth in the *Cylinder* for some days, but always for some time, before the change of the *Weather*, whether for *Snow* or *Rain*: So, that then it presently falleth again, even before the *Snow* or *Rain* falls. Whereas, if the *Weight* of the *Aer*, were the only, or the chief *Cause* of the ascent of the *Mercury*; than as it riseth all the while the *Weather* is gathering, so it would keep its standing or heighth, until the *Weather* breaks and falleth down: which yet it never doth, but always falls before it; sometimes no less than a whole day. The *Cause* whereof is, in that all the while the *Mercury* riseth in the *Cylinder*, the *Aer* is *crowded* with more and more *Saline Parts*, which by

the

the *Winds*, or otherwife, are carryed into it; and fo caufeth it to prefs upon the *Mercury* in the *Box*: but after that in fome time the *Salts* are diffolved or incorporated in the *Aqueous* Parts of the *Aer*, as in *Rain* or *Snow*; fo foon as that is done, there is fome *space* gaind; and fo, before any *Weather* falleth, the *Aer* is lefs crowded, and prefieth lefs upon the *Mercury* in the *Box*, which gives way to its defcent in the *Cylinder*.

5. §. From hence alfo it is, that the *Mercury* rifeth higher with *Cold Winds*, than it doth with thofe which are *Warme*. Both becaufe that in *cold Winds* there is the greateft quantity of *Nitre*: and that the *coldeft Winds*, are ufually the dryeft. So that the *Nitre* wanting *Moyfture* fully to diffolve it; it takes up fo much the greater fpace, and fo caufeth a greater preffure in the *Aer*, as hath been faid.

6. §. *Laftly*, For the fame reafon it comes to pafs, that the *Mercury* firft rifeth higher, and then falleth lower before *Snow*, than it doth before *Rain*. Becaufe that for the production of *Snow*, the *Aer* is crowded with a greater quantity of *Nitre*, or fome other like *Salts*; which before they are diffolved, take up fo much the more fpace; and afterwards fo much the lefs, even before the *Snow* falls: as hath been proved.

F I N I S.

AN INDEX OF THE Chief Matters,

In which, Id. *signifies* Idea. An. Anotmy. *The Figuers before* §. *the Page. The Figures following* §. *the Section in that Page.*

A.

ACid, *commonly the predominant Principle in Plants,* 240.§.8. *That is of the Parenchyma.* Id.§.48.

Aer, *how to be examined, as relating to Vegetation,* Id.§.60.

Aer, *in Plants, How made,* An.93. §.61.
Where it enters the Plant, 127. §. 1.
Its Motion and Courſe in Plants, ibid.

Aereal *Salt,* Id.§.60.

Aer-Veſſels, *their Structure,* 115. §.16, &c. *See* Root *and other Parts.*

Affinities *of Plants,* Id.6.§.11.

Age *of Roots, see* Roots.

Agitation, *a Cauſe of Mixture,* 230.§.6.

Akern, 186.

Albumen, *see* Seed.

Alkaline *Salt, in many Plants in their natural eſtate.* 240.§.9.
This the predominant Principle of the true Wood of a Plant, Id. §.52.

Anagallis, *of what Taſte,* 284. §. 10.

Angellica *Roots, when dry, full of Roſin,* Id.§.41.

Anatomy *of Plants, why fit to be made,* Id.§.17.
In what manner, §. 18.
What to be obſerved thereby, §. 19.
Of what uſe, §. 20.

Animals, *their Parts mixed with ſeveral Menſtruums,* 247. *to* 253.
Cantharides, *of what nature,* 249.
Antimony, *of what nature,* 245. §.23.

Apertures *of Seeds,* An.2. §. 5. & 200.§.1.

Apple *described,* An. 40. §. 2. & 179.

Aprecock, 148.

Arſmart, *coded, how its Seed ejaculated,*

Y y

The Index.

lated, 188. §. 18.
Arenulæ *in Pears*, An. 41. §. 4. & 241. §. 20.
Arum-Root, *of what Taste*, 281. §. 21
The Pestil of what Scent, Id. §. 28.
Aqua-fortis *double, mixed with Spirit of Wine, what remarquable thereupon*, 242. §. 26.
With Steel, 244 §. 22.
With Tin. 245. §. 27.
Asa fœtida, *of what nature*, 258. Query, 2.
Ascent *of the Trunk, how made*, An. 22. §. 21.
A Magnetick Motion, 136.
Ascent *of the Sap, how made*, An. 24. §. 29. & 126. §. 13.
Asparagus, *of what Taste*, 284. §. 10.
Attire *of Plants see* Flower.

B.

Barbado *Nut*, Id §. 30.
Barque *of the Root, see* Root.
Of the Trunk, see Trunk.
Bawme, *its Tincture in Water*, 274. §. 11.
In Spirit of Wine, 275. §. 14.
Beams *of the Sun, different from the Heat of Common Fire*, Id. §. 61.
Bean *dissected*, An. §. 1.
Beech-Wood, An. 20.
Berry, *see* Fruits.
Bezoar, *its nature*, 252. §. 49.
Bezoardicum minerale, 245. §. 25.
Bleeding *of Plants*, Id. §. 23. An. 124. §. 3.
Bolus, *what*, 242. §. 2.
Bonus Henricus, *of what Taste.* 284. §. 10.
Bones, *their different nature*, 249. §. 18.
Branch, *how made*, An. 28. §. 3. *Its Claspers*, An. 27. *see* Trunks.
Bud *of a Branch, how originated, nourished, and kept*, An. 28. §. 1. *How kept*, 145. §. 2.
Bud *of the Seed, see* Seed.
Butyr *of Flax*, Id. §. 51.

C.

Calamus Aromaticus, *of what Tast*, 283. §. 6.
Cantharides, *their nature*, 249. §. 14.
Case *of the seed, of several manners*, An. 45. §. 2. & 186.
Carduus *green Leavs, their scent*, Id. §. 28.
Castor, 250. §. 28.
Celandine, *little, where tasted*, 284. §. 10.
Cherry, 185.
Circulation *of the Sap*, An. 17. §. 30.
Claspers, An. 27.
Clematis peregrina, *the Seed-Case of what Tast*, 283. §. 3.
Coats *of the Seed, see* Seed.
Colocynthis. *Its nature*, 240. §. 13. & 257. Query 5.
Where tasted, 284. §. 8.
Colours *of Plants, To what Parts of Plants they belong*, Id. 26.
How to be observed, Id. §. 27.
Colours *of Roots*, An. 94. §. 65. & 270. §. 5.
Of Leavs, 270. §. 6.
Of Flowers, 271. §. 15.
By Infusion, in Oyl, 273. *In Water*, 274. *In Spirit of Wine*, ibid.
By their Mixture with other Bodies, 375.
By Cultivation. 277.
Their Causes summed up, 276.
Compression *a Cause of Mixture; and of Dissolution*, 229. §. 3. 232. §. 4. & 237. §. 3, 4.
Contents *of Plants, in what Method to be examined*, Id. §. 21 to 26. & 31 to 47. *Of what kind*, §. 21.
Their Receptacles, §. 22. *Motions*, §. 23. *Quantities*, §. 24. *Consistence.* §. 25.
How made in the several Parts of a Plant, An. 92. §. 57.
What in the Seed, 208. §. 15.
Contrayerva, *of what Taste*, 283. §. 6.
Convolution *of the Trunk a Magnetick*

The Index.

netick *Motion,* 137.
Copper, *its nature.* 245. §. 28.
Copperas, 246. §. 38.
Coral, *the Magistery,* 244. §. 15. *Tincture,* Id. §. 28.
Corin *white,* 285.
Cortical *Body, see* Barque.
Covers *of the seed, see* Seed.
Cucumer, An. 181.
  Leavs of the wild, of what Taste, 280. §. 11. & 283. §. 6. *Where tasted,* 284. §. 4, 8.

D.

Daisy *Leaves, where tasted,* 284. §. 10.
Descent *of the Root, how made,* An. 34. §. 3.
Diametral *Rays, see* Roots.
Digester, *the nature of that invented by Monf.* Pappin, 237.
Dissolution *of Bodies promoted by Compression.* 237.
Dissimilar *Leavs, see* Leavs.
Dragon *Root,* 59. §. 13.
Dung *of Pigeons,* 251. §. 37.

E.

Earth, *how to be examin'd, as relating to Vegetation,* Id. §. 57.
  How nature prepares it for the growth of Plants, 11. §. 8.
Empalement, *see* Flower.
Emulsionss, *sometimes for Glysters,* Id. 39.
Enula, *of what Taste,* 283. §. 3, 34.
Essential *Salts of Plants, see* Salts.
Evergreen, 156. §. 2.
Euphorbium, *of what nature,* 200. §. 1. 241. §. 19. & 258. Query 2. *Of what Taste,* 283. §. 6.

F.

Fat, *how made by Art or Nature,* 233. §. 3.
Fermentation, 253. §. 55.
Fibers *of the Leaf, see* Leavs.
  Of the Seed, see Seed.
Figures, *of Plants,* Id. §. 11.
  Of Roots, An. 58. §. 4.
  Of Trunks, 135. *Of Leavs,* 150. §. 1.
  Of Seeds, 195.
Figs, *their Sugar.* Id. §. 41.
Flax, *its nature,* 258. Query 1.
Flower, *its Empalement,* An. 35. §. 2. & 163.
  Foliature, its Foulds, An. 36. §. 5. & 164. §. 1. *Protections,* An. 36. §. 7. *Hairs,* An. 36. §. 8. & 168. §. 8. *Globulets,* An. 37. §. 10. & 165. §. 9. *Number of Leavs,* 165. §. 11. *Parts of the Leavs,* 166. §. 15. *Use,* An. 37. §. 11. & 166. §. 18. *Shape or Figure,* 167. §. 20.
  Attire, Seminiform, An. 37. §. 13. & 167. *Florid,* An. 38. §. 17. & 170. *Globulets or Sperme of both,* An. 38. §. 15. 39. §. 21. 168. §. 9. & 170. §. 5.
  Use of the Atire, An. 39. §. 22. & 171.
Flower, *when formed,* 173.
Colours *of the Flower,* 271.
  How by the Flower to find out to what sort a Plant belongeth, 175. §. 13.
Fœtus, *see* Seed.
Foulds *of Leavs, see* Leavs.
Formation *of the Root, see* Root.
Fruits; *Apple,* An. 40. §. 2. & 179. *Limon,* 180. *Cucumer,* 181. *Pear,* An. 41. §. 3. & 182. *Quince,* 183. *Plum,* An. 42. §. 5. & 183. *Aprecock,* 184. *Peach,* 184. *Cherry,* 185. *Walnut,* ibid. *Grape,* ibid. *Gooseberry,* An. 43. §. 9. & 185. *White Corin,* 185. *Filbert,* An. 43. §. 8. & 186. *Akern,* 186.
  The Use of the Fruit, An. 44. §. 10.

The Index.

§.10. *Of its Parts to its self,* 189. *To the Seed,* 191, & 209. *When the Fruit formed,* 192.§.9.
Furr *of a Hare,* 247.§.3.

G.

Gall-Stones, 252.§.47.
Generation *of the Seed, and other Parts, see the Seed, and other Parts.*
Gentian *Root, where tasted,* 284.§.5.
Germen, *see Bud.*
Glysters, *sometimes best made of Emulsions,* Id. §.39.
Globulets, *see Leavs and Flowers.*
Gold, *its nature,* 245.§.31.
Gooseberry, 185.
Grape, 185.
Gravel, *its nature,* 251.§.40.
Gums, *of three kinds,* 134.§.15.

H.

Hairs, *see Leavs and Flowers.*
Hares *Furr,* 247.§.3.
Harts-Horn, 248.§.8.
Hazel *Nut, see Fruits.*
Hellebore *black, of what Taste,* 280.§.12. & 283.§.3. &c. *where tasted,* 284.§.4.
Hoglice, 249.§.15.
Horses *Hoofe,* 247.§.5.

I.

Jalap, *of what Taste,* 283.§.6. 284.§.10.
Insertions, *in the Root, and other Parts, see Root and other Parts.*
Iris *Root described,* 60.§.14.
Irish *slate, its nature,* 243.§.4.
Iron, *see Steel.*

L.

Lapis Calaminaris, *its nature,* 243.§.9.
Lapis Lazuli, 243.§.7.

Lapis Tuthiæ, 243.§.9.
Lead, *its nature,* 244.§.16.
Lead Spar, 244.§.12.
Leavs, *the two first which come of every Seed, what,* An.8.§.42, &c. *Their Use,* An.10.§.46.
Leavs; *their Protections,* An. 32.§.17. & 145.§.2. *Foulds,* An. 31.§.14. & 147.§.9. *Shapes and Measures,* An.30.§.17. & 150. §.1. *Globulets,* An.34.§.7. & 148.§.1. *Hairs,* An. 34.§.4. 149.§.8. *Spots,* 148.§.4. *Thorns,* 148.§.6.
Their Compounding Parts, An. 29.§.7. *Skin,* 153.§.1. *Parenchyma or Pulp,* 153.§.5. *Fibers or Vessels, Their Position In the Body of the Leaf,* 152.§.19. *In the Stalk,* 154.§.9. *The Lignous Vessels.* 155.§.16. *The Aer-Vessels,* 155.§.19. *Texture of a Palm Leaf or Bag,* 156.§.20. *Duration of the Leaf,* 156.§.2. *Time and manner of its Generation,* 156.§.4. & 174.
Colour of the Leaf, 270.
How by this to find out to what sort a Plant belongeth, 174.§.1.
Lignous *Body, see Trunk and other Parts*
Lilium convalle, *its nature,* Id. §.30.
Limor, *described,* 180.
Lithosperm *the Seed, its nature,* 241.§.21.
Lixivial *Salts, see Salts.*
Lobes *of the Seed, see Seed.*
Lympha *out of which the Seed is first nourished, see Seed.*
Lympheducts *their Structure,* 111. §.30.

M.

Magistery *of Corals,* 244.§.15 *Of Pearls,* 252.§.43.
Mallow, *its nature,* 257. Query 5.
Marine *Salt of Plants, see Salts.*
Mastick, *its nature,* 258. Query. 2.
Measures *of Leavs,* 150.§.1.
Mechanick *uses of Timber,* 137.

Membranes

The Index.

Membranes *of the Seed, fee* Seed.
Menstruum *of the Stomach,* 253. § 53.
Metals; *Lead,* 244. § 16. *Mercury,* 244. § 16. *Steel,* 244. § 20. *Antimony,* 245. § 23. *Tin,* 245. § 26. *Copper,* 245. § 28. *Silver,* 245. § 29. *Gold,* 245. § 31.
Milks *of Plants,* Id. § 21, & 26.
How made, An. 67. § 19. & 93. § 60. & 133. § 12.
Milk-Vessels, *their Structure,* 112. § 35.
Millipedes, 249. § 15.
Minerals *of all sorts, how easily tryed,* 247. § 48.
Mixture; *the received Doctrine hereof,* 222.
Its nature explained, 225. Causes, 229. Power and Use, 231.
Mixture of the Parts of Plants with several Menstruums, 239, &c. Of Minerals. 247, &c. Of Animals, 247, &c.
Motions, *Of Plants,* Id. § 16.
Of Roots, and other Parts, see Roots, and other Parts.
Of the Sap, see Sap.
Of the Aer, see Aer.
Muciducts, An. 66. § 18.
Mucilages, Id. § 21. & An. 210. § 4.
Musk, *its nature,* 250. § 29.

N.

Nature of Bodies, *how discoverable,* 235
Navel-*Fibers, see* Seed.
Nightshade *deadly, of what Tast,* 284. § 9.
Nitre, *of what Tast,* 280. § 6.
Noli me tangere, *how the Seed ejaculated,* 188. § 18.
Number *of Leavs in Flowers.* 165. § 11.
Number of Seeds, 198.
Nut Barbado, Id. § 30. & 205. § 17. Hazel Nut, 43. § 8. & 186.

O

OAK-Wood, *described,* An. 20, & 21.
Odors *of Plants, how to be observed,* Id. § 28.
Some Instances how made, An. 44 § 46. Imitated, 235.
Olibanum, *its nature,* 258. Query, 2.
Oyls *stillatitious, how mingled with Water &c.* 232. § 7, & 237.
Oyly Sap, *how made,* 132. § 6.

P

PArenchyma *or Cortical, Pithy, and Pulpy parts of a Plant, their predominant Principle,* Id. § 48. Described in the Root and other parts, see Root, &c.
How formed, see Roots and Leavs.
Peach, see Fruits.
Pear, see Fruits
Pearls, *their magistery,* 252. § 43.
Philosophy, *begins and ends with Theology.* 79. § 1.
Pimpinel, *where tasted,* 284. § 10.
Pith, *its structure,* 76, § 7. & 120 § 11, &c.
Plants, *their Natural History how far cultivated,* Id. § 3. *Wherein defective,* § 2. *Fit to be further improved,* § 3. & § 63. *What to be enquired of,* § 6. *The usefulness hereof* § 8.
Plants, *their Nature and Virtue how judged of, see* Virtues.
Plants, *their places of Growth,* Id. 15.
Propotions, § 13.
Plants, *their Parts only Two Essentially distinct,* 47. § 14.
Plants; *the general structure of their Parts,* 120. § 11, &c.
Plants, *their Pinciples how to be observed,* Id. § 48.
For what purpose, § 53. What predo-

The Index.

predominant therein, 240. § 8.
Plants, *how to find out to what kind any one belongs*, 174
Plum, *see* Fruits.
Principles *of Bodys*, 223. *which predominant in the true wood of a Plant*. Id. § 52.
Principles *of Principles*. Id. § 62.
Protections *of the Leaf and Flower*. See Leaf and Flower.
Pyrethrum *the Root, of what Taft*, 281. § --- & 284. §7.

R.

Radicle, *see* Seed.
Raisins, *their Sugar*, Id. § 41.
Rings *annual in the Trunk*, An. 19. § 6.
Roots; *their Original*, 57. § 1. *Shapes or Figures, & Sizes* 58. § 4, & 89. § 41, *Motions*, An. 15. § 24. &c. 34. § 3. 59. § 9. &c. 90. § 48 &c. *Ages*, 60. § 16. & 91. § 54.
Parts, *the Barque, its Skin*, An. 11, § 2. & An. 61.
Parenchyma *described*, An. 11. § 3. & 63. § 2. *How form'd*, 87. § 34. *Its Diametral rays*, 64. § 7. *Veffels*, 65, 66, 67.
The Wood, *Herein the Infertions*, An. 12, § 10 & 17. § 28. & 70. § 2. *Lignous Fibers or Veffels*, 70. § 4, 8, 9. *Aer Veffels*, An. 12. §. 7. & 71. § 5, 6, 10, &c.
The Pith, An. 13. § 16. & An. 16. § 27. & An. 75, 76. § 7.
Root, *how it grows*, An. 14. § 23.
The Sap, *how imbibed and diftributed to its feveral Parts*, 82. § 15 &c. *How circulated*, An. 17. § 29.
  How all the parts are form'd, 85. § 26. &c. *And differently difposed*, 88. § 36, &c. *The Colours of Roots*, 170. § 5. *How made*, 94. § 65.
Root *of Wormwood, where tafted*, 285. § 12.

Rofin, *how made by Art*, 233. § 4.
Rofin *in dryed Roots of Angelica*, Id. § 41.
Common Rofin, *its nature*, 258. Query 2.

S.

Salt aereal, Id. § 60.
Salt Alkaline, *in many Plants in their natural eftate*, 240. § 9.
Salt ammoniac, 246, § 44.
Salt, *effential of Plants, How made*, 262. § 3, &c. & 265. § 3. &c. *Of feveral forts*, Id. § 48. *Inftanced alfo in thofe of Rofemary, Black Thorn, Scurvey-Grafs, Wormwood, Afh*, 865, § 6 &c. *Taftable in good Rhubarb*, Id. § 41.
Salt *fixed, of what ufe in Purgation*, 260.
Salt Lixivial *of Plants, how imitated*, 233. § 6. *Of different nature*, 264. § 2.
Salt *of Afh, of what nature*, 167. § 22. *Of Tartar*. Ib. *Yielded in different quantitys by the Barque of Afh, Rofemary, Black-Thorn, Agrimony*, 256. Query 1. *Garden and Sea Scurvy-grafs*, 256. Query 2. *Mint diftill'd, and not*, 256, Query 3. *Majorane, Oak-Barque, Liquorifh, Anifeeds, Sorrel, Garden Scurveygrafs, Mint, Sea Scurveygrafs*, 256. Query 4. *Majorane, Agrimony, Mugwort, Mint, Mallow, Rhubarb, Sena, Jalap, Colocynthis*, 257, Query 5. *Flax*, 258. Qu. 1. *Gum Arabick, Euphorbium, Myrrh, Opium, Aloe, Scammouy, Gutta gamba*, 258. Query 2.
Salt Marine, *its nature*, 246. § 43.
Salt Marine *of Plants, how made by Nature or Art*, 234. § 8. 263 § 12, &c. 266. § 16. *Of feveral forts, inftanced in thofe of Rofemary, Scurvey grafs, Black Thorn, Wormwood*, 266. § 17 &c.
Salt *of the dead Sea*, 263. § 14.

Saps

The Index.

Saps of Plants, *how to be observed,* Id. § 21. to 26. and 31 to 47. *Their several kinds,* §. 21. *Receptacles,* § 22. *Motions,* § 23. *Quantitys,* § 24. *Consistence,* § 25.

Sap, *how imbibed, and distributed to the several parts of the Root.* An. 82. § 15 &c. *Its Circulation therein.* An. 17. § 29. *where, and how it ascends in the Trunk,* An. 24. § 29. & An. 124.

Sap *and other contents of the several Parts how made,* An. 92. § 57 &c. & 131. *How a Milky Sap,* An. 67. § 19. & 93. § 60. & 133. § 12 &c. *How a Winy,* 93 § 62. & 132. § 3 &c. *How one very Oyly,* 132. § 6 &c.

Scurveygrass Garden, *of what Tast,* 283. § 6.

Scurvey-grass Sea, *its Nature,* 256. Query 4.

Seasons of Plants, Id. § 14.

Secundine *see Seed.*

Seeds; *their Case or Uterus,* An. 45. § 2. *of several manners,* 186. *Figures,* An. 45. § 3. &c. 195. *Numbr.* 198. *Motions,* 188. § 18 & 199 § 3, &c. *Stones,* 201. § 2. & 209. *Mucilages,* 201, § 4. *Coats or Membranes,* An. 2. § 3. 45. § 3, &c. 46 § 10. 47. § 15 & 201. § 6. 210 &c. *Apertures,* An. 2. § 5. & 200 § 1.

Vitellum, 20. 2. § 9. *The Fœtus, or true Seed, its Radicle and Lobes,* An. 2 & 3. § 9 to 12. & 203. *Plume or Bud,* An. 3. § 13 & 206. *skin,* An. 4. § 16. & 207. § 9. *Parenchyma,* An. 4. § 18. & 207. § 10. *Seminal Root or Vessels,* An. 5. § 21, &c. & 207. § 11, &c. *Navle Fiber,* An. 48. § 17. & 212. *Content.* 208. § 15. *The manner of its Vegetation,* An. 6. § 30, &c. *Of its Generation,* An. 48. § 18 &c. & 209 &c.

Shape of Roots, and other Parts. *see Roots and other parts.*

Shells, *their Nature,* 248. § 9, 10, 11.

Skin, *see Seed and other Parts.*

Silver, *its Nature,* 245 § 29.

Smell *of green Carduus,* Id. § 28. *Of the Pestil of Arum.* Ib.

Soyl. *see Earth.*

Sperme of Plants. *see Flower.*

Spirit of Salt, 247. § 46.

Spirit of Salt Armoniac, 247. § 47.

Spirit of Peas-Cods, Id. § 30.

Spirits urinous, *how made less offensive,* Id. § 45.

Spirit of Wine *mixed with Aqua fortis, what thereupon remarkable,* 242. § 26.

Stalks. *see Trunks and Branches.*

Steel, *its nature,* 244. § 20. *Mixed with double Aqua fortis, what thereupon remarkable,* 244. §. 22.

Stillatitious Oyls, *how mixed with water,* 232. § 7. & 237.

Stomachick menstruum, 253. § 53.

Stones; *a strange one bred in the Stomach,* 252. § 48. *Others probably bred there,* 253. § 51. *Of the Kidneys or Bladder, of what nature,* 251. § 32. *How prevented,* 251. § 41. & 252. § 42. Gall Stone, *its nature,* 252. § 47. Bezoar, *its nature,* 252. § 49.

Lead-Spar, 244. § 12.

Lapis Calaminaris, 243. § 9.

Tuthiæ, ibid.

Lazuli, 244 § 12.

Structure of a Plant. 120. § 11, &c.

Sugar of Raisins and Figs. Id. § 41.

Sulphur *predominant in the true wood of a Plant,* id. § 52.

Sun, *its Influence on Plants how to be examined.* Id. § 61.

Tastes.

The Index.

T.

TAsts of Plants; *how to be obſerved*, Id. § 29. *ſimple*, 280. § 6. *Compounded*, 281. *Their Degree*, 282. *Motions or Terms*, 283. *Seat*, 284. *Cauſes*, An. 95. § 68 &c. & 286 &c.

Taſt *of Arum Root*, 281. § *Aſparagus Root*, 284. §. 10. *Bonus Henricus*, 284. § 10. *Calamus Aromaticus*, 283. §. 6. *Celandine little*, 284. § 10. *Clematis peregrina the Seed-Caſe*, 283. § 3. *Colocynthis*, 284. § 8 *Contrayerva*, 283. § 6. *Cucumer wild, the Leavs*, 284. § 4, 8. *Daiſy Leavs*, 284. § 10. *Enula*, 283. § 3, 4. *Euphormium*, 283. § 6. *Gentian Root*, 284. § 5, 8. *Hellebore black, the Root*, 280. § 12. 283. § 3. 284. § 4. *Jalap*, 283. § 6. 284. § 10. *Nightſhade deadly*, 284. § 9. *Nitre* 280. § 6. *Pimpinel*, 284. §. 10. *Pyrethrum the Root*, 281. §. & 284. § 7. *Tamarisk-Leavs*, Id. § 29. *Wormwood-Root*, 285. § 12. *Yarrow*, 283. § 6.

Texture *of a Plant*, 120. § 11, &c.

Thorns, *their kinds*. An. 33. § 1.

Timber, *ſee* Trunk.

Tin, *its nature*, 245. § 26.

Tin *mixed with ſtrong Aqua fortis, what thereupon obſervable*, 245. § 27.

Tincture *of Bawme in Water*, 274. § 11. *In Spirit of Wine*, 275. §. 14.

Tincture *of Corals, a cheat*, Id. § 28.

Trunks, *and Branches ſeveral deſcribed as they appear to the naked Eye, ſc. of Indian Wheat, Dandelyon, Borage, Colewort, Holyoak, wild Cucumer, Scorzonera,* *Burdock, Endive, Vine, Sumach* 103, &c.

Trunk, *skin*, An. 19. § 2. & 107. § 2 to 5.

The Barque, An. 19. § 3.

Its Parenchyma, 108. §. 7.

Veſſels, 108. § 8. to p. 113.

The Lympheducts their Structure, 111. §. 30, &c.

Milk-Veſſels, their Structure, 112. § 35, &c.

Different Surface of the Barque how made, 129. § 4.

How united to the Wood, 129, §. 2, 3. *How this always keeps moyſt, not the Pith*, An. 20. § 7. & 93. § 58. & 124. §. 2.

The Wood, An. 19. § 4. to 11. & An. 20 & 21.

Its Annual Rings, An. 19. § 6.

Inſertions, An. 19. §. 5. 12. to 15. 17. & 128. § 8. &c.

True wood, 114. § 10, &c.

How dilated, An. 22. § 22, 23. *And why*, § 24, &c.

Aer-Veſſels, An. 20. § 8, 9. & 115. § 16, &c.

How leſs in the Trunk, than in the Root, and whence formed late in the year &c. 130. § 10. & 131. § 16.

The Pith, An. 19. § 5, 18, 19, 20. & 119. to 122 & 129. § 5. &c.

Trunks, *their different Structure whence*, 129.

Shapes, whence, 135.

Motions, whence, An. 22. § 21. & 136.

Trunks, *how fitted for Mechanick uſe*, 137.

Trunks, *of their Bleeding*, Id. § 23. & 124. § 3, &c.

Trunk-Roots, An. 27, 28.

Turnep, *deſcribed*, An. 13.

Valves,

The Index.

V.

Valves, *no where in Plants*, An. p. 21. §.16.
Vegetables, *see* Plants.
Vegetation *of the Seed, see* Seed.
The manner of Vegetation, how judged of, Id. §. 53.
Vessels *of the Root and other Parts, see* Root *and other Parts*.
Virtues *of Plants, how to be observed and judged of,* Id. §.12,30, 47. & p. 236. 290.
Imitable, 235.
The reason of them, how knowable, Id. §. 55.
Vitriols, *their nature,* 246. §.38.
Uterus *of Plants, fleshy or membraneous,* 186.

W.

Wallnut, 185.
Water, *how to be examined as relating to Vegetation*.
Water, *how mingled with stillatitious or other Oyls,* 237.
Wood *of the Root and other Parts, see the* Parts.
Wood *of Beech,* An. 20. & 21.
Of Oak, Ibid.
Its predominant Principle, Id. §. 52.

Y.

Yarrow *Leavs, their Tast,* 283 §. 6.

THE
EXPLICATION
OF THE
TABLES,

Reduced to a narrow compaſs; as ſerving to clear thoſe Particulars, chiefly, which the *Deſcriptions* before given, have not reached.

The TABLES to the Firſt BOOK, *are Four.*

TAB. I. Figure 1, a, *The Foramen.*
F. 2, a, *the Radicle lodged in the Body of the Iner Coat.*
F. 3, a, *the Radicle,* b, *the Plume or Bud.*
F. 4, *the Seed covered;* c, *the Seed open;* e, *the ſame magnified.*
F. 5. a, *the Corn covered;* c, *naked and a little magnified.*
F. 6. a, b, *the two Lobes;* e, *the Radicle;* c, *the Radicle and Bud;* d, *the Hollow in which the Bud lies.*
F. 7. a, *the Seed covered;* c, *naked;* e, *open.*
F. 8. a, *one Lobe;* --b, *the Bud;* b, *magnified.*
F. 9. *the Slice a little magnified.*
F. 10. *The Radicle* d, *cut tranſverſly* c.
F. 11. *The Plume or Bud* a, *cut tranſverſly* c.
F. 12. *Cut by the Length.*
F. 13. *A Lobe cut tranſverſly.*
F. 14. *Both the Lobes pared by the Length, to ſhew the Seminal Root.*
F. 15. a, *the convex ſide of one Lobe, ſhewing the Seminal Root without cuting;* c, *the flat ſide.*

TAB. II. F. 1, 2, & 3. *ſhew the gradual converſion of the Lobes of the Seed, into Leavs.*
F. 4. a, *the Radicle cut by the length;* b, *tranſverſly.*
F. 5. *The white Wedges, are the Inſertions; the black, are the Wood; the pricks are the Aer-Veſſels; and the black half ovals, the Lympheducts in the Barque.*
F. 6. *The three black Rings, are the terms of three years growth.*
F. 7. a, *the upper part;* b, *the lower.*
F. 8. *A Turnep cut tranſverſly, and part of the Rind cut off.*
F. 9. *ſheweth the gradual growth of the Pith.*

TAB. III. F. 1. *The Bud cut tranſverſly, and part of the Radicle*

by

The Explication of the Tables.

by the Length, in a Bean newly sprung up.

F. 2. sheweth the Wood as it appears to the naked Eye.

F. 3. the Cane split down.

F. 4. the Corn newly sprouted.

F. 5. A Branch of five years growth. From the Circumference, to the utmost black Ring, goes the Barque.

F. 6. a, a piece of the Stalk; b, magnified.

F. 7. a, a piece of Oak-wood cut transversly; b, the same magnified. The white Lines are the lesser and greater Insertions. The Pricks, are the Wood. The little and great Holes two sorts of Aer-Vessels.

F. 8. Part of a Branch ten years old, with the Barque stripped off, and cut both transversly and down the length, to shew how the Barque is inserted into the Wood.

TAB. IV. F. 1. shewing how the Insertions appear, in a piece of Beech-Tree split down, to be braced or woven in together with the Wood.

F. 2. to 11. shew the different position and Figure of the Lignous Fibers.

F. 12. a, one of the Thecæ Seminiformes in a Lily, with the spermatick Powder therein, as apparent to the naked Eye.

F. 13. a, one of the suits in the Florid Attire, as it appears to the naked Eye; b, the Floret; c, the Sheath; d, the Blade.

F. 14. Wherein the white Pentangular Acetary is bounded by the Calculary.

F. 15. The Branches which run through the Stone to the Flower and Seed.

F. 16. The Innermost Cover of the Seed, as shaped when it is ripe.

F. 17. The Coats cut open.

F. 18. The Seminal Root.

The *TABLES* to the Second *BOOK* are Thirteen.

TAB. V. sheweth the generation of Roots out of the Descending Trunk. So F. 6. is a treble Root of three years descent; the lowermost, half-roted off.

TAB. VI. F. 1. sheweth the Surface of the Barque.

F. 2. the midle part.

F. 3. the Barque striped.

F. 4. the Root cut down the length.

F. 5. the Barque striped off.

F. 6. the Network both of the Lympheducts, and of the Aer-Vessels.

F. 7. the Generation of a Bud.

F. 8, 9, 10, 11. The Root split down, to shew the Position of the Vessels, and the Figure of the Pith at the top of the Root.

TAB. VII. The Roots all cut transversly, and their Varieties described, in the second Book, as they appear to the naked Eye.

TAB. VIII. Other Roots cut transversly, and the varieties of their Parts also described in the second Book.

TAB. IX. More Roots cut transversly.

TAB. X. F. 1. A Slice of the Root cut transversly; but a little too big for the life.

F. 2. AA, One half of a like Slice.

b b, The Skin.

AADD, The Barque or all that part of the Root analogous to it.

GD, The Lympheducts on the inner edge of the Barque.

GG, The Wood.

GT, The Aer-Vessels therein.

TT, The Pith.

TAB. XI. F. 1. The Neck of
the

The Explication of the Tables.

the Root cut transversly.

F. 2. One half of the same split down.

F. 3. Magnified.

A B, The Skin.

A E, The Barque.

E E, The Lympheducts.

The black Columns under them, are the Wood.

The Holes in the Columns are the Aer-Vessels.

The white Columns E L, are Insertions betwixt the Barque and the Pith.

L e, The Pith.

c e, The angular Bladders of the Pith.

T A B. XII. A, one half of F. 1. magnified.

A b, The Skin.

A G, The Barque, or all that part of the Root which answers to it.

In which the round black spots, are the Muciducts.

D G, The common Lympheducts.

D T, The Pithy Part of the Root.

T T, More Lympheducts.

In both which, the black Holes are the Aer-Vessels.

T A B. XIII. A, One half of F. 1. magnified.

A C, The Skin.

A G, The Barque, or that part of the Root which answers to it.

D D, The Milk-Vessels placed in Rings.

E E, The Parenchymous Rings betwixt them.

G T, The Bladders streaming in Rays, by the mixture of the Lympheducts with the Lacteals.

G G, To the Centre, the Wood.

In which the Holes are the Aer-Vessels.

T A B. XIIII. A b., The Skin, which should have been thicker.

A F, The Barque.

G b, The Bladders in the outer part of the Barque, oblong and postured circularly.

S S The Bladders in the inner part, standing in Arches.

F F, A Ring of Sap-Vessels.

d d, Parenchymous Insertions.

d L d, The Wood.

In which, the Holes edged with white Rings are the Aer-Vessels.

T A B. XV. A A, The Skin.

A B, The Barque.

B L, The Sap-Vessels in the form of a Glory.

B E, The Wood.

In which, the Holes are the Aer-Vessels.

G E, A Ring of more Sap-Vessels.

E E, The Pith.

T A B. XVI. A b, The Skin.

A C, The Barque.

In which the round Holes B, are Balsame-Vessels.

B. C. Parcels of Lympheducts.

In which there are more Balsame-Vessels.

C D, Parenchymous Insertions.

D E, Parcels of Wood,

In which the Holes are the Aer-Vessels.

T A B. XVII. A, the Skin.

A B, The Barque.

L S, A parcel of Sap-Vessels.

L l, A Parcel of Wood.

In which the Holes great and small are Aer-Vessels.

B B, Parenhcymous Insertions betwixt the parcels of Wood.

D D, Others within them.

The *TABLES* to the *Third BOOK* are 23.

T A B. XVIII. Hereof see the Description in the Third Book, Chap. 1.

T A B. XIX. F. 1. A Branch of Corin Tree.

A, sheweth the surface of the Barque. B.

The Explication of the Tables.

B, *Of the wood.*
F. 2. *Stalk of Sonchus split down.*
F. 3. *Branch of Vine split down.*
In both, the several Storys or Chambers of the Pith.
F. 4. *Branch of Walnut.* A. *an older.* B, *a younger: in both, the Pith parted into transvers Membrans.*

TAB. XX. F. 1. *Sheweth the Surface of a Walking Cane.*
And the Clusters of Aer-Vessels, surrounded with Rings of Succiferous.
F. 2. *The surface of the Skin of Borage Stalk.*
F. 3. *The Turpentine Vessels running through the length of the Barque; one of them cut down the middle, the other entire.*
F. 4. *The Milk-Vessels shewed in the same manner.*

TAB. XXI. *Sheweth the Woody and Aer-Vessels by the length of the Branch, part of the Barque, and wood, being taken away.*

TAB. XXII. A B, *The Skin.*
A C, *the Barque.*
Q, *the Parenchymous part.*
H I, *Parcels of Muciducts in a Ring.*
D C, *Common Lympheducts.*
C D E F, *the Wood of 3 years growth.*
K L M N, *The second years growth.*
O ϱ, *the great Insertions.*
P ϱ, *the smaller.*
X X, *Lignous parcels.*
Within which the Holes are the Aer-Vessels.
E F G, *the Pith.*

TAB. XXIII. A B, *the Skin.*
A C, *the Barque.*
Q, *the simple Parenchyma.*
H I, *a Ring of special Vessels.*
P, *common Sap-Vessels.*
C D E F, *the Wood of 3 years growth.*
K L M N, *one years growth.*
X, *great Insertions.*
P O, *lesser between them.*

The black parcels are the wood. In which the Holes are the Aer-Vessels.
E F G, *the Pith.*

TAB. XXIV. A B, *the Skin.*
A B C D, *the Bark.*
N N, *the Parenchyma.*
H I, *a Ring of special Sap-Vessels.*
D M C, *Parcels of Lympheducts.*
C D E F, *the Wood.*
E F L K, *one years growth.*
K P Q L, *the larger Aer-vessels in the several parcels of Wood.*
ϱ ϱ, *the lesser Aer-vessels.*
M T, *the insertions.*
E F G, *the Pith.*

TAB. XXV. A B, *the Skin.*
A B C D, *the Barque.*
H I, *special Sap Vessels in arched parcels.*
O O, *the common Sap-vessels which begin to turn into Wood.*
C D E F, *the Wood.*
K L M N, *one years growth.*
The Holes are the Aer-vessels in the wood.
ϱ ϱ, *the true wood.*
O z, O y, *the Insertions.*
E F, *other Sap-vessels.*
E F G, *The Pith.*

TAB. XXVI. A B, *the skin.*
A B C D, *the Barque.*
Q Q, *the Parenchyma.*
H I, *special sap-vessels in arched parcels.*
D C, *a Ring of common Lympheducts.*
D C F E, *the Wood.*
K L M N, *one years growth.*
The Holes are the Aer-Vessels.
O O, *the greater Insertions.*
P O, *the smaller.*
E F, *other Sap-vessels.*
E F G, *the Pith.*

TAB. XXVII. A B, *the Skin.*
A B C D, *the Barque.*
W V, *the Parenchyma.*
H I, *round parcels of Sap-Vessels.*
D C, *the common Sap Vessels.*
D C E F, *the Wood of 5 years growth.*

The Explication of the Tables.

QRFE, *one years growth.*
XX, *the true wood.*
The Holes both great and small are the Aer-Vessels.
SS. *The great Insertions.*
TS, *the smaller.*
EFG, *the Pith.*
TAB. XXVIII. AB, *the Skin.*
ABCD, *the Barque.*
HTI, *special Sap-Vessels in round Parcels.*
DSC, *common Sap-Vessels.*
DCEF, *the Wood of five years growth.*
ee, *the true wood.*
KL &c. *the great Aer-Vessels.*
DC, *the smaller.*
SS, *the Insertions.*
EFS, *the Pith.*
TAB. XXIX. ABCD, *the Barque.*
AB, *a Ring of Sap-Vessels in round parcels next the Skin.*
HI, *the Parenchyma.*
Another Ring of round parcels.
DOC, *Common Lympheducts.*
DCEF, *the wood.*
MNEF, *one years growth.*
SS, *the true wood.*
KL, *the great Aer Vessels.*
PQ, *the lesser.*
OO, *the Insertions.*
EFG, *The Pith.*
e, *the Bladders of the Pith.*
TAB. XXX. AB, *the Skin.*
ABCD, *the Barque.*
RR, *the Parenchyma.*
HRI, *two Rings of special Sap-Vessels.*
DC, *Common Lympheducts.*
DCEF, *the wood of four years growth.*
dd, *the true wood.*
Qd, *part of it whiter, by the mixture of special Sap-Vessels represented by the transvers Lines.*
MN, *the great Aer-Vessels.*
ce, *parcels of lesser ones.*
EF, *a Ring of other Sap-Vessels.*
EFG, *the Pith.*
TAB. XXXI. ABCD, *the Barque.*

mm, *the Parenchyma.*
HmI, *Milk Vessels in arched parcels.*
DKC, *Lympheducts.*
DCEF, *the wood of one years growth.*
ST, *probably milk Vessels heretofore.*
The Holes in the Aer-Vessels.
KK, *the Insertions.*
EvF, *other Milk-Vessels.*
EFG, *the Pith.*
TAB. XXXII. ABCD, *the Barque.*
MN, *The Parenchyma.*
DLC, *the Lympheducts.*
HI, *The Vessels which carry the Turpentine.*
DCFE, *the Wood.*
LL, *the Insertions.*
EFG, *the Pith.*
The greater Holes both in the Wood and Pith, are more Turpentine Vessels.
TAB. XXXIII. ABCD, *the Barque.*
XY. *The Parenchyma.*
KXYL. *Special Vessels in round parcels.*
HI, *others in a Ring.*
DC, *Common Lympheducts.*
DCEF, *the wood.*
SZT, *probably one sort of Sap-Vessels heretofore in the Barque.*
QMQN, *small Aer-Vessels.*
MN, *great Aer-Vessels.*
RQ, *the small Insertions.*
QQ, *the great ones.*
EFG, *the Pith.*
TAB. XXXIV. AB a a, *the hairy Skin.*
ABCD, *the Barque.*
HwI, *the Parenchyma.*
DMC, *the common Lympheducts.*
KL, *the Milk Vessels.* v, *one Vessel.*
HI, *Another sort of Lympheducts, arched over the Milk Vessels.*
XX, *seems to be a third sort of Lympheducts.*
DCFE, *the Wood.*

MM

The Explication of the Tables.

M M, *the Insertions.*
X X, *the true Wood.*
The Holes therein are the Aer-Vessels.
E F, *a Ring of Lympheducts.*
E F G, *the Pith.*
TAB. XXXV. A B C D, *the Barque.*
A M B, *the Parenchyma.*
H M I, *Balsam Vessels.*
K L, *another sort of Sap-Vessels in parcels.*
K L D C, *Lympheducts.*
D C E F, *The Wood.*
In which the Holes are the Aer-Vessels.
M M, *the Insertions.*
E F, *more Balsame-Vessels.*
E F G, *the Pith.*
TAB. XXXVI. a a, *part of a Vine-Branch cut transversly, and also split half way down the midle.*
B B, *The same magnified. Shewing the Position of the Bladders in the Barque and Pith in perpendicular Rows; in the Insertions, in Horizontal Rows.*
And the Vessels or Parcels of Wood not raced as in many other Trees.
TAB. XXXVII. *Sheweth the bracing of the Vessels. And how the several Parcels of Vessels or Wood are interwoven with the Insertions.*
TAB. XXXVIII. A B C D, *the Barque.*
H I, *The Parenchyma.*
e e, *A sort of Sap-Vessels.*
a a, *Another sort.*
c c, *Milk Vessels.*
D C E F, *the Wood.*
V V, *the Aer-Vessels.*
t t, *More Lympheducts.*
ſſ, *More Milk-Vessels.*
a t, *The Insertions.*
E F G, *The Pith, composed of angular Bladders, the Bladders of Threds, and the Threds of single Fibers.*
e, *One of the single Fibers.*
TAB. XXXIX. *Sheweth the Structure of the Lympheducts or of the Lignous Fibers both in the Barque, and the Wood.*
F. 1. a, & F. 2, *A single Vessel in the Barque of Flax, composed of a great number of other Lignous Fibers; with which also the Parenchymous are intermixd. Not visible, except very highly magnified.*
F. 3. *A parcel of the same Vessels in Wood.*
F. 4, & F. 5. *shew the manner of the Ascent of the Sap, both in the Lympheducts, and in the Lactiferous and other larger Vessels.*
TAB. XL. *The Fibers which hang down from the Barque are the Lympheducts; one of which is composed of a great many other smaller Fibers.*
The large Tubes are the Milk-Vessels composed of Bladders.
The Fibers which hang down from the wood, are some of them the old Lympheducts turn'd to wood.
And some, Aer-Vessels unroav'd.
The thin Plate between the two wedges of wood, is one of the Insertions, composed of Bladders, and those Bladders of Threds.
The remainder, is part of the Pith, composed of Thredy or Fibrous Bladders.

The *TABLES* to the Fourth *BOOK* are 42.

TAB. XLI. b, *a Dock-Leaf covered with the Veil.*
d, *the Leaf naked.*
a c, *the Veil spred open.*
In Clary, the Bud is embraced by the Curled Leavs.
In Sumach, the Bud lies within the Stalk, as an Egg or Kernel within a Shell.
TAB. XLII. F. 1. *sheweth how the Pipes are inclosed one within another.*
F. 2. a, *the Leaf foulded up.* b, *opened.*
F. 3. a b, *the Bud.* b, *a little magnified*

The Explication of the Tables.

magnified.

F. 4. a b, *the Leaf rowled up, inward.* c, *a little magnified and cut transversly, to shew the Rowl.*

F. 5. a, *the Leaf rowled up.* b, *magnified and cut transversly.*

F. 6. a, *the Leaf rowled backward.* b, *magnified and cut transversly.*

TAB. XLIII. F. 1. *sheweth the Tenter-Hooks, by which the Leaf climbs.*

F. 2. *sheweth the Globulets, turned to a white powder.*

The Leaf of Jerusalem Cowslip. sheweth the Way of the Insect under the Skin.

TAB. XLIV. & XLV. *sheweth the Measures of Leavs by the Circumference.*

TAB. XLVI & XLVII. *sheweth the proportion between the chief Fibers; and also the Angles they usually make together.*

TAB. XLVIII. F. 1, 2, & 3. *shew the Apertures in several Leavs.*

F. 4. *sheweth the same. And likewise, the peculiar composure of the Bladders and Fibers of the Leaf.*

TAB. XLIX. *sheweth the difference in the Bladders, and in the Position of the Lignous Fibers in the Stalks of Leavs.*

TAB. L. *sheweth the Pulp of a Borage-Leaf and many others composed of Bladders; the sides of which Bladders, are made of other smaller ones.*

And the distribution of the Lignous Fibers (and of the Aer-Vessels sheathed within them) not like that of Veins in Animals, but of the Nervs, &c. See the description of the Leaf.

TAB. LI. F. 1. *The appearance of the Aer-Vessels like Cobwebs to the naked Eye, upon breaking the Leaf.*

F. 2. *A small peice cut off of the Leaf.*

F. 3. *The same magnified in which the same Vessels look like spiral wyers stretched out.*

F. 4. *The same as they stand entire within the Wood.*

TAB. LII. *Representeth the Aer-Vessels of Scabious, as in* Tab. LI.

TAB. LIII. *Sheweth the manner of the Generation of the Leaf, chiefly, by the help of several Salts, wherewith the Sap is impregnated.*

F. 5. (1) *The Foundation of the work.*

F. 6. (1 & 2) *strengthned.*

F. 7. (1 & 3); *in which* (3) *is set with the square end to end: and with the poynt-side of one, to that of another.*

F. 8. *The same, directing the Position of the Lignous Fibers at very Acute Angles.*

F. 9. *At less Acute Angles.*

F. 10. *The greater Fibers at Acute, and the smaller at Right Angles.*

F. 11. *The greater at Right Angles with the help of* (1) (2) *or* (3).

F. 12. (3) *directing the Fiber in the Edge of the Leaf into a greater Circle.*

F. 13. *Into a less, and with divers Diameters.*

F. 14. (4) *derecting the Parenchymous Fibers in making the Bladders.*

F. 15. *In winding from one Bladder to another.*

F. 17. *Or about the Lignous Fibers.*

F. 16. *In making the Aer-Vessels.*

TAB. LIV. *sheweth how Nature manages the Folds of Flowers according to their Shape.*

TAB. LV. F. 1. *sheweth the Edges of the Leaf fastned by their Indented Hairs.*

F. 2. *The Balsamick Knobs in the place of Hairs.*

F. 3. *The number 5 running 3 times into its self in 13.*

F. 4. *And five times in 21.*

F. 5. &c. *The Seminiform Attire in* Clematis Austriaca. *With one of the*

The Explication of the Tables.

the *Thecæ* magnified; of which, there are about 30 or 40 in one Flower.

F. 8, &c. The same in *Blattaria*, with one of the *Thecæ* magnified; of which are there about 5 in one Flower.

TAB. LVI. The same in yellow Henbane.

With one of the *Thecæ* magnified; of which there are about 5 in one Flower.

And the Column on the top of the Seed-Case.

TAB. LVII. The same in St. *Johns wort*, entire, together with the Seed-Case or *Uterus*.

TAB. LVIII. The Varieties of the Spermatick Particles in the Seminiform Attire.

TAB. LIX. The Florid Attire of Golden Rod;

In which, the several suits consist but of two pieces. And of which Attire, the Flower doth almost wholly consist.

TAB. LX. F. 1, &c. The same Attire in French Marigold or *Flos Africanus*, with one suit magnified. Of which, there are about 12, in one Flower; and every suit consisting of 3 Pieces.

F. 5. One of another Flower, consisting also of 3 Pieces.

TAB. LXI. One suit of the same Attire in Marigold, and Knapweed, each of them consisting of three Pieces.

F. 5. a, The Attire of one Piece, proper to each Leaf in a Marigold Flower, besides that in the bosome of the Flower.

F. 8. a b, the Seed-Case or *Uterus* at the bottom of every suit.

TAB. LXII. The Attire (of 3 Pieces) proper to each Leaf in the Flower of Cichory.

TAB. LXIII. sheweth the Flower of Mezereon perfectly formed in all its Parts, in the year before it appears. But differs in Shape, as a Fœtus doth when newly formed.

TAB. LXIV. sheweth the same in the Flower of *Asarum*.

TAB. LXV. sheweth the position of the 20 chief Branches in an Apple.

Their Production from the Stalk to the Seeds and Flower.

And a part of the *Parenchyma* magnified, sc. that which is pricked out from the Coar to the Skin; shewing the oblong Figure of the Bladders, and the Divisions in every Bladder.

TAB. LXVI. F. 1,&c. sheweth the Bladders in the Rind of a Limon conteining the Oyl.

The Bags and Bladders of the Pulp, conteining the sower Juyce.

And the Position of the Vessels belongeth to the Fruit, Seed, and Flower.

F. 5. shews the same Vessels, and treble Parenchyma in a Cucumer.

TAB. LXVII. Representeth the Parts of a Pear.

The position and production of the Vessels.

The Chanel from the top of the Pear to the botome of the Coar.

The Tartareous Knots.

And the Bladders radiated to them.

TAB. LXVIII. See the Descriptions of Fruits; and the last Chapter of the Generation of the Seed.

TAB. LXIX. F. 5. Sheweth the Parts of a Goosberry.

The darker part is the sower Rind. Consisting of two sorts of Bladders, of which some very small, and others very great.

The white pieces on the circumference of the Berry, are the Lignous Fibers.

The two opposite white and radiated Bodys are the Middle Parenchyma.

And the oblong Bages round about the several Seeds or Seed-Cases, are the sweet Pulp.

TAB. LXX. Sheweth the seed-Case of Radish opened, and the Seeds hanging on two Ropes.

That of Poppey both entire, and

The Explication of the Tables.

split down the midle.

A slice of the Cod of Garden-Bean, while very young; and therein the Bladders and Threds of the Spongy Parenchyma.

And the gradual ripening and opening of that of yellow Henbane.

TAB. LXXI. sheweth the Seed Case of Tulip entire, cut transversly, and split downe.

A slice of Thorn-Apple, or of th' Seed-Case of Stramonium, while young.

That of Pimpinel naturally divided into two Hemisphers; with the Button, on which the Seeds grow, crested in the middle.

The manner of the ejaculation of the Seed, in Coded Arsmart.

And the Coat of the Seed of Azarum formed the year before it ripens.

TAB. LXXII. sheweth the measures of Plum-stones.

The Apertures, and Divisions, of the covers of the Seed.

The Seed and Seed-Case of Harts Tongue, opened with a Spring.

And other contrivances both for the Motion, and Arrest of other Seeds.

TAB. LXXIII & LXXIV. See the Descriptions.

In Tab. 74. the corners and edges of that of Fox-glove should have been rounder.

The Figures are all done pretty near a Scale.

TAB. LXXV. The Belly and Back of a Datestone, and the small sprouting Node taken out of the Hole in the back cut open.

The Shapes and Foulds of divers Seeds.

The Vitellum of Orach, and Rhapontick.

In great blew-Lupine, d, the Navle; b, the descending part of the Radicle.

TAB. LXXVI. Flag. 1, the Seed. 2, split open. 3, the true Seed which lies in the hollow made in the Cover (2) 4, one half of (2) magnifyd. 5, the Seed (3) magnifyd.

Purging Angola Nut. 1, with the shell on. 2. taken off. 3, the soft Cover split down. 4, the Seed which lies in it; the Lobes hereof answerable to two Leaves, and Radicle to the Stalk.

And so in the rest.

TAB. LXXVII. Coffee Berry stone. 1, The belly of the Stone. 2, the Black. 3, pared a little. 4, the Kernel taken out of it. 5, the same magnifyd.

Goosgrass. 1, the entire Seed. 2, the back of the hard Cover. 3, the belly. 4, cut in two. 5, the same magnifyd. 6, the true Seed taken out of it.

Staphisagria. 1, the entire Seed. 2, the hard Cover. 3, Split in two. 4, the true Seed taken out of it. 5, The same magnifyd.

Peony, 1, the Seed commonly so call'd. 2, one half of it split down. 3, the other half. 4. the true seed taken out of it. 5, the same magnifyd.

Stramonium. 1, the Seed entire. 2, the iner thick cover. 3, the same split in two. 4, the true seed taken out of it. 5. half the thick Cover (3) magnifyd. 6, the seed (4) magnifyd.

TAB. LXXVIII. Some examples of the Buds of Seeds before they are sown.

Sena. 1, the naked Seed. 2, the Lobes divided to shew the Bud. 3, one Lobe with the Bud magnifyd.

Carduus Benedictus, 1, the entire Seed. 2, with the outer Covers off. 3, naked. 4, divided. 5, that half with the Bud, magnifyd.

Hemp. 1, the naked Seed divided. 2, 3, the same magnifyd.

Almond. 1. one half of the Kernel. 2, the Radicle and Bud at the bottom of it. 3, the same broken off. 4, magnifyd. 5, opened.

TAB. LXXIX. F. 1. a b, Part of the outer Coat.

c d,

The Explication of the Tables.

c d, *Part of the Inner Coat.*

c d e, *one Lobe cover'd with the Skin.*

f g, *the other, with the skin and part of the Parenchyma pared off.*

f f, *the Skin.*

h h. *the Parenchyma.*

i i. *the Seminal Root.*

k k, *the Radicle.*

k l, *where it is cut off from the Lobes.*

M, *the Plume or Bud.*

N, *The Cavitys in which it is lodg'd.*

F. 2. *Sheweth the Barque, Vessels and Pith of the Radicle.*

TAB. LXXX. F. 1. *A Slice of a young Apricock, cut transversly, near the lower end; shewing the duplicature of the Skin half way through the Stone.*

F. 2. *A Slice, cut near the upper end; shewing the duplicature of the Skin quite throw the Stone*

F. 3. *A well-grown Apricock cut by the length.*

F. 4, 5, *The Membranes of a Filbert full ripe.*

F. 6, *The Membranes of a young Apricock, with part of the Seed-branch.*

F. 7. *the two Membranes cut by the length.*

TAB. LXXXI. F. 1, *The outer and midle Coats or Membranes; with the Chanel, oval at both ends, now formed in the latter.*

f. 2, *Part of the same, with the upper Oval grown larger, and the inmost Cover now also formed therein.*

f. 3. *the same with the inmost cover grown larger.*

f. 4. *the Inmost Cover more magnified, and the hollow in the smaller end, laid open, to shew the Seed it self, newly begun in a round Node.*

f. 5. *the same; in which the Node begins to be divided into two Lobes.*

f. 6, 7, 8, *the gradual forming of the Lobes.*

f. 9, *next the forming of the Radicle.*

f. 10, 11, *Its gradual contraction at the point, into a short and slender Navle string.*

Which in the further growth of the Seed, breaks and disappears.

TAB. LXXXII. a a, *the Pulp, or open Parenchyma.*

b b, *the close Parenchyma or ground of the Stone.*

c c, *the Flower-Branch running through the body of the Stone.*

d d, *the Seed-Branch striking into the hollow of the Stone, and so running round the outer Membrane* e e.

f f, *the middle Membrane.*

g g, *the Chanel.*

h, *the inner Membrane, in which lies the Seed.*

TAB. LXXXIII. f. 1. *the manner of the generation of the Essential Salts of Plants.*

f. 2, *a Crystal of the Essential Salt of Rosemary, a little magnifyd.*

f. 3, a b, *two of Wormwood,* a, *upon the second Solution;* b *upon the first.*

f. 4, *one of G. Scurvygrass;* a, *one side;* b *the other.*

f. 5. *a Crystal of the Marine Salt of Rosemary.*

f. 6, *of Garden Scurvy-grass;* a *the upper side;* b. *the nether.*

f. 7, *of Wormwood.*

f. 8. *of Black Thorne.*

f. 9. *another of the same.*

f. 10, *of Firne.*

f. 11, *another of Wormwood.*

FINIS.

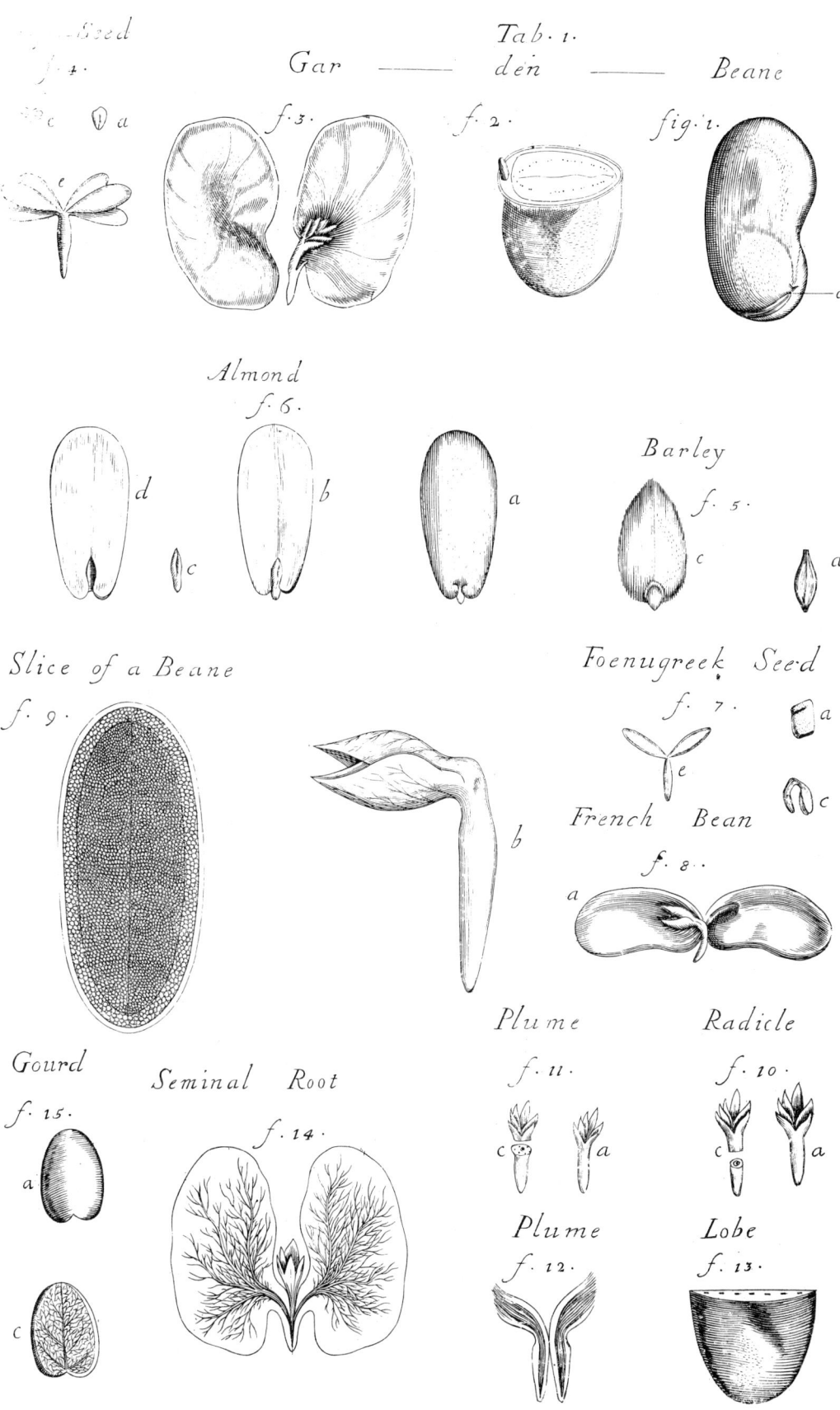

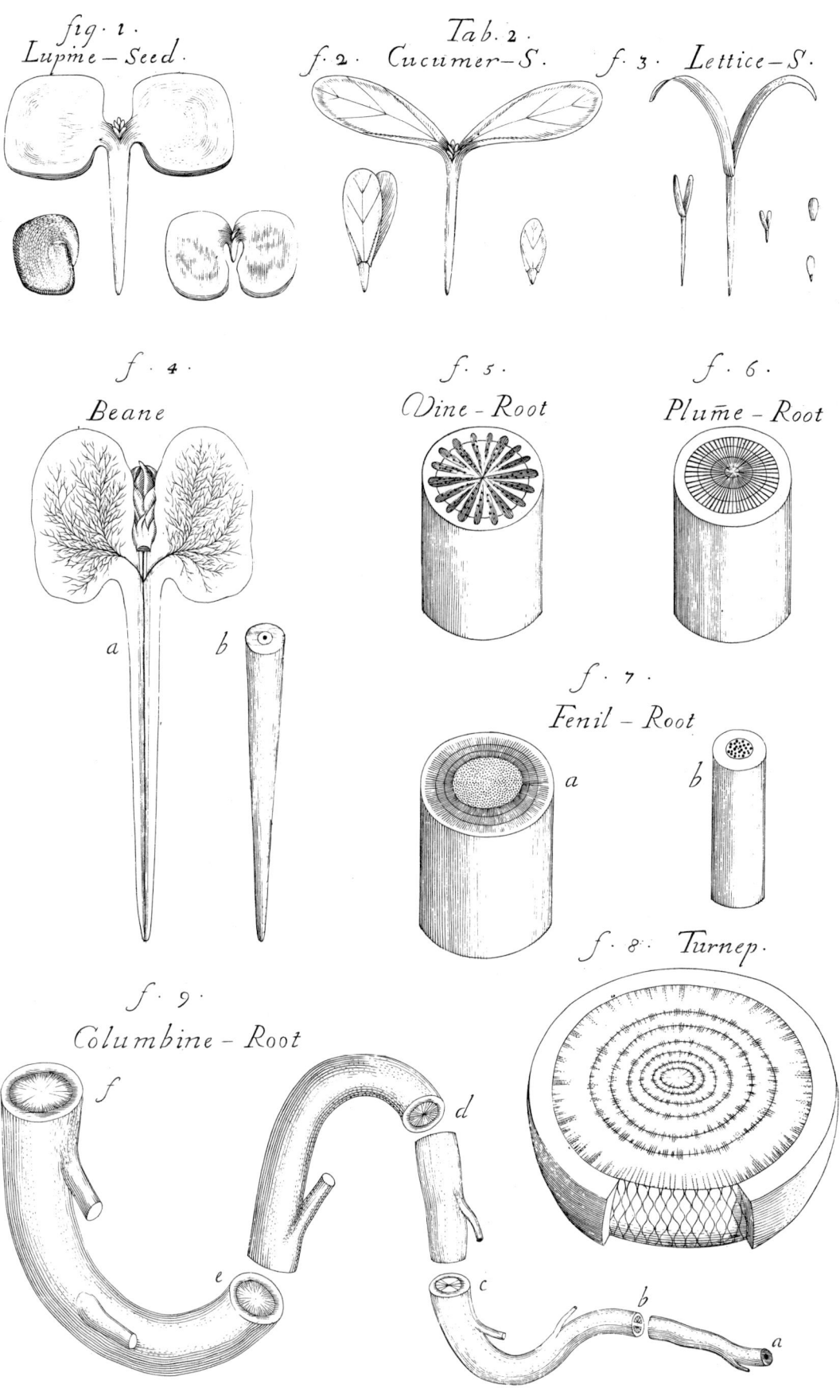

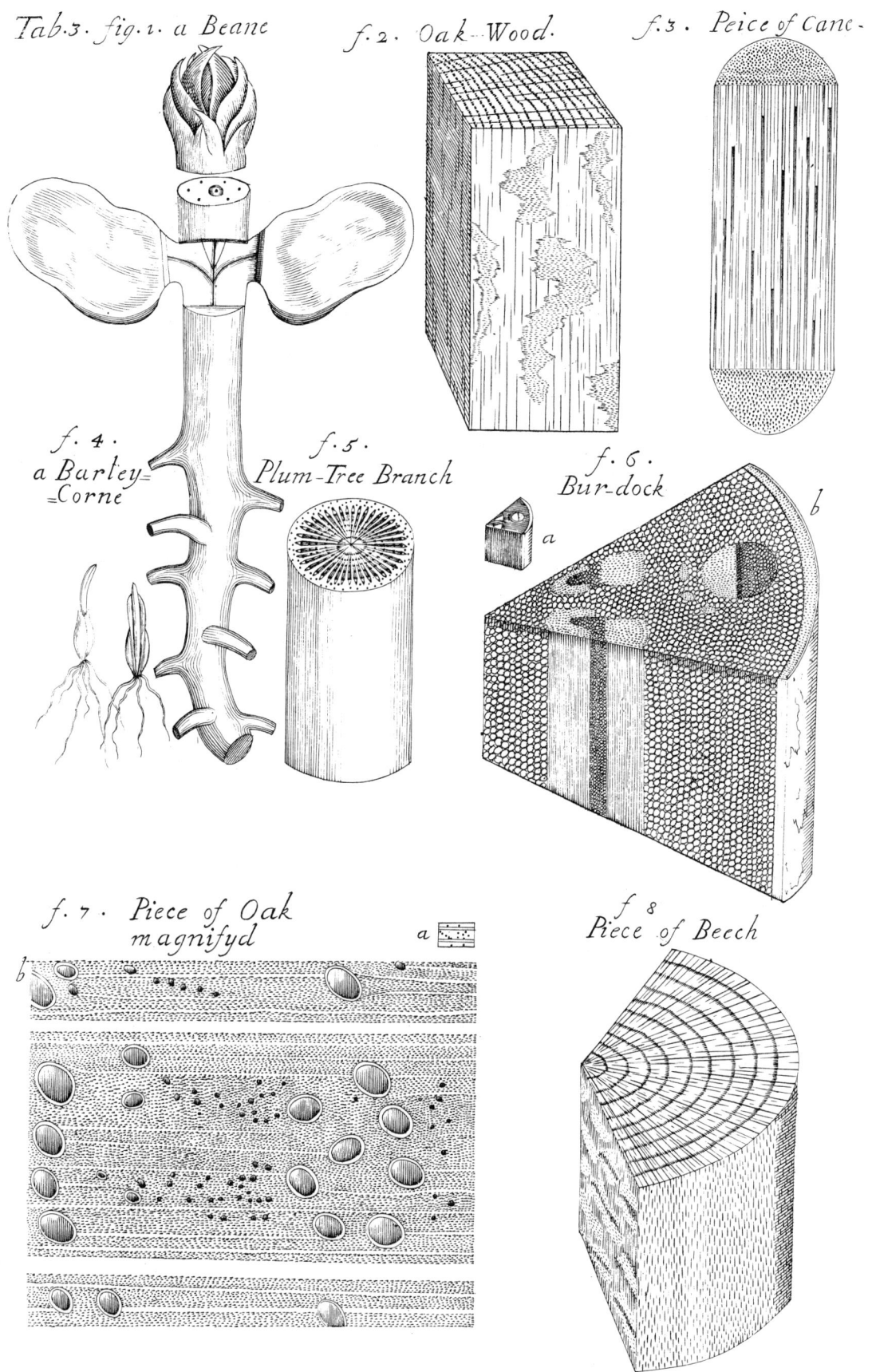

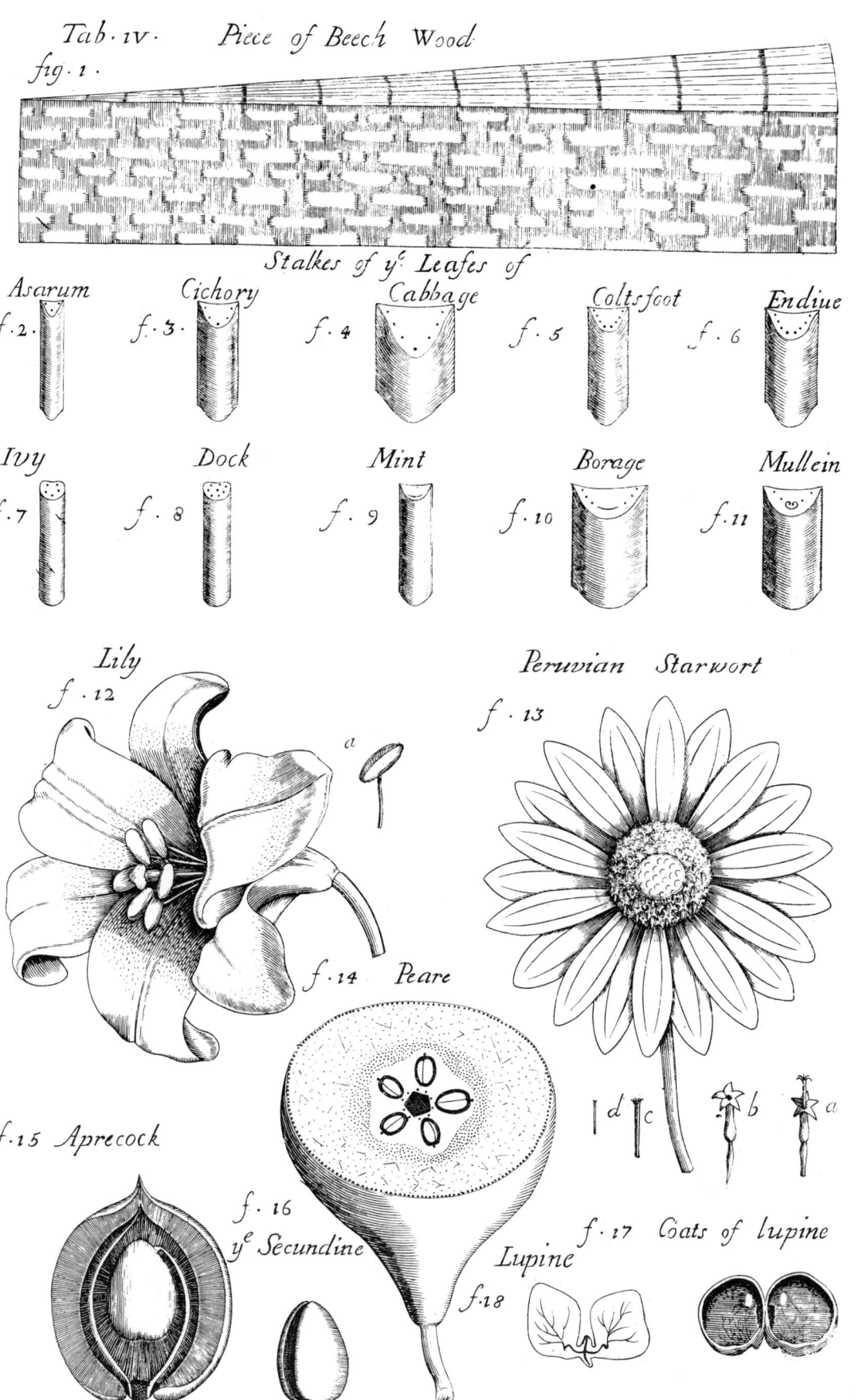

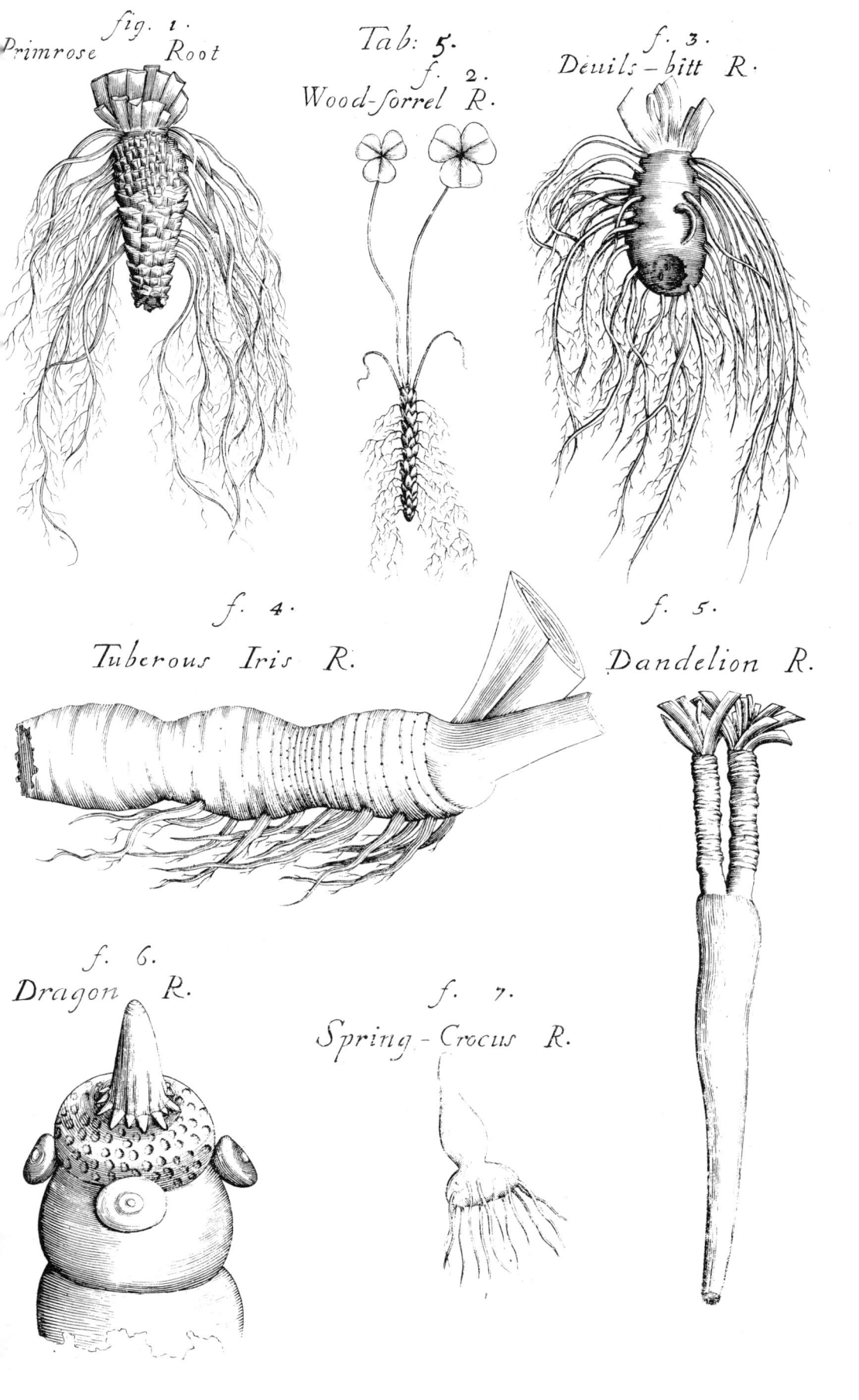

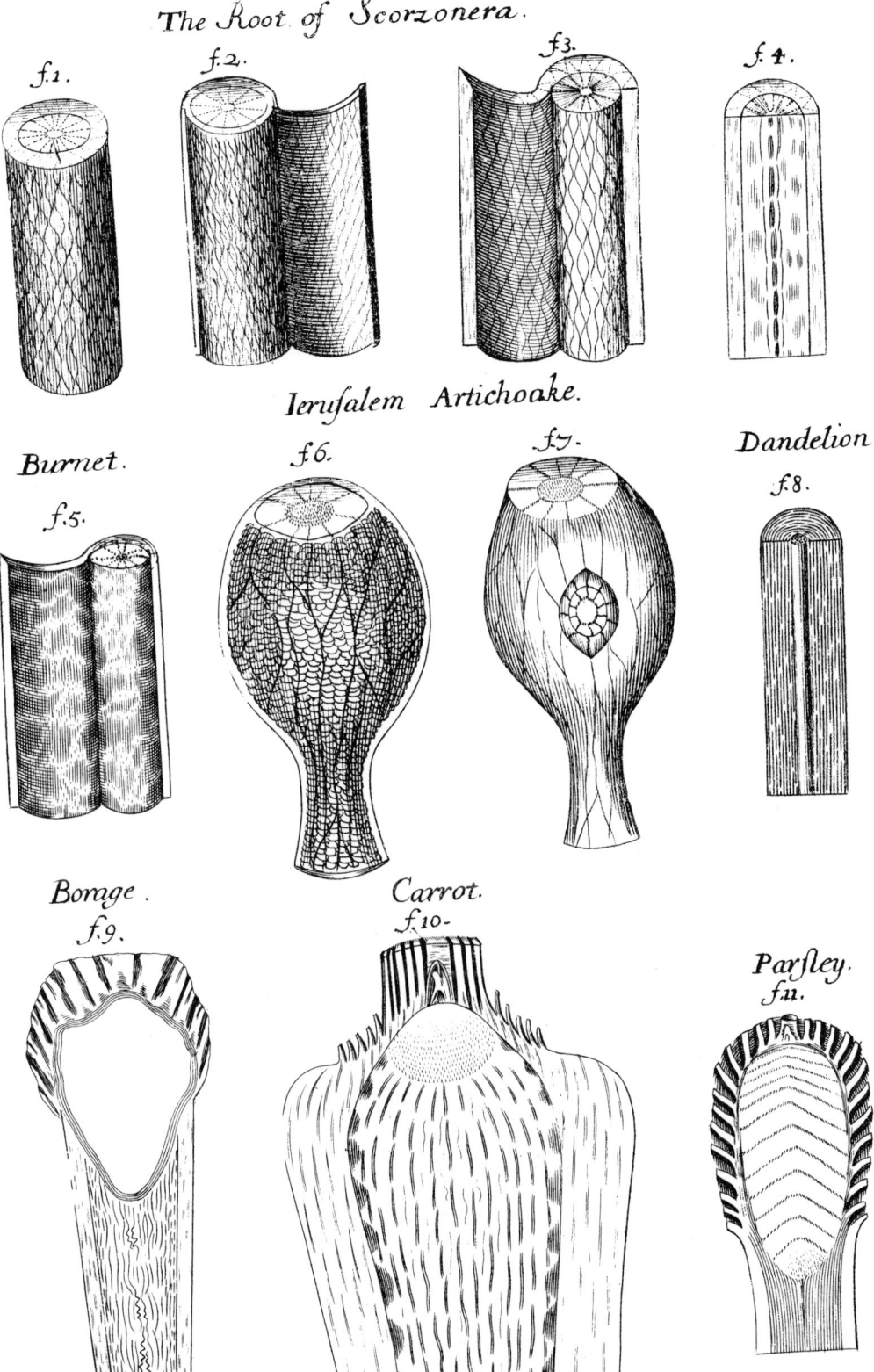

Tab: VI.

Tab. VII. Roots w.ch Bleed little or nothing.

Marsh Mallow.

Patience.
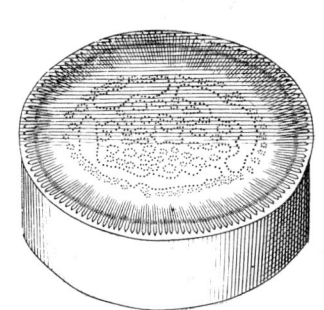

Iris.
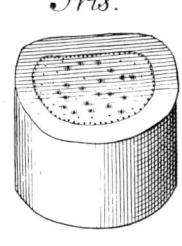

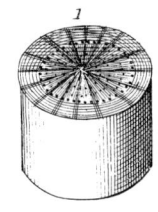

Peony.
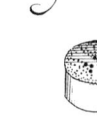

Bistord.
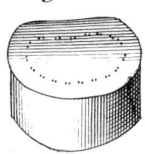

Roots which Bleed a Lympha.

Bugloss.
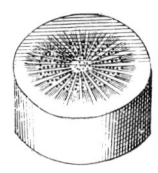

Bryony.
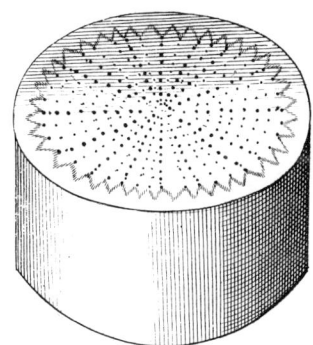

Borage.
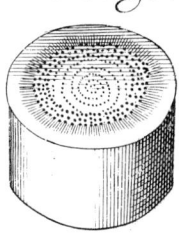

Black Henbeane.
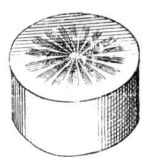

Horse Radish.

Deadly Nightshade.
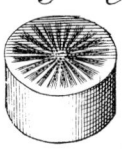

Brownwort.
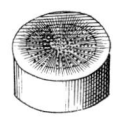

Non Bulbous Lily.

Asparagus.

Columbine.
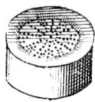

Tab. VIII.   Roots which Bleed a Lympha.

f. 1. Parsnep   f. 2. Carrot   f. 3. Beet

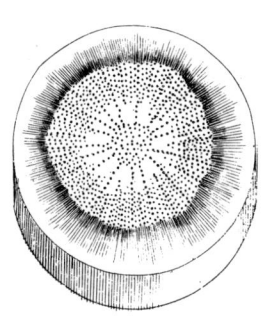
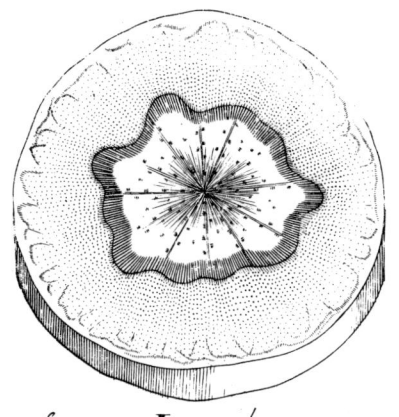
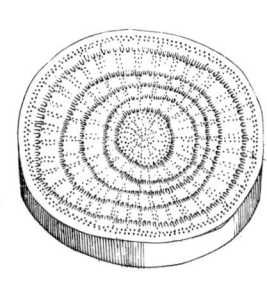

f. 5. Nettle   f. 4: Jerusalem Artichoak   f. 7. Dropwort

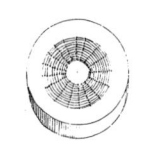
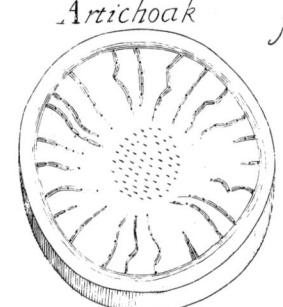

f. 6. Valerian   f. 8. Lychnis

Roots which Bleed a Milk or Oyly Sap

f. 10. Butyr Bur   Lovage f. 9.   f. 11. Dandelyon

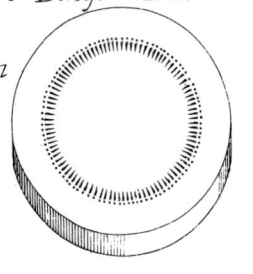
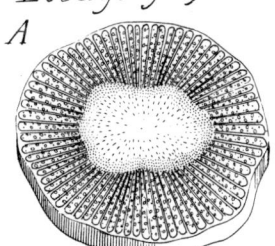
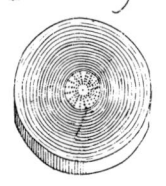

f. 12. Great Celaudine   B   f. 13. Cychory

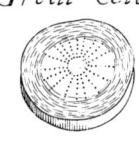
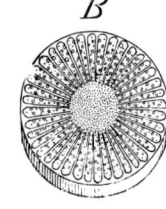

f. 15. Bishopsweed   f. 14. Goatsbeard

Tab: 9.

Roots with Milky or Balsamick Vessels, and Lymphæducts, both apparent

F: 1. Fenil.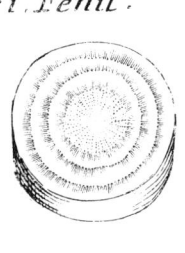

F: 2. Enula camp: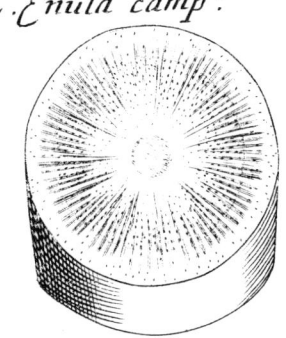

F: 3. Trachelium.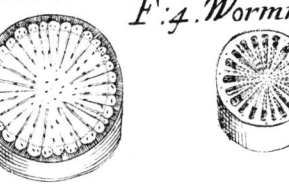

F: 4. Wormwood.

Roots with two sorts of Lymphaticks; in some, Aquæducts and Muciducts.

F: 5 Gard: Scurvygrass.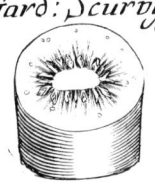

F: 6. Wild Cucumer.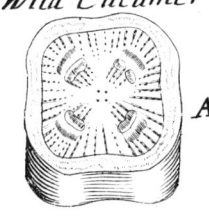
A

F: 6. B.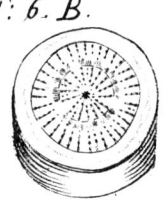

F: 7 Melilote.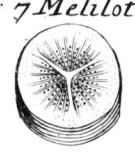

F: 8. Potato.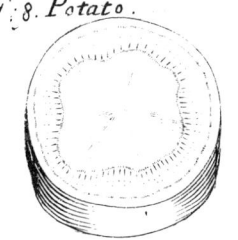

F: 9. Carduus beued.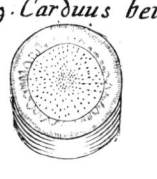

F: 10. Eryngo.

F: 11 Cumfrey.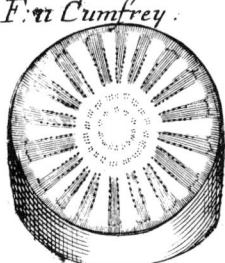

F: 12. Monks Hood.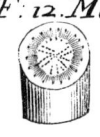

F: 14. Cinquefoyle.

F: 13. Valerian.

F: 15 Burnet

Tab. x.

Fig: 2. *The same magnify'd*

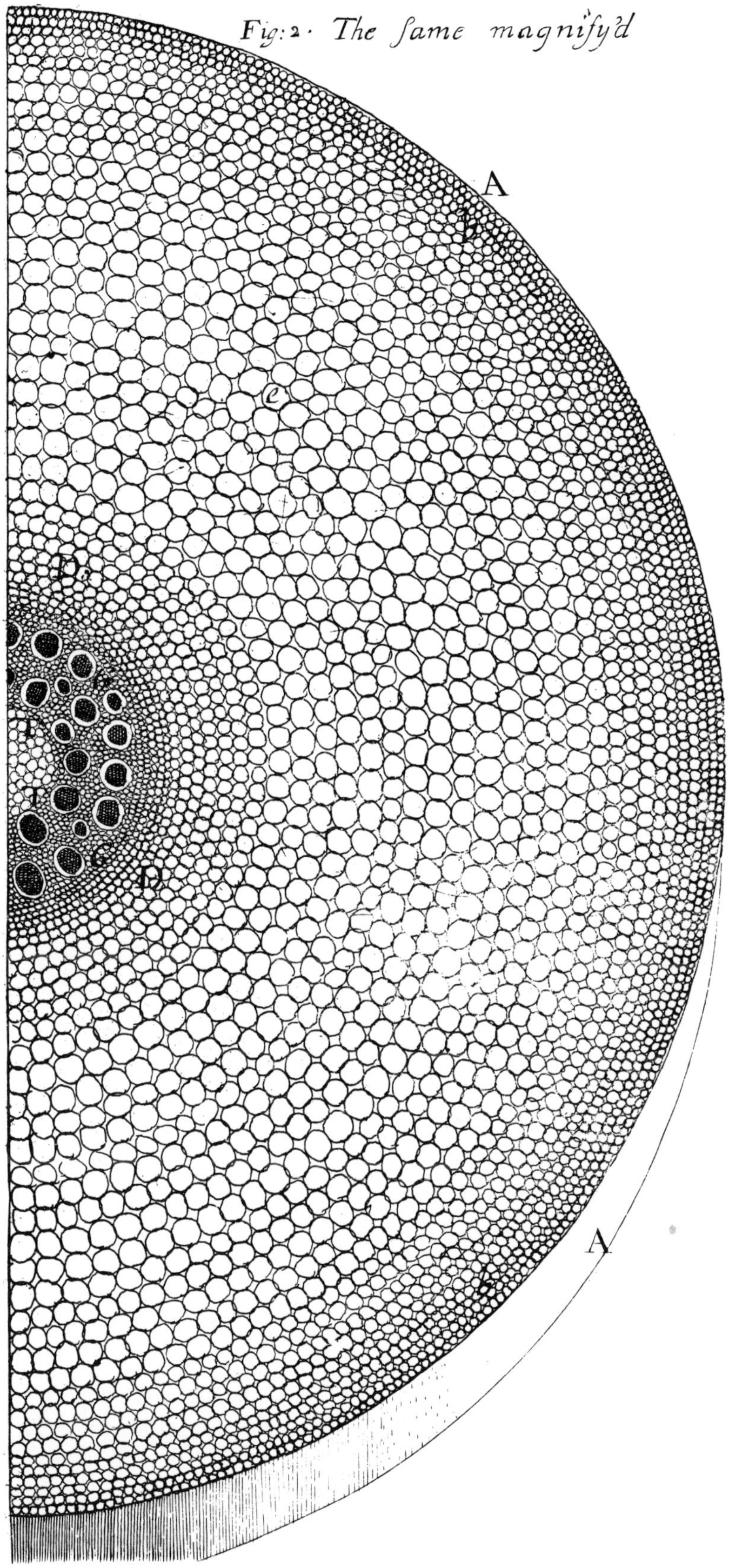

Fig 1
Small Root of
Asparagus

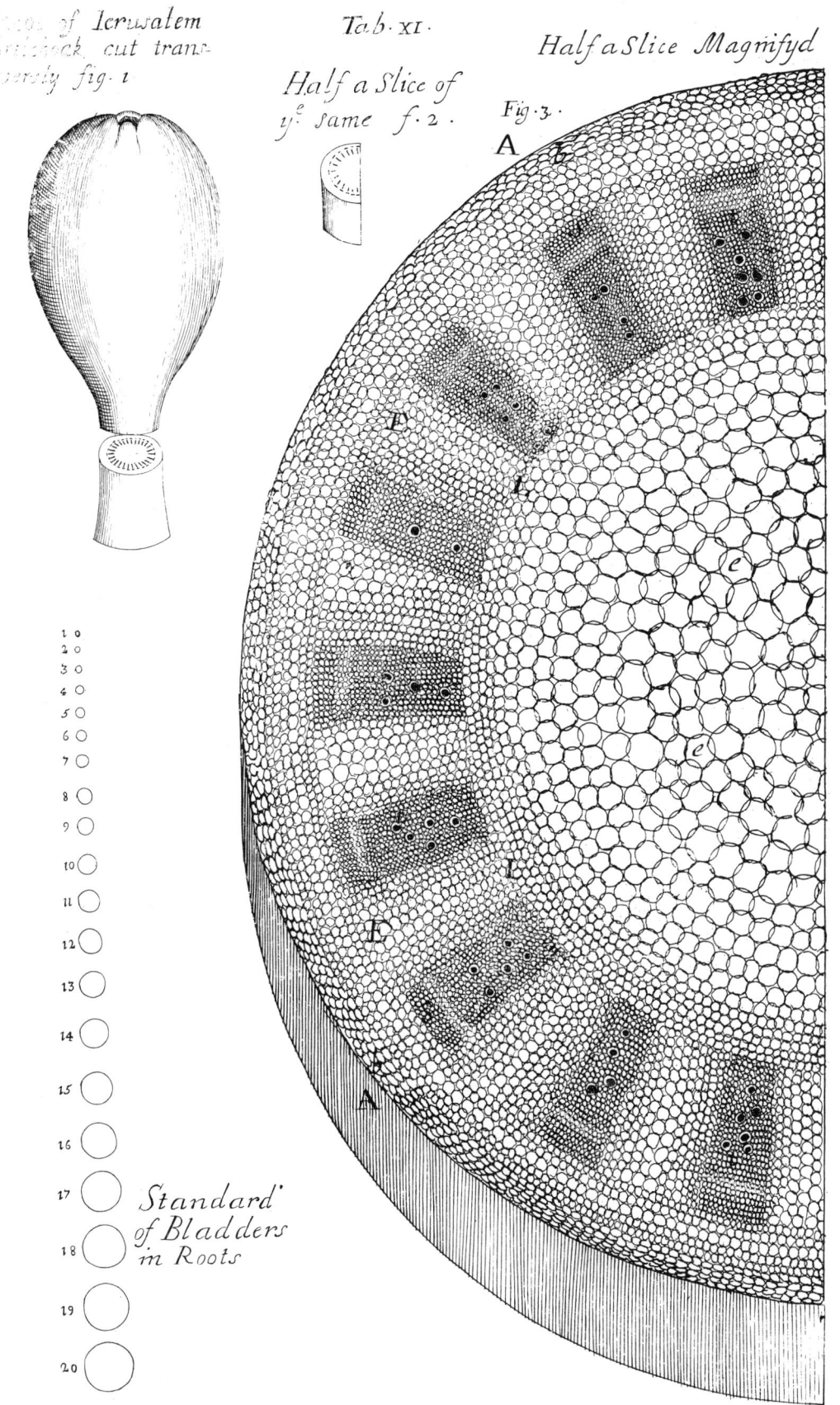

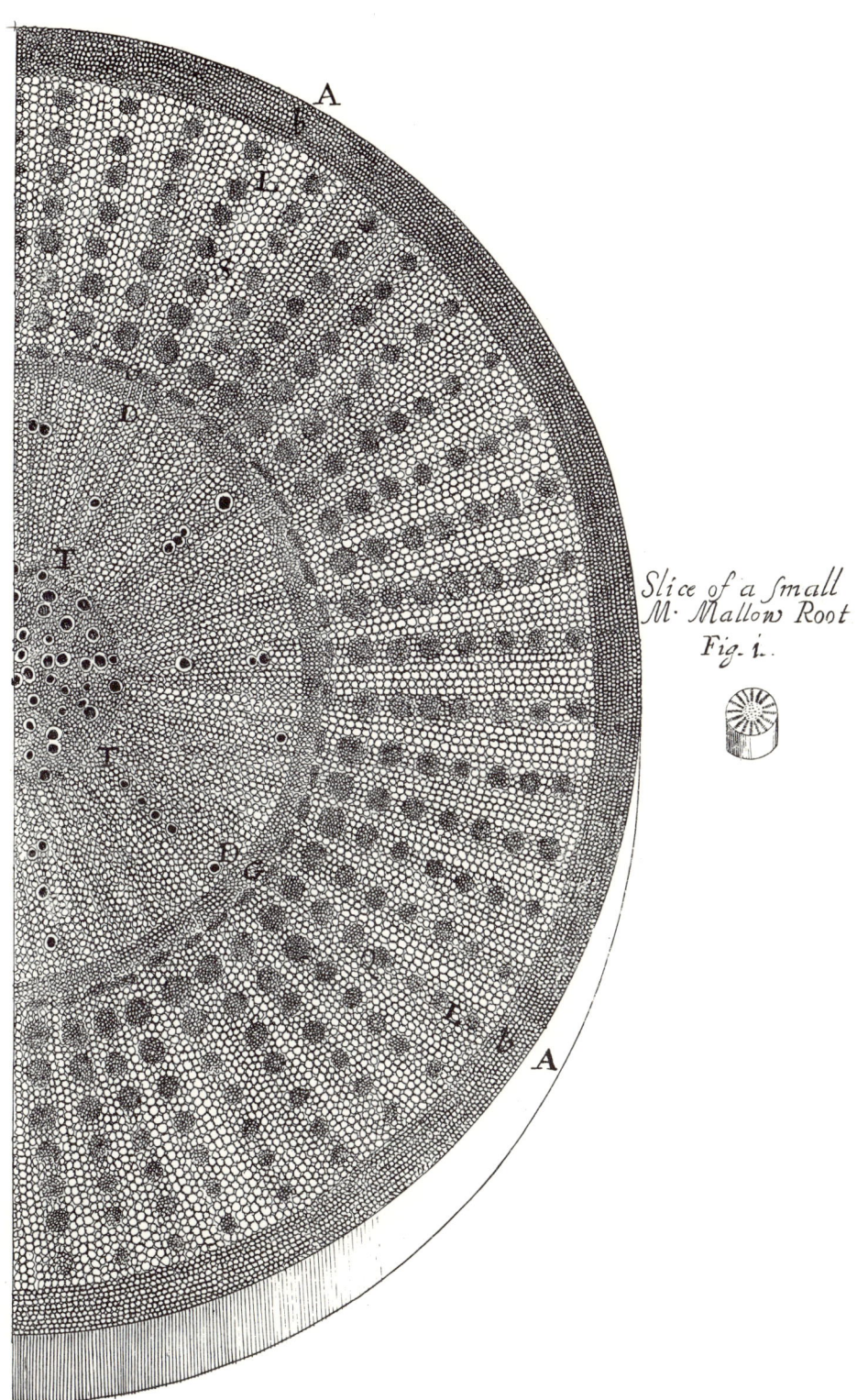

Tab: XII.

The same Magnifid Fig. 2.

Slice of a small M. Mallow Root Fig. 1.

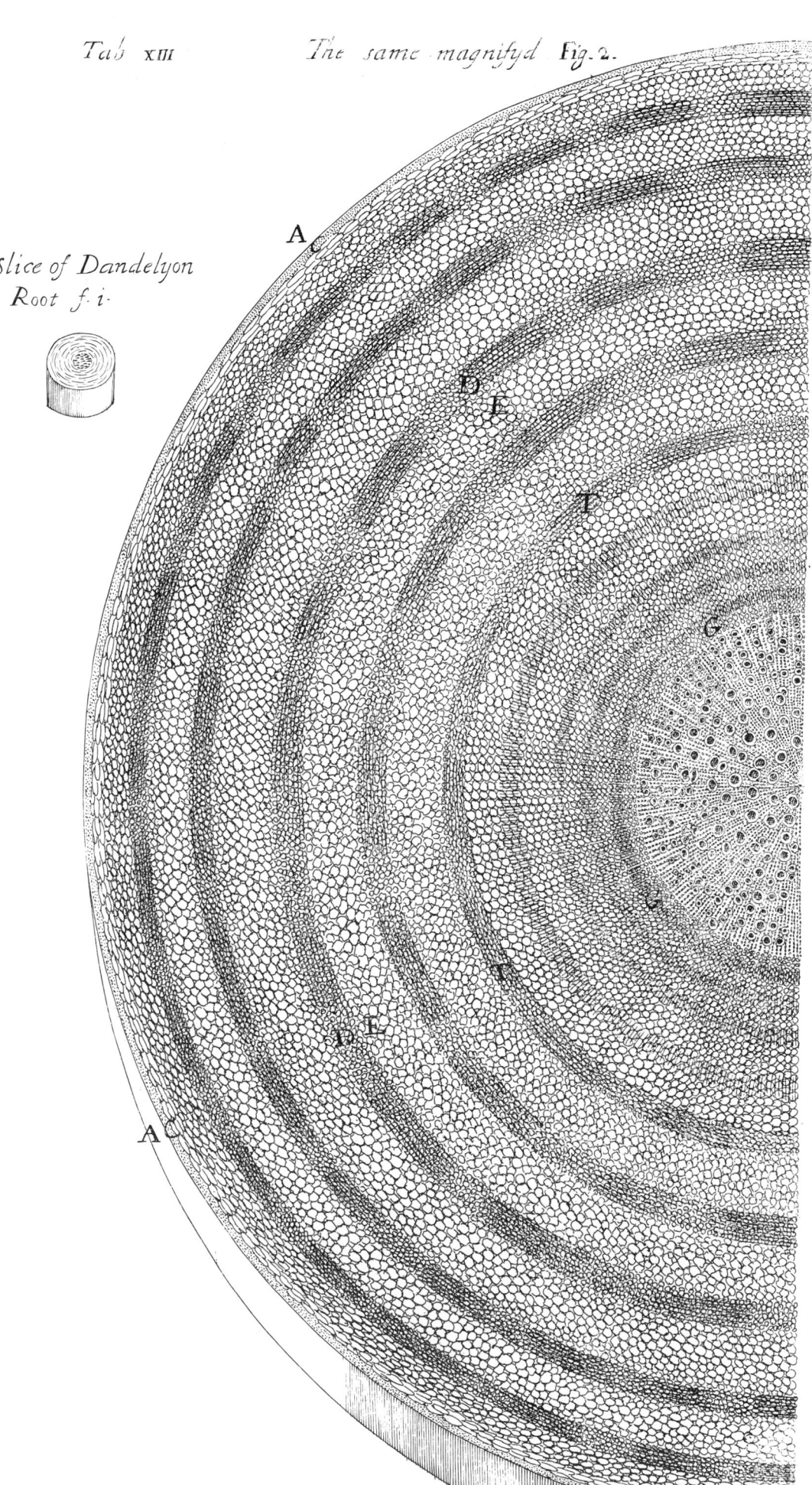

Tab 19.

F.3. Vine.

F.2. Sonchus.

F.1. Corke.

F.4. Walnut.

with ẏ Barque off.

A.

B.

Tab. 20.

A. Magnify'd.

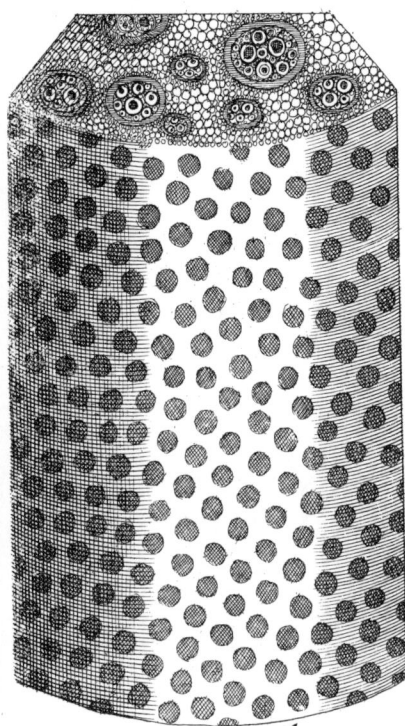

B. Magnify'd.

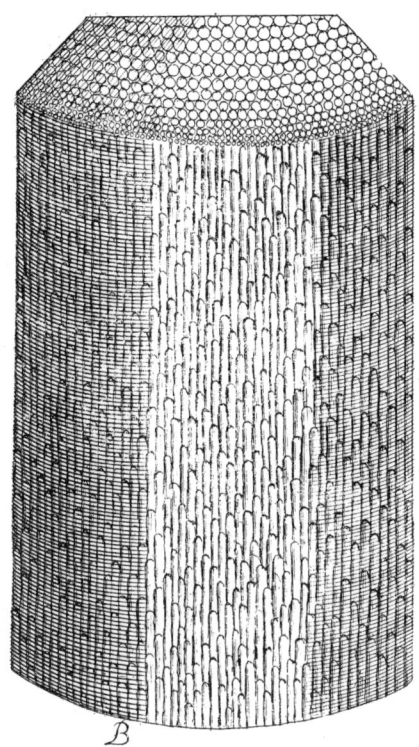

the Surface of Walking-Cane. Fig. 1.

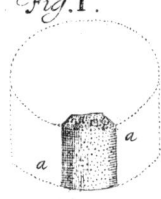

the Skin of Borage-Stalk. F. 2.

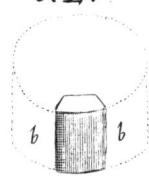

Turpentine-Vessels in the Barque of Pine. F. 3.

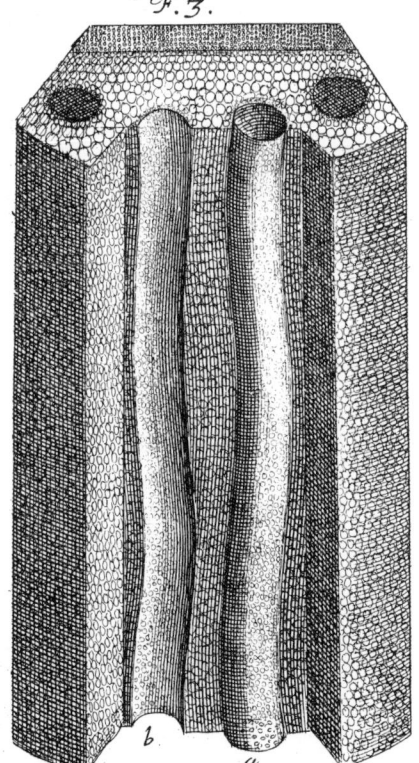

Milk-Vessels in the Barque of Sumach. F. 4.

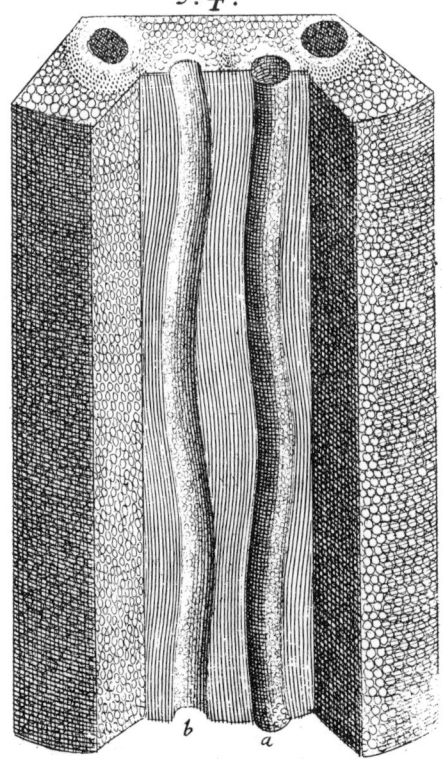

Tab. 21.

AB Piece Cut out of (ab) and Magnify'd to Shew y^e Lymphæducts & Aer-vessels.

a.b. Part of a Vine Branch Cut transversly

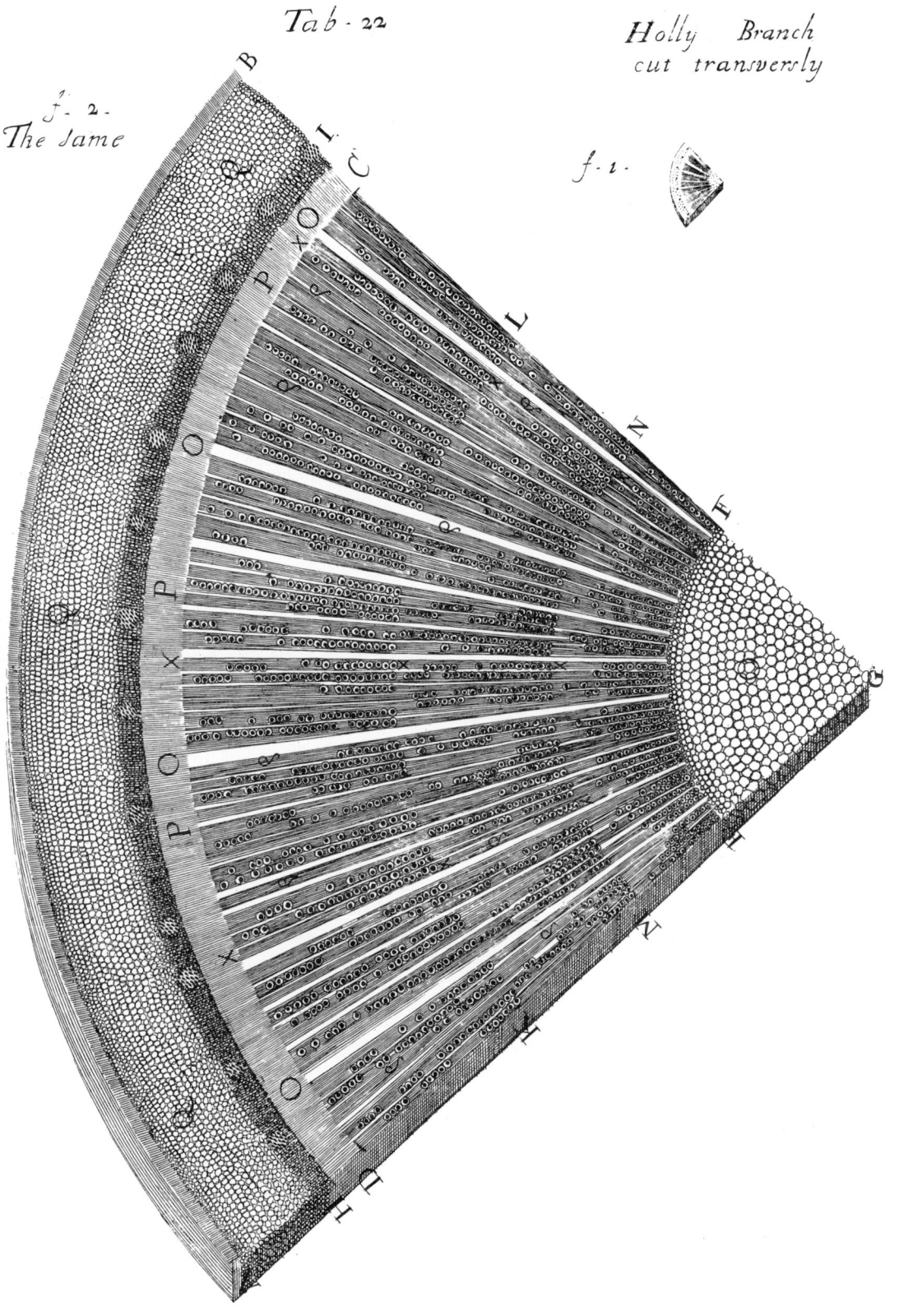

Tab. 22

Holly Branch cut transversly

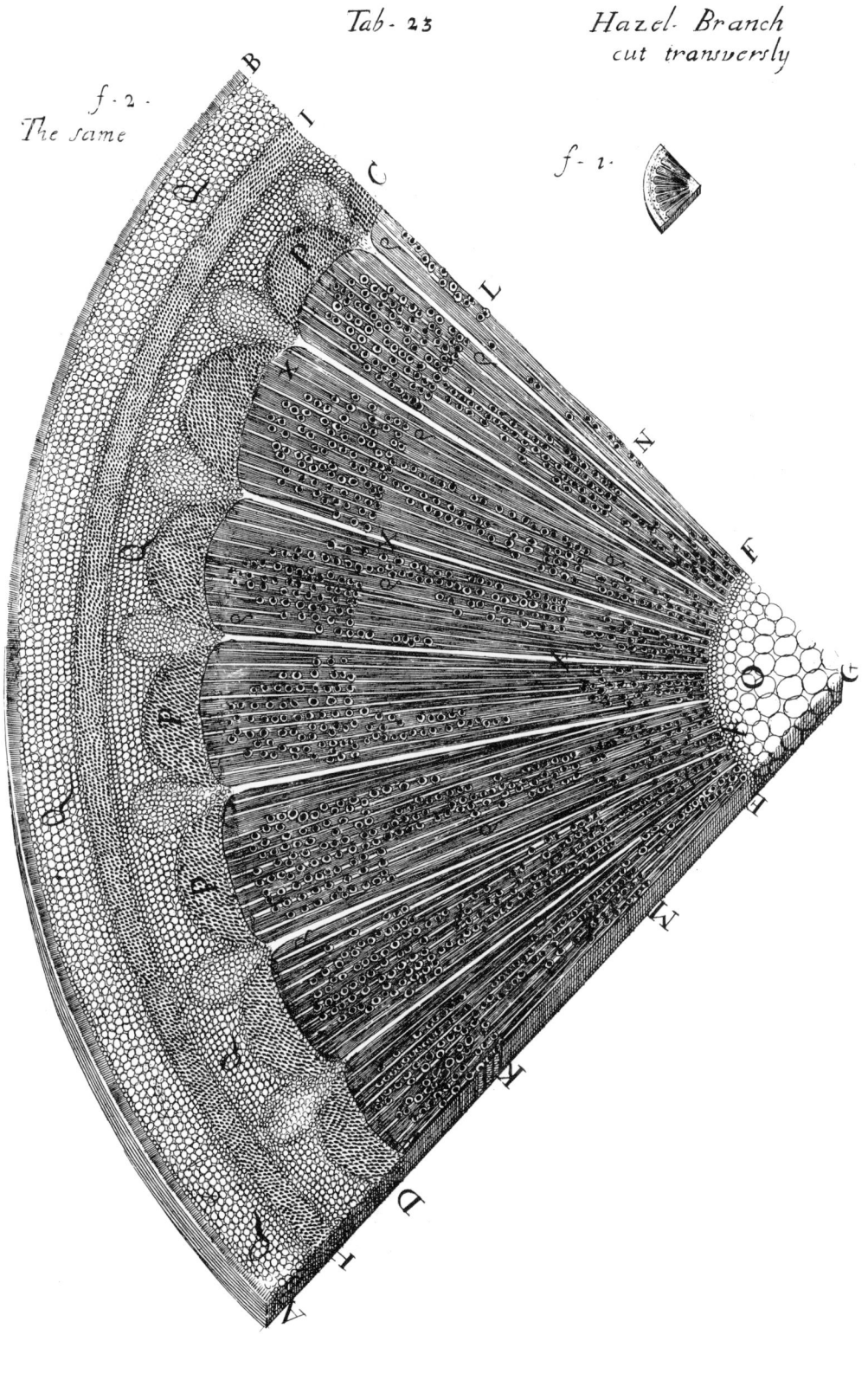

Tab. 23 Hazel Branch cut transversly

f. 2. The same

f. 1.

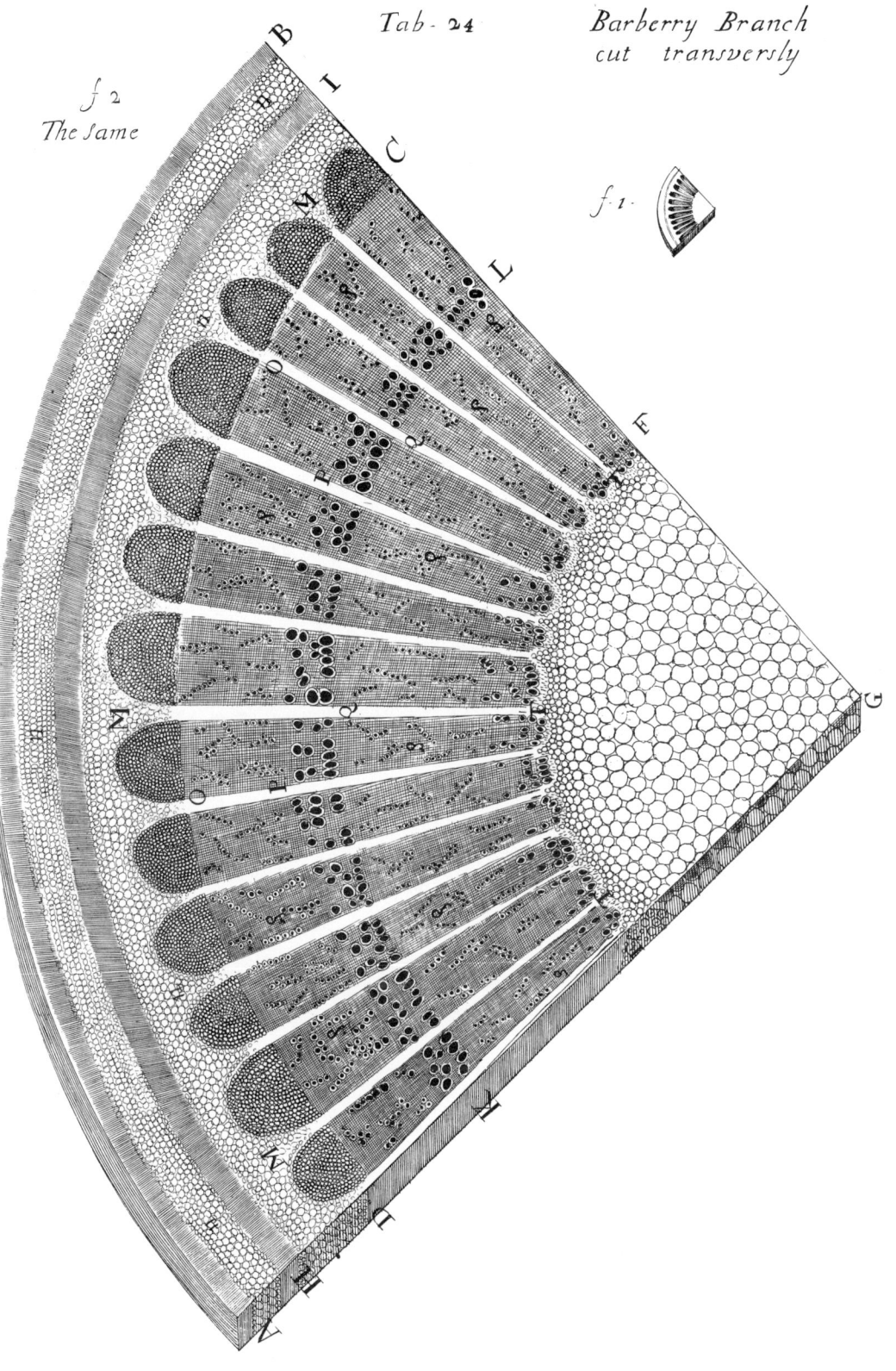

Tab. 24. Barberry Branch cut transversly

f. 2 The Same

f. 1.

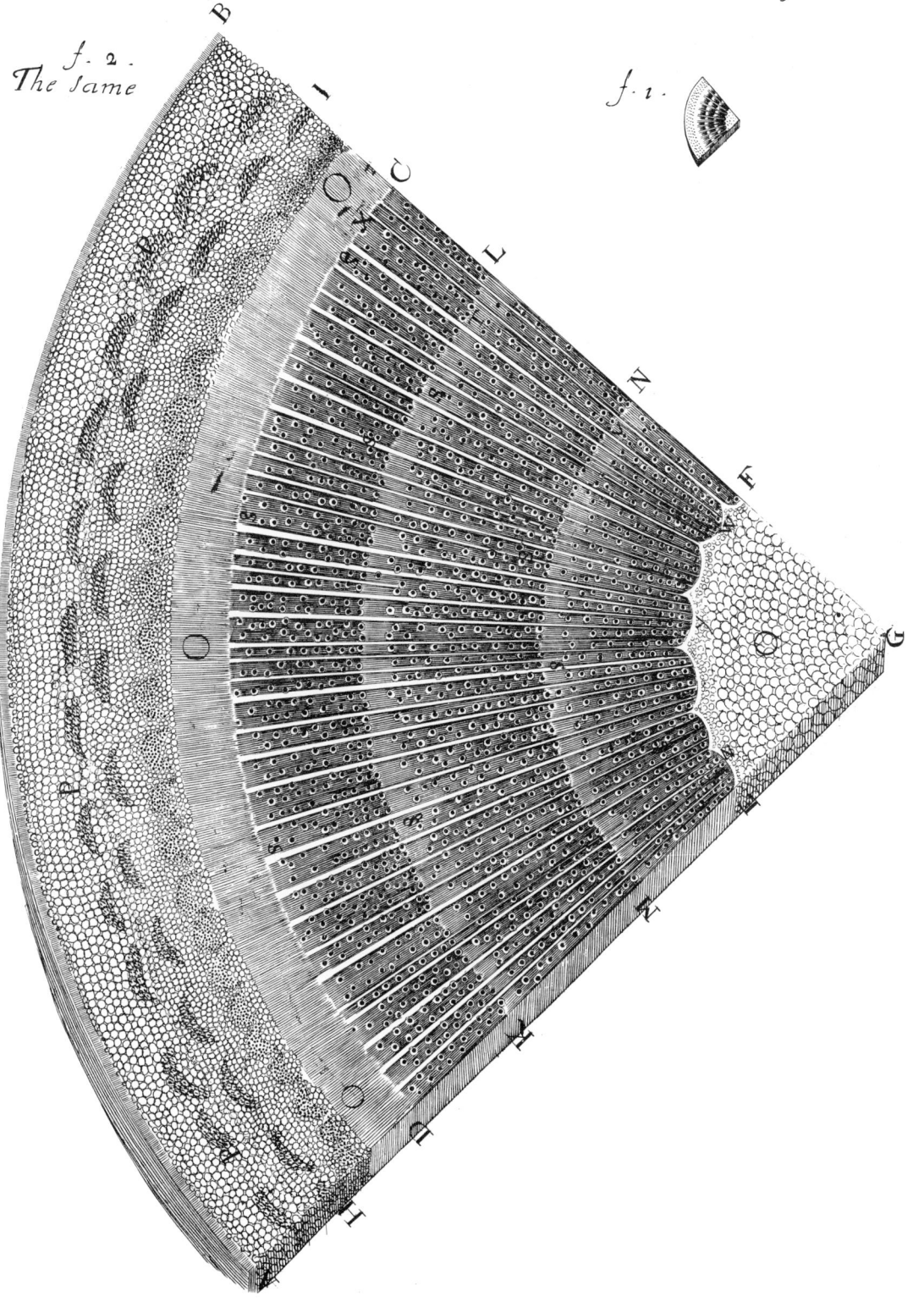

Tab. 25

Apple Branch cut transversly

f. 1.

f. 2. The same

Tab. 26   Pear Branch cut transversly

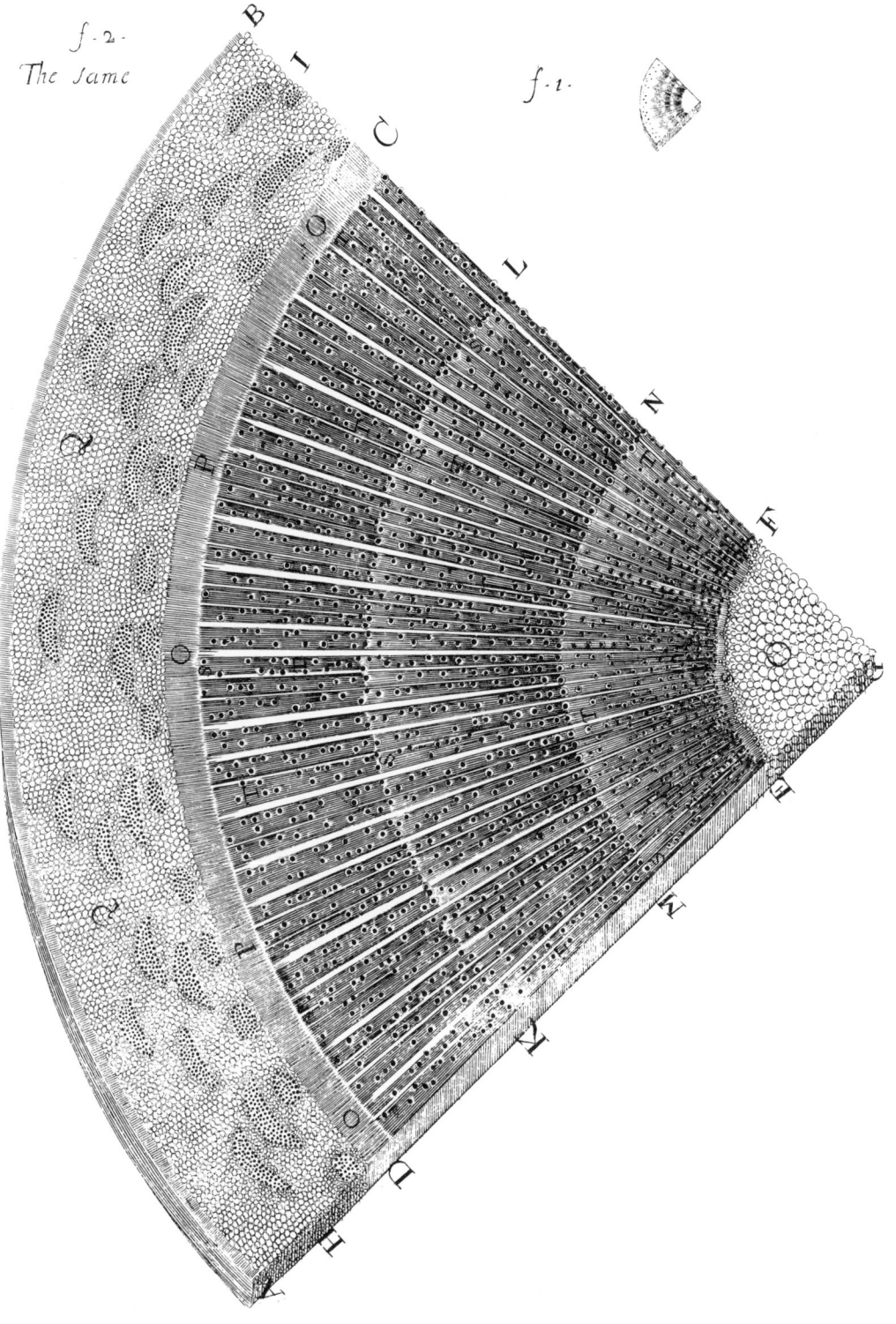

f. 2. The same   f. 1.

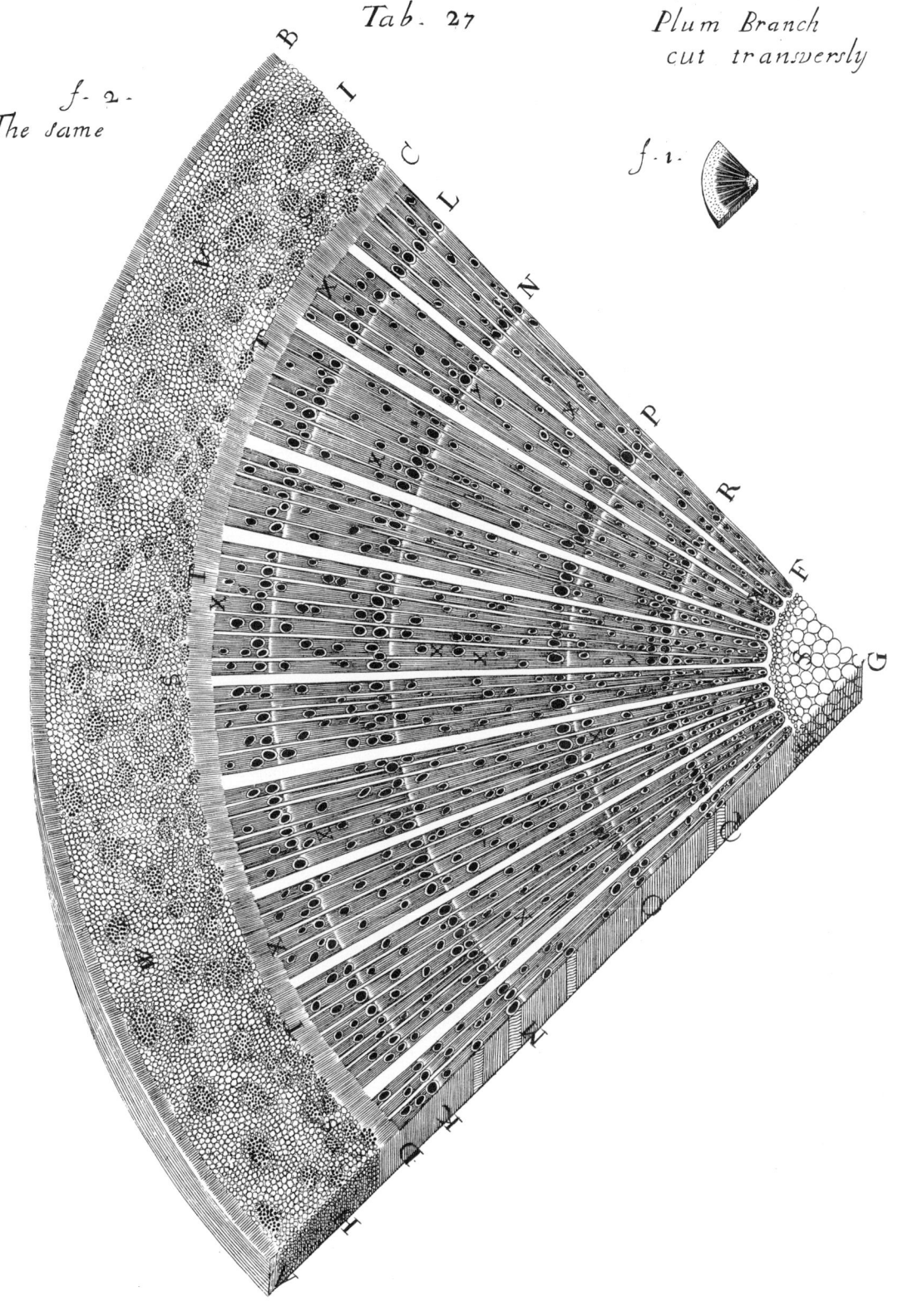

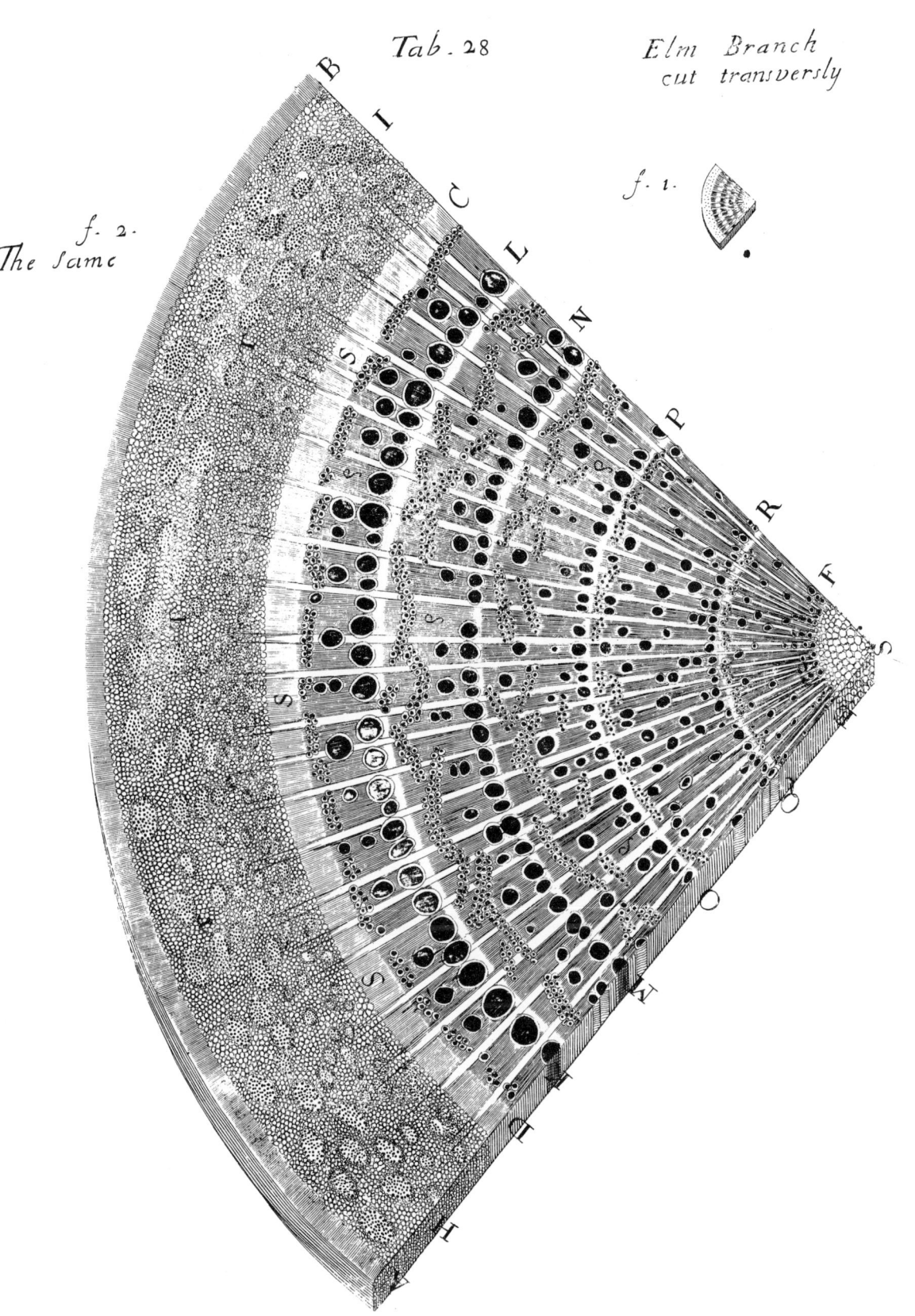

Tab. 28 Elm Branch cut transversly

f. 2. The same

f. 1.

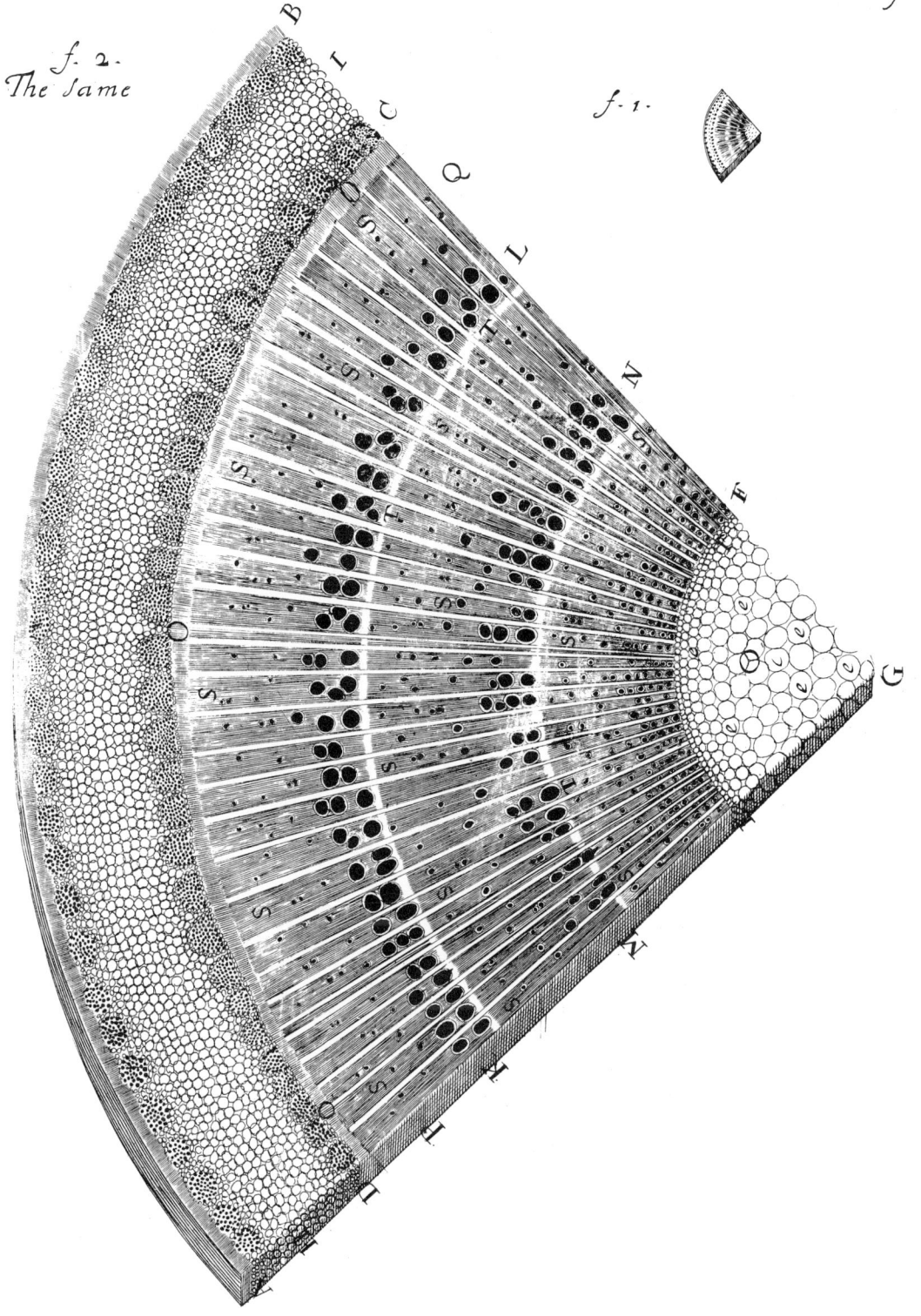

Tab. 29 — Ash Branch cut transversly

f. 2. The Same

f. 1.

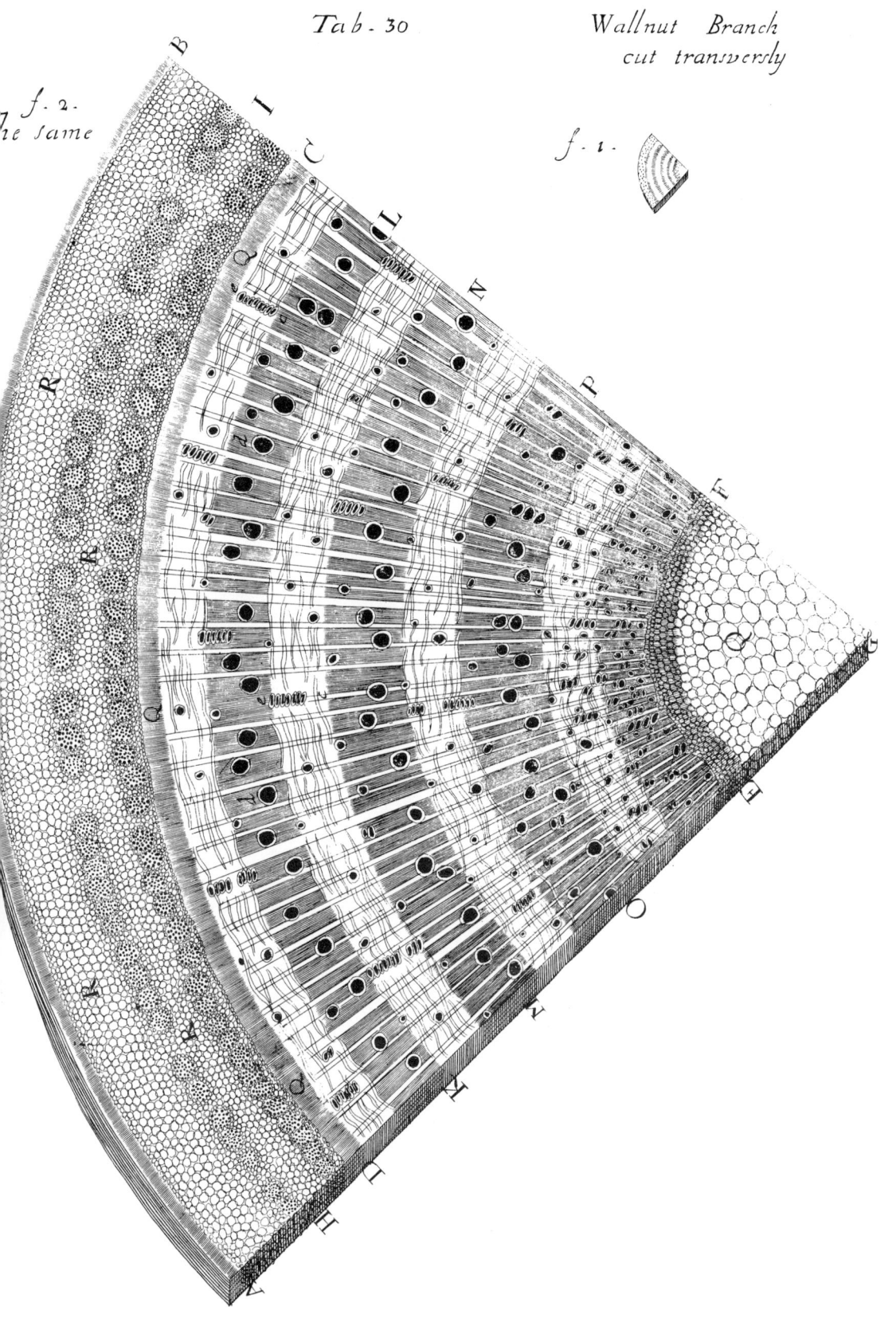

Tab. 30. Wallnut Branch cut transversly. f. 1. f. 2. the same

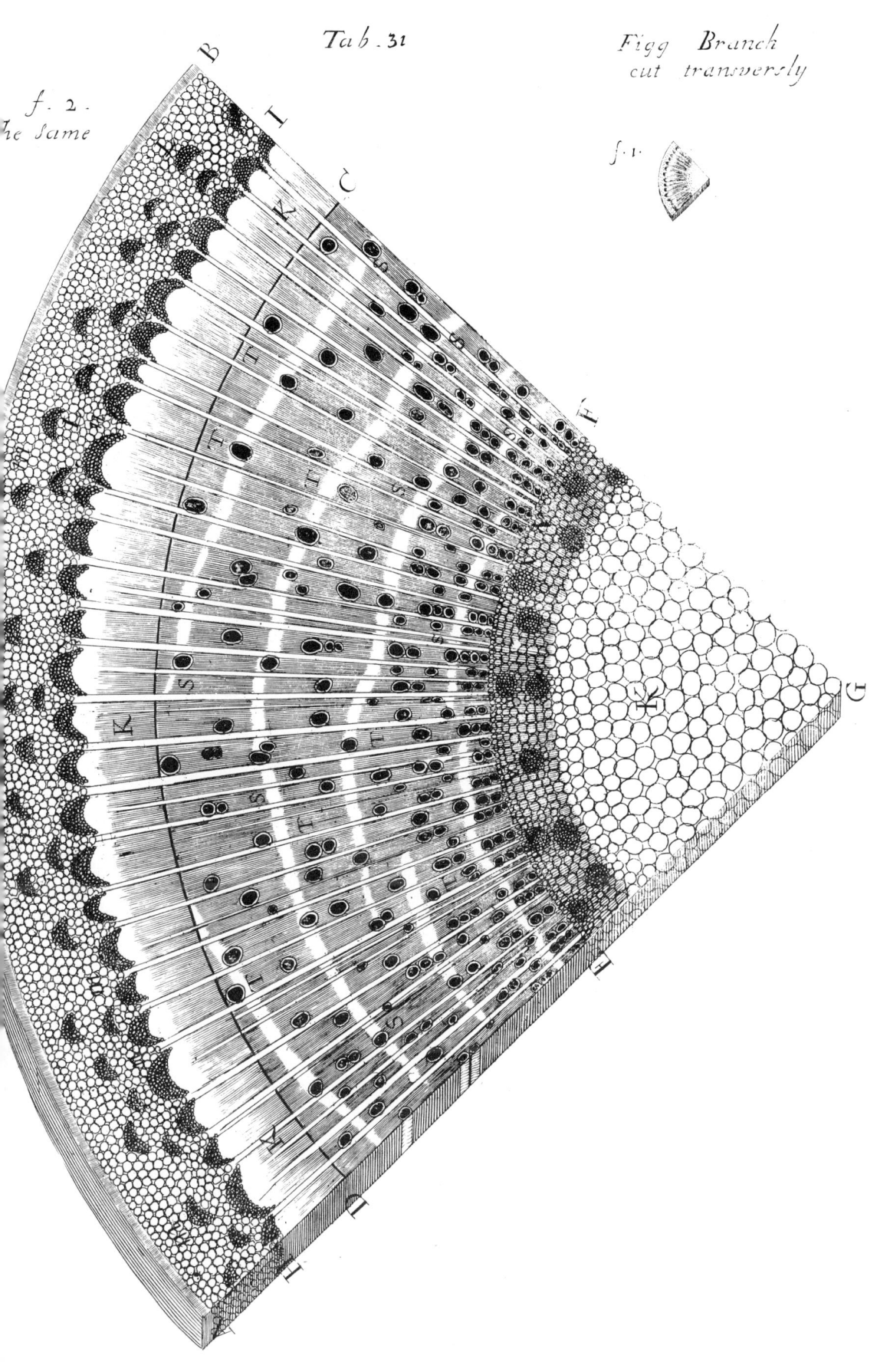

Tab. 31 Figg Branch cut transversly

f. 2. The Same

f. 1.

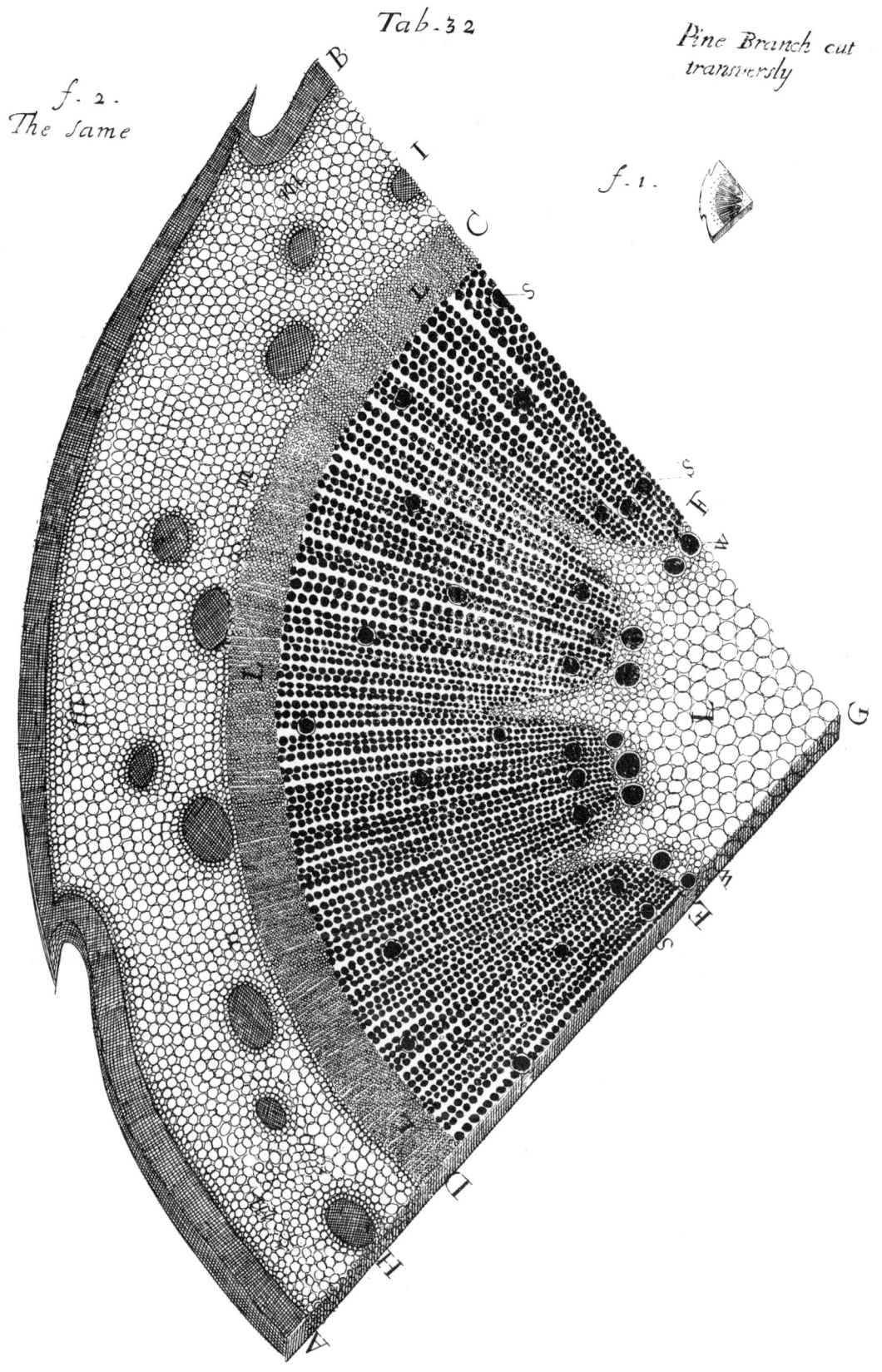

Tab. 32 Pine Branch cut transversly

f. 2. The same f. 1.

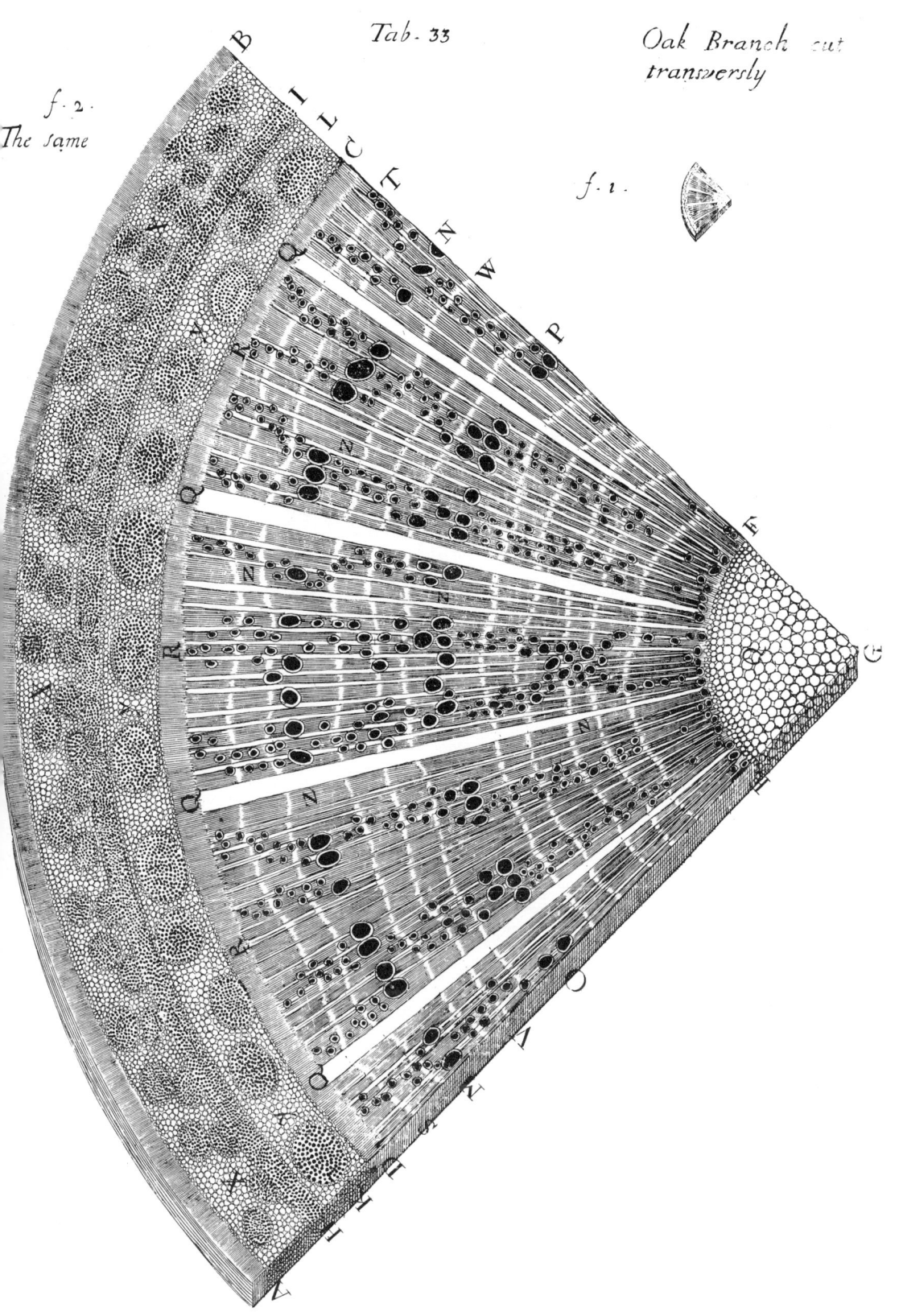

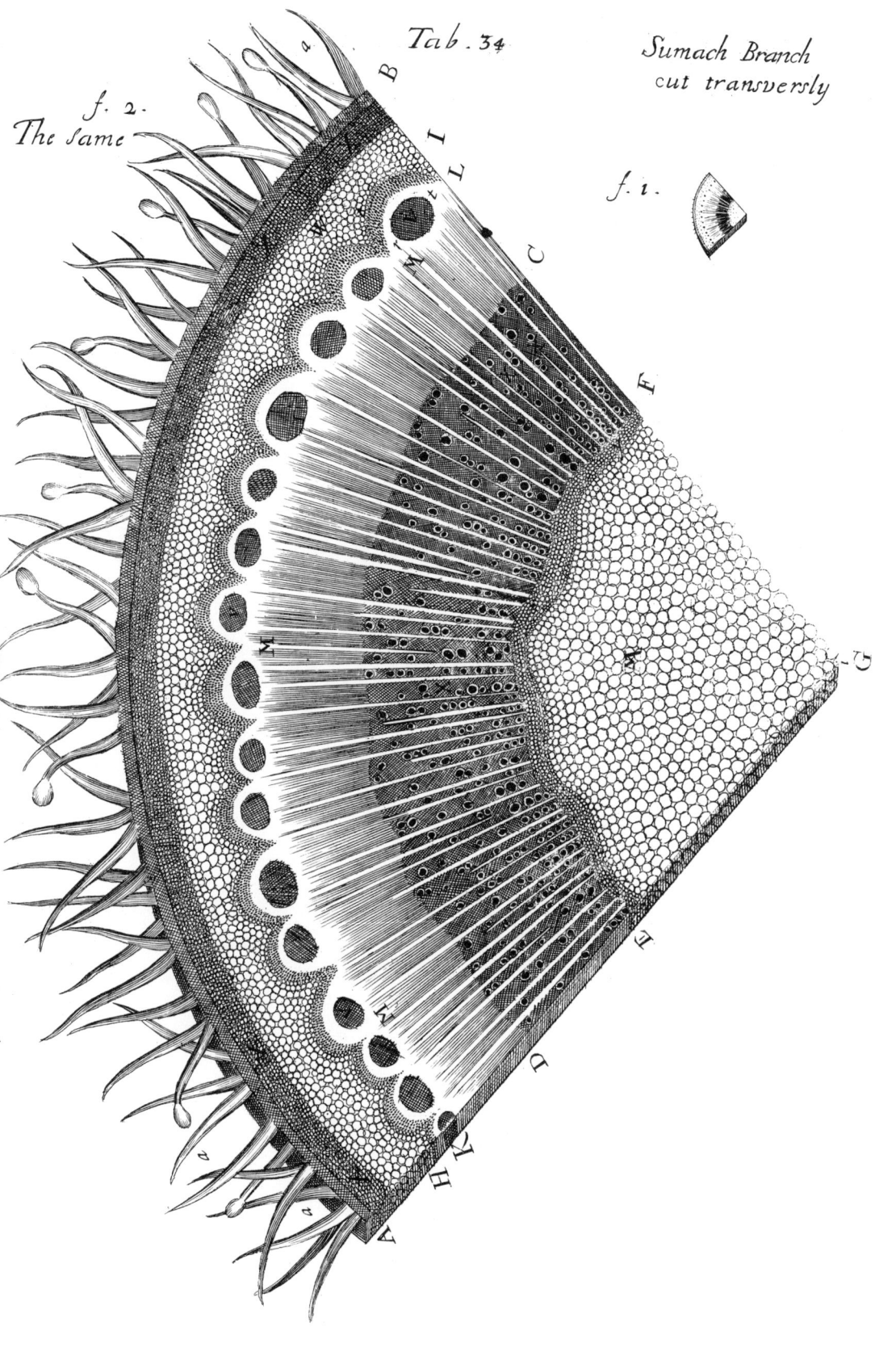

Tab. 34 Sumach Branch cut transversly

f. 2. The same

f. 1.

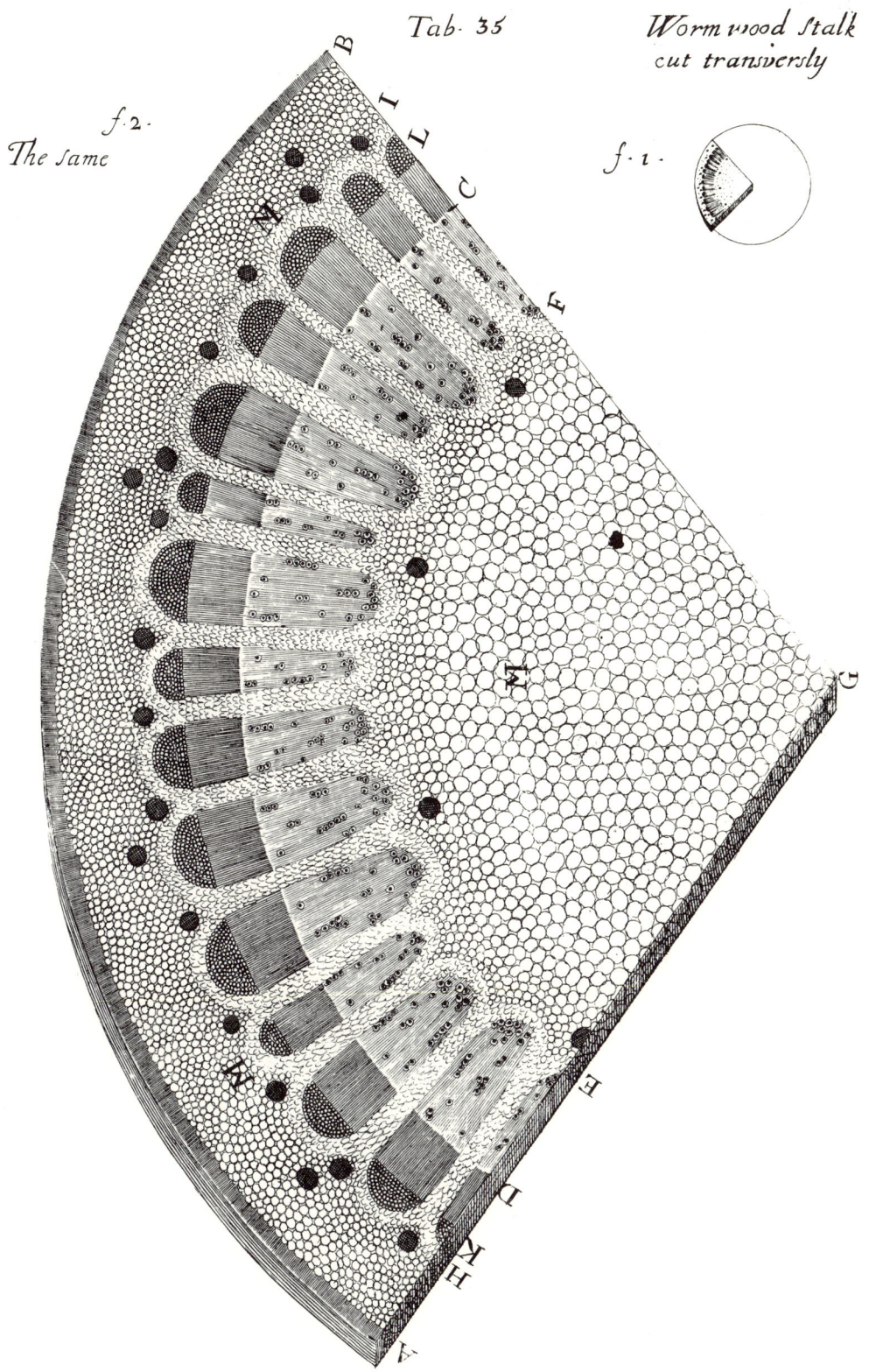

TAB. XXXVI. *Part of a Vine Branch cut transversly, and splitt half way downe ye midle.*

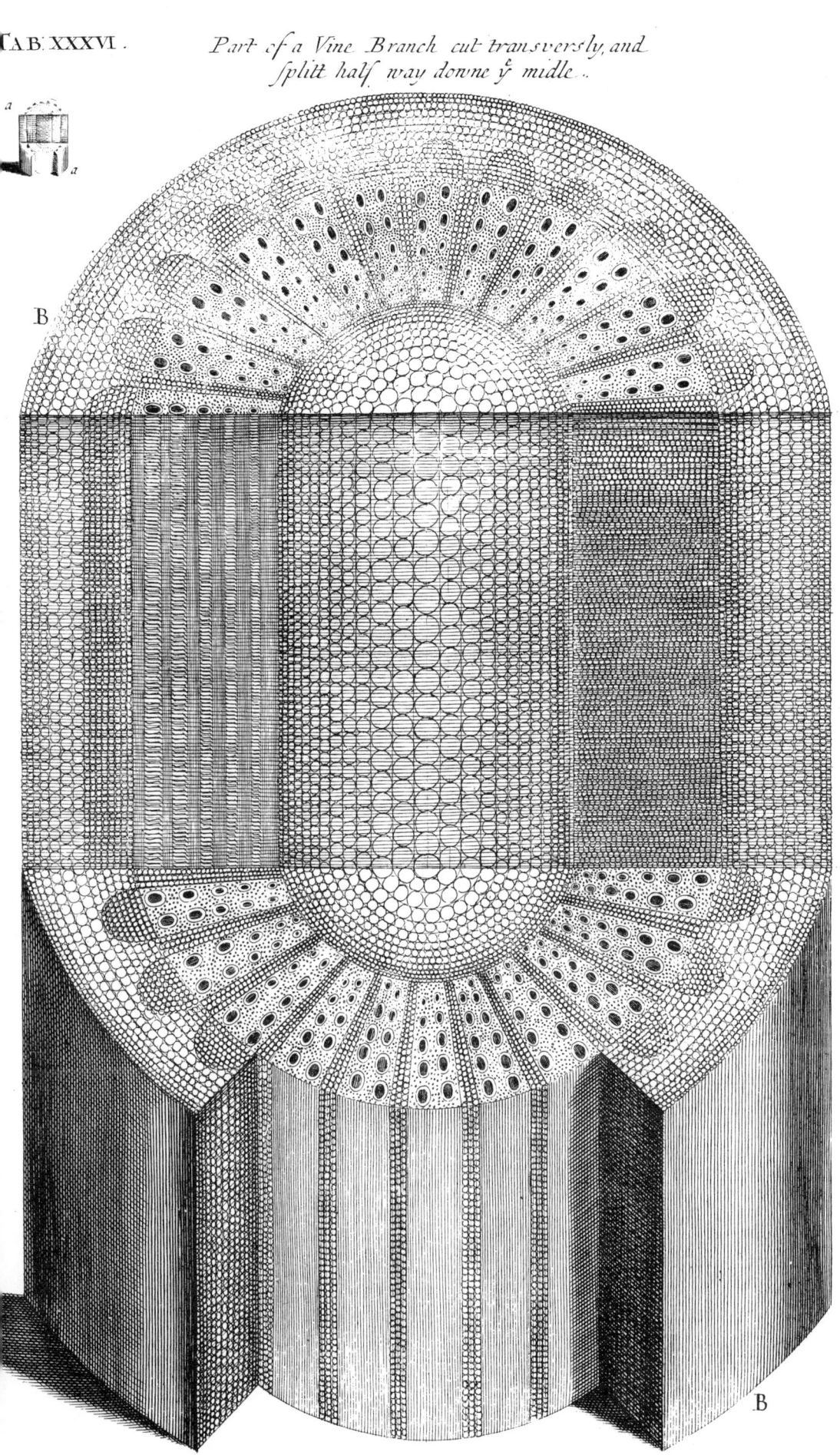

Tab:37. *Part of a Corin Branch Cut as in Tab. 36.*

A

a

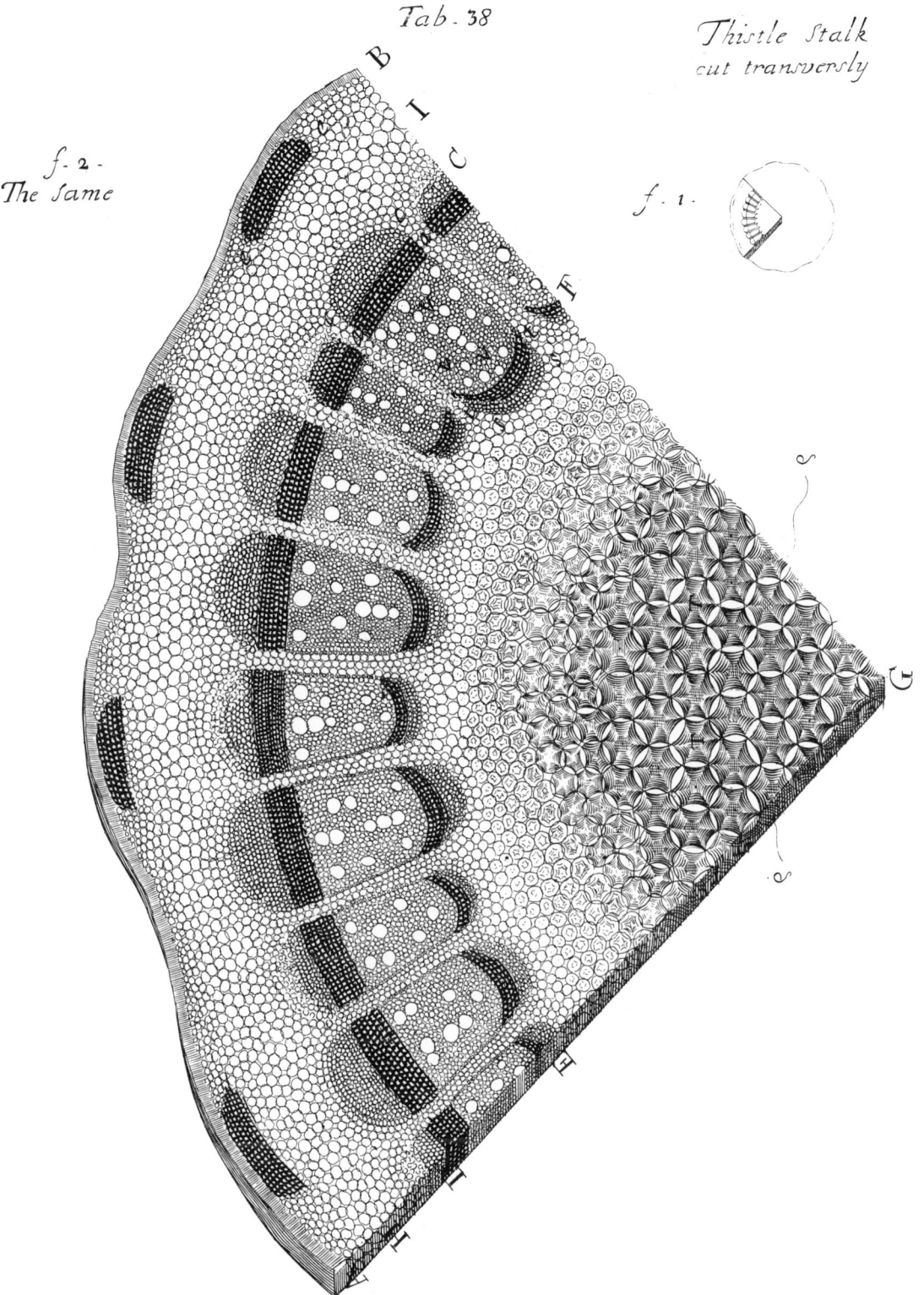

Tab. 38

Thistle Stalk cut transversly

f. 2. The Same

f. 1.

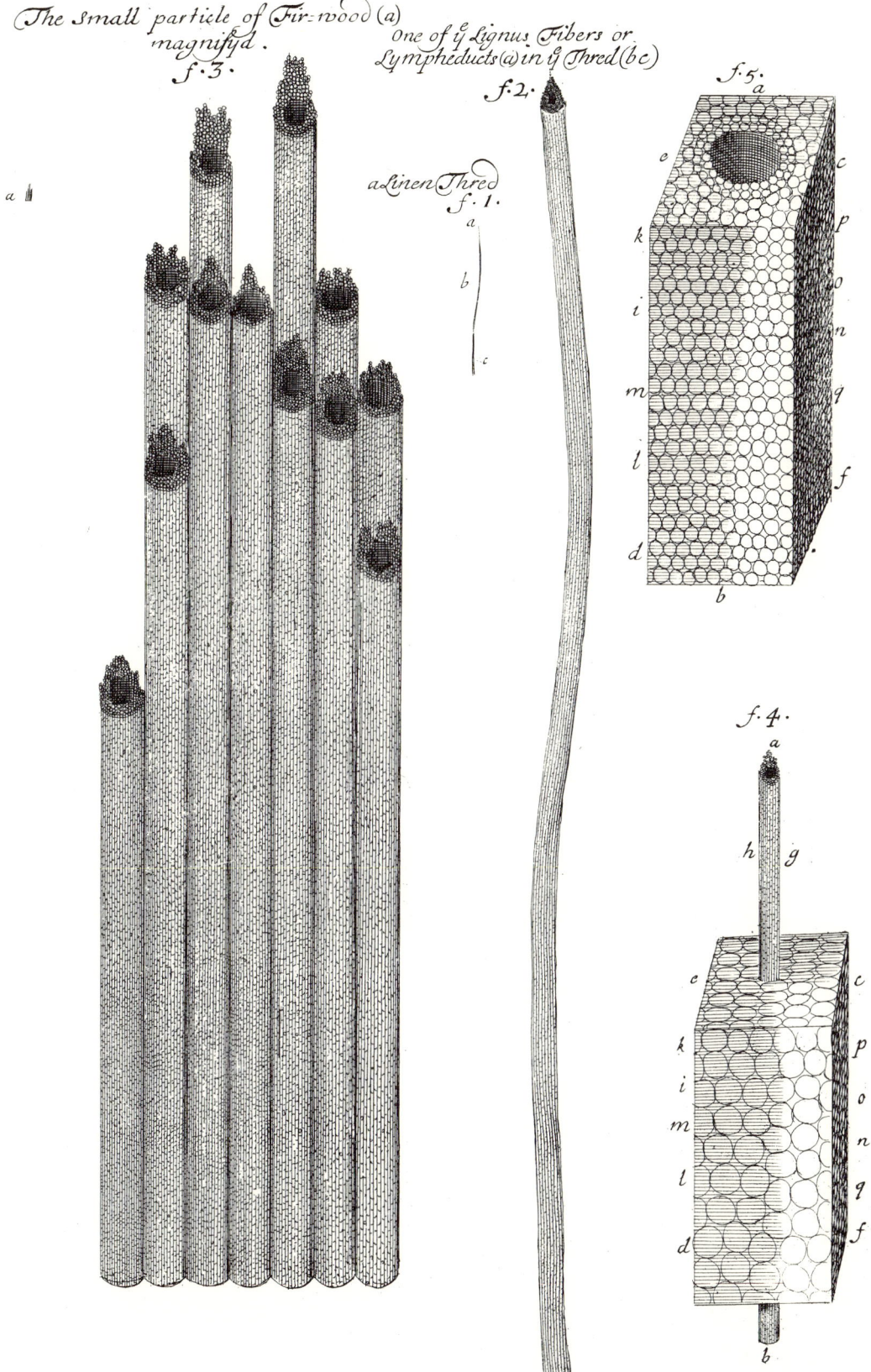

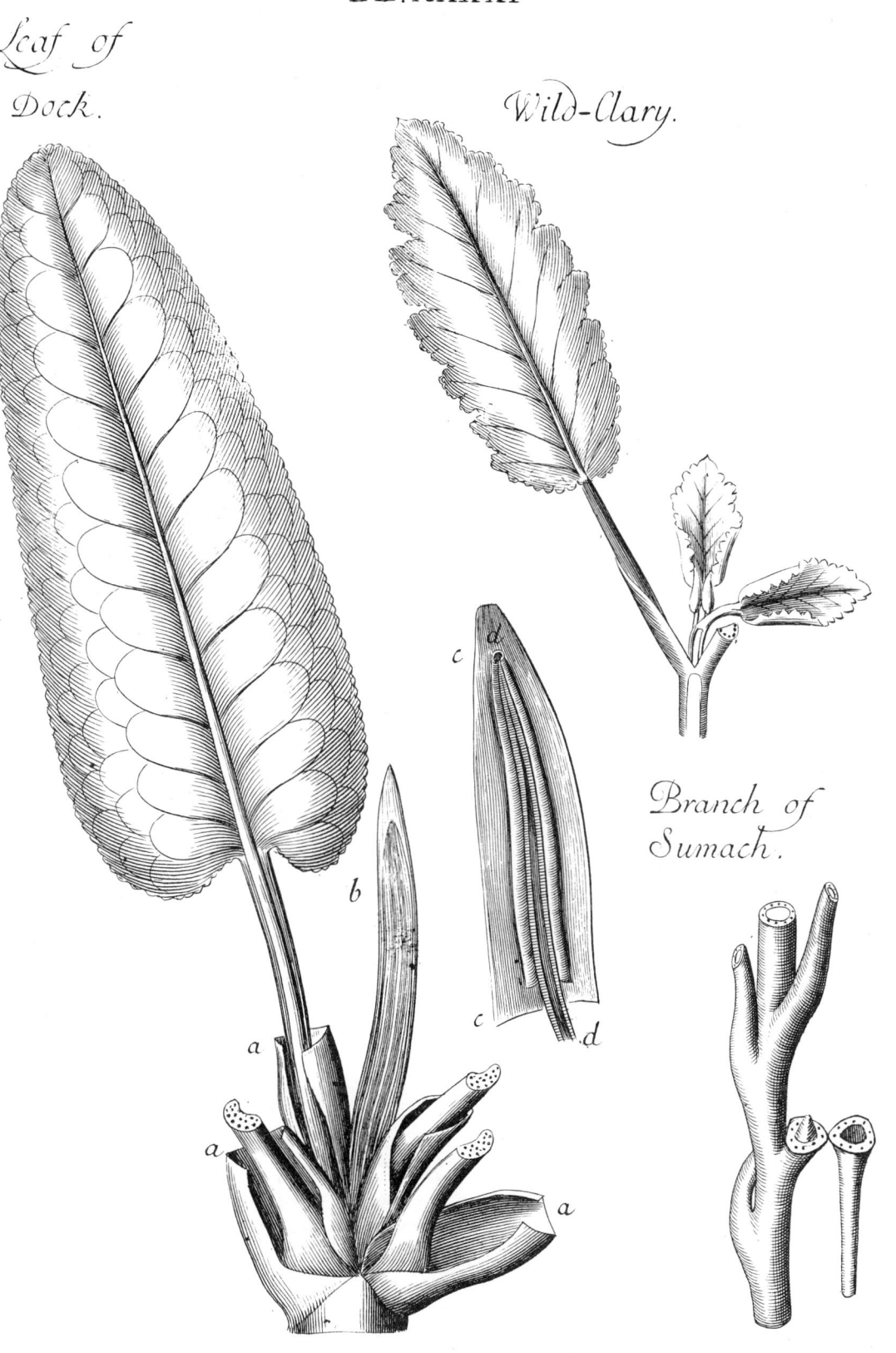

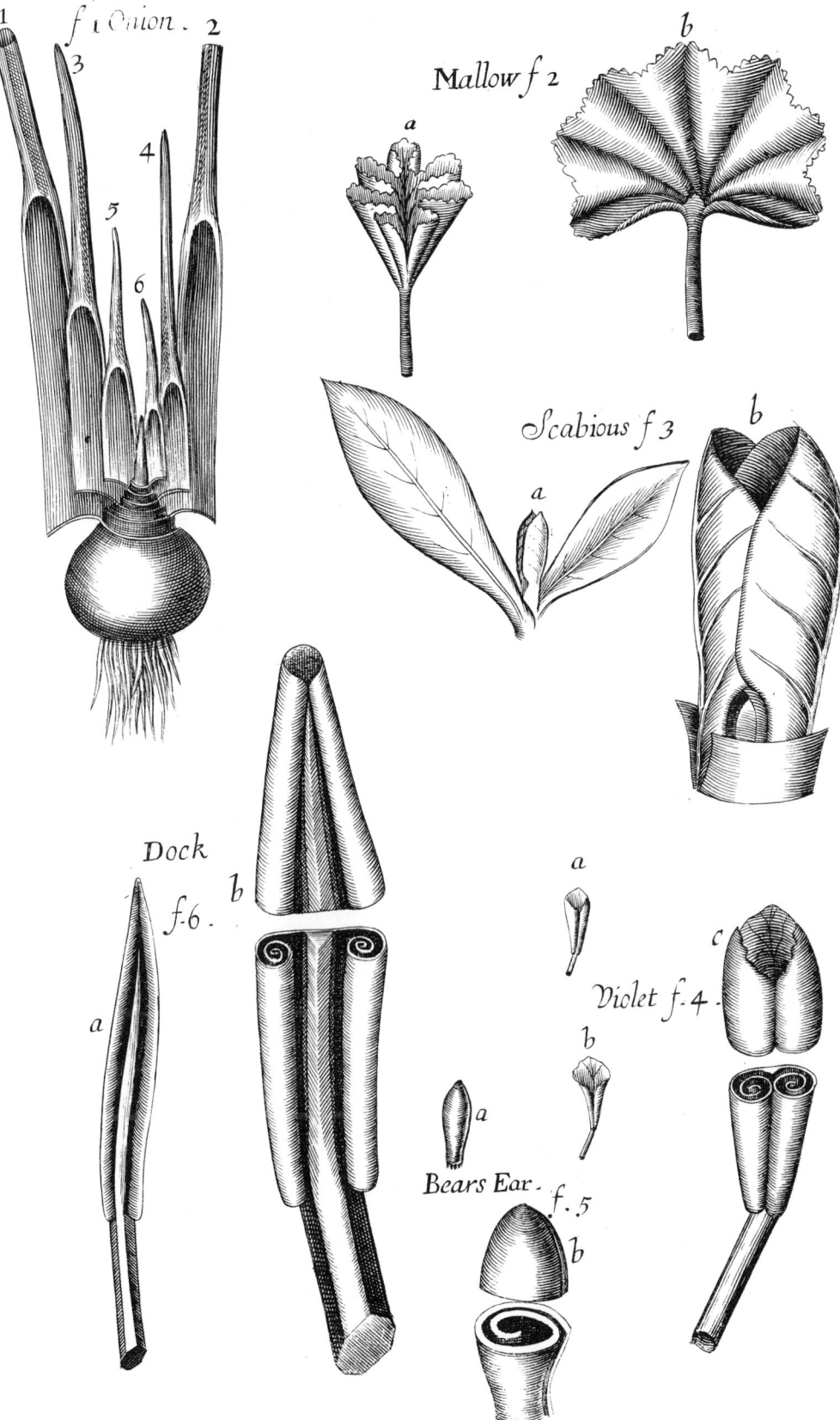

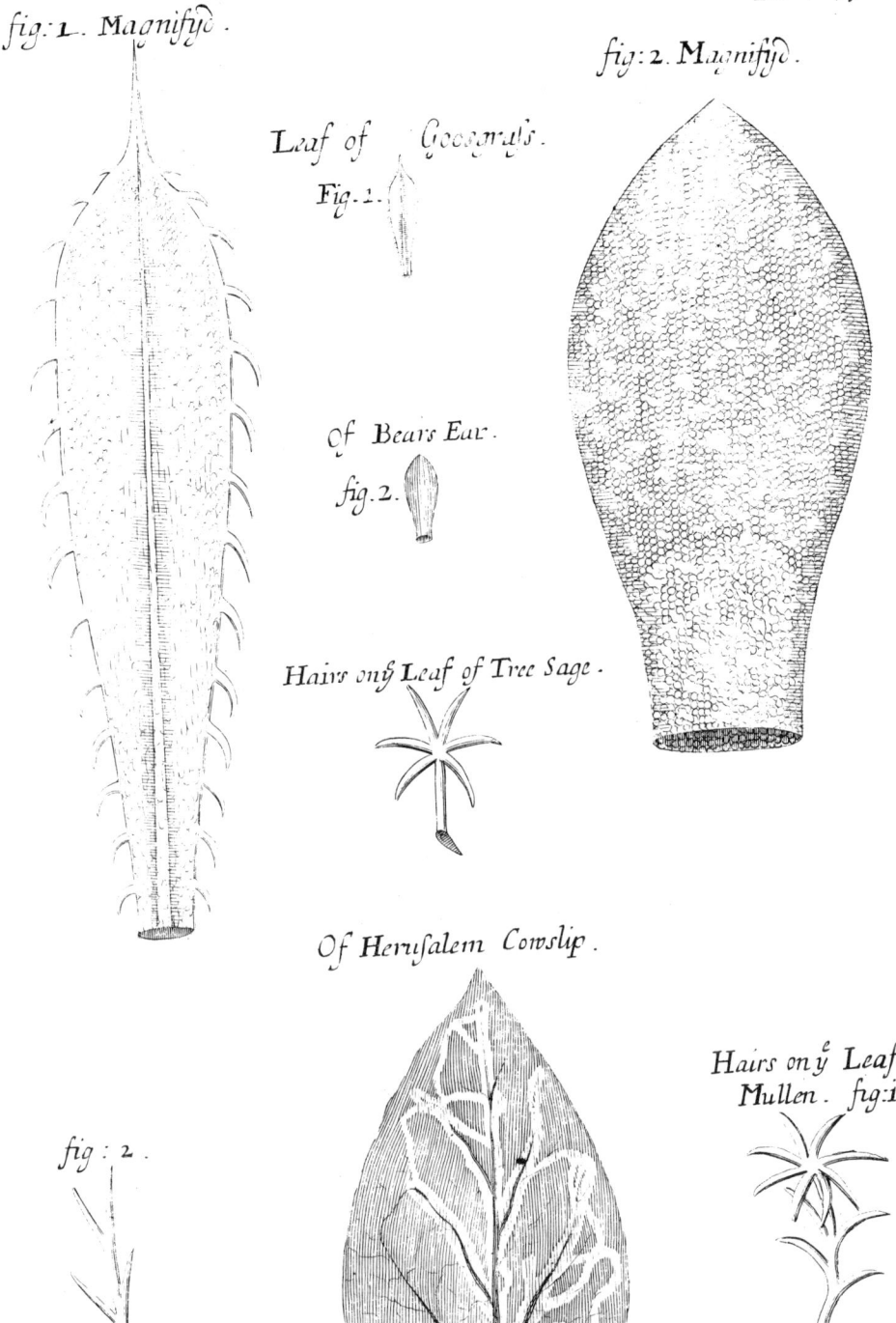

Tab XLIII.

Leaf of Venetian Vetch

Orange

Sage-Leav'd Ironwort

Cornelian Cherry

Broad Leav'd Laser-wort

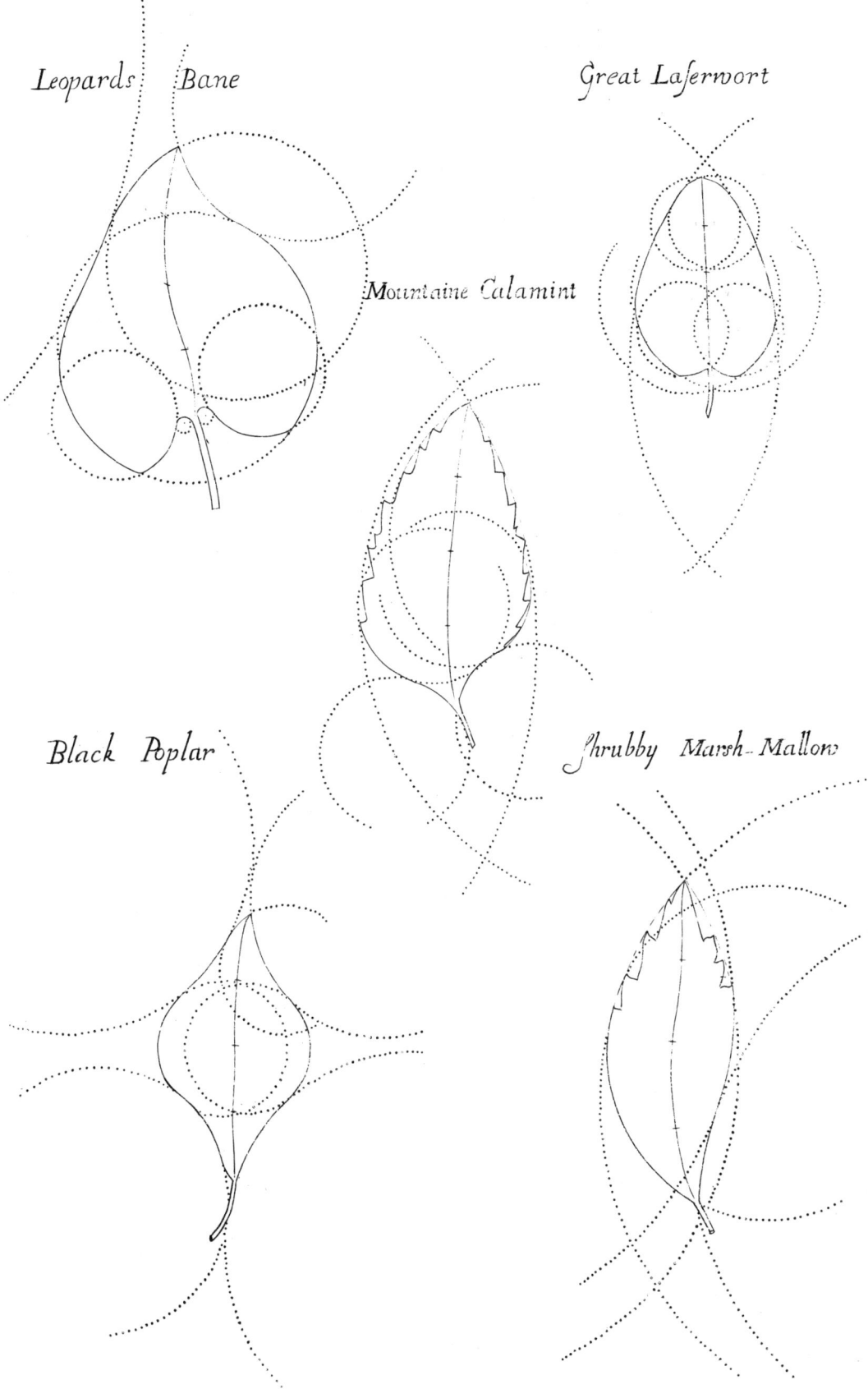

Tab. 46. Leaves of
Clematis Syl. m. Mallow.
Vine.

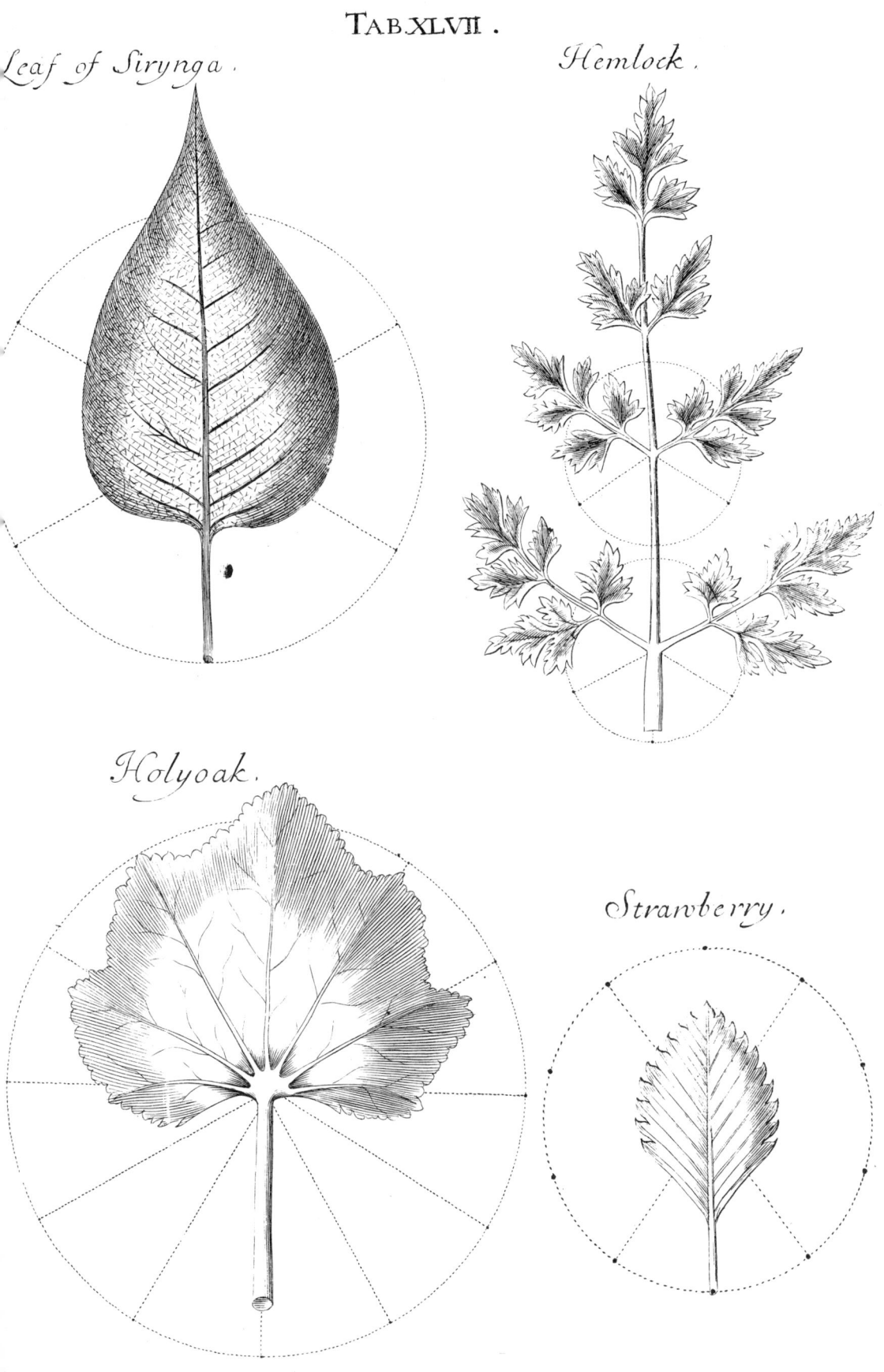

TAB. XLVIII.

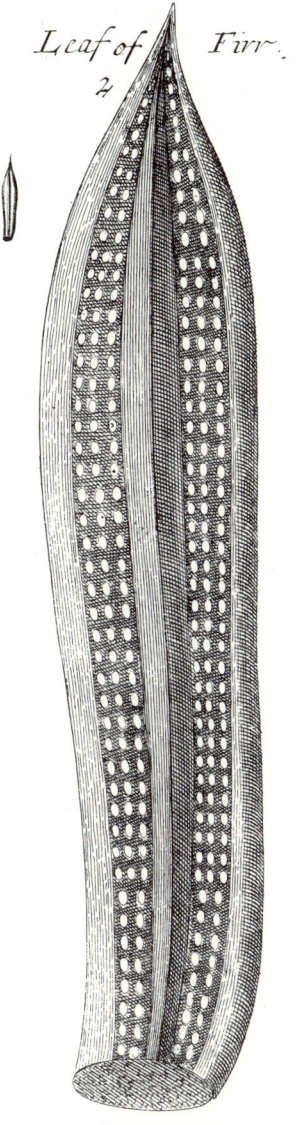

Leaf of Firr.

Rhamnus Salicis Folio.

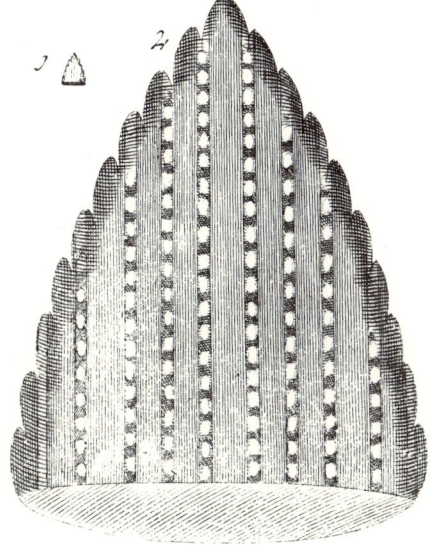

The top of Pine Leaf.

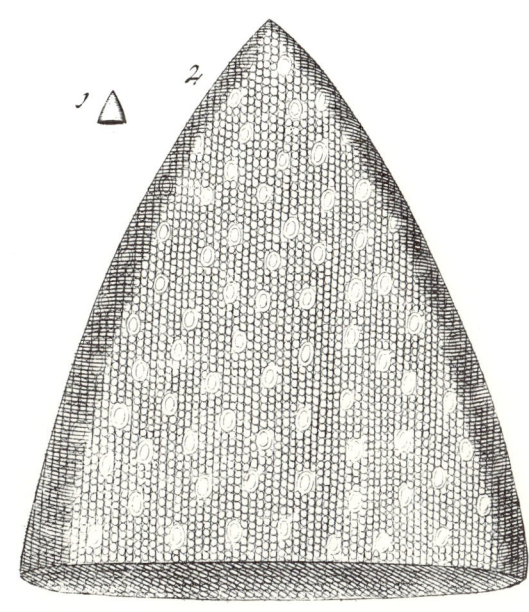

The top of Lilly leaf.

TAB. XLIX.   Stalks of

Mallow.
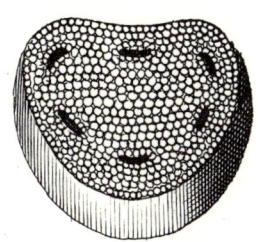

Dock.

Dandelion.

Wild-Clary.

Borage.
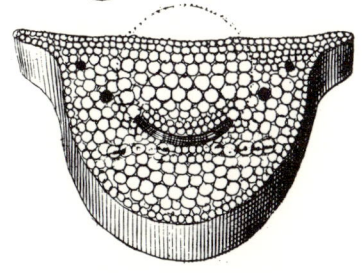

Mullen.
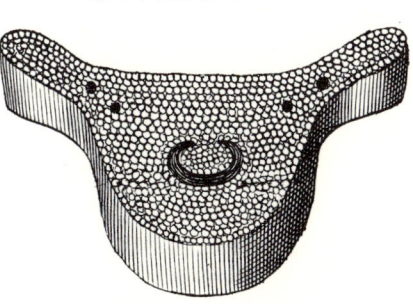

Tab. 50.

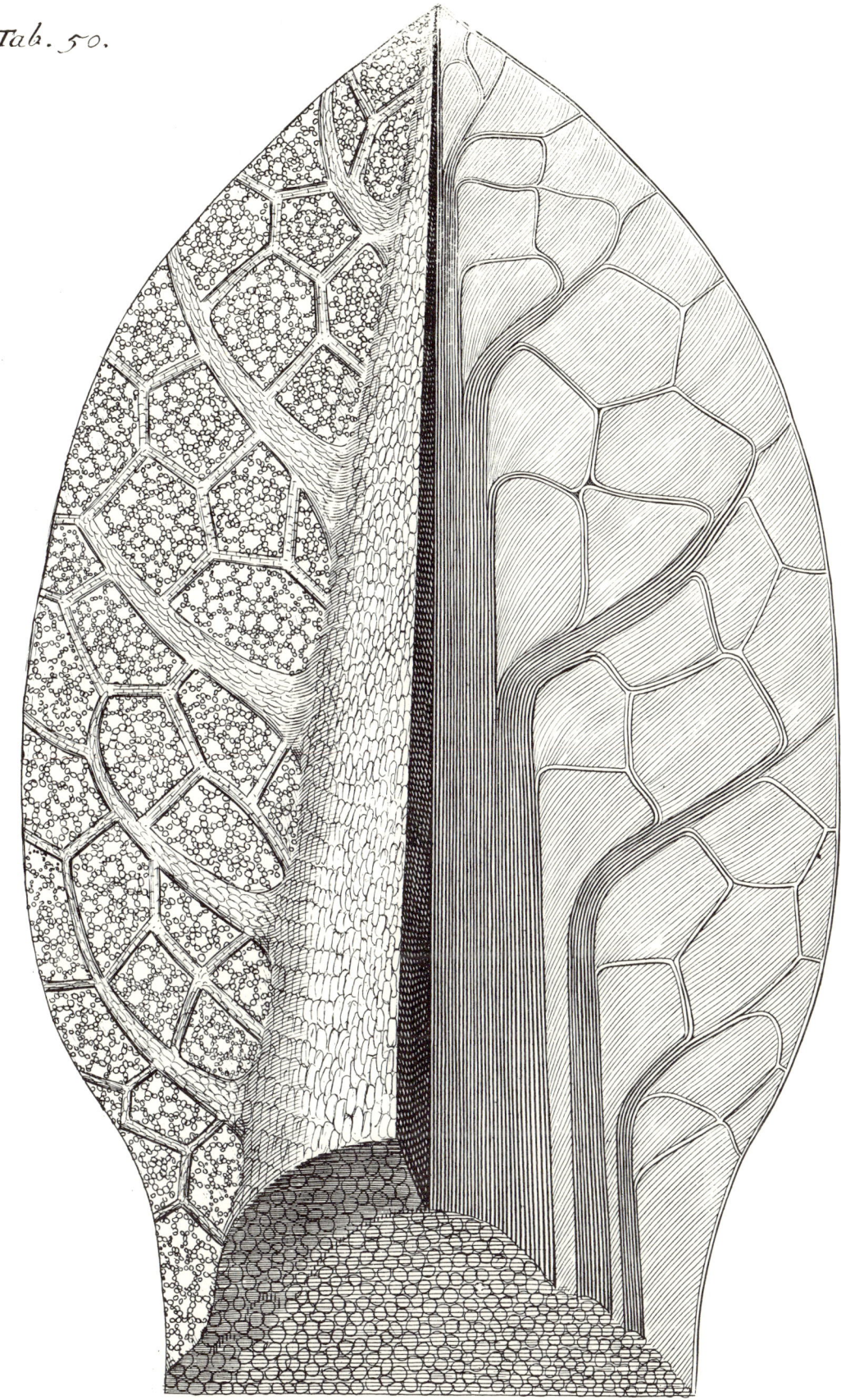

A young Borage Leaf

Tab: 51.

The Aer-Vessels unroaved in a Vine Leafe.

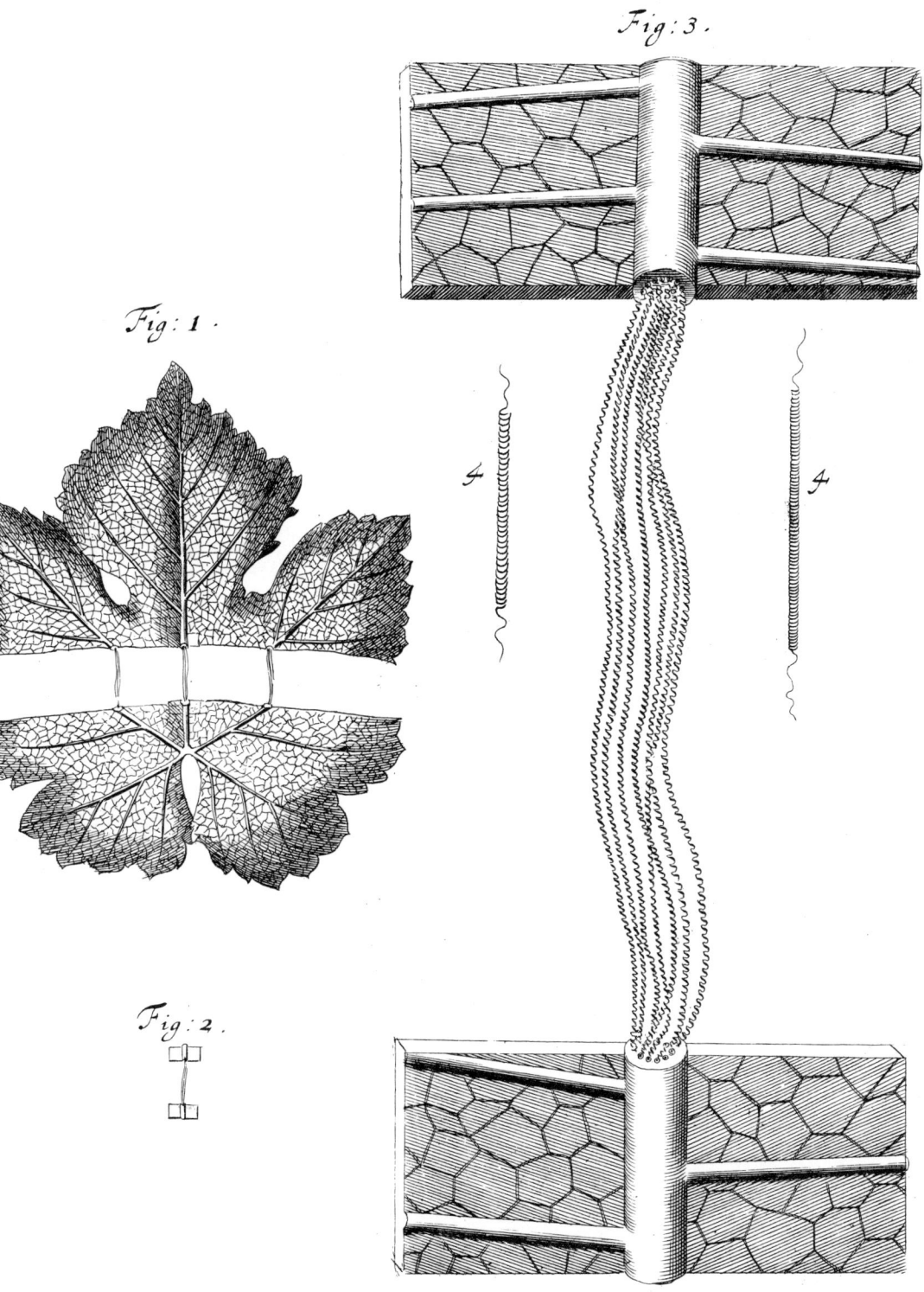

Tab. LII

The Aer-Vessels unroaved in a Scabions Leafe.

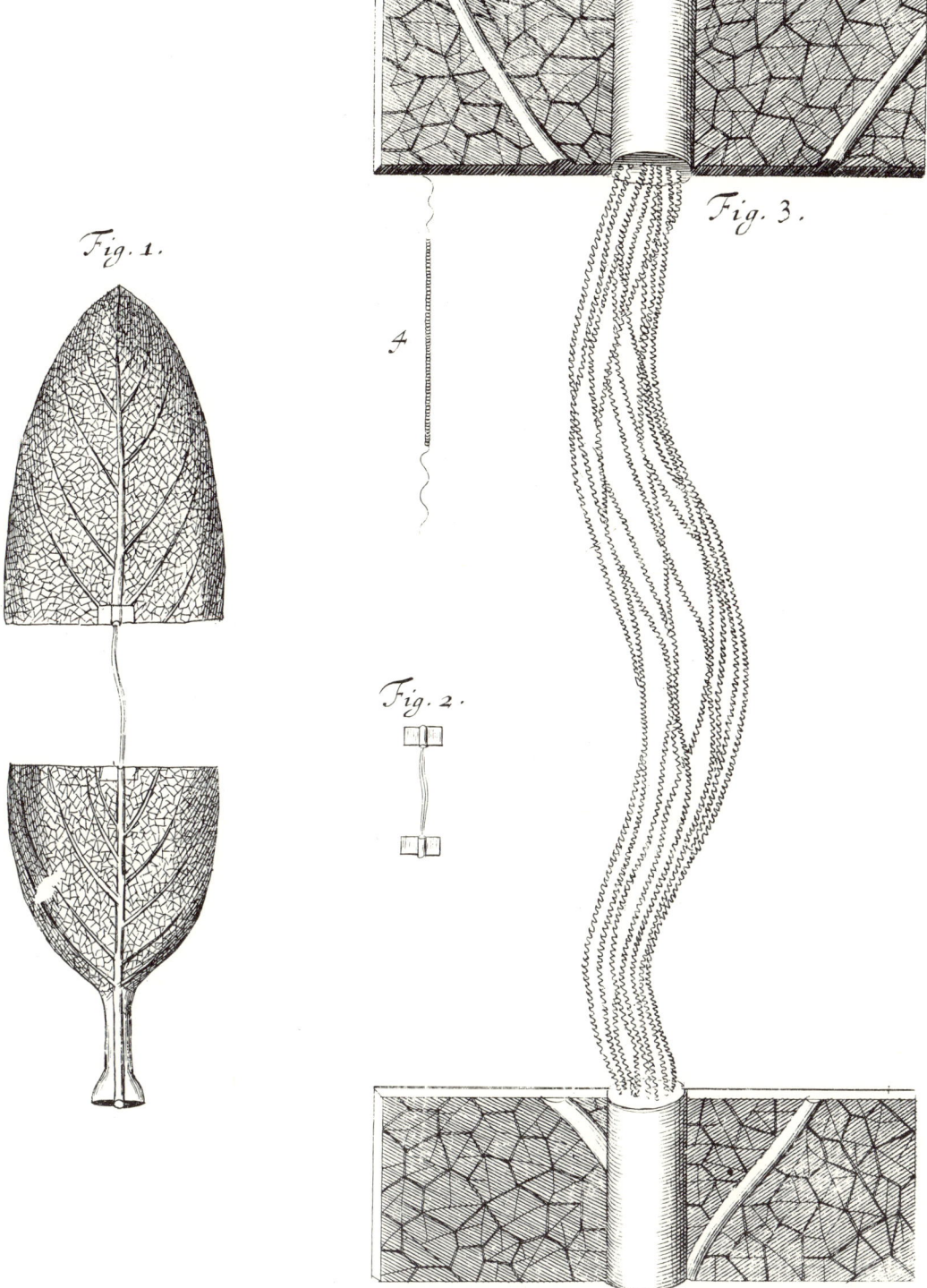

Fig. 1.

Fig. 2.

Fig. 3.

4

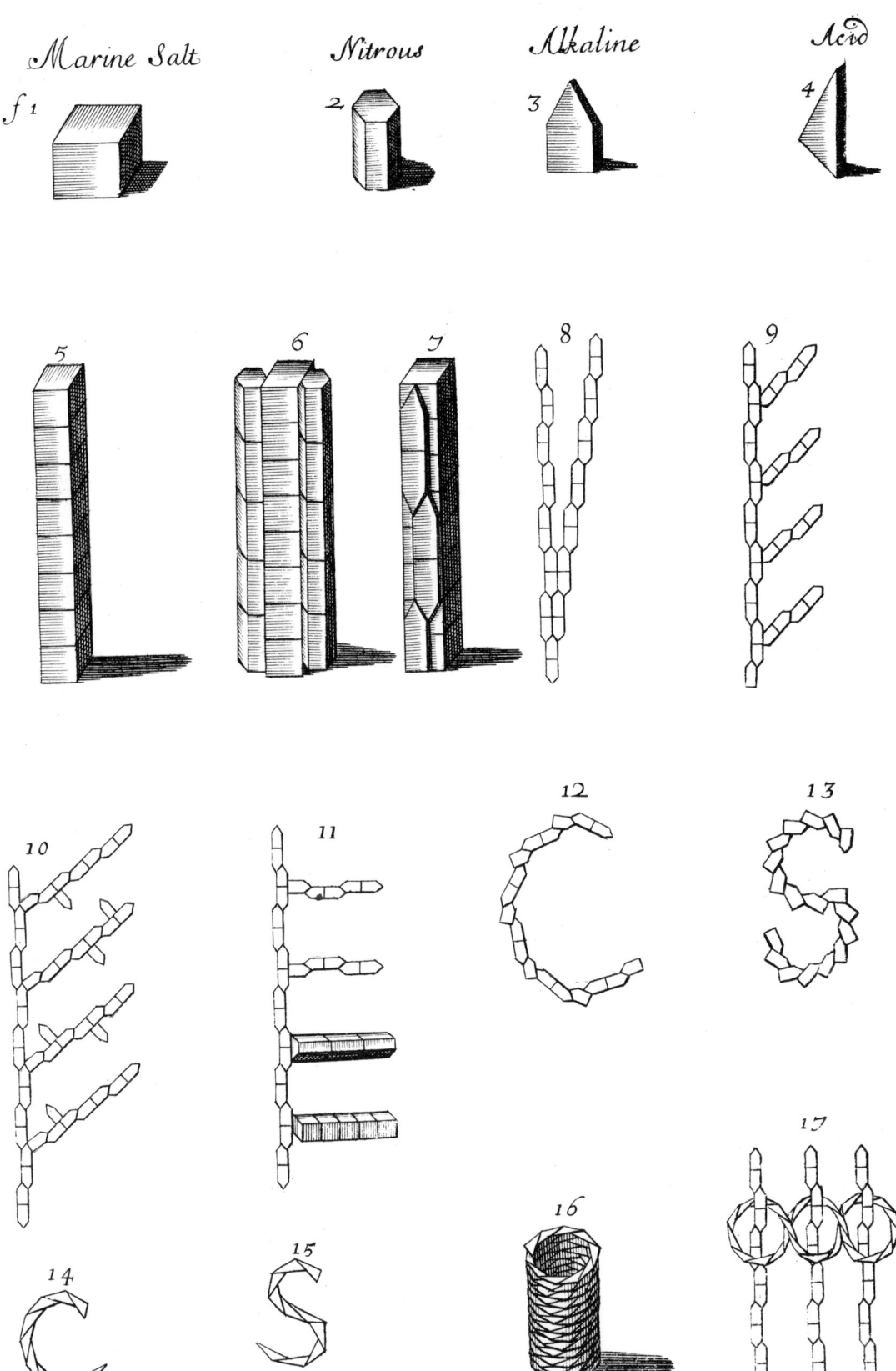

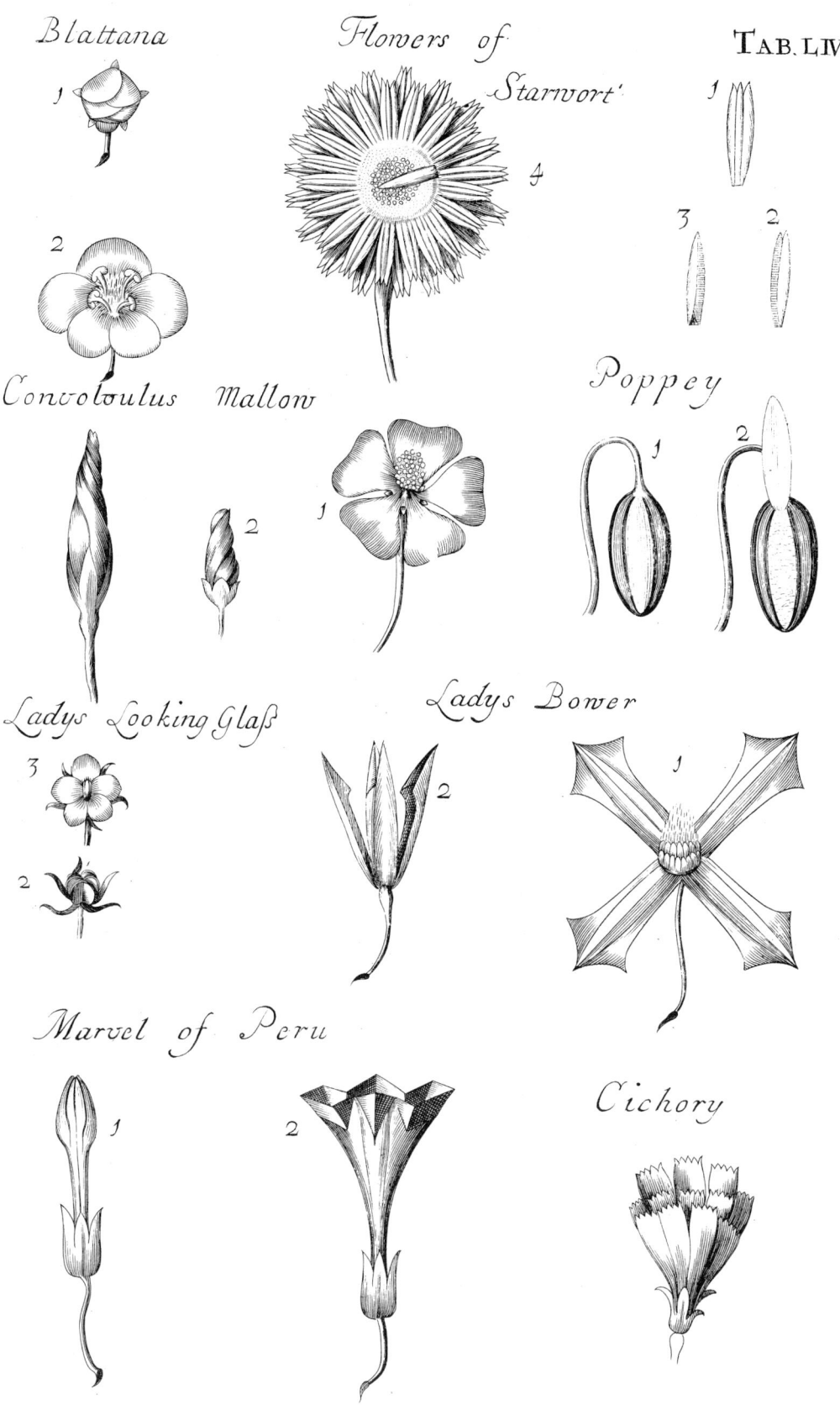

Tab. 55.

f. 1.
Flower of Dulcamara magnifi'd.

f. 2.
Flower of Colus Jovis magnifi'd.

f. 3.
Flower of St James's wort.

12345. 2345. 2345.

f. 7.
Theca (a) magnifi'd.

f. 4.
Fl. of Chamemile.

f. 10.
The Theca (c) magnifi'd.

12345. 2345. 2345. 2345.

f. 5.
Fl. of Clematis Austriaca.

f. 8.
Fl. of Blattaria.

f. 9.
One of ye Thecæ.

f. 6.
One of ye Thecæ in ye flower.

a

Tab. 56.

f. 3.
The Backside of ẏ Theca. (a)

f. 4.
The Belly of the Theca. (a)

f. 7.
The Column (e) Magnifi'd.

f. 1.
Flower of Hyoscyanus.

f. 2.
One of ẏ Spermatick Thecæ.

f. 6.
The Column in ẏ Midle of ẏ flower.

f. 5.
The Edges of the Theca (a) open.

f. 1.
Flower of
St. Johnswort.

f. 2.
The same a little magnifid.

Tab. 57.

f. 3.

f. 4.

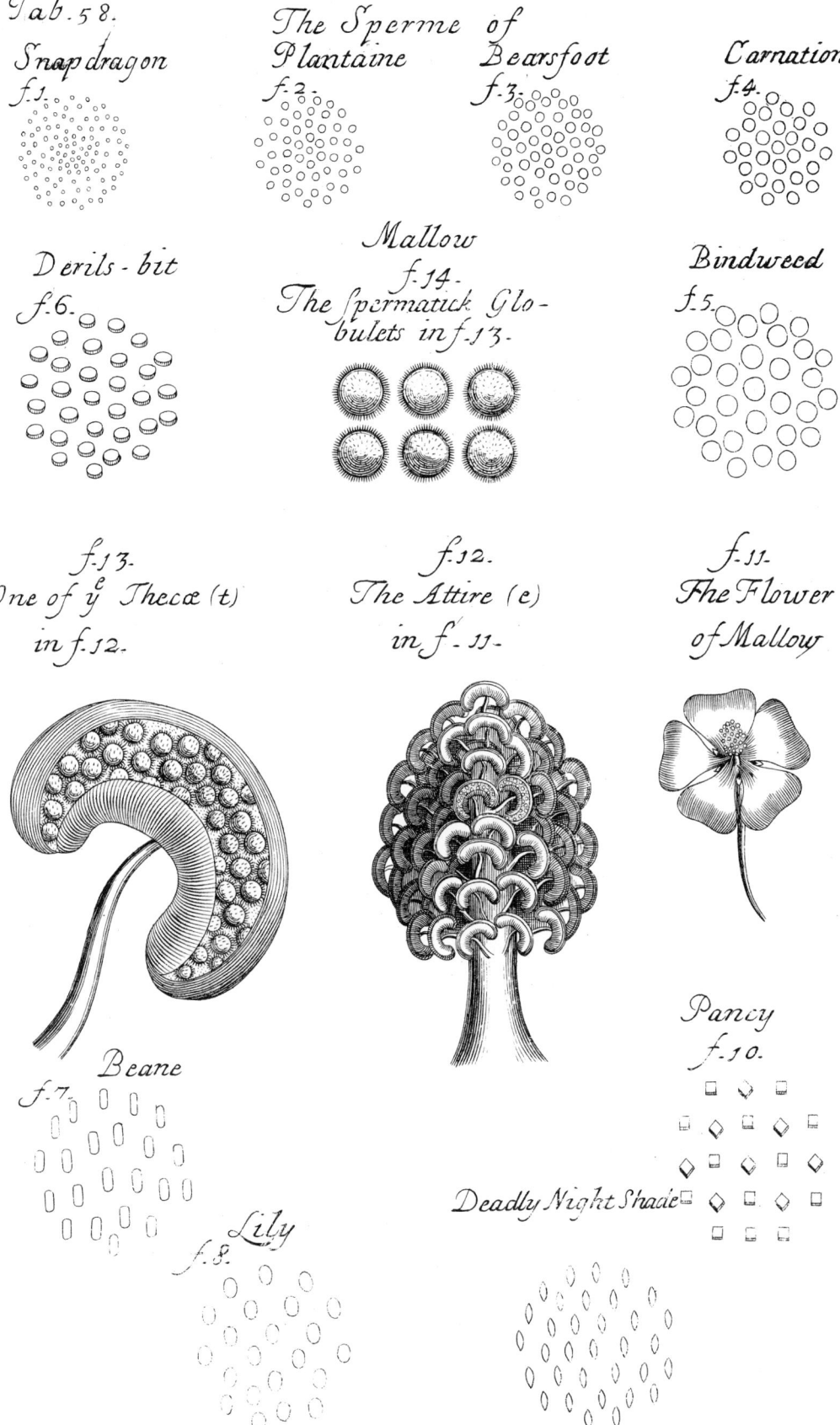

Tab. 58.

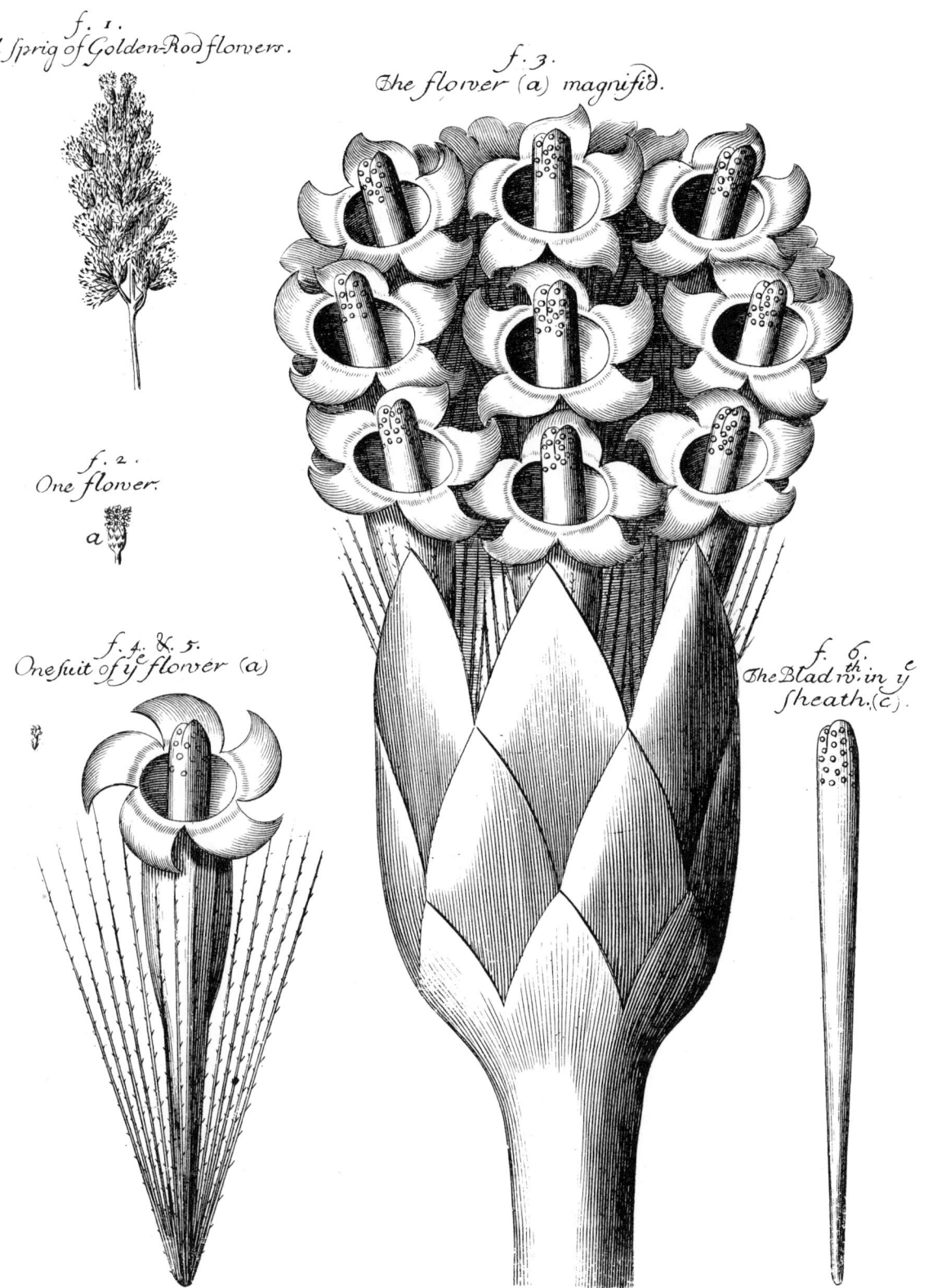

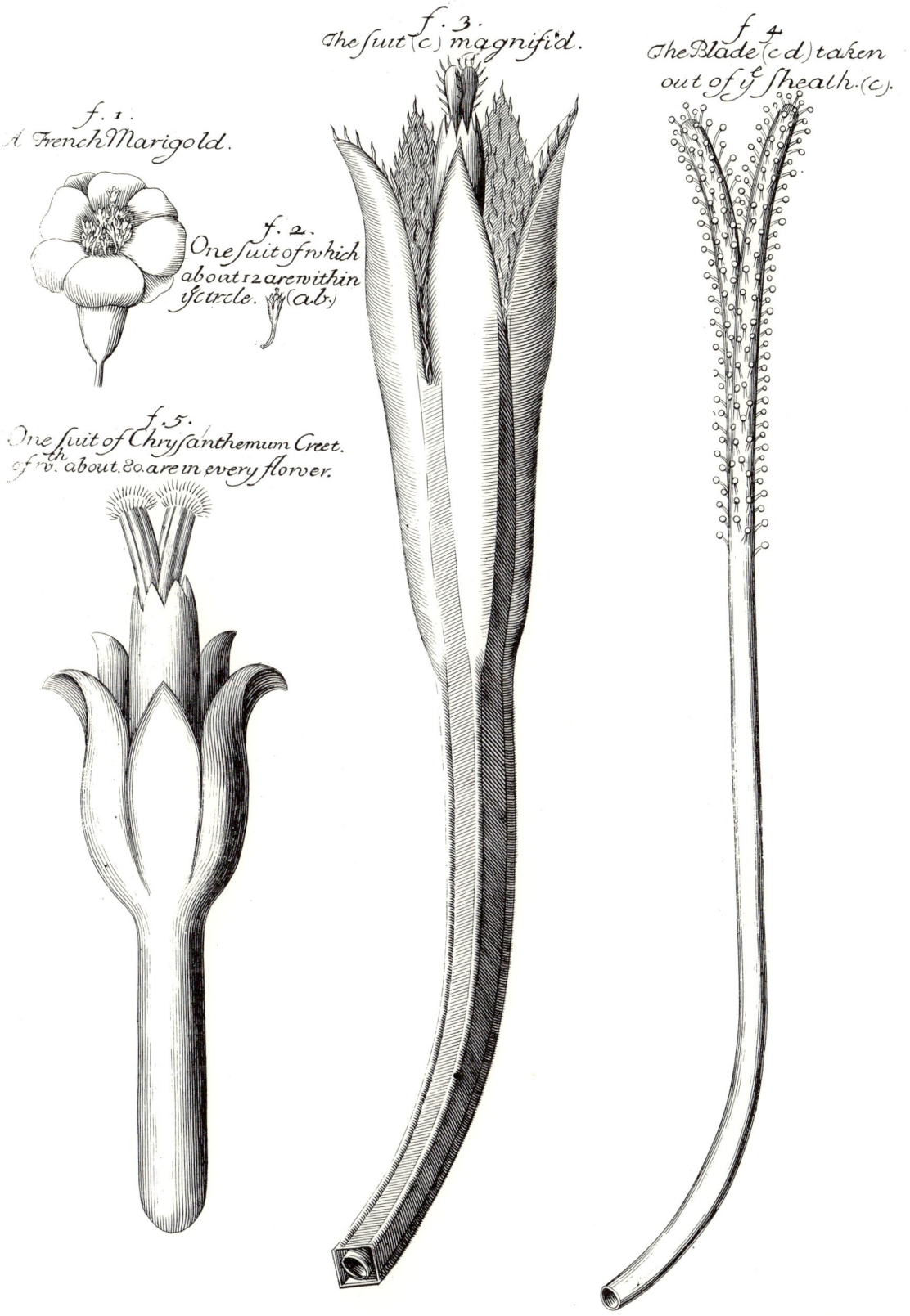

Tab. 61.

f. 1. Marigold.

f. 8. The suit (d) Magnifi'd.

f. 2. One of y⁰ suits, of w̄ are about 40. in y⁰ circle. (a b)

f. 7. One suit of Knapweed.

f. 3. The suit (c) Magnifi'd.

f. 5. Is (b) Magnifi'd.

f. 6. The flower of Knapweed.

f. 4. A Marigold leaf.

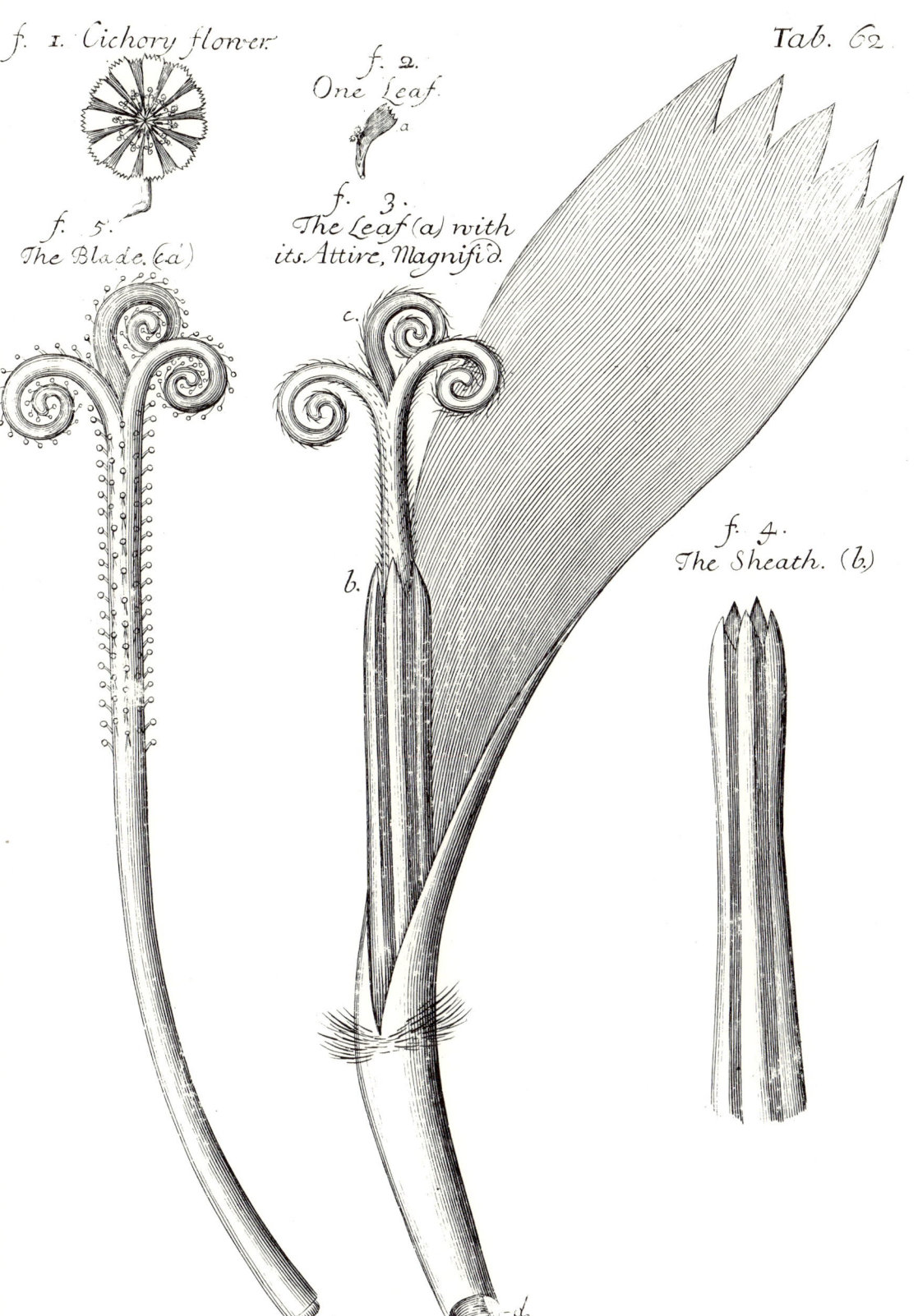

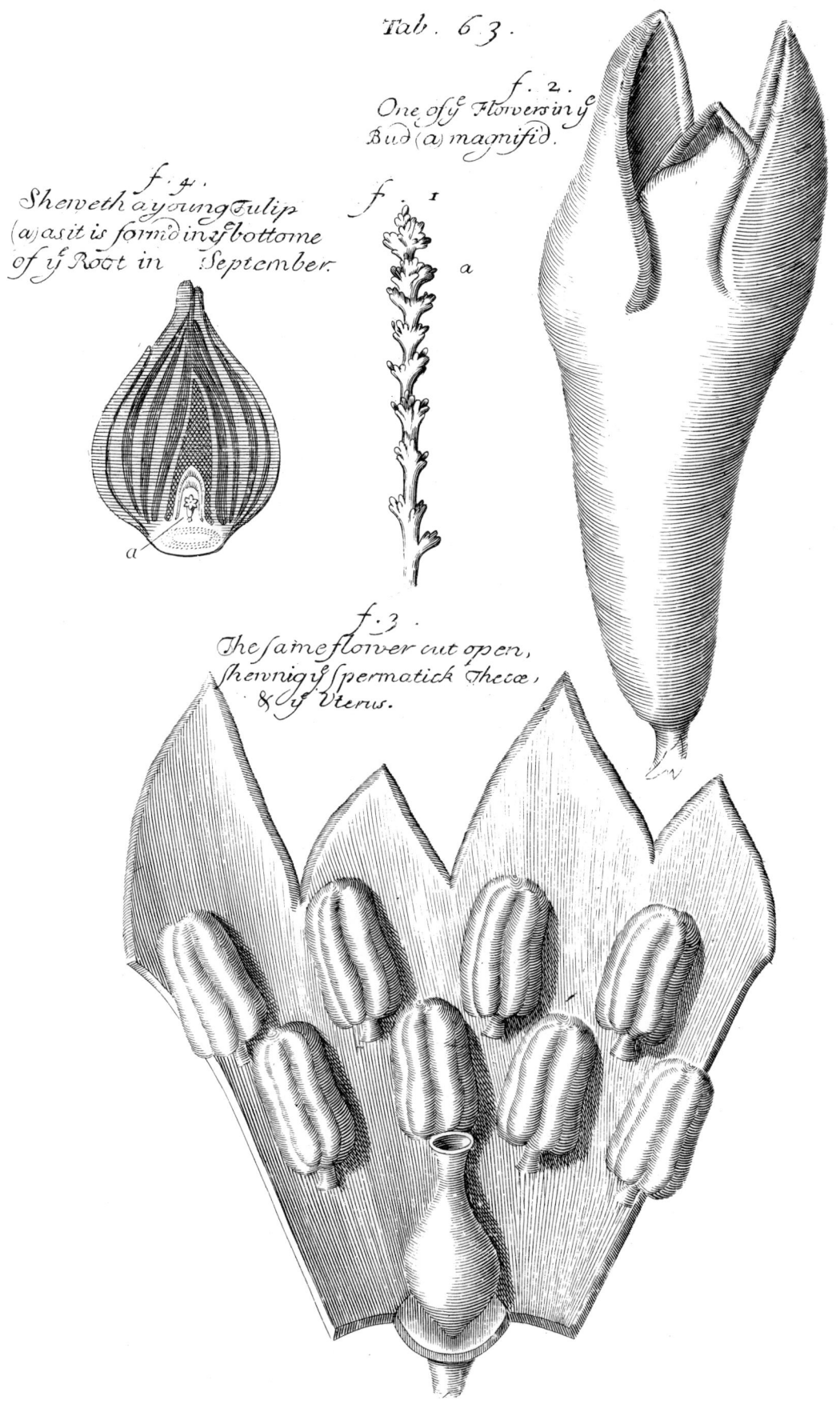

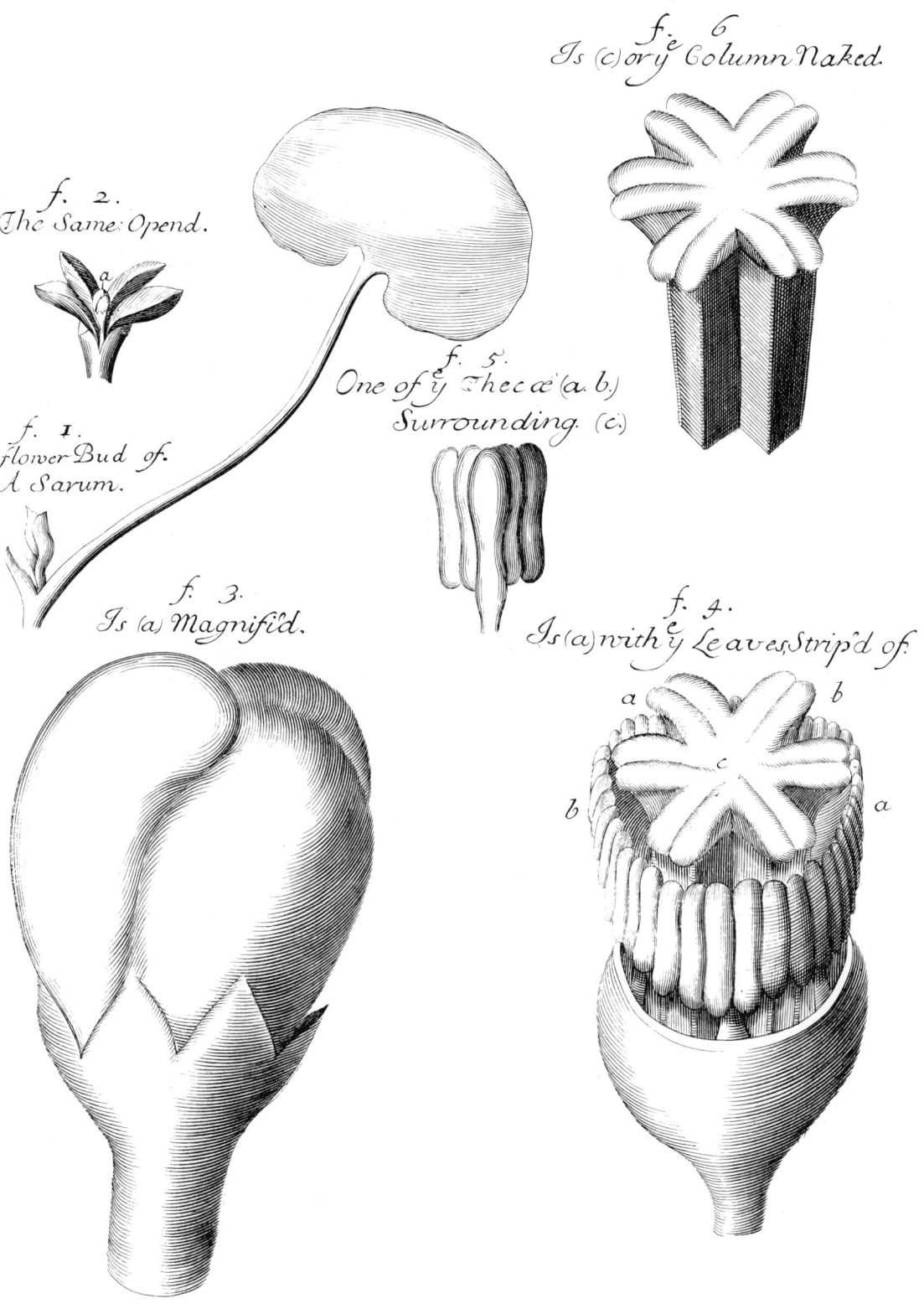

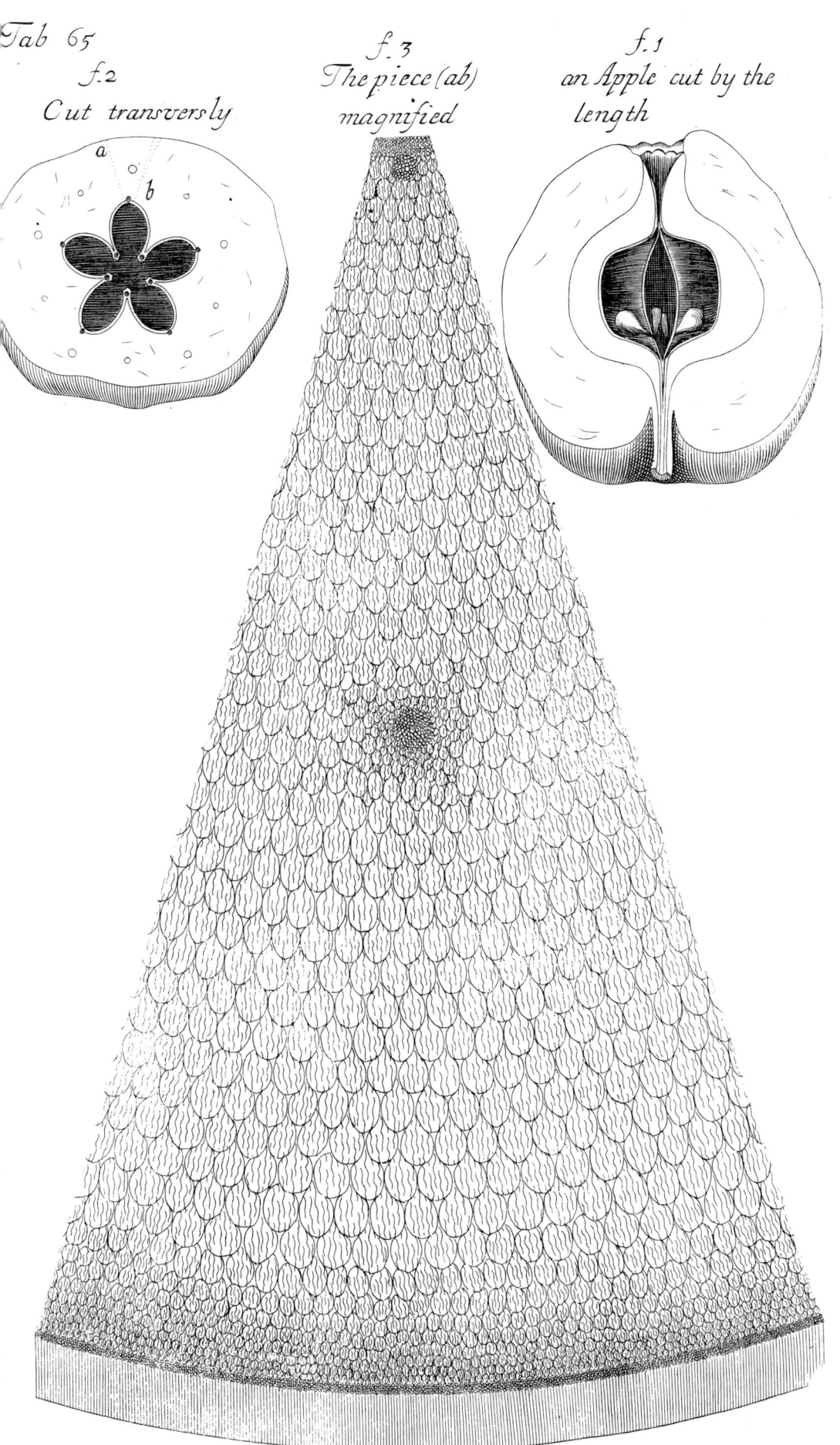

Tab. 66.

f. 3 great
One of the Baggs, a.c.e.

f. 4
One of the little Baggs
c.e. cut transversly

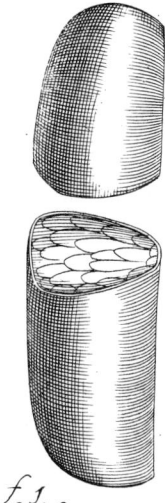

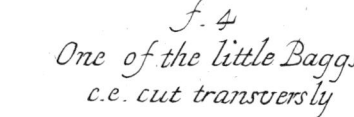

f. 1.
a Limon cut downe

f. 2.
cut transversly

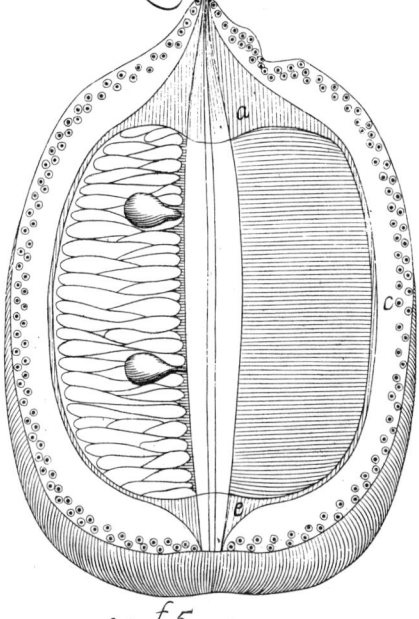

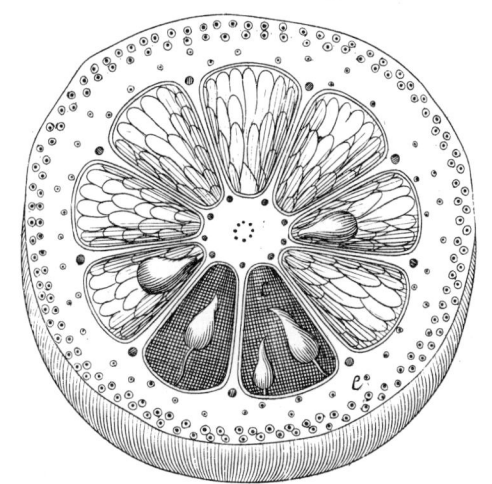

f. 5 Garden Cucumer

f. 6 Wile Cucumer

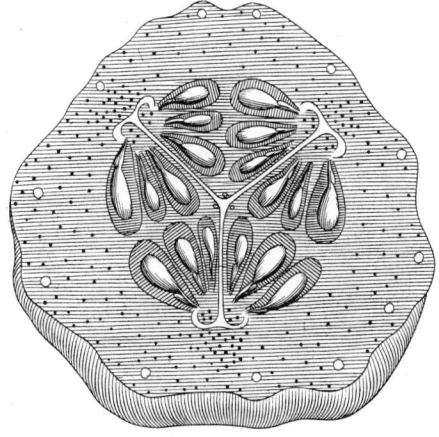

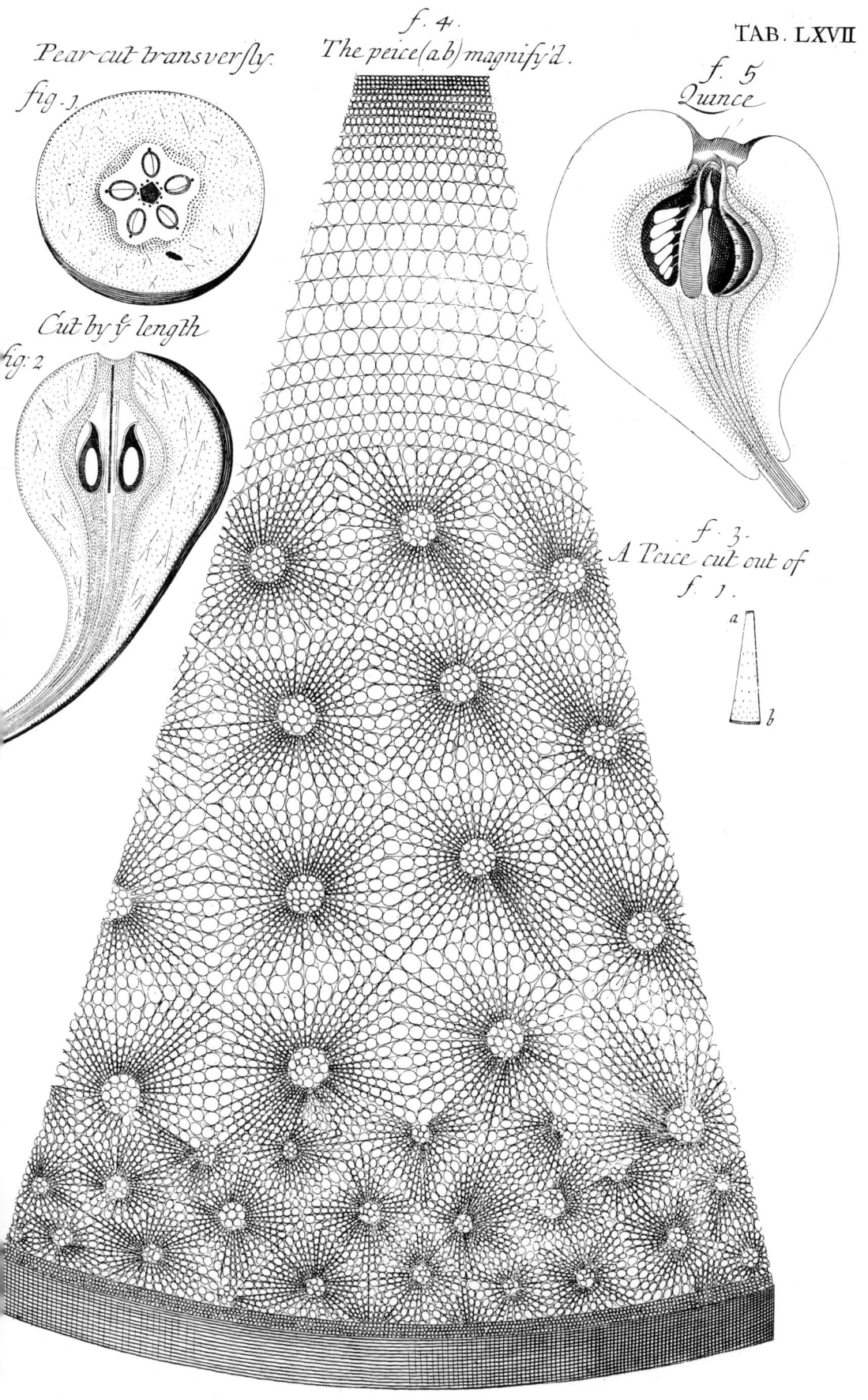

f. 1.
a plum cut
transversly

Tab. 68.
(See Tab. 80. &c)
f. 3.
a piece taken
out of *f. 2.*

f. 2.
an Aprecock cut
transversly

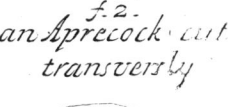
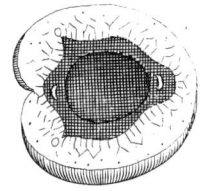

f. 4.
the piece a,
Magnify'd

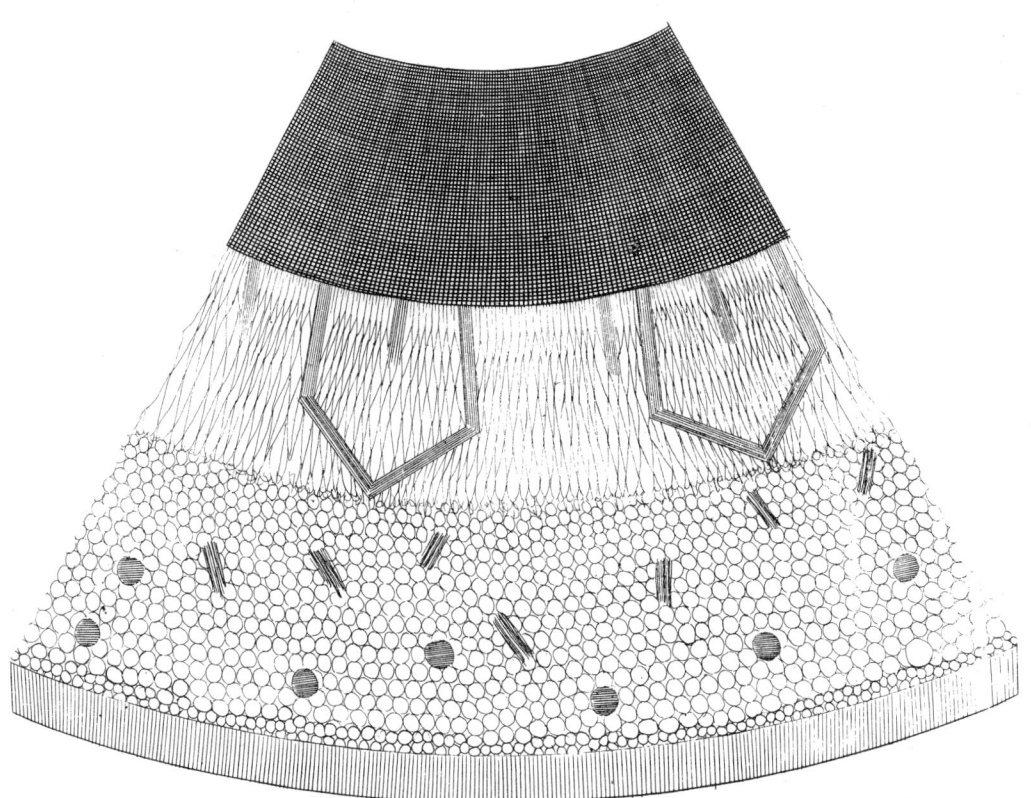

f. 5.
A young one wth ye now
bulky Coats of ye Seed

f. 6.
the same by
ye Length

f. 7.
One with the
kernell full grown

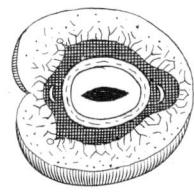
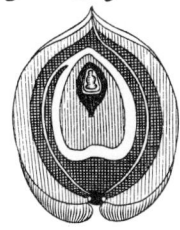
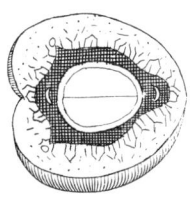

Tab. 69.

f.3.
a Goosberry cut downe.

f.2.
a Grape.
a   b

f.4.
a Goosberry cut transversly.

f.5.
Is (f.4.) magnifyd.

f.1.
a Cherry wthkind

f.6.
a Filbert
a   b

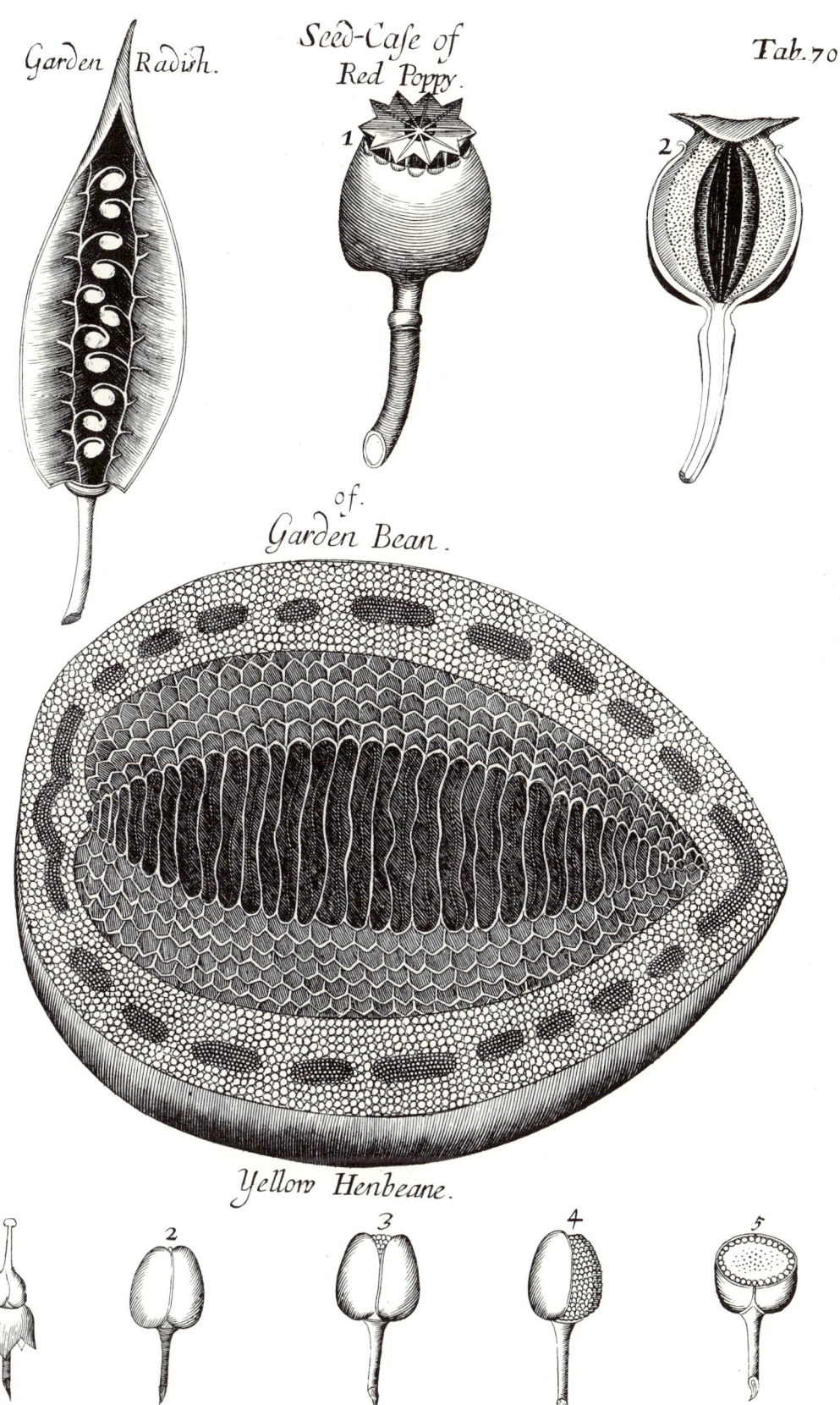

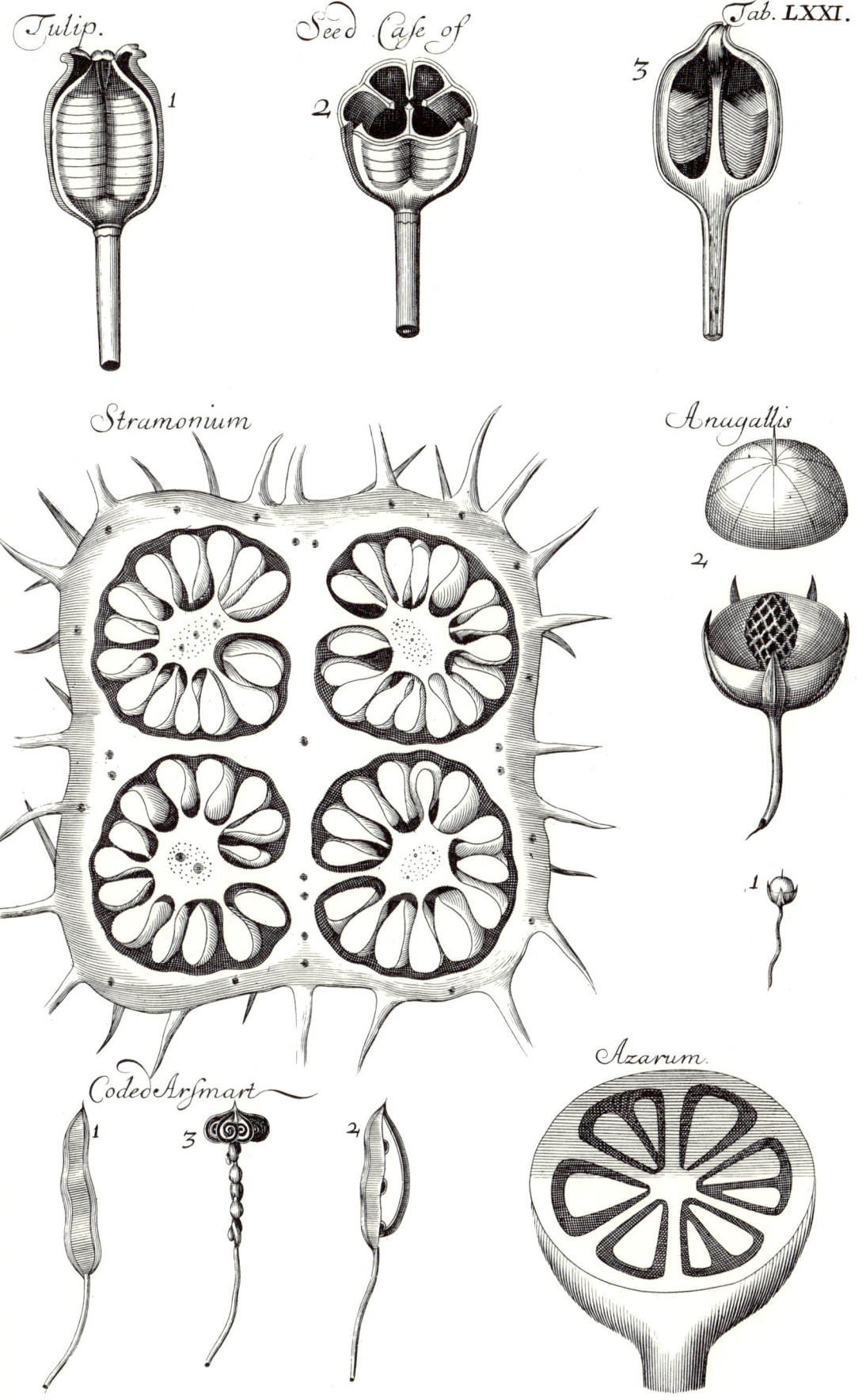

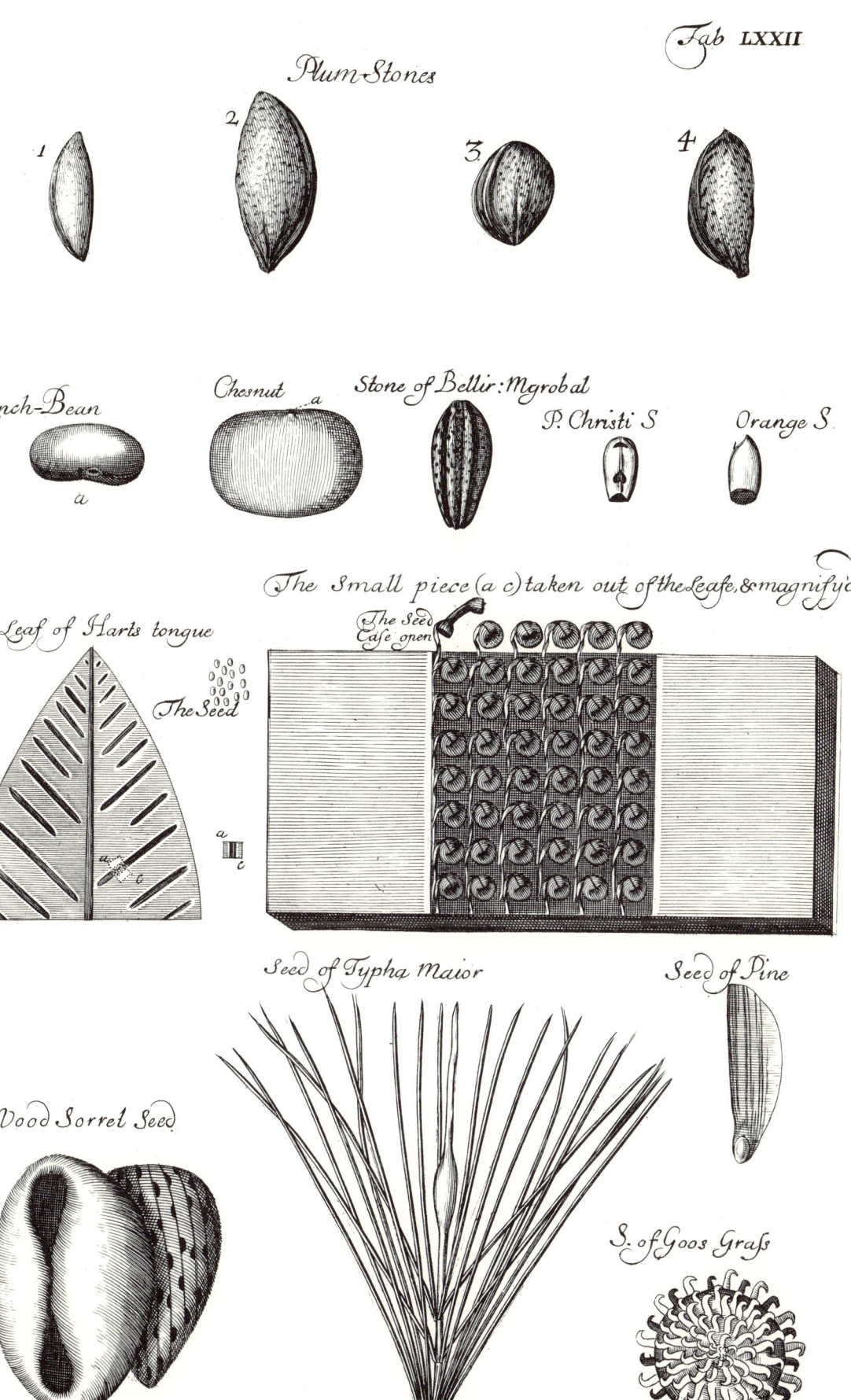

TAB. LXXIII.

Seeds of:

Poppy. Little Century. Spergula. Little Celandine. Lychnis.

Great Celandine. Chickweed. Ben. Pentaphyll. fragif.

Doves-foot. Brooklime. Rush. Little Bell. Sed. maj.

Vervaine. Ragwort. Barley Grass. S.t Iohns wort.

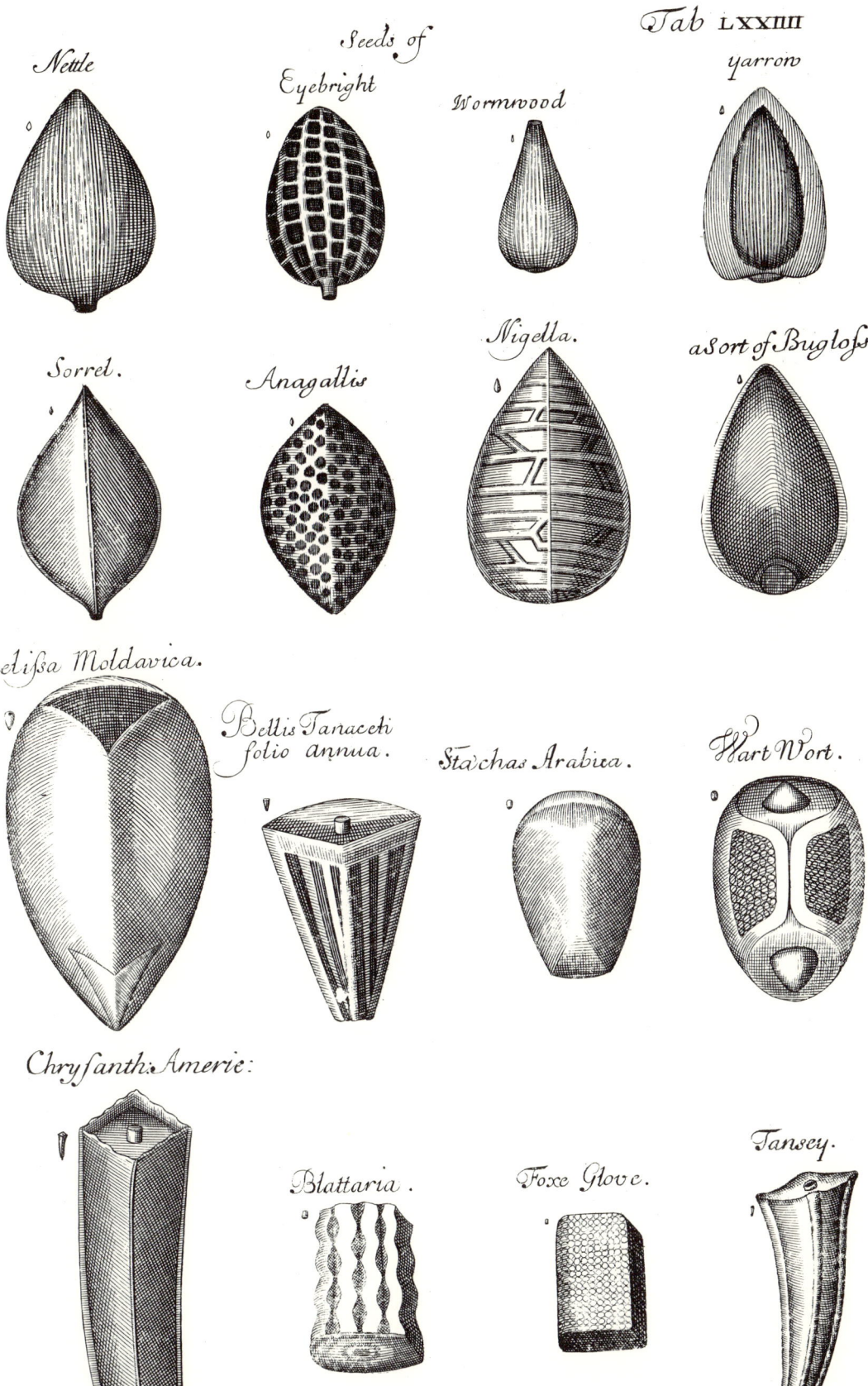

Seeds of — Tab: 75.

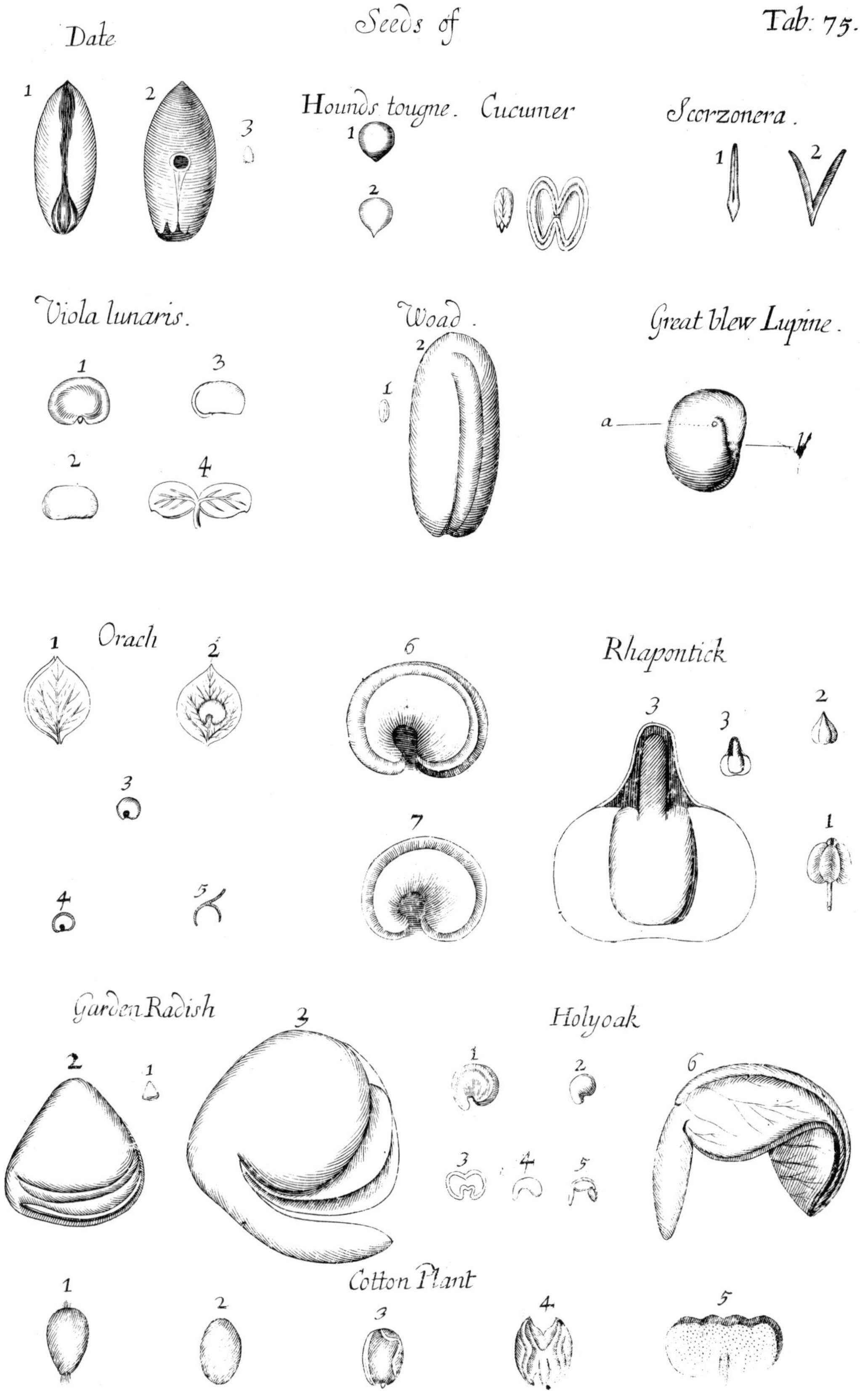

Date

Hounds tougne. Cucumer

Scorzonera.

Viola lunaris.

Woad.

Great blew Lupine.

Orach

Rhapontick

Garden Radish

Holyoak

Cotton Plant

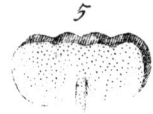

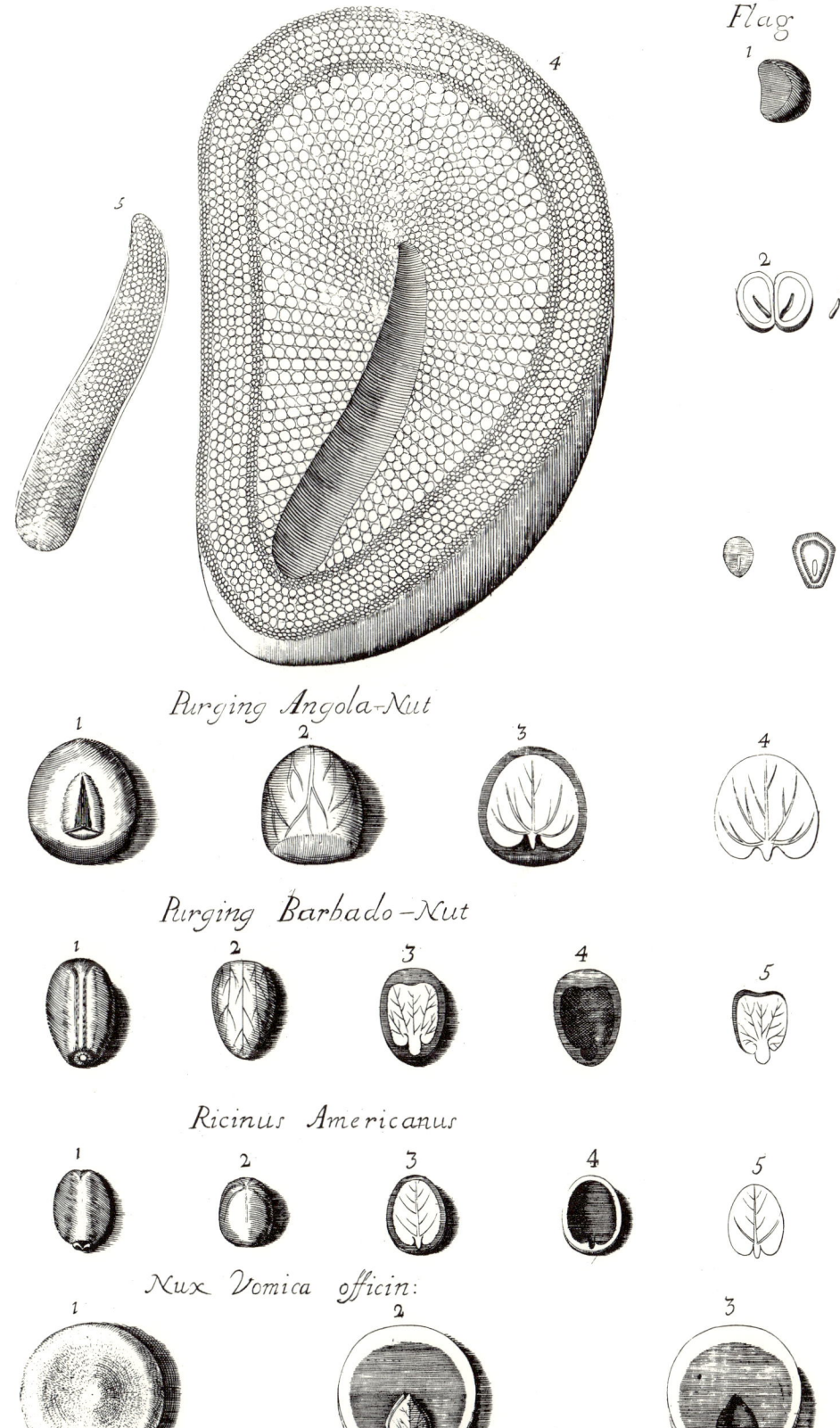

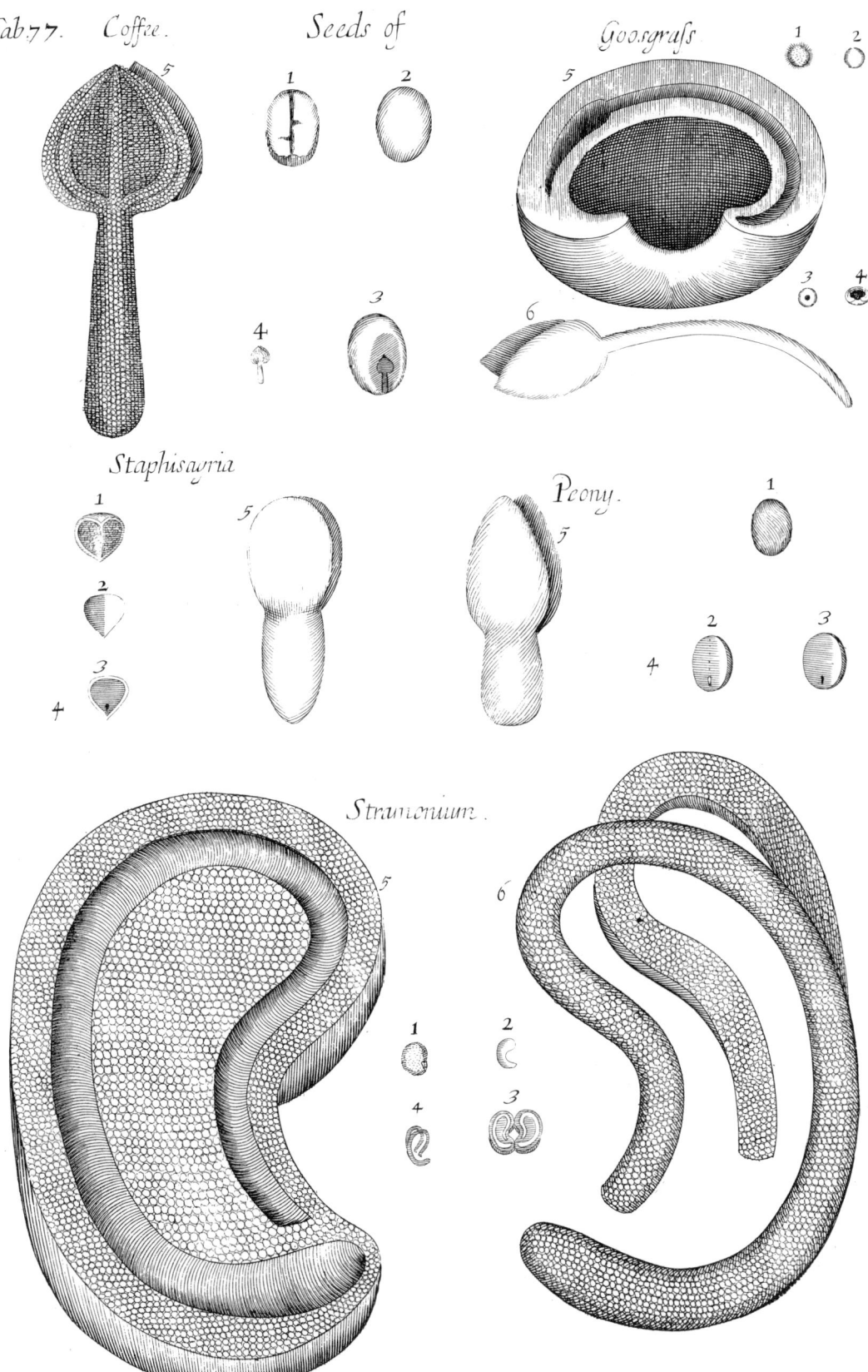

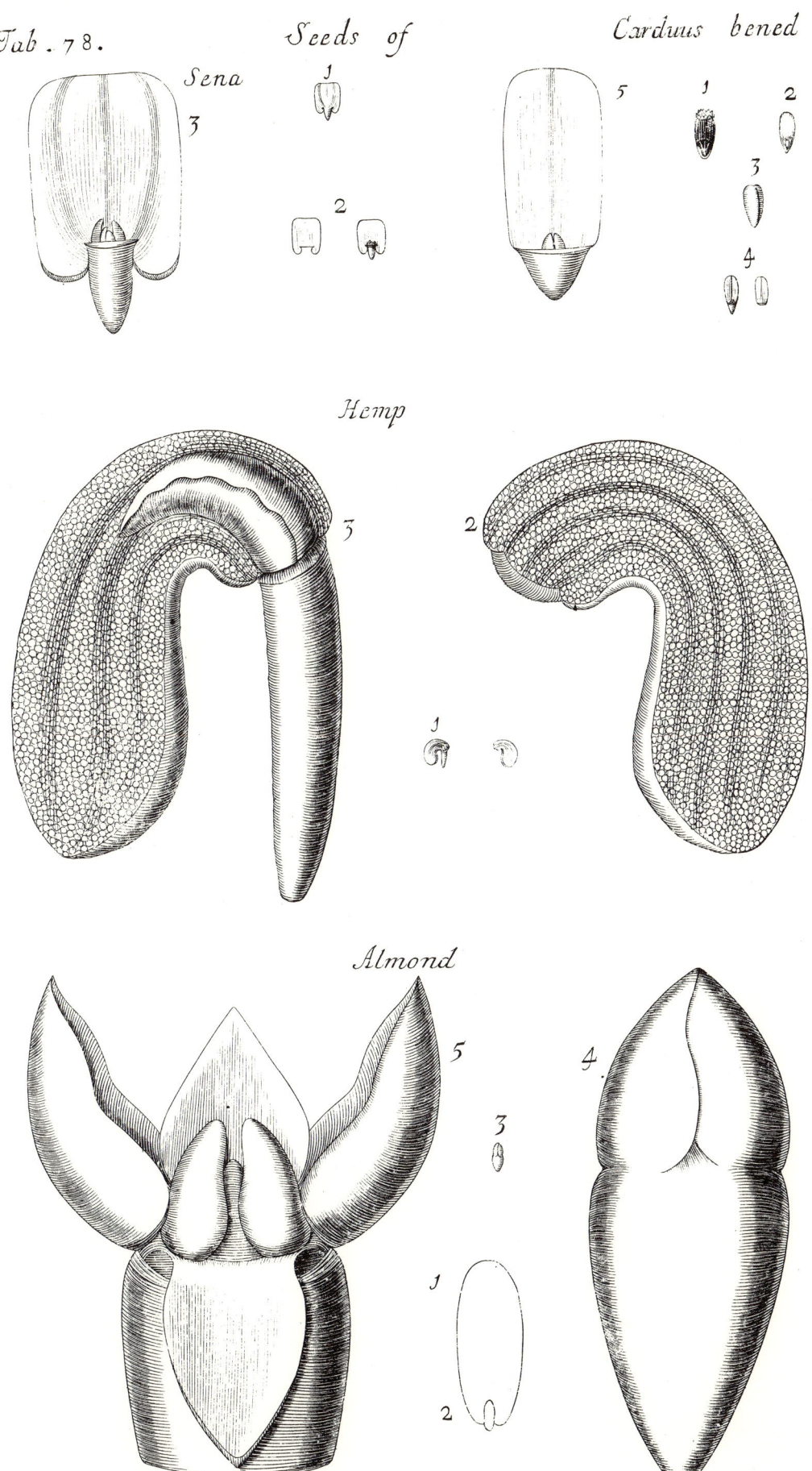

Tab: LXXIX.

A Garden Bean, in one Lobe of w.ch the Seminal Root is layd bare.

f. 1.

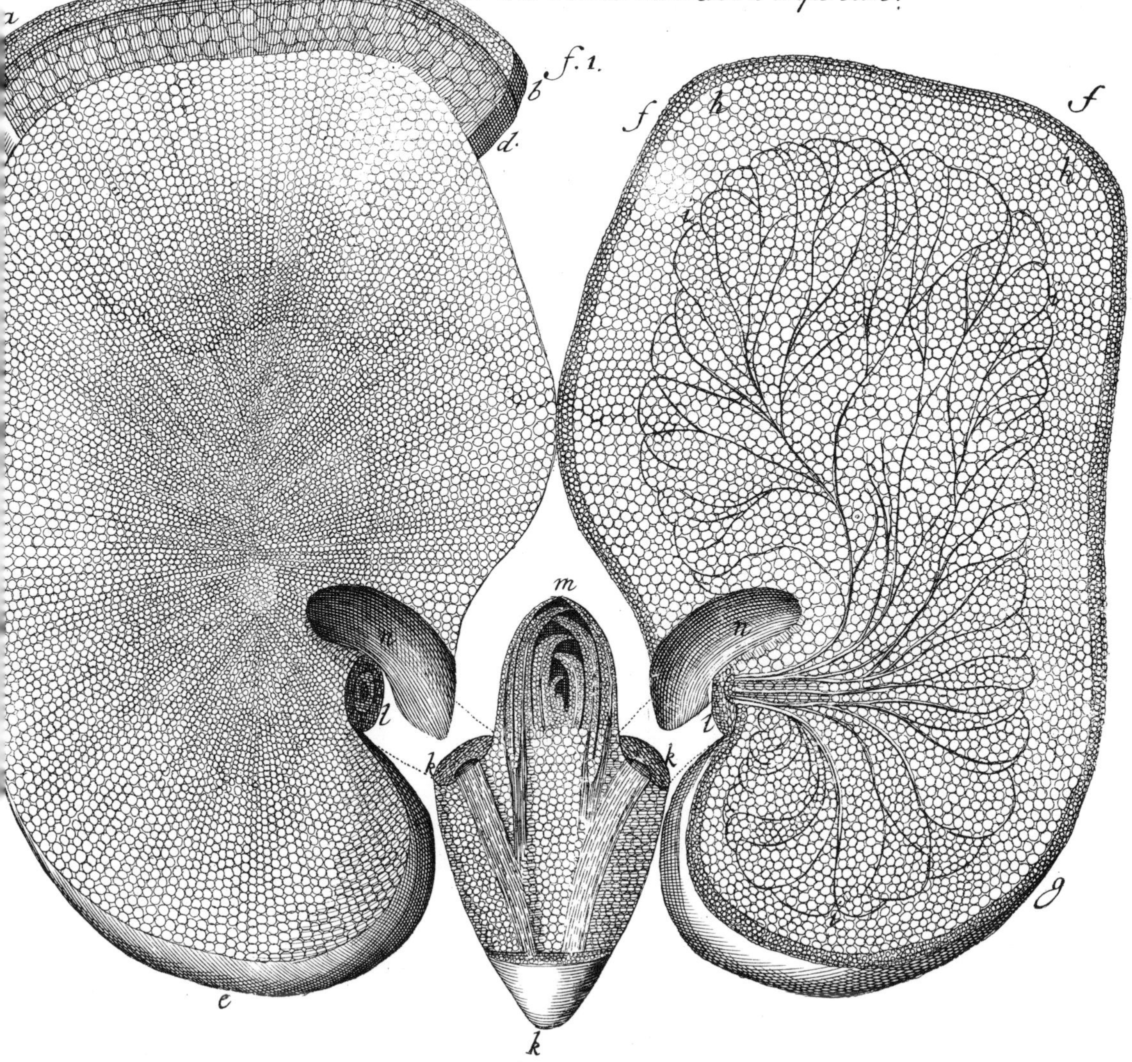

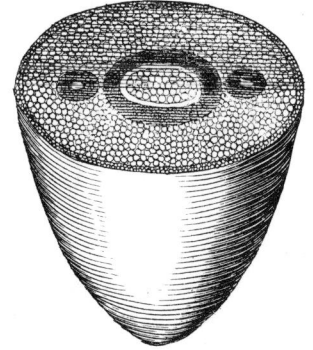

The Radicle cut transversly.
f. 2.

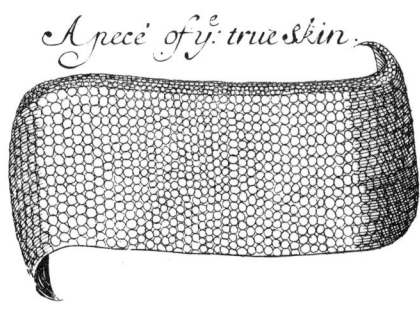

A peece of y.e true skin.

Tab. 80. Sheweth ye Structure of ye Stone & ye two uper Membranes of ye Seed

f. 1. f. 2. f. 4. f. 5. f. 3. f. 6. f. 7.

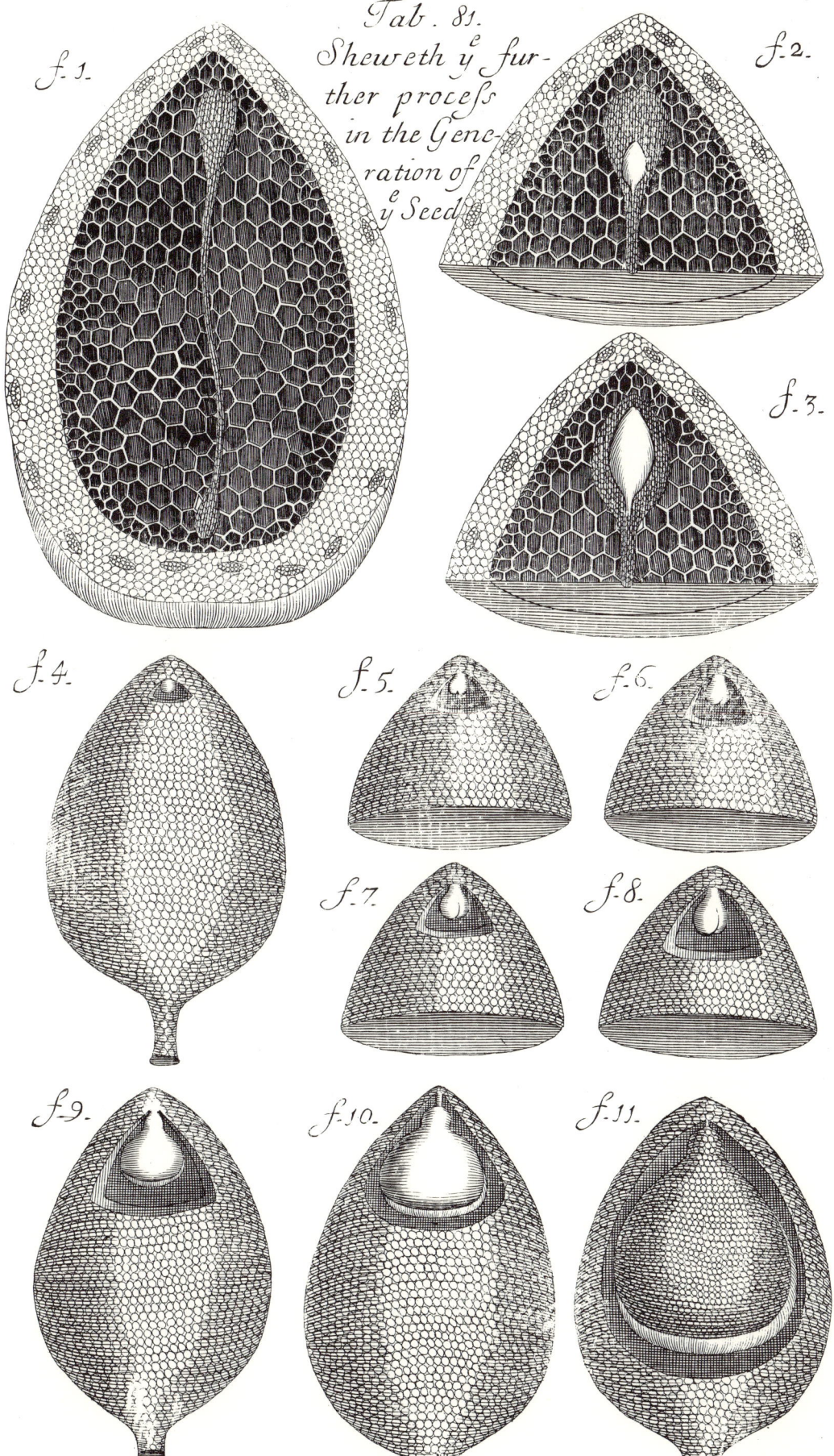

The young Fruit, Three Membranes, & Seed now loose,

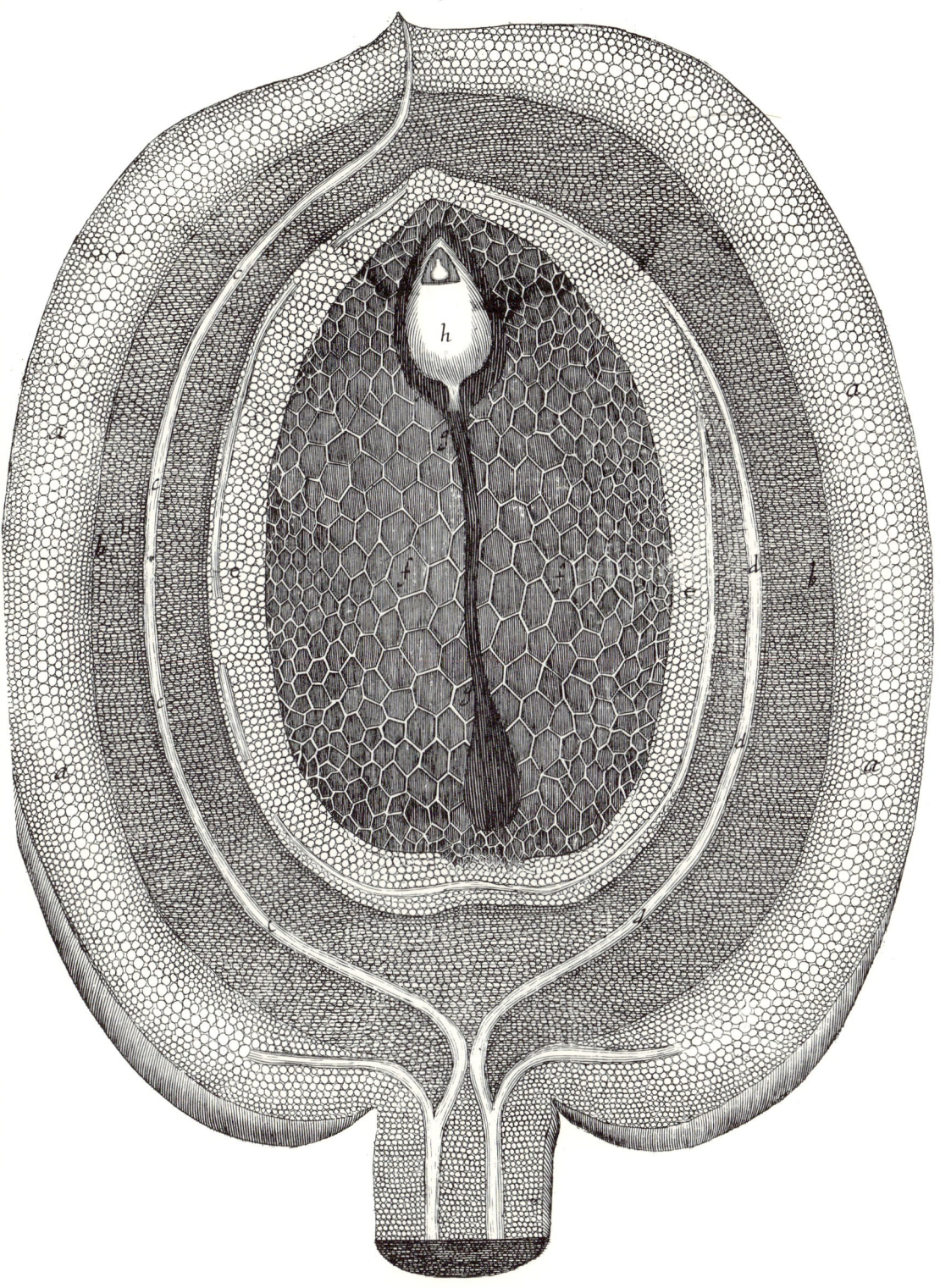

Tab. 83.   Essent: Salts of Plants.

 f.2.
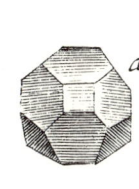 f.3. a b

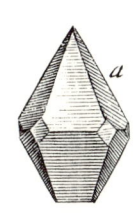 f.4. a
 b

f.1.
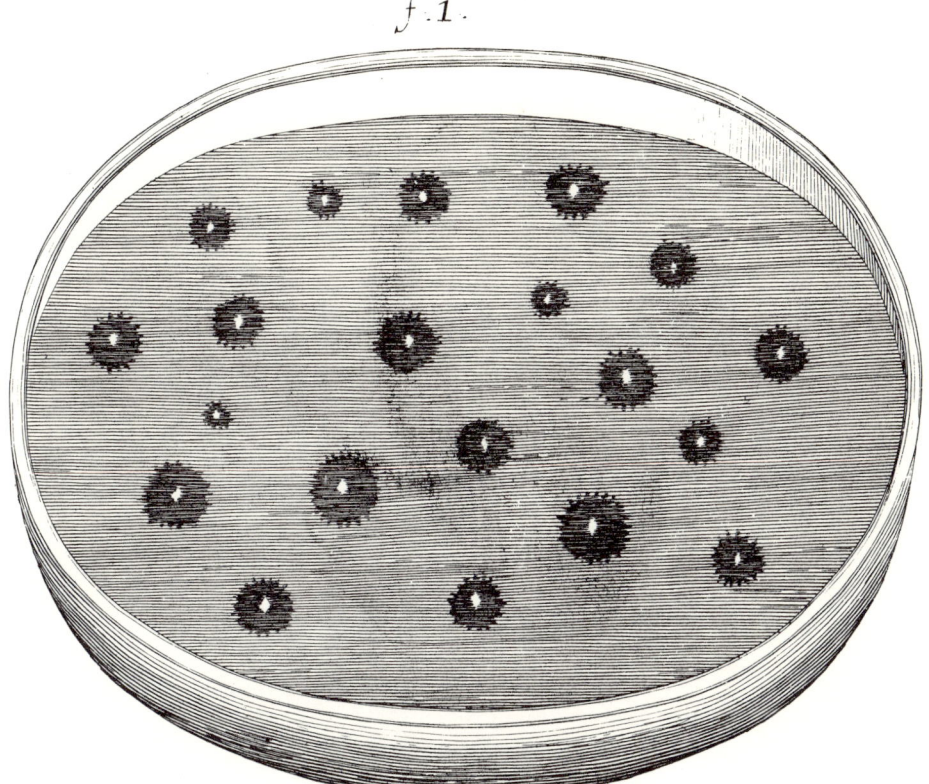

Manne Salts of Plants.

f.5.
f.6.  a   b
f.7.

f.8.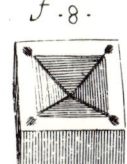
f.9.
f.10.
f.11

DATE DUE